Evolving Approaches to Fuel Desulfurization With Nanocomposites

Tawfik A. Saleh
https://orcid.org/0000-0002-3037-5159
King Fahd University of Petroleum and Minerals, Saudi Arabia

IGI Global
Scientific Publishing
Publishing Tomorrow's Research Today

Published in the United States of America by
IGI Global Scientific Publishing
701 East Chocolate Avenue
Hershey, PA, 17033, USA
Tel: 717-533-8845 | Fax: 717-533-7115
Website: https://www.igi-global.com E-mail: cust@igi-global.com

Library of Congress Cataloging-in-Publication Data

Names: Saleh, Tawfik A. editor
Title: Evolving approaches to fuel desulfurization with nanocomposites / edited by Tawfik A. Saleh.
Description: Hershey, PA : IGI Global Scientific Publishing, [2026] | Includes bibliographical references and index. | Summary: "This book explores the evolving approaches to fuel desulfurization using advanced nanomaterials and nanocomposites, highlighting their enhanced catalytic performance and efficiency in industrial applications"-- Provided by publisher.
Identifiers: LCCN 2025052971 (print) | LCCN 2025052972 (ebook) | ISBN 9798337359663 hardcover | ISBN 9798337359670 paperback | ISBN 9798337359687 ebook
Subjects: LCSH: Fossil fuels--Desulfurization--Materials | Petroleum chemicals--Desulfurization--Materials | Nanocomposites (Materials)--Industrial applications
Classification: LCC TP156.D43 E96 2026 (print) | LCC TP156.D43 (ebook)
LC record available at https://lccn.loc.gov/2025052971
LC ebook record available at https://lccn.loc.gov/2025052972
British Cataloguing in Publication Data
A Cataloguing in Publication record for this book is available from the British Library.

Table of Contents

Detailed Table of Contents

Humanity's need for energy is an undeniable fact, and this reality is increasing day by day. To meet its growing energy needs, humanity has turned to fossil fuels. Fossil fuels are fuels derived from animal and plant fossils formed over millions of years. However, the excessive use of fossil fuels for energy needs has created a dangerous situation that threatens to deplete fossil fuel reserves. Furthermore, the burning of fossil fuels releases toxic emissions. These emissions cause serious health problems for living organisms and also harm the ecosystem. These gases include sulfate (SO_2) and sulfur derivatives, carbon dioxide (CO_2), carbon monoxide, and nitrogen oxide derivatives (NOx). This chapter focuses on catalytic techniques using nanocomposite materials for the removal of SO_2 and sulfur derivatives, with detailed references to relevant sources. SO_2 and sulfur gases are harmful gases released by the combustion of fossil fuels. Nanocomposite production methods include sol-gel formation, gel shaping, chemical vapor deposition (CVD), and physical vapor deposition (PVD), layered production, electrospinning, mechanical and ultrasonic techniques, additive manufacturing, and melting mixture. Desulfurization processes using these methods are as follows: (i) oxidative desulfurization, photocatalytic desulfurization, and absorptive-catalytic desulfurization. All techniques are explained with references to the relevant sources. This chapter provides theoretically important information for researchers working on nanotechnological catalytic studies in desulfurization.

The removal of sulfur from fossil fuels is crucial for achieving clean energy and complying with stringent environmental regulations. Although traditional desulfurisation methods are effective, they are often energy-intensive, costly, and inefficient for addressing recalcitrant sulfur compounds. Plant-based nanocomposites present a promising alternative owing to their tunable structures, high porosity, and unique surface properties that facilitate sulfur removal. This review examines the utilisation of agricultural and forestry wastes, such as rice husks, coconut shells, sugarcane bagasse, and fruit peels, in the synthesis of these materials. Eco-friendly techniques, including pyrolysis, hydrothermal carbonisation, and steam activation, can be used to produce materials with high surface areas and advantageous properties. Additionally, it discusses the necessity of balancing the hydrophobic and hydrophilic properties for effective application with hydrocarbons.

Green and sustainable nanomaterials are the new hopeful solutions on how to lighten the environmental impact of the sophisticated fuel purification process. The topic of this chapter is bio-derived carbon nanomaterials and low-temperature synthesis paths as alternatives to the traditional energy-intensive nanomaterial fabrication. Agricultural waste, algae, and industrial biowaste are biomass-based precursors that can provide renewable, cheap, and scalable sources of carbon. There is a focus on hydrothermal and microwave-assisted, and low-temperature chemical vapor deposition technologies, which consume less energy and contribute to fewer dangerous emissions. The chapter is a critical discussion of physicochemical properties, functionalization strategies, and integration with the nanocomposites to desulfurize fuel efficiently. Green nanotechnology is also considered in terms of environmental, economic, and regulatory perspectives to emphasize the viability of the industry.

Abdullah Ozkan, Iskenderun Technical University, Turkey
Vildan Ozkan, Iskenderun Technical University, Turkey
Pinar Bozbeyoglu, Biomedical Device Technology Program, Electronics and Automation Department, Gümüşhane University, Turkey
Elif Esra Altuner, Kocaeli Health and Tecnology University, Turkey

The mounting global demand for energy signifies that we must depend on fossil fuels, which engenders environmental degradation and precipitates the emission of substantial quantities of sulfur oxides (SOx). Protecting themselves from environmental pollution is a top priority for many countries. Proper removal of sulfur-containing compounds from fossil fuels is becoming a major focus. Fuels such as diesel, fuel oil, and gasoline are particularly important in this regard. Fuels such as diesel, fuel oil, and gasoline are particularly important in this regard. The numerous negative consequences of traditional hydrodesulfurization (HDS) methods are what researchers focus on, and these methods are limited. Some examples of these negative effects include higher costs, the use of too much pressure and temperature, and using too much hydrogen. Scientific studies show that biosourced nanocomposite materials are good for desulfurization methods because they can reduce the potential negative consequences. Using raw materials such as microorganisms, biopolymers, and plant extracts is one of the most important ways to increase energy efficiency and eliminate chemical emissions. The large surface areas, enhanced active site homogeneity, and diverse functional groups of nanocomposite materials obtained from such biosourced raw materials make them promising candidates for use in adsorptive and oxidative desulfurization processes. Metal-metal oxide (Fe, Ni, Cu, TiO_2) and carbon-based (graphene, carbon nanotube, and biochar-supported) nanocomposites are particularly preferred for achieving high sulfur removal with low energy consumption. The provision of information on the production methods, chemical properties, sulfur removal methods, and recent scientific research on biosourced nanocomposites is the purpose of this section. In addition, the environmental benefits, economic benefits, and potential roles of biosourced materials in alternative energy systems are discussed.

Nayab Ahmad, The University of Agriculture, Peshawar, Pakistan
Sidra Ahmad, The University of Agriculture, Peshawar, Pakistan
Micheal Ayodeji Sokan-Adeaga, Department of Community Health and Primary Care, Faculty of Clinical Sciences, College of Medicine, University of Lagos, Lagos, Nigeria
Irshad Ali Khan, Department of Agriculture, The University of Swabi, Khyber Pakhtunkhwa, Pakistan
Adewale Allen Sokan-Adeaga, Public Health Program, College of Health Sciences, Bowen University, Iwo, Nigeria
Sezai Ercisli, Department of Horticulture, Faculty of Agriculture, Ataturk University, Erzurum, Turkey & HGF Agro, Ata Teknokent, Erzurum, Turkey

This chapter explores magnetic and hybrid bionanocomposites as a novel class of recyclable, eco-friendly materials for green fuel desulfurization. It covers the structural and functional engineering of bio-based core–shell frameworks and the incorporation of magnetic cores (for example, Fe_3O_3) into biogenic supports such as biochar, cellulose, or graphene oxide. These materials demonstrate superior efficiencies in both adsorptive and oxidative desulfurization under mild conditions, making them suitable for the removal of refractory sulfur compounds. The chapter also discusses green synthesis methods, magnetic separation, and reusability of materials over multiple cycles. Environmental and safety considerations are addressed from a life cycle sustainability perspective, including biodegradability, toxicity, and waste management. In addition, this chapter analyses the challenges and scalability of economic and process integration.

Rupam Hazra, Global Institute of Management and Technology, Krishnanagar, India

This chapter examines the evolving landscape of fuel purification and emission control, emphasizing innovative desulfurization approaches using nanocomposites and advanced materials. As environmental regulations tighten and global efforts to combat pollution and climate change intensify, cleaner fuel technologies have become critical. Conventional methods like hydrodesulfurization, though widely used, often fail to remove refractory sulfur compounds and involve high energy and operational costs. In response, researchers are turning to nanostructured materials—such as metal–organic frameworks (MOFs), graphene derivatives, mesoporous silica, and functionalized zeolites—for their superior selectivity and reactivity. The

Preface

Nanotechnology in desulfurization is a relatively recent and rapidly emerging field. The development of nanomaterials has become a central focus of research and technological advancement across disciplines including chemistry, physics, applied sciences, petroleum engineering, and materials engineering. Research in nanotechnology has primarily concentrated on the synthesis of nanomaterials with unique chemical, thermal, and mechanical properties that enable a wide range of applications. Numerous properties and phenomena have been extensively investigated, with many studies aimed at understanding the structure–property relationships and practical applications of nanomaterials.

Due to their enhanced chemical and mechanical properties, the nanomaterials play promising roles in enhancing the desulfurization. Nanomaterials have properties that are useful for enhancing surface-to-volume ratio, reactivity, strength, and durability. In pursuit of the same goal, this book offers detailed, up-to-date chapters on the synthesis, properties and technological developments of nanomaterials, and their applications in petroleum.

ORGANIZATION OF BOOK

The first chapter of the book encompasses the occurrence of organosulfur compounds in petroleum and various techniques for the removal of these compounds. The next two chapters encompass the nanocomposites, hybrid materials and carbon-based nanomaterials for adsorptive desulfurization including preparation and evaluation.

The fourth chapter discussed the polyoxometalates as efficient catalysts for the activation of oxidants in desulfurization processes with the main goal to design heterogeneous polyoxometalate based catalysts. This chapter pretends to inform the research society about the scientific directions that have been taken using heterogeneous polyoxometalate catalysts in oxidative desulfurization of simulated and real liquid fuels. In chapter five, the relation of catalytic properties with components and structures of Ni/ZnO sorbent is discussed. Based on detailed characterization and reaction results, the dynamic change of Ni/ZnO sorbents during RADS is highlighted, the mechanisms of desulfurization, and the sulfur transfer and sulfur adsorption, etc..

In the next chapter, several scientific and technological advances reported in the literature for the desulfurization of fuels are reviewed and discussed. Amongst these techniques are oxidative desulfurization (ODS) and adsorptive desulfurization (ADS) which are proposed as additional steps to complement HDS in meeting the mandated ultra-low sulfur levels (10 ppmS).

Advances in the reduction of the costs inherent to fossil fuels' biodesulfurization towards its potential industrial applications are discussed in chapter seven. In the next chapter, environmental concerns and the importance of desulfurization is discussed.

In this book, we have tried to cover many aspects of nanomaterials for desulfurization, which is of current interest. This book is written for a large readership, including university students and researchers from diverse backgrounds such as chemistry, petroleum, materials science, physics, and engineering. It can be used not only as a textbook for both undergraduate and graduate students but also as a review and reference book for researchers in these fields. We hope that the chapters of this book will provide the readers with valuable insight into state-of-the-art advanced and functional nanomaterials and technologies.

However, it is possible that some topics have been left out owing to constraints on the size of the book and possible errors in judgment. We trust that the preface will be useful to students, teachers, and researchers.

The Editor,

Tawfik A. Saleh

Department of Chemistry, King Fahd University of Petroleum and Minerals, Saudi Arabia

Acknowledgment

I would like to acknowledge the support provided by King Fahd University of Petroleum & Minerals. I would like to express my gratitude to the many people who saw me through this book; to all those who provided support, read, wrote, offered comments, allowed me to quote their remarks, and assisted in the editing, proofreading, and design. Above all, I want to thank my parents, relatives, wife, children, and the rest of my family, who supported and encouraged me in spite of all the time it took me away from them. Last but not least: I beg forgiveness of all those who have been with me over the course of the months and whose names I have failed to mention.

Many thanks.

Tawfik Abdo Saleh

Department of Chemistry, King Fahd University of Petroleum and Minerals, Saudi Arabia

Chapter 1
Catalytic Nanocomposites in Oxidative on Adsorptive Delsulfurization Process: A Comprehensive Review

Elif Esra Altuner
https://orcid.org/0000-0001-7663-6898
Kocaeli Health and Tecnology University, Turkey

Pinar Bozbeyoglu
Gümüşhane Vocational School, Department of Electronics and Automation, Biomedical Device Technology Program, Gümushane University Gümüşhane University, Turkey

Vildan Özkan
https://orcid.org/0000-0001-8719-9099
Iskenderun Technical University, Turkey

Abdullah Ozkan
https://orcid.org/0000-0002-6799-2396
Iskenderun Technical University, Turkey

ABSTRACT

Humanity's need for energy is an undeniable fact, and this reality is increasing day by day. To meet its growing energy needs, humanity has turned to fossil fuels. Fossil fuels are fuels derived from animal and plant fossils formed over millions of years. However, the excessive use of fossil fuels for energy needs has created a dangerous

DOI: 10.4018/979-8-3373-5966-3.ch001

situation that threatens to deplete fossil fuel reserves. Furthermore, the burning of fossil fuels releases toxic emissions. These emissions cause serious health problems for living organisms and also harm the ecosystem. These gases include sulfate (SO_2) and sulfur derivatives, carbon dioxide (CO_2), carbon monoxide, and nitrogen oxide derivatives (NOx). This chapter focuses on catalytic techniques using nanocomposite materials for the removal of SO_2 and sulfur derivatives, with detailed references to relevant sources. SO_2 and sulfur gases are harmful gases released by the combustion of fossil fuels. Nanocomposite production methods include sol-gel formation, gel shaping, chemical vapor deposition (CVD), and physical vapor deposition (PVD), layered production, electrospinning, mechanical and ultrasonic techniques, additive manufacturing, and melting mixture. Desulfurization processes using these methods are as follows: (i) oxidative desulfurization, photocatalytic desulfurization, and absorptive-catalytic desulfurization. All techniques are explained with references to the relevant sources. This chapter provides theoretically important information for researchers working on nanotechnological catalytic studies in desulfurization.

1. INTRODUCTION

Humanity's need for energy is increasing day by day (Darabi et al., 2023). Energy types are needed in a wide variety of areas, including heating (Guelpa & Verda, 2019; Jamar et al., 2016; Ürge-Vorsatz et al., 2015), electricity (H. Chen et al., 2009; Dillon, 2010), and systems to increase clean water resources (Süme et al., 2024). Especially for essential needs like heating, humanity has turned to fossil fuels. Energy derived from fossil fuel sources such as coal (Longwell et al., 1995; Strielkowski et al., 2021), oil (Abas et al., 2015; Chew, 2014; Shafiee & Topal, 2009), and natural gas(Shafiee & Topal, 2010) largely meets this need. However, because humanity is so heavily reliant on fossil fuels to meet its energy needs, energy reserves are nearing depletion, and finding a new energy source is crucial (Ritchie & Rosado, 2017). Furthermore, sulfur-derived gases and toxic emissions from fossil fuels are extremely harmful to living organisms, polluting the environment and damaging ecosystems (Glass, 1979; J. Liu et al., 2024). Therefore, researchers have focused on eliminating toxic emissions from fossil fuels.

Fossil fuels are formed millions of years ago when animal and plant remains undergo chemical changes under high temperatures and pressures (Armaroli & Balzani, 2011a; Ritchie & Rosado, 2017; Soeder, 2021). Their high density and easy accessibility are key advantages (Wuebbles & Jain, 2001). Fossil fuels take various forms depending on the chemical reactions they undergo, their pressure and temperature, and their evolution (Burnham, 2017). These forms include coal (Lockwood, 2017), natural gas (Stejskal et al., 2021), and petroleum (Peixoto et al., 2011). Coal is classified as anthracite (J. Liu et al., 2019), lignite (Ali & Zhao, 2020), and hard coal (Kapusta & Stańczyk, 2011); natural gas is classified as methane (Howarth, 2014; J. Wang et al., 2020), ethane (Schwietzke et al., 2014), and propane (Q. Liu et al., 2019), depending on its carbon content (Seewald et al., 1998); and gasoline (Wadsley et al., 2004) is classified as diesel, Liquefied Petroleum Gas (LPG) (Ashok et al., 2015), gasoline (Wadsley et al., 2004), fuel oil (Blumer & Sass, 1972), and kerosene (Lam et al., 2012). All of these fossil fuels are crucial to human energy needs and have become integral to our lives. The excessive and unbalanced use of these fossil fuels leads to the release of toxic emissions (Armaroli & Balzani, 2011b). These emissions threaten both the health of living creatures and the ecosystem (Ritchie & Rosado, 2017). The most common emission gases can be listed as follows: (i) carbon dioxide (CO_2) (van Droffelaar & Jacobs, 2018), carbon monoxide (CO) (Yang et al., 2019), sulfur dioxide (SO_2) (H. Zhao et al., 2018) and its derivatives, nitrogen oxides (NO_x) (Dignon, 1992; Zou et al., 2017), methane gas (CH4) (Schwietzke et al., 2016), and various volatile organic compounds (VOCs) (Y. Wang et al., 2022). All of these emission gases are harmful. In addition to being harmful, some emission gases are poisonous. Emission gases that are both toxic and harmful are CO (van Droffelaar & Jacobs, 2018), SO2 (H. Zhao et al., 2018), NOx, and their derivatives (Dignon, 1992; Zou et al., 2017), VOCs (Y. Wang et al., 2022). This chapter addresses the desulfurization study for the removal of SO2. Catalytic processes are used in desulfurization studies. Nanotechnology is of great importance in catalytic processes.

Nanotechnology is a branch of science that deals with particles between 1 and 100 nm (Altuner, Bekmezci, et al., 2021; Altuner et al., 1 C.E., 2023; Altuner, Gur, et al., 2021). 1 nm is one-billionth of a meter, and the term nano means "dwarf" in Latin (Altuner, Bekmezci, et al., 2022; Altuner, Sen, et al., 2025). Therefore, because of such small particles, nanotechnology is intertwined with almost all branches of science and has a multidisciplinary approach. For example, it is used in a wide variety of fields, including sensors (Altuner, Ozalp, Yilmaz, Bekmezci, et al., 2022; Altuner, Ozalp, Yilmaz, Sudagidan, et al., 2022), catalytic systems (Altuner, 2025; Ni et al., 2023), medicine and health (Ozdemir et al., 2022; Özdemir et al., 2023; Torlak et al., 2025), chemistry (Zare et al., 2024), physics (Miwa et al., 2006), biology (Kocak et al., 2024), optics (S. Chen et al., 2021; Nur Tatli et al., 2024), and electronics (Cui & Lieber, 2001). In the literature, bimetallic iron-cobalt nanoparticles (FeCo NPs) have been synthesized, and a dopamine sensor has been developed (Esra Altuner et al., 2025). Another study reported the removal of methylene blue (MB) by synthesizing zinc nanoparticles (Zn NPs) (Altuner, 2025). Another study reported a catalytic study for hydrogen energy production based on copper-iron nanoparticles (Al Janabi et al., 2024). Catalytic studies and the study of nanoparticles are crucial.

Catalytic processes are processes that involve the use of a catalyst to increase the rate and activity of a chemical reaction (Altuner, Tatli, et al., 2025). A wide variety of catalysts exist, including heat (Doner et al., 2023), metals(Sloan et al., 2010), pressure (Rogalinski et al., 2008), and metal oxides (Kojima et al., 2002). However, in addition to all these catalysts, nanocomposite materials are a key element of catalytic systems because the very small size of nanoparticles significantly enhances their catalytic effectiveness.

This chapter provides a detailed discussion, with references to relevant sources, of catalytic systems developed using nanocomposite materials for the removal of toxic and harmful sulfur derivatives released into the environment from fossil fuels. This study provides detailed theoretical information on the fundamentals of catalytic systems developed using nanoparticles for the removal of toxic wastes, particularly for desulfurization studies.

2. FOSSIL FUELS

Fuels such as coal(Y. Wang et al., 2017), petroleum (Peixoto et al., 2011), oil (Chew, 2014), and natural gas (Howarth, 2014) are fossil fuels (figure 1). Much of the energy we use today comes from fossil fuels. Fossil fuels have been formed over millions of years by the decay of plants and animals (Friedemann, 2021; Martins et al., 2018). Fossil fuels can be extracted from the earth's surface by drilling or excavation (Süto, 2010). Currently, these fuels are formed underground by the effects of heat and pressure (Oparin et al., 2017). However, the rate at which fossil fuels are formed is much slower than the rate at which they are consumed (Fulkerson et al., 1990; Inman, 2013). Therefore, fossil fuels can be considered non-renewable energy sources in the short term (Nyambuu & Semmler, 2014; Radišić, 2024; Ragazzi et al., 2017). Population growth, urbanization, and industrialization, in particular, are causing the energy needs met by these fuels to increase (Avtar et al., 2019; Mudakkar et al., 2013). In addition to the depletion and the constant rise in prices of fossil fuels, the environmental damage caused by their combustion and their impact on human health are also significant.

Figure 1. The type of fossil fuels.

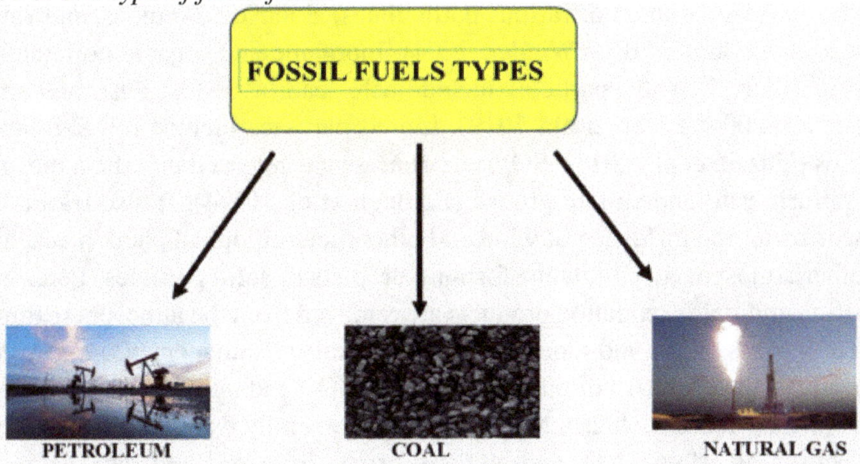

The primary component of fuels is organic carbon (C) (Huang et al., 2015). Heat is released when organic carbon reacts with oxygen (O2) (Khvorostyanov et al., 2008). The heat of reaction released in this way is utilized in various processes (Lyon, 2000). The energy value per unit mass or volume of fuel is called the calorific value (Demirbaş & Demirbaş, 2004). The calorific value of fuels is an important property. The upper calorific value is the calorific value obtained from the fuel when the combustion products released by the combustion of a unit amount of fuel are cooled back to their initial temperature and the water present in the fuel is condensed (Demirbas, 2004; Kaya et al., 2021). The lower calorific value is the energy released by the combustion of the fuel (Chomiak et al., 1989). The lower calorific value is approximately 8-9% lower than the upper calorific value, equal to the heat of condensation of the water vapor released by combustion (Wu et al., 2011). Fuels release their energy in two ways: chemical (combustion) or nuclear (fusion).

3. TOXIC EMISSION GASES

This section examines toxic gases emitted into the ecosystem by fossil fuels (Nayak et al., 2019). Toxic emission gases can be categorized into two groups based on their chemical structure: inorganic and organic (Borah et al., 2020). Inorganic gases include nitrogen oxides (NOx) (Muzio & Quartucy, 1997), carbon dioxide (CO_2) (Maginn, 2010), carbon monoxide (CO) (Ferry, 1995), and sulfur dioxide

(SO_2) (Srivastava et al., 2001) (T. M. Chen et al., 2007). Organic gases include fluoride, chloride, ammonia, various particulates, aldehydes, ketones, and various hydrocarbons (Gupta, 2020; Winder, 2004). Inorganic and organic pollutants are shown in figure 2. Solid particles include ash, lead, asbestos, dust, and smoke (Feng et al., 2004; Klebe et al., 2019). Liquid particles include oil, fog, smoke, and acids (Husain et al., 2014). SO_2 is a colorless gas released into the atmosphere from sulfuric acid and sulfate groups (Eatough et al., 1994). It also travels long distances under the influence of wind and other meteorological phenomena, along with other atmospheric pollutants, forming droplets or solid particles (Pérez et al., 2020). Sulfur dioxide oxidation products are removed from the atmosphere through the effectiveness of dry and moist deposits (acid rain) (Ramadan, 2004; Yun et al., 2019). NO_x species consist of nitrogen dioxide (NO_2) and nitrogen monoxide (NO) nitrites (NO) (Mousavi, 2020). NO_2 species are generally exhaled at a rate of 90% (Lundberg et al., 1996). NO_2 and NO are formed by ozone and radicals and are very dangerous to human health. Toxic emission gases are produced by industrial processes, combustion boilers, and air, sea, and land traffic (Turkay, 2018). SO_2 and its types are discussed in the next subheading of this chapter on emission gases.

Figure 2. The pollutants of fossil fuels.

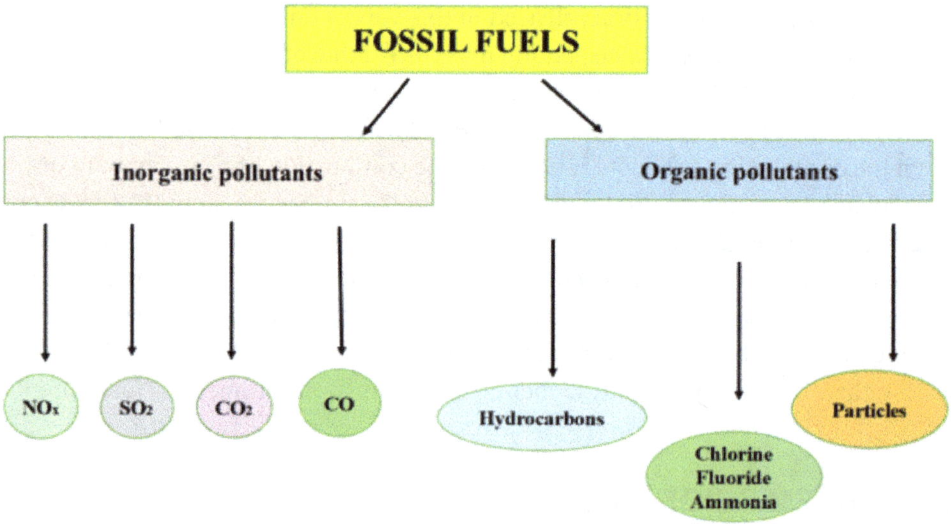

4. TOXIC SO2 GASES AND THEIR DERIVATES

SO2 gas is produced by the combustion of sulfur-containing fossil fuels. While SO2 gas is generally released in industrial processes, it is also released into the atmosphere during natural events such as volcanic eruptions. The biggest disadvantage of this gas is that its release into the atmosphere is invisible and undetectable. This harms the ecosystem and the health of living organisms. In industrial applications, SO2 gas is typically released from metal-melting furnaces and power plants. From the moment SO2 gas is released into the atmosphere, it interacts with oxygen and water molecules. This interaction results in the formation of sulfuric acid aerosols. The release of sulfuric acid particles creates acid rain. Acid rain negatively affects the structure of nature, reduces agricultural and soil fertility, and accelerates chemical accumulation. Figure 3 shows an illustration of the release of SO2 gas and its harmful effects. On the other hand, this gas also reduces air quality and makes the air heavier. Sulfate particles cause conditions such as asthma, COPD, shortness of breath, and bronchitis. Children and the elderly are particularly affected by this.

Figure 3. The schematic illustration of the formation of SO_2 and the hazards of SO_2.

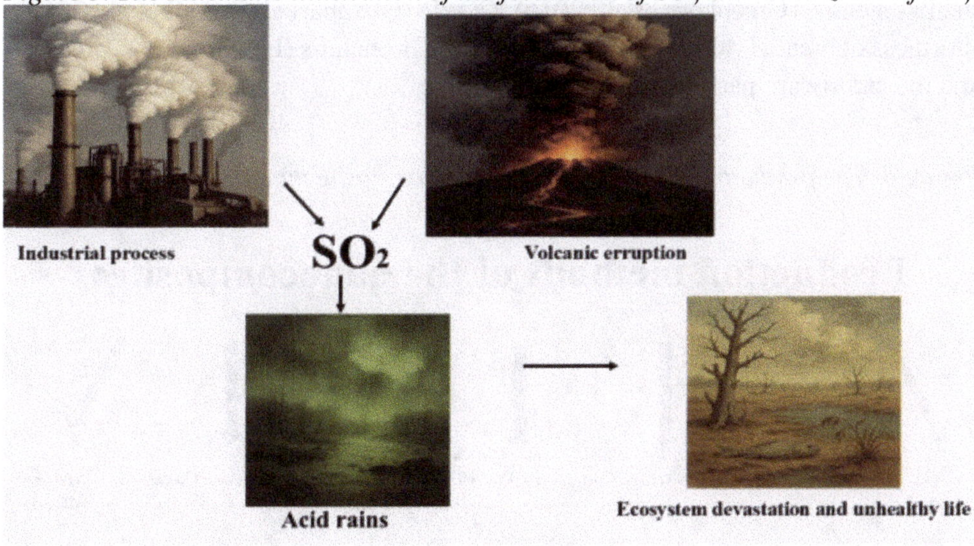

5. NANOCOMPOSITION TECHNIQUES

SO_2 gas is produced by burning sulfur-containing fossil fuels (Perraud et al., 2015; Sarbassov et al., 2018). It is generally released during industrial processes (Hung et al., 2020). It can also enter the atmosphere during natural events, such as volcanic eruptions (von Glasow et al., 2009). The biggest disadvantage of this gas is that its release into the atmosphere is invisible and undetectable (Saxena & Shukla, 2023). This harms the ecosystem and the health of living organisms.

In industrial applications, SO_2 gas is typically released from metal-melting furnaces and power plants (W. Liu et al., 2019). From the moment SO_2 gas is released into the atmosphere, it interacts with oxygen and water molecules (He et al., 2020). This interaction results in the formation of sulfuric acid aerosols (Schobesberger et al., 2015; Temelso et al., 2012; Xu & Zhang, 2012). The release of sulfuric acid particles creates acid rain. Acid rain negatively affects the environment's structure, reduces agricultural and soil fertility, and accelerates the accumulation of chemicals (Mohnen, n.d.; Sivalingam et al., 2024). Figure 3 illustrates the release of SO_2 gas and its harmful effects. On the other hand, this gas also reduces air quality and makes the air heavier (Sivalingam et al., 2024). Sulfate particles cause conditions such as asthma (Greenberg et al., 2016), COPD (Ghanbari Ghozikali et al., 2016), shortness of breath (van Thriel et al., 2010), and bronchitis (Lewin, 2011). Children and the elderly are particularly affected by this.

Figure 4. The production methods of the nanocomposite materials.

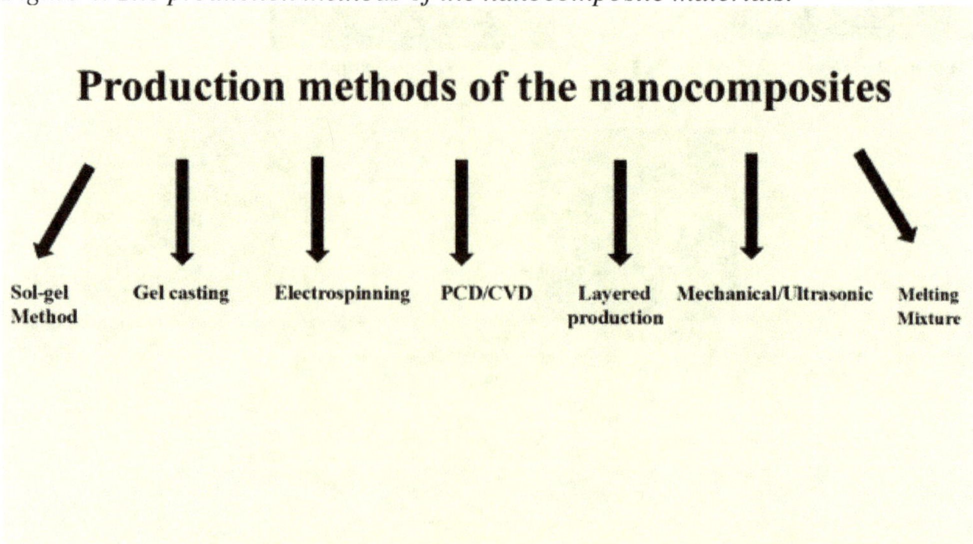

The production techniques for nanocomposites are as follows:

. The method of sol-gel

This method is generally used to produce ceramic matrix nanocomposites (Livage, 1997). The sol-gel method refers to the controlled conversion of a liquid solution into a gel and then into a solid (Danks et al., 2016; X. Guo et al., 2016). The sol-gel method is commonly used in the production of ceramic matrix nanocomposites (Suib, 2017; Zanurin et al., 2022). This method is based on the controlled conversion of a liquid solution (sol) into a solid structure (gel). Nano-sized components are homogeneously distributed in the matrix material, and then the material is brought to its final form by drying or heat treatment (Gill & Ballesteros, 2000).

. Gel casting

In the gel-forming method, the gel, once it has been prepared, is shaped and transformed into a solid form (Tallon & Franks, 2011). Drying, sintering, and baking techniques are applied to achieve this solid form (Abou Hammad et al., 2022).

. PVD and CVD vapor gas deposition methods

This technique is used to produce support materials such as graphene or carbon nanotubes. It has two methods: Chemical Vapor Deposition (CVD) (Suman et al., 2023; Vahlas et al., 2006; M. Wang et al., 2017) and Physical Vapor Deposition (PVD) (Baptista et al., 2018; Inspektor & Salvador, 2014; Mattox, 2002). In the CVD method, nano-components are deposited as a thin layer on a substrate (Shen et al., 2018). In the PVD method, solid materials are vaporized to form a nano-sized coating (Panjan et al., 2020).

. The method of electrospinning:

Electrospinning is a method used in nanofiber production. In this method, thin fibers are drawn from a liquid solution using high voltage. The nanofibers, integrated into a polymer matrix, increase the strength and flexibility of the material (Ji et al., 2024).

. Melting and Mixing Techniques

In the melt-and-mix method, a thermoplastic polymer matrix is used (Advani & Hsiao, 2012; Yousefpour et al., 2004). Then, a nanoscale filler is added. The resulting mixture is mixed until homogenized. The homogeneous mixture is then heated to the desired form (Albdiry, 2024).

. Additive Manufacturing (3D Printing and Additive Production)

The design flexibility resulting from the production of nanocomposite materials using 3D printing technologies can be utilized in the design of the material. This method is achieved by creating multiple layers of filaments modified with nanofillers (Jiménez et al., 2019).

. Mechanical and Ultrasonic Mixing Techniques

Ultrasonic waves or mechanical stirrers are used to ensure even distribution of nanocomponents within the matrix. This method is ideal for polymer-based nanocomposites (Kundalwal & Rathi, 2020).

6. DESULFURIZATION PROCESS

As mentioned in previous sections, SO_2 gas is a gas produced by the combustion of fossil fuels such as coal and fuel oil. This gas is toxic (Hannun & Abdul Razzaq, 2022; Moroń & Rybak, 2015). Therefore, its harmful effects are mitigated through desulfurization methods. Sulfur dioxide is absorbed chemically by bringing the waste gas and the scrubbing solution into contact in three different arrangements (Hanif et al., 2020; Pandey et al., 2005). In the first arrangement, perforated tubes are immersed in the scrubbing solution, and high pressure is applied to them (W. Liu et al., 2025). As a result of the pressure, air bubbles are released. If the depth of the scrubbing solution is sufficient and the air bubbles are as small as possible, this type of application ensures that the waste gas reaches a very close chemical equilibrium with the scrubbing solution (Rene et al., 2012). However, the disadvantage of this type of device is the high pressure loss (W. Liu et al., 2025; Rene et al., 2012). The second arrangement is the spray dryer stage. Here, as the waste gas rises from the bottom to the top in the tower, the scrubbing solution sprayed from the top moves downwards within the gas (Filková & Mujumdar, 2020; Lerotholi et al., 2022). In this technique, it is more difficult and complicated for the gas to reach equilibrium with the spray particles compared to the first stage (Aggarwal, 1998; Sánchez et al., 2015). This method is preferred to prevent sludge formation and other problems, such as clogging (Sánchez et al., 2015). At the third level, there is an absorption column (Pérez Sánchez et al., 2016). In this absorption column, waste gases are sent to a packed bed, and contact with a film layer is ensured on the packing (Flagiello et al., 2021; Sovilj et al., 2019). In practice, gas velocities in desulfurization devices vary between 0.3 m/s in packed bed columns and 3 m/s in spray dryers (Z. Li et al., 2022; P. Liu et al., 2025; Scala et al., 2004). The most commonly used method for controlling SO_2 emissions released into the atmosphere from the combustion of sulfur in fossil fuels is the absorption of sulfur dioxide gas with a scrubbing liquid containing finely ground limestone powder ($CaCO_3$) (T. Zhang et al., 2022). The investment and operating costs of such treatment plants constitute one of the main expenses of thermal power plants that generate electricity (Bihari et al., 2010; Elio & Milcarek, 2022). However, accelerating desulfurization methods with nanocomposite materials is more practical and convenient (Boshagh et al., 2022; Irshad et al., 2024; Morshedy et al., 2021; Zhou et al., 2021a). Techniques using nanocomposite materials with catalytic techniques can be summarized as oxidative desulfurization (ODS) (Muhammad et al., 2024), adsorptive-catalytic desulfurization (Botin et al., 2023), and photocatalytic desulfurization (X. Li et al., 2017; Shafiq et al., 2022; Zhou et al., 2021b).

6.1. Oxidative Desulfurization (ODS)

Sulfur compounds are oxidized to sulfones or sulfoxides under mild conditions in the presence of nanocomposite catalysts . These products can be easily removed from the fuel phase (Rajabi et al., 2021). Substances such as H_2O_2 and tert-butyl hydroperoxide are used as oxidizers. Examples of catalysts include MoO_3/rGO, Fe_3O_4@CNT, and Pd–Ag/rGO nanocomposites (Ghanbari Kermanshahi & Bahrami, 2019; Mirfakhraei et al., 2018). ODS is considered an environmentally friendly method because it takes place at low temperatures and without the need for hydrogen (Mashkina, 1990).

Figure 5. The illustration of the oxidative desulfurization nanocpomposite technique.

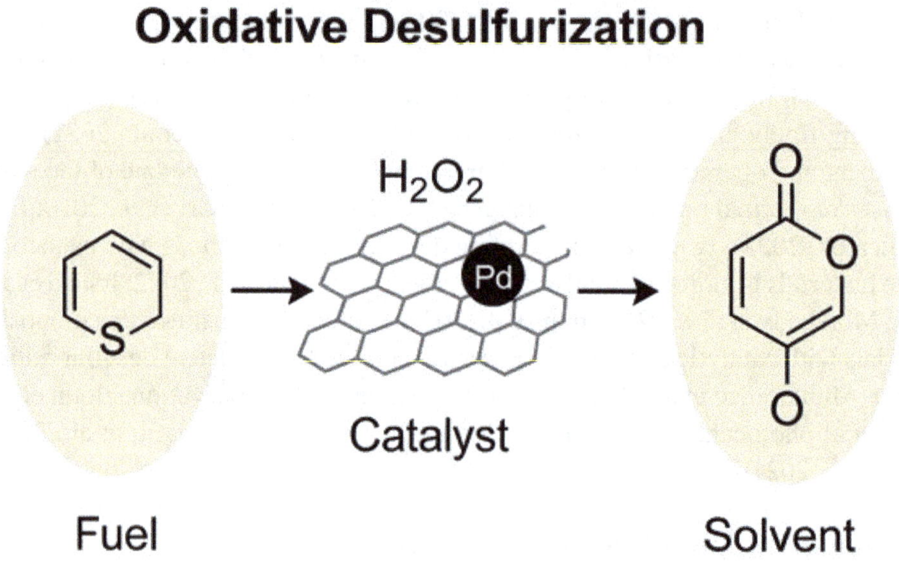

6.2. Adsorptive–Catalytic Desulfurization

In this method, nanocomposites act as both adsorbents and catalysts (Bazyari et al., 2016). Sulfur compounds are retained on the active surfaces and simultaneously decomposed (H. Q. Guo et al., 2014; C. Wang et al., 2020). Carbon-based nanocomposites and metal oxide-supported hybrid structures provide high selectivity (figure 6). Moreover, the dual functionality of these nanocomposites enables a synergistic interaction between adsorption and catalytic transformation, minimizing mass transfer limitations (Abdullah et al., 2025; De Clippel et al., 2013; J. Shi, 2012). The strong affinity of sulfur-containing molecules toward the nanocomposite surface enhances local concentration at the catalytic sites, thereby accelerating reaction kinetics (P. Wang et al., 2024; M. Zhao et al., 2024). Additionally, the structural tunability of carbon-based and metal oxide-supported hybrids allows optimization of pore size, surface chemistry, and active site distribution (Gao et al., 2023; Pérez-Mayoral et al., 2016). This results in improved desulfurization efficiency and catalyst reusability under mild operating conditions (Zaman et al., 2022).

Figure 6. The adsorptive–catalytic desulfurization using nanocomposites.

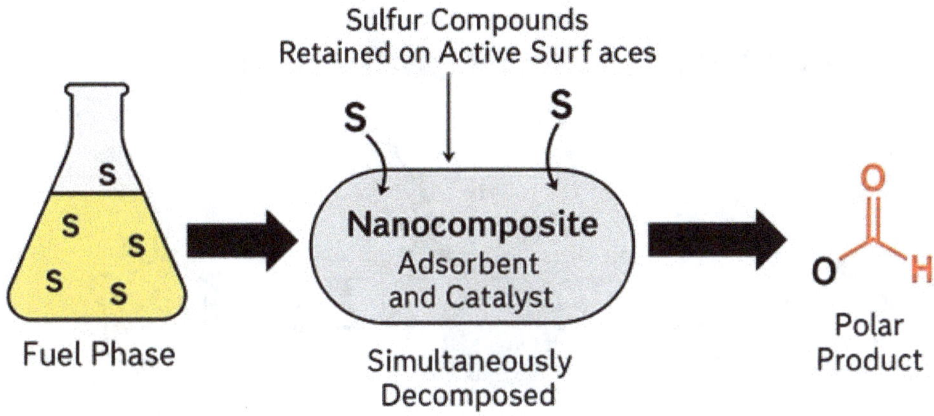

6.3. Photocatalytic Desulfurization

Sulfur compounds are broken down using semiconductor nanocomposites activated under UV or visible light (X. Li et al., 2016; Zaman et al., 2022). The catalysts used are TiO_2/graphene and ZnO-metal NP hybrids (figure 7). Moreover, the incorporation of graphene or metal nanoparticles into semiconductor matrices enhances charge separation and suppresses electron–hole recombination, leading to improved photocatalytic efficiency (X. Li et al., 2016; Marschall, 2014). These nanocomposite structures also provide a higher density of active sites and facilitate interfacial electron transfer (Liang et al., 2018; Lu et al., 2017; Z. Zhang et al., 2019). As a result, sulfur-containing compounds are oxidized more rapidly into polar products under mild reaction conditions (Q. Shi & Wu, 2021; Xi et al., 2023).

Figure 7. The illustration of the desulfurization of nanocomposite photocatalytric technique.

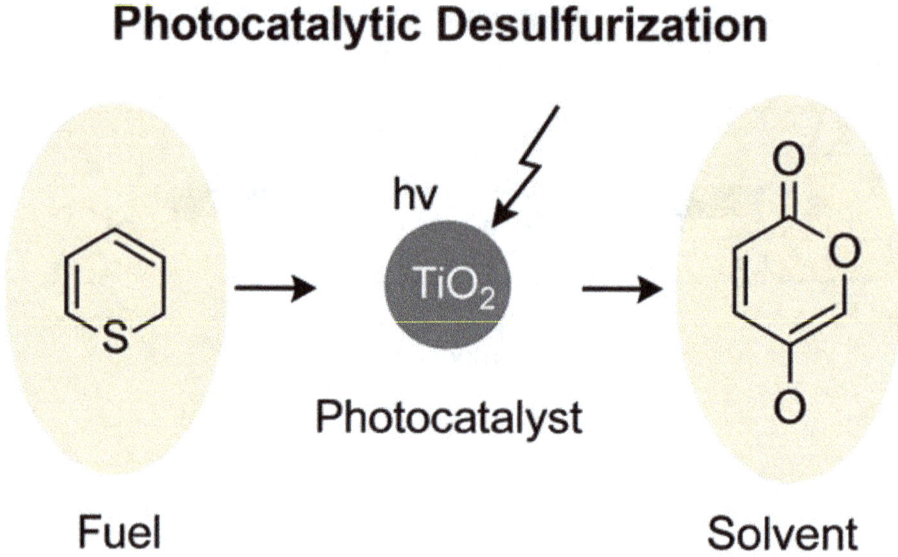

7. CONCLUSION

Given humanity's ever-increasing needs, the necessity for energy is an undeniable fact. Energy is essential for heating, electricity, and countless other needs. Humanity has turned to fossil fuels to meet its heating needs. Fossil fuels are fuels formed from animal and plant fossils that have been stored for millions of years, through various pressures, temperatures, and chemical reactions. Depending on the type of chemical reaction, temperature, pressure, environment, and the layers in which they are found, fossil fuels are classified into various types. The best known of these are petroleum, natural gas, and coal. Fossil fuels are a great necessity for heating. However, humanity's reliance on fossil fuels to meet its energy needs has created an excessive demand for them. This demand and excessive use have brought fossil fuel resources to the point of depletion, and the use of fossil fuels is also harmful to the ecosystem and the health of living beings. The release of toxic and harmful emissions from fossil fuels is an undesirable situation. Emission gases such as carbon monoxide (CO), carbon dioxide (CO_2), NO and its derivatives, sulfate (SO_2) ions, and sulfur derivatives are among the most commonly known. This chapter provides detailed information on SO_2 removal, or desulfurization. It explains how desulfurization methods can be made more effective with nanocomposite materials and the types of catalytic techniques used. Nanocomposite materials have various production techniques, including: sol-gel method, gel forming, PVD, CVD, melting mixture, electrospinning, ultrasonic and mechanical techniques, and layer production. Desulfurization processes with nanocomposite materials are: oxidative desulfurization, photocatalytic desulfurization, and adsorptive-catalytic desulfurization. Oxidative desulfurization; Sulfur compounds are oxidized to sulfones or sulfoxides under mild conditions in the presence of nanocomposite catalysts. Photocatalytic desulfurization techniques utilize sunlight. Sulfur compounds are broken down using semiconductor nanocomposites activated under UV or visible light. Adsorptive-catalytic desulfurization techniques involve nanocomposites acting as both adsorbents and catalysts. Sulfur compounds are retained on the active surfaces and simultaneously decomposed. This chapter provides detailed theoretical information on catalytic and desulfurization studies.

REFERENCES

Abas, N., Kalair, A., & Khan, N. (2015). Review of fossil fuels and future energy technologies. *Futures*, *69*, 31–49. DOI: 10.1016/j.futures.2015.03.003

Abdullah, N., Zumra, N., & Ahmed, S. (2025). Recent innovations in nanocomposite beads for the removal of pollutants from water: A critical review. *Environmental Science. Water Research & Technology*, *11*(6), 1428–1445. DOI: 10.1039/D5EW00215J

Abou Hammad, A. B., Mansour, A. M., Elhelali, T. M., & El Nahrawy, A. M. (2022). Sol-Gel/Gel Casting Nanoarchitectonics of Hybrid Fe2O3–ZnO/PS-PEG Nanocomposites and Their Optomagnetic Properties. *Journal of Inorganic and Organometallic Polymers and Materials 2022 33:2, 33*(2), 544–554. DOI: 10.1007/s10904-022-02519-2

Advani, S. G., & Hsiao, K. T. (2012). Introduction to composites and manufacturing processes. *Manufacturing Techniques for Polymer Matrix Composites (PMCs): A Volume in Woodhead Publishing Series in Composites Science and Engineering*, 1–12. DOI: 10.1533/9780857096258.1.1

Aggarwal, S. K. (1998). A review of spray ignition phenomena: Present status and future research. *Progress in Energy and Combustion Science*, *24*(6), 565–600. DOI: 10.1016/S0360-1285(98)00016-1

Al Janabi, M. A. Y., El Houda Tiri, R. N., Cherif, A., Altuner, E. E., Lee, C. J., Sen, F., Dragoi, E. N., Karimi, F., & Kalikeri, S. (2024). Hydrogen Generation by Methanolysis of NaBH4 via Efficient CuFe2O4 Nanoparticle Catalyst: A Kinetic Study and DNN Model. *Topics in Catalysis*, *67*(9–12), 843–852. DOI: 10.1007/s11244-024-01904-0

Albdiry, M. (2024). Effect of melt blending processing on mechanical properties of polymer nanocomposites: A review. *Polymer Bulletin*, *81*(7), 5793–5821. DOI: 10.1007/s00289-023-05012-z

Ali, A., & Zhao, C. (2020). Direct liquefaction techniques on lignite coal: A review. *Chinese Journal of Catalysis*, *41*(3), 375–389. DOI: 10.1016/S1872-2067(19)63492-3

Altuner, E. E. (2025). Synthesis, characterisation and photocatalytic activation of zinc nanoparticles via. biogenic methods. *International Journal of Environmental Analytical Chemistry*, *105*(19), 1–11. DOI: 10.1080/03067319.2025.2451356

Altuner, E. E., Bekmezci, M., Bayat, R., Akin, M., Isik, I., & Şen, F. (2021). Dendrimer-based nanocomposites for alcohol fuel cells. *Nanomaterials for Direct Alcohol Fuel Cells*, 337–352. DOI: 10.1016/B978-0-12-821713-9.00007-X

Altuner, E. E., Bekmezci, M., & Sen, F. (2022). Manufacturing Techniques of Magnetic Polymer Nanocomposites. *Handbook of Magnetic Hybrid Nanoalloys and Their Nanocomposites*, 1–16. DOI: 10.1007/978-3-030-34007-0_12-1

Altuner, E. E., Gur, T., & Şen, F. (2021). Ternary/quaternary nanomaterials for direct alcohol fuel cells. *Nanomaterials for Direct Alcohol Fuel Cells*, 157–172. DOI: 10.1016/B978-0-12-821713-9.00001-9

Altuner, E. E., Ozalp, V. C., Yilmaz, M. D., Bekmezci, M., & Sen, F. (2022). High-efficiency application of CTS-Co NPs mimicking peroxidase enzyme on TMB(ox). *Chemosphere*, *292*, 133429. DOI: 10.1016/j.chemosphere.2021.133429

Altuner, E. E., Ozalp, V. C., Yilmaz, M. D., Sudagidan, M., Aygun, A., Acar, E. E., Tasbasi, B. B., & Sen, F. (2022). Development of electrochemical aptasensors detecting phosphate ions on TMB substrate with epoxy-based mesoporous silica nanoparticles. *Chemosphere*, *134077*, 134077. Advance online publication. DOI: 10.1016/j.chemosphere.2022.134077

Altuner, E. E., Sen, F., & Dar, U. A. (2025). Commercialization of quinone-based drugs. *Quinone-Based Compounds in Drug Discovery: Trends and Applications*, 293–300. DOI: 10.1016/B978-0-443-24126-0.00017-2

Altuner, E. E., Tatli, H. N., Aldoori, B., & Dar, U. A. (2025). Waste Dye Removal with Nanotechnological Methods: Photocatalytic Processes. *Nanotechnology in the Life Sciences*, (Part F525), 249–270. DOI: 10.1007/978-3-031-82794-5_11

Armaroli, N., & Balzani, V. (2011a). The legacy of fossil fuels. *Chemistry, an Asian Journal*, *6*(3), 768–784. DOI: 10.1002/asia.201000797

Armaroli, N., & Balzani, V. (2011b). The legacy of fossil fuels. *Chemistry, an Asian Journal*, *6*(3), 768–784. DOI: 10.1002/asia.201000797

Ashok, B., Denis Ashok, S., & Ramesh Kumar, C. (2015). LPG diesel dual fuel engine – A critical review. *Alexandria Engineering Journal*, *54*(2), 105–126. DOI: 10.1016/j.aej.2015.03.002

Avtar, R., Tripathi, S., Aggarwal, A. K., & Kumar, P. (2019). Population–Urbanization–Energy Nexus: A Review. *Resources 2019, Vol. 8, Page 136, 8*(3), 136. DOI: 10.3390/resources8030136

Baptista, A., Silva, F., Porteiro, J., Míguez, J., & Pinto, G. (2018). Sputtering Physical Vapour Deposition (PVD) Coatings: A Critical Review on Process Improvement and Market Trend Demands. *Coatings 2018, Vol. 8, Page 402, 8*(11), 402. DOI: 10.3390/coatings8110402

Bazyari, A., Khodadadi, A. A., Haghighat Mamaghani, A., Beheshtian, J., Thompson, L. T., & Mortazavi, Y. (2016). Microporous titania–silica nanocomposite catalyst-adsorbent for ultra-deep oxidative desulfurization. *Applied Catalysis B: Environmental*, *180*, 65–77. DOI: 10.1016/j.apcatb.2015.06.011

Bihari, P., Gróf, G., & Gács, I. (2010). Efficiency and cost modelling of thermal power plants. *Thermal Science*, *14*(3), 821–834. DOI: 10.2298/TSCI1003821B

Blumer, M., & Sass, J. (1972). Oil pollution: Persistence and degradation of spilled fuel oil. *Science*, *176*(4039), 1120–1122. DOI: 10.1126/science.176.4039.1120

Borah, P., Kumar, M., & Devi, P. (2020). Types of inorganic pollutants: metals/metalloids, acids, and organic forms. *Inorganic Pollutants in Water*, 17–31. DOI: 10.1016/B978-0-12-818965-8.00002-0

Boshagh, F., Rahmani, M., & Zhu, W. (2022). Recent Advances and Challenges in Developing Technological Methods Assisting Oxidative Desulfurization of Liquid Fuels: A Review. *Energy & Fuels*, *36*(21), 12961–12985. DOI: 10.1021/acs.energyfuels.2c03218

Botin, A. A., Boldushevskii, R. E., Mozhaev, A. V., Ghambarian, M., Balar, M., Ghashghaee, M., & Nikulshin, P. A. (2023). Reactive adsorption desulfurization of model FCC gasoline on Ni-based adsorbents: Effect of active phase dispersion on activity and HDS/HYD selectivity. *Applied Catalysis B: Environmental*, *337*, 122946. DOI: 10.1016/j.apcatb.2023.122946

Burnham, A. K. (2017). Global Chemical Kinetics of Fossil Fuels: How to Model Maturation and Pyrolysis. *Global Chemical Kinetics of Fossil Fuels: How to Model Maturation and Pyrolysis*, 1–315. https://doi.org/DOI: 10.1007/978-3-319-49634-4/COVER

Chen, H., Cong, T. N., Yang, W., Tan, C., Li, Y., & Ding, Y. (2009). Progress in electrical energy storage system: A critical review. *Progress in Natural Science*, *19*(3), 291–312. DOI: 10.1016/j.pnsc.2008.07.014

Chen, S., Liu, Z., Du, H., Tang, C., Ji, C. Y., Quan, B., Pan, R., Yang, L., Li, X., Gu, C., Zhang, X., Yao, Y., Li, J., Fang, N. X., & Li, J. (2021). Electromechanically reconfigurable optical nano-kirigami. *Nature Communications 2021 12:1*, *12*(1), 1–8. DOI: 10.1038/s41467-021-21565-x

Chen, T. M., Gokhale, J., Shofer, S., & Kuschner, W. G. (2007). Outdoor Air Pollution: Nitrogen Dioxide, Sulfur Dioxide, and Carbon Monoxide Health Effects. *The American Journal of the Medical Sciences*, *333*(4), 249–256. DOI: 10.1097/MAJ.0b013e31803b900f

Chew, K. J. (2014). The future of oil: Unconventional fossil fuels. *Philosophical Transactions of the Royal Society A: Mathematical, Physical and Engineering Sciences, 372*(2006). https://doi.org/DOI: 10.1098/RSTA.2012.0324/58963

Chomiak, J., Longwell, J. P., & Sarofim, A. F. (1989). Combustion of low calorific value gases; Problems and prospects. *Progress in Energy and Combustion Science, 15*(2), 109–129. DOI: 10.1016/0360-1285(89)90012-9

Cui, Y., & Lieber, C. M. (2001). Functional Nanoscale Electronic Devices Assembled Using Silicon Nanowire Building Blocks. *Science, 291*(5505), 851–853. DOI: 10.1126/science.291.5505.851

Danks, A. E., Hall, S. R., & Schnepp, Z. (2016). The evolution of 'sol–gel' chemistry as a technique for materials synthesis. *Materials Horizons, 3*(2), 91–112. DOI: 10.1039/C5MH00260E

Darabi, R., Gu, Q., Mehrizi, A. A., Altuner, E. E., Alhrasishawi, W., Gulbagca, F., Tiri, R. N. E., Tekeli, Y., Seyrankaya, A., Kaynak, I., & Sen, F. (2023). Synthesis, characterization, and enhanced hydrogen generation from NaBH4 methanolysis of highly dispersed bimetallic PdNi nanoparticles supported on Vulcan carbon. *Molecular Catalysis, 547*, 113245. DOI: 10.1016/j.mcat.2023.113245

De Clippel, F., Dusselier, M., Van De Vyver, S., Peng, L., Jacobs, P. A., & Sels, B. F. (2013). Tailoring nanohybrids and nanocomposites for catalytic applications. *Green Chemistry, 15*(6), 1398–1430. DOI: 10.1039/c3gc37141g

Demirbas, A. (2004). Combustion characteristics of different biomass fuels. *Progress in Energy and Combustion Science, 30*(2), 219–230. DOI: 10.1016/j.pecs.2003.10.004

Demirbaş, A., & Demirbaş, A. H. (2004). Estimating the calorific values of lignocellulosic fuels. *Energy Exploration & Exploitation, 22*(2), 135–143. DOI: 10.1260/0144598041475198

Dignon, J. (1992). NOx and SOx emissions from fossil fuels: A global distribution. *Atmospheric Environment. Part A, General Topics, 26*(6), 1157–1163. DOI: 10.1016/0960-1686(92)90047-O

Dillon, A. C. (2010). Carbon Nanotubes for Photoconversion and Electrical Energy Storage. *Chemical Reviews, 110*(11), 6856–6872. DOI: 10.1021/cr9003314

Doner, N., Burhan, H., Bayat, R., Esra Altuner, E., & Sen, F. (2023). Energy Generation through Thermopower Waves using Multi-walled Carbon Nanotube Yarn. *Fuel, 331*, 125906. DOI: 10.1016/j.fuel.2022.125906

Eatough, D. J., Caka, F. M., & Farber, R. J. (1994). The Conversion of SO2 to Sulfate in the Atmosphere. *Israel Journal of Chemistry*, *34*(3–4), 301–314. DOI: 10.1002/ijch.199400034

Elio, J., & Milcarek, R. J. (2022). Techno-economic analysis and case study of combined heat and power systems in a wastewater treatment plant. *Energy*, *260*, 125106. DOI: 10.1016/j.energy.2022.125106

Esra Altuner, E., Yousif, M., & Al Janabi, A. (2025). Synthesis and Characterization of Iron-Cobalt@Activated Carbon Nanoparticles for Electrochemical Sensing of Dopamine Neurotransmitters. *Chemistry Africa*, *2025*(3), 1–9. DOI: 10.1007/s42250-025-01197-0

Feng, Q., Lin, Q., Gong, F., Sugita, S., & Shoya, M. (2004). Adsorption of lead and mercury by rice husk ash. *Journal of Colloid and Interface Science*, *278*(1), 1–8. DOI: 10.1016/j.jcis.2004.05.030

Ferry, J. G. (1995). CO dehydrogenase. *Annual Review of Microbiology*, *49*(1), 305–334. https://go.gale.com/ps/i.do?p=AONE&sw=w&issn=00664227&v=2.1&it=r&id=GALE%7CA17715866&sid=googleScholar&linkaccess=fulltext. DOI: 10.1146/annurev.mi.49.100195.001513

Filková, I., & Mujumdar, A. S. (2020). Industrial Spray Drying Systems. *Handbook of Industrial Drying*, 263–307. DOI: 10.1201/9780429289774-9

Flagiello, D., Parisi, A., Lancia, A., & Di Natale, F. (2021). A Review on Gas-Liquid Mass Transfer Coefficients in Packed-Bed Columns. *ChemEngineering 2021, Vol. 5, Page 43, 5*(3), 43. DOI: 10.3390/chemengineering5030043

Friedemann, A. J. (2021). *Life after Fossil Fuels. 81*. DOI: 10.1007/978-3-030-70335-6

Fulkerson, W., Judkins, R. R., & Sanghvi, M. K. (1990). Energy from Fossil Fuels. *JSTOR, 263*(3), 128–135. https://www.jstor.org/stable/24996937?seq=1

Gao, M., Wang, L., Yang, Y., Sun, Y., Zhao, X., & Wan, Y. (2023). Metal and Metal Oxide Supported on Ordered Mesoporous Carbon as Heterogeneous Catalysts. *ACS Catalysis*, *13*(7), 4060–4090. DOI: 10.1021/acscatal.2c05894

Ghanbari Ghozikali, M., Heibati, B., Naddafi, K., Kloog, I., Oliveri Conti, G., Polosa, R., & Ferrante, M. (2016). Evaluation of Chronic Obstructive Pulmonary Disease (COPD) attributed to atmospheric O3, NO2, and SO2 using Air Q Model (2011–2012 year). *Environmental Research*, *144*, 99–105. DOI: 10.1016/j.envres.2015.10.030

Ghanbari Kermanshahi, M., & Bahrami, K. (2019). Fe 3 O 4 @BNPs@SiO 2 –SO 3 H as a highly chemoselective heterogeneous magnetic nanocatalyst for the oxidation of sulfides to sulfoxides or sulfones. *RSC Advances*, *9*(62), 36103–36112. DOI: 10.1039/C9RA06221A

Gill, I., & Ballesteros, A. (2000). Bioencapsulation within synthetic polymers (Part 1): Sol-gel encapsulated biologicals. *Trends in Biotechnology*, *18*(7), 282–296. DOI: 10.1016/S0167-7799(00)01457-8

Glass, N. R. (1979). Environmental effects of increased coal utilization: Ecological effects of gaseous emissions from coal combustion. *Environmental Health Perspectives*, *33*, 249–272. DOI: 10.1289/ehp.7933249

Greenberg, N., Carel, R. S., Derazne, E., Bibi, H., Shpriz, M., Tzur, D., & Portnov, B. A. (2016). Different effects of long-term exposures to SO2 and NO2 air pollutants on asthma severity in young adults. *Journal of Toxicology and Environmental Health. Part A.*, *79*(8), 342–351. DOI: 10.1080/15287394.2016.1153548

Guelpa, E., & Verda, V. (2019). Thermal energy storage in district heating and cooling systems: A review. *Applied Energy*, *252*, 113474. DOI: 10.1016/j.apenergy.2019.113474

Guo, H. Q., Xie, L. L., Wang, X. L., Liu, F. R., Wang, M. J., & Hu, R. S. (2014). Sulfur removal and release behaviors of sulfur-containing model compounds during pyrolysis under inert atmosphere by TG-MS connected with Py-GC. *Journal of Fuel Chemistry & Technology*, *42*(10), 1160–1166. DOI: 10.1016/S1872-5813(14)60047-5

Guo, X., Zhang, Q., Ding, X., Shen, Q., Wu, C., Zhang, L., & Yang, H. (2016). Synthesis and application of several sol–gel-derived materials via sol–gel process combining with other technologies: a review. *Journal of Sol-Gel Science and Technology 2015 79:2*, *79*(2), 328–358. DOI: 10.1007/s10971-015-3935-6

Gupta, P. K. (2020). Solvents, Gasses, and Vapors. *Problem Solving Questions in Toxicology:*, 181–193. DOI: 10.1007/978-3-030-50409-0_13

Hanif, M. A., Ibrahim, N., & Abdul Jalil, A. (2020). Sulfur dioxide removal: An overview of regenerative flue gas desulfurization and factors affecting desulfurization capacity and sorbent regeneration. *Environmental Science and Pollution Research 2020 27:22*, *27*(22), 27515–27540. DOI: 10.1007/s11356-020-09191-4

Hannun, R. M., & Abdul Razzaq, A. H. (2022). Air Pollution Resulted from Coal, Oil and Gas Firing in Thermal Power Plants and Treatment: A Review. *IOP Conference Series. Earth and Environmental Science*, *1002*(1), 012008. DOI: 10.1088/1755-1315/1002/1/012008

He, G., He, H., He, H., & He, H. (2020). Water Promotes the Oxidation of SO2 by O2 over Carbonaceous Aerosols. *Environmental Science & Technology, 54*(12), 7070–7077. DOI: 10.1021/acs.est.0c00021

Howarth, R. W. (2014). A bridge to nowhere: Methane emissions and the greenhouse gas footprint of natural gas. *Energy Science & Engineering, 2*(2), 47–60. DOI: 10.1002/ese3.35

Huang, Y., Shen, H., Chen, Y., Zhong, Q., Chen, H., Wang, R., Shen, G., Liu, J., Li, B., & Tao, S. (2015). Global organic carbon emissions from primary sources from 1960 to 2009. *Atmospheric Environment, 122*, 505–512. DOI: 10.1016/j.atmosenv.2015.10.017

Hung, Y.-T., Wang, L. K., & Shammas, N. K. (2020). *Handbook of Environment and Waste Management. 03*. DOI: 10.1142/11470

Inman, M. (2013). The True Cost pf Fossil Fuels. *JSTOR, 308*(4), 58–61. https://www.jstor.org/stable/26018067

Inspektor, A., & Salvador, P. A. (2014). Architecture of PVD coatings for metalcutting applications: A review. *Surface and Coatings Technology, 257*, 138–153. DOI: 10.1016/j.surfcoat.2014.08.068

Irshad, T., Saeed, M., Munir, M., Kousar, R., Intisar, A., Din, M. I., Ahsan, H., Abid, M. A., Irfan, M., & Osman, S. M. (2024). Excellent catalytic performance over reduced graphene-boosted novel nanoparticles for oxidative desulfurization of fuel oil. *Nanotechnology Reviews, 13*(1), 20240116. Advance online publication. DOI: 10.1515/ntrev-2024-0116

Jamar, A., Majid, Z. A. A., Azmi, W. H., Norhafana, M., & Razak, A. A. (2016). A review of water heating system for solar energy applications. *International Communications in Heat and Mass Transfer, 76*, 178–187. DOI: 10.1016/j.icheatmasstransfer.2016.05.028

Ji, D., Lin, Y., Guo, X., Ramasubramanian, B., Wang, R., Radacsi, N., Jose, R., Qin, X., & Ramakrishna, S. (2024). Electrospinning of nanofibres. *Nature Reviews Methods Primers 2024 4:1, 4*(1), 1-. DOI: 10.1038/s43586-023-00278-z

Jiménez, M., Romero, L., Domínguez, I. A., Espinosa, M. D. M., & Domínguez, M. (2019). Additive Manufacturing Technologies: An Overview about 3D Printing Methods and Future Prospects. *Complexity, 2019*(1), 9656938. DOI: 10.1155/2019/9656938

Kapusta, K., & Stańczyk, K. (2011). Pollution of water during underground coal gasification of hard coal and lignite. *Fuel, 90*(5), 1927–1934. DOI: 10.1016/j.fuel.2010.11.025

Kaya, D., Çanka Kılıç, F., & Öztürk, H. H. (2021). *Fuels and Combustion*. 227–263. DOI: 10.1007/978-3-030-25995-2_8

Khvorostyanov, D. V., Krinner, G., Ciais, P., Heimann, M., & Zimov, S. A. (2008). Vulnerability of permafrost carbon to global warming. Part I: Model description and role of heat generated by organic matter decomposition. *Tellus, Series B: Chemical and Physical Meteorology, 60 B*(2), 250–264. https://doi.org/DOI: 10.1111/J.1600-0889.2007.00333.X;WGROUP:STRING:PUBLICATION

Klebe, S., Leigh, J., Henderson, D. W., & Nurminen, M. (2019). Asbestos, Smoking and Lung Cancer: An Update. *International Journal of Environmental Research and Public Health 2020, Vol. 17, Page 258, 17*(1), 258. DOI: 10.3390/ijerph17010258

Kocak, Y., Aygun, A., Altuner, E. E., Gonca, S., Berikten, D., Tiri, R. N. E., & Sen, F. (2024). *Eco-friendly production of platinum nanoparticles: physicochemical properties, evaluation of biological and catalytic activities. 21*, 51–62. https://doi.org/DOI: 10.1007/S13762-023-05232-W/FIGURES/7

Kojima, Y., Suzuki, K. I., Fukumoto, K., Sasaki, M., Yamamoto, T., Kawai, Y., & Hayashi, H. (2002). Hydrogen generation using sodium borohydride solution and metal catalyst coated on metal oxide. *International Journal of Hydrogen Energy, 27*(10), 1029–1034. DOI: 10.1016/S0360-3199(02)00014-9

Kundalwal, S. I., & Rathi, A. (2020). Improved mechanical and viscoelastic properties of CNT-composites fabricated using an innovative ultrasonic dual mixing technique. *Journal of the Mechanical Behavior of Materials, 29*(1), 77–85. DOI: 10.1515/jmbm-2020-0008

Lam, N. L., Smith, K. R., Gauthier, A., & Bates, M. N. (2012). Kerosene: A review of household uses and their hazards in low-and middle-income countries. *Journal of Toxicology and Environmental Health. Part B, Critical Reviews, 15*(6), 396–432. DOI: 10.1080/10937404.2012.710134

Lerotholi, L., Everson, R. C., Koech, L., Neomagus, H. W. J. P., Rutto, H. L., Branken, D., Hattingh, B. B., & Sukdeo, P. (2022). Semi-dry flue gas desulphurization in spray towers: a critical review of applicable models for computational fluid dynamics analysis. *Clean Technologies and Environmental Policy 2022 24:7, 24*(7), 2011–2060. DOI: 10.1007/s10098-022-02308-y

Lewin, A. (2011). *Hospitalisation pour problèmes respiratoires et exposition aux émissions d'une aluminerie chez l'enfant* [Universite de Montreal]. DOI: 10.71781/725

Li, X., Yu, J., Wageh, S., Al-Ghamdi, A. A., & Xie, J. (2016). Graphene in Photocatalysis: A Review. *Small, 12*(48), 6640–6696. DOI: 10.1002/smll.201600382

Li, X., Zhu, W., Lu, X., Zuo, S., Yao, C., & Ni, C. (2017). Integrated nanostructures of CeO2/attapulgite/g-C3N4 as efficient catalyst for photocatalytic desulfurization: Mechanism, kinetics and influencing factors. *Chemical Engineering Journal, 326*, 87–98. DOI: 10.1016/j.cej.2017.05.131

Li, Z., Ma, X., Liao, Y., & Zhao, N. (2022). Characteristics analysis and parameters optimization of desulfurization wastewater evaporation in a rotary spray drying tower. *Powder Technology, 399*, 117211. DOI: 10.1016/j.powtec.2022.117211

Liang, Y., Zhang, W., Wu, D., Ni, Q. Q., & Zhang, M. Q. (2018). Interface Engineering of Carbon-Based Nanocomposites for Advanced Electrochemical Energy Storage. *Advanced Materials Interfaces, 5*(14), 1800430. DOI: 10.1002/admi.201800430

Liu, J., Burciaga, R., Tang, S., Ding, S., Ran, H., Zhao, W., Wang, G., Zhuang, Z., Xie, L., Lyu, Z., Lin, Y., Du, A., Yuan, A., Fu, J., Song, B., Zhu, J., Sun, Z., Jin, X., Huo, Z. Y., & Zhu, W. (2024). Heterogeneous catalysis for the environment. *The Innovation Materials, 2*(3), 100090. DOI: 10.59717/j.xinn-mater.2024.100090

Liu, J., Jiang, Y., Yao, W., Jiang, X., & Jiang, X. (2019). Molecular Characterization of Henan Anthracite Coal. *Energy & Fuels, 33*(7), 6215–6225. DOI: 10.1021/acs.energyfuels.9b01061

Liu, P., Yang, S., & Wang, H. (2025). Mass transfer behavior and desulfurization characteristics in the three-dimensional wet flue gas desulfurization tower. *Energy, 319*, 134840. DOI: 10.1016/j.energy.2025.134840

Liu, Q., Wu, X., Wang, X., Jin, Z., Zhu, D., Meng, Q., Huang, S., Liu, J., & Fu, Q. (2019). Carbon and hydrogen isotopes of methane, ethane, and propane: A review of genetic identification of natural gas. *Earth-Science Reviews, 190*, 247–272. DOI: 10.1016/j.earscirev.2018.11.017

Liu, W., Gao, W., Zhang, H., Wei, B., Jiang, R., Liang, L., Zhang, L., & Xie, Y. (2025). A New Clean Completion Technology for Soluble Screen Tubes in Ultra-Deep Well with High Temperature and High Pressure. *International Petroleum Technology Conference, IPTC 2025*. DOI: 10.2523/IPTC-25094-MS

Liu, W., Xu, H., Liao, Y., Quan, Z., Li, S., Zhao, S., Qu, Z., & Yan, N. (2019). Recyclable CuS sorbent with large mercury adsorption capacity in the presence of SO2 from non-ferrous metal smelting flue gas. *Fuel, 235*, 847–854. DOI: 10.1016/j.fuel.2018.08.062

Livage, J. (1997). Sol-gel processes. *Current Opinion in Solid State and Materials Science, 2*(2), 132–138. DOI: 10.1016/S1359-0286(97)80057-5

Lockwood, T. (2017). A Compararitive Review of Next-generation Carbon Capture Technologies for Coal-fired Power Plant. *Energy Procedia, 114*, 2658–2670. DOI: 10.1016/j.egypro.2017.03.1850

Longwell, J. P., Rubin, E. S., & Wilson, J. (1995). Coal: Energy for the future. *Progress in Energy and Combustion Science, 21*(4), 269–360. DOI: 10.1016/0360-1285(95)00007-0

Lu, Z., Zeng, L., Song, W., Qin, Z., Zeng, D., & Xie, C. (2017). In situ synthesis of C-TiO2/g-C3N4 heterojunction nanocomposite as highly visible light active photocatalyst originated from effective interfacial charge transfer. *Applied Catalysis B: Environmental, 202*, 489–499. DOI: 10.1016/j.apcatb.2016.09.052

Lundberg, J. O. N., Weitzberg, E., Lundberg, J. M., Alving, K., Lundberg, E. J. O. N., Weitzberg, J. M., & Lundberg, K. A. (1996). Nitric oxide in exhaled air. *The European Respiratory Journal, 9*(12), 2671–2680. DOI: 10.1183/09031936.96.09122671

Lyon, R. E. (2000). *Heat Release Kinetics. Fire and Materials An International Journal.* Wiley., DOI: 10.1002/1099-1018(200007/08)24:4

Maginn, E. J. (2010). What to Do with CO2. *The Journal of Physical Chemistry Letters, 1*(24), 3478–3479. DOI: 10.1021/jz101582c

Marschall, R. (2014). Semiconductor composites: Strategies for enhancing charge carrier separation to improve photocatalytic activity. *Advanced Functional Materials, 24*(17), 2421–2440. DOI: 10.1002/adfm.201303214

Martins, F., Felgueiras, C., & Smitková, M. (2018). Fossil fuel energy consumption in European countries. *Energy Procedia, 153*, 107–111. DOI: 10.1016/j.egypro.2018.10.050

Mashkina, A. V. (1990). Catalytic Synthesis of Sulfoxides and Sulfones Via Oxidation of Sulfides by Molecular Oxygen. *Catalysis Reviews. Science and Engineering, 32*(1–2), 105–161. DOI: 10.1080/01614949009349941

Mattox, D. M. (2002). Physical vapor deposition (PVD) processes. *Metal Finishing, 100*(1, SUPPL.), 394–408. DOI: 10.1016/S0026-0576(02)82043-8

Mirfakhraei, S., Hekmati, M., Eshbala, F. H., & Veisi, H. (2018). Fe3O4/PEG-SO3H as a heterogeneous and magnetically-recyclable nanocatalyst for the oxidation of sulfides to sulfones or sulfoxides. *New Journal of Chemistry*, *42*(3), 1757–1761. DOI: 10.1039/C7NJ02513K

Miwa, K., Aoki, M., Noritake, T., Ohba, N., Nakamori, Y., Towata, S. I., Züttel, A., & Orimo, S. I. (2006). Thermodynamical stability of calcium borohydride. *Physical Review B: Condensed Matter and Materials Physics*, *74*(15), 155122. DOI: 10.1103/PhysRevB.74.155122

Mohnen, V. A. (n.d.). The Challenge of Acid Rain on JSTOR. *Scientific American, JSTOR, 1988*, 30–39. Retrieved December 12, 2025, from https://www.jstor.org/stable/24989192

Moroń, W., & Rybak, W. (2015). NOx and SO2 emissions of coals, biomass and their blends under different oxy-fuel atmospheres. *Atmospheric Environment*, *116*, 65–71. DOI: 10.1016/j.atmosenv.2015.06.013

Morshedy, A. S., Ali, H. R., Nada, A. A., Rabie, A. M., & El-Maghrabi, H. H. (2021). Highly efficient Imprinted Polymer Nanocomposites for photocatalytic desulfurization of real diesel fuel. *Environmental Technology & Innovation*, *21*, 101206. DOI: 10.1016/j.eti.2020.101206

Mousavi, A. (2020). An Investigation of How Nitrogen Monoxide (NO) and Nitrogen Dioxide (NO2) Are Produced in Oxidations by Nitric Acid (HNO3). *Comments on Modern Chemistry. Part A, Comments on Inorganic Chemistry : a Journal of Critical Discussion of the Current Literature*, *40*(6), 307–313. DOI: 10.1080/02603594.2020.1803296

Mudakkar, S. R., Zaman, K., Khan, M. M., & Ahmad, M. (2013). Energy for economic growth, industrialization, environment and natural resources: Living with just enough. *Renewable & Sustainable Energy Reviews*, *25*, 580–595. DOI: 10.1016/j.rser.2013.05.024

Muhammad, T., Ahmad, W., Ahmad, I., & Yaseen, M. (2024). Integrated oxidative-adsorptive desulfurization of model and real fuel oils over a Zn-impregnated hydroxyapatite-activated carbon (Zn/HA-AC) composite. *Journal of Sulfur Chemistry*, *45*(2), 217–238. DOI: 10.1080/17415993.2023.2264439

Muzio, L. J., & Quartucy, G. C. (1997). Implementing NOx control: Research to application. *Progress in Energy and Combustion Science*, *23*(3), 233–266. DOI: 10.1016/S0360-1285(97)00002-6

Nayak, S., Das, M., Pradhan, B., & Kundu, S. N. (2019). Environmental Impacts of Fossil Fuels. *Sustainable Energy and Environment*, 245–270. DOI: 10.1201/9780429430107-13

Ni, K., Wu, Y., Karimi, F., Gulbagca, F., Seyrankaya, A., Esra Altuner, E., Kocak, Y., & Sen, F. (2023). Palladium based bimetallic nanocatalysts: Synthesis, characterization and hydrogen fuel production. *Fuel, 341*, 127577. DOI: 10.1016/j.fuel.2023.127577

Nur Tatli, H., Beniz Gunduz, S., Sahin, M., Esra Altuner, E., & Ali Dar, U. (2024). Fluorimetric methods for determination of aluminum in water resources utilizing newly synthesized N,N'-bis(2,5-dihydroxybenzylidene)-4,4′-diamino diphenyl ether. *Spectrochimica Acta. Part A: Molecular and Biomolecular Spectroscopy, 321*, 124726. DOI: 10.1016/j.saa.2024.124726

Nyambuu, U., & Semmler, W. (2014). Trends in the extraction of non-renewable resources: The case of fossil energy. *Economic Modelling, 37*, 271–279. DOI: 10.1016/j.econmod.2013.11.020

Oparin, V. N., Kiryaeva, T. A., Gavrilov, V. Y., Tanashev, Y. Y., & Bolotov, V. A. (2017). Initiation of underground fire sources. *Journal of Mining Science 2016 52:3, 52*(3), 576–592. DOI: 10.1134/S1062739116030850

Ozdemir, S., Turkan, Z., Kilinc, E., Altuner, E. E., & Sen, F. (2022). Anoxybacillus flavithermus loaded ɣ-Fe2O3 magnetic nanoparticles as an efficient magnetic sorbent for the preconcentrations of Cu(II) and Mn(II). *Food and Chemical Toxicology, 168*, 113334. DOI: 10.1016/j.fct.2022.113334

Özdemir, S., Turkan, Z., Kılınc, E., Altuner, E. E., & Sen, F. (2023). Bioaccumulation, resistance, and remediation of Mn(II) and Cu(II) and their impacts on antioxidant enzymes of Anoxybacillus flavithermus. *International Journal of Environmental Science and Technology 2023 20:10, 20*(10), 10823–10834. DOI: 10.1007/s13762-023-05133-y

Pandey, R. A., Biswas, R., Chakrabarti, T., & Devotta, S. (2005). Flue gas desulfurization: Physicochemical and biotechnological approaches. *Critical Reviews in Environmental Science and Technology, 35*(6), 571–622. DOI: 10.1080/10643380500326374

Panjan, P., Drnovšek, A., Gselman, P., Čekada, M., & Panjan, M. (2020). Review of Growth Defects in Thin Films Prepared by PVD Techniques. *Coatings 2020, Vol. 10, Page 447, 10*(5), 447. DOI: 10.3390/coatings10050447

Peixoto, R. S., Vermelho, A. B., & Rosado, A. S. (2011). Petroleum-Degrading Enzymes: Bioremediation and New Prospects. *Enzyme Research, 2011*(1), 475193. DOI: 10.4061/2011/475193

Pérez, I. A., García, M. Á., Sánchez, M. L., Pardo, N., & Fernández-Duque, B. (2020). Key Points in Air Pollution Meteorology. *International Journal of Environmental Research and Public Health 2020, Vol. 17, Page 8349, 17*(22), 8349. DOI: 10.3390/ijerph17228349

Pérez-Mayoral, E., Calvino-Casilda, V., & Soriano, E. (2016). Metal-supported carbon-based materials: Opportunities and challenges in the synthesis of valuable products. *Catalysis Science & Technology, 6*(5), 1265–1291. DOI: 10.1039/C5CY01437A

Pérez Sánchez, A. J., Pérez Sánchez, E. J., & Segura Silva, R. M. (2016). Design of a packed-bed absorption column considering four packing types and applying MATLAB. *Nexo Revista Científica, ISSN-e 1995-9516, Vol. 29, No. 2, 2016 (Ejemplar Dedicado a: Nexo Revista Científica), Págs. 83-104, 29*(2), 83–104. DOI: 10.5377/nexo.v29i2.4577

Perraud, V., Horne, J. R., Martinez, A. S., Kalinowski, J., Meinardi, S., Dawson, M. L., Wingen, L. M., Dabdub, D., Blake, D. R., Gerber, R. B., & Finlayson-Pitts, B. J. (2015). The future of airborne sulfur-containing particles in the absence of fossil fuel sulfur dioxide emissions. *Proceedings of the National Academy of Sciences of the United States of America, 112*(44), 13514–13519. DOI: 10.1073/pnas.1510743112

Radišić, B. (2024).. . *Non-Renewable and Renewable Energy Sources., 2024*(1), 1–5. DOI: 10.3997/2214-4609.202449BGS49

Ragazzi, M., Ionescu, G., & Cioranu, S. I. (2017). Assessment of environmental impact from renewable and non-renewable energy sources. *International Journal of Energy Production and Management, 2*(1), 8–16. DOI: 10.2495/EQ-V2-N1-8-16

Rajabi, F., Vessally, E., Luque, R., & Voskressensky, L. (2021). Highly efficient and selective aqueous aerobic oxidation of sulfides to sulfoxides or sulfones catalyzed by tungstate-functionalized nanomaterial. *Molecular Catalysis, 515*, 111931. DOI: 10.1016/j.mcat.2021.111931

Ramadan, A. E. K. (2004). *Acid Deposition Phenomena.*

Rene, E. R., Veiga, M. C., & Kennes, C. (2012). Combined biological and physicochemical waste-gas cleaning techniques. *Journal of Environmental Science and Health. Part A, Toxic/Hazardous Substances & Environmental Engineering, 47*(7), 920–939. DOI: 10.1080/10934529.2012.667289

Husain, I. A. F., Alkhatib, M. F., Jami, M. S., Mirghani, M. E. S., Zainudin, Z. Bin, & Hoda, A. 2014. Problems, Control, and Treatment of Fat, Oil, and Grease FOGReview, A. (•••).. . *Journal of Oleo Science*, *63*(8), 747–752. DOI: 10.5650/jos.ess13182

Ritchie, H., & Rosado, P. (2017). Fossil fuels. *Our World in Data*. https://ourworldindata.org/fossil-fuels

Rogalinski, T., Ingram, T., & Brunner, G. (2008). Hydrolysis of lignocellulosic biomass in water under elevated temperatures and pressures. *The Journal of Supercritical Fluids*, *47*(1), 54–63. DOI: 10.1016/j.supflu.2008.05.003

Sánchez, A. L., Urzay, J., & Liñán, A. (2015). The role of separation of scales in the description of spray combustion. *Proceedings of the Combustion Institute*, *35*(2), 1549–1577. DOI: 10.1016/j.proci.2014.08.018

Sarbassov, Y., Duan, L., Manovic, V., & Anthony, E. J. (2018). Sulfur trioxide formation/emissions in coal-fired air- and oxy-fuel combustion processes: A review. *Greenhouse Gases: Science and Technology*, *8*(3), 402–428. DOI: 10.1002/ghg.1767

Saxena, P., & Shukla, P. (2023). A review on recent developments and advances in environmental gas sensors to monitor toxic gas pollutants. *Environmental Progress & Sustainable Energy*, *42*(5), e14126. DOI: 10.1002/ep.14126

Scala, F., D'Ascenzo, M., & Lancia, A. (2004). Modeling flue gas desulfurization by spray-dry absorption. *Separation and Purification Technology*, *34*(1–3), 143–153. DOI: 10.1016/S1383-5866(03)00188-6

Schobesberger, S., Franchin, A., Bianchi, F., Rondo, L., Duplissy, J., Kürten, A., Ortega, I. K., Metzger, A., Schnitzhofer, R., Almeida, J., Amorim, A., Dommen, J., Dunne, E. M., Ehn, M., Gagné, S., Ickes, L., Junninen, H., Hansel, A., Kerminen, V. M., & Worsnop, D. R. (2015). On the composition of ammonia-sulfuric-acid ion clusters during aerosol particle formation. *Atmospheric Chemistry and Physics*, *15*(1), 55–78. DOI: 10.5194/acp-15-55-2015

Schwietzke, S., Griffin, W. M., Matthews, H. S., & Bruhwiler, L. M. P. (2014). Natural Gas Fugitive Emissions Rates Constrained by Global Atmospheric Methane and Ethane. *Environmental Science & Technology*, *48*(14), 7714–7722. DOI: 10.1021/es501204c

Schwietzke, S., Sherwood, O. A., Bruhwiler, L. M. P., Miller, J. B., Etiope, G., Dlugokencky, E. J., Michel, S. E., Arling, V. A., Vaughn, B. H., White, J. W. C., & Tans, P. P. (2016). Upward revision of global fossil fuel methane emissions based on isotope database. *Nature 2016 538:7623*, *538*(7623), 88–91. DOI: 10.1038/nature19797

Seewald, J. S., Benitez-Nelson, B. C., & Whelan, J. K. (1998). Laboratory and theoretical constraints on the generation and composition of natural gas. *Geochimica et Cosmochimica Acta*, *62*(9), 1599–1617. DOI: 10.1016/S0016-7037(98)00000-3

Shafiee, S., & Topal, E. (2009). When will fossil fuel reserves be diminished? *Energy Policy*, *37*(1), 181–189. DOI: 10.1016/j.enpol.2008.08.016

Shafiee, S., & Topal, E. (2010). A long-term view of worldwide fossil fuel prices. *Applied Energy*, *87*(3), 988–1000. DOI: 10.1016/j.apenergy.2009.09.012

Shafiq, I., Shafique, S., Akhter, P., Abbas, G., Qurashi, A., & Hussain, M. (2022). Efficient catalyst development for deep aerobic photocatalytic oxidative desulfurization: Recent advances, confines, and outlooks. *Catalysis Reviews. Science and Engineering*, *64*(4), 789–834. DOI: 10.1080/01614940.2020.1864859

Shen, P. C., Lin, Y., Wang, H., Park, J. H., Leong, W. S., Lu, A. Y., Palacios, T., & Kong, J. (2018). CVD Technology for 2-D Materials. *IEEE Transactions on Electron Devices*, *65*(10), 4040–4052. DOI: 10.1109/TED.2018.2866390

Shi, J. (2012). On the Synergetic Catalytic Effect in Heterogeneous Nanocomposite Catalysts. *Chemical Reviews*, *113*(3), 2139–2181. DOI: 10.1021/cr3002752

Shi, Q., & Wu, J. (2021). Review on Sulfur Compounds in Petroleum and Its Products: State-of-the-Art and Perspectives. *Energy & Fuels*, *35*(18), 14445–14461. DOI: 10.1021/acs.energyfuels.1c02229

Sivalingam, S., Vishal, G., & Anush, B. (2024). Environmental and health effects of acid rain. *Health and Environmental Effects of Ambient Air Pollution: Volume 1: Air Pollution, Human Health, and the Environment, 1*, 91–107. DOI: 10.1016/B978-0-443-16088-2.00007-7

Sloan, M. E., Staubitz, A., Clark, T. J., Russell, C. A., Lloyd-Jones, G. C., & Manners, I. (2010). Homogeneous Catalytic Dehydrocoupling/Dehydrogenation of Amine–Borane Adducts by Early Transition Metal, Group 4 Metallocene Complexes. *Journal of the American Chemical Society*, *132*(11), 3831–3841. DOI: 10.1021/ja909535a

Soeder, D. J. (2021). Fossil Fuels and Climate Change. *Fracking and the Environment*, 155–185. DOI: 10.1007/978-3-030-59121-2_9

Sovilj, M. N., Nikolovski, B. G., Spasojević, M., & Mauhar, S. M. (2019). Packed bed absorption column: Hydrodynamics and mass transfer. *Acta Periodica Technologica*, *50*(50), 260–267. DOI: 10.2298/APT1950260S

Srivastava, R. K., Jozewicz, W., & Singer, C. (2001). SO2 scrubbing technologies: A review. *Environmental Progress*, *20*(4), 219–228. DOI: 10.1002/ep.670200410

Stejskal, V., Vendl, T., Aulicky, R., & Athanassiou, C. (2021). Synthetic and Natural Insecticides: Gas, Liquid, Gel and Solid Formulations for Stored-Product and Food-Industry Pest Control. *Insects 2021, Vol. 12, Page 590, 12*(7), 590. DOI: 10.3390/insects12070590

Strielkowski, W., Lutsenko, E., & Pavlov, D. (2021). Fossil fuel industry development in the 21st century: a case of coal. *Fossil Fuel Industry Development in the 21st Century: A Case of Coal, 128*, 02004. DOI: 10.1051/shsconf/202112802004

Suib, S. L. (2017). A Review of Nanoceramic Materials for Use in Ceramic Matrix Composites. *Sol-Gel Based Nanoceramic Materials: Preparation, Properties and Applications*, 185–230. DOI: 10.1007/978-3-319-49512-5_7

Suman, S., Pravalika, J., Manjula, P., & Farooq, U. (2023). Gender and CVD- Does It Really Matters? *Current Problems in Cardiology, 48*(5), 101604. DOI: 10.1016/j.cpcardiol.2023.101604

Süme, V., Daneshfaraz, R., Kerim, A., Abbaszadeh, H., & Abraham, J. (2024). Investigation of Clean Energy Production in Drinking Water Networks. *Water Resources Management 2024 38:6, 38*(6), 2189–2208. DOI: 10.1007/s11269-024-03752-9

Süto, L. (2010). Mining: Extraction of Fossil Fuels. *Anthropogenic Geomorphology: A Guide to Man-Made Landforms*, 131–154. DOI: 10.1007/978-90-481-3058-0_10

Tallon, C., & Franks, G. V. (2011). Recent trends in shape forming from colloidal processing: A review. *Journal of the Ceramic Society of Japan, 119*(1387), 147–160. DOI: 10.2109/jcersj2.119.147

Temelso, B., Phan, T. N., & Shields, G. C. (2012). Computational Study of the Hydration of Sulfuric Acid Dimers: Implications for Acid Dissociation and Aerosol Formation. *The Journal of Physical Chemistry A, 116*(39), 9745–9758. DOI: 10.1021/jp3054394

Torlak, Y., Altuner, E. E., Tekeli, Y., & Tekeli, T. (2025). Advanced Nanocomposite-based Electrochemical Biosensors for Sensitive Detection of Pharmaceutical Drugs. *Journal of Pharmaceutical Research International, 37*(11), 34–58. DOI: 10.9734/jpri/2025/v37i117765

Turkay, M. (2018). *KARAYOLU ULAŞIMINDAN KAYNAKLANAN SERA GAZI EMİSYONUNUN (KARBON AYAK İZİNİN) HESAPLANMASI: ESKİŞEHİR İLİ ÖRNEĞİ* [Fen Bilimleri Enstitüsü]. https://tez.yok.gov.tr/UlusalTezMerkezi/tezDetay.jsp?id=h_oND-zFpEPniNFRp8KCtw&no=z7z8rI4ek1lfdPBfXH2JGQ

Ürge-Vorsatz, D., Cabeza, L. F., Serrano, S., Barreneche, C., & Petrichenko, K. (2015). Heating and cooling energy trends and drivers in buildings. *Renewable & Sustainable Energy Reviews*, *41*, 85–98. DOI: 10.1016/j.rser.2014.08.039

Vahlas, C., Caussat, B., Serp, P., & Angelopoulos, G. N. (2006). Principles and applications of CVD powder technology. *Materials Science and Engineering R Reports*, *53*(1–2), 1–72. DOI: 10.1016/j.mser.2006.05.001

van Droffelaar, B., & Jacobs, M. (2018). Nature-Based Training Program Fosters Authentic Leadership. *Journal of Leadership Studies*, *12*(3), 7–18. DOI: 10.1002/jls.21569

van Thriel, C., Schäper, M., Kleinbeck, S., Kiesswetter, E., Blaszkewicz, M., Golka, K., Nies, E., Raulf-Heimsoth, M., & Brüning, T. (2010). Sensory and pulmonary effects of acute exposure to sulfur dioxide (SO2). *Toxicology Letters*, *196*(1), 42–50. DOI: 10.1016/j.toxlet.2010.03.013

von Glasow, R., Bobrowski, N., & Kern, C. (2009). The effects of volcanic eruptions on atmospheric chemistry. *Chemical Geology*, *263*(1–4), 131–142. DOI: 10.1016/j.chemgeo.2008.08.020

Wadsley, J. W., Stadel, J., & Quinn, T. (2004). Gasoline: A flexible, parallel implementation of TreeSPH. *New Astronomy*, *9*(2), 137–158. DOI: 10.1016/j.newast.2003.08.004

Wang, C., Chen, S., Chen, Y., Zi, F., Hu, X., Qin, X., Zhang, Y., Yang, P., He, Y., He, P., Lin, Y., & Zhang, G. (2020). Modification of activated carbon by chemical vapour deposition through thermal decomposition of thiourea for enhanced adsorption of gold thiosulfate complex. *Separation and Purification Technology*, *241*, 116632. DOI: 10.1016/j.seppur.2020.116632

Wang, J., Tchapmi, L. P., Ravikumar, A. P., McGuire, M., Bell, C. S., Zimmerle, D., Savarese, S., & Brandt, A. R. (2020). Machine vision for natural gas methane emissions detection using an infrared camera. *Applied Energy*, *257*, 113998. DOI: 10.1016/j.apenergy.2019.113998

Wang, M., Wang, X., Moni, P., Liu, A., Kim, D. H., Jo, W. J., Sojoudi, H., & Gleason, K. K. (2017). CVD Polymers for Devices and Device Fabrication. *Advanced Materials*, *29*(11), 1604606. DOI: 10.1002/adma.201604606

Wang, P., Xi, B., & Xiong, S. (2024). Insights into the Optimization of Catalytic Active Sites in Lithium–Sulfur Batteries. *Accounts of Chemical Research*, *57*(15), 2093–2104. DOI: 10.1021/acs.accounts.4c00244

Wang, Y., Ding, L., Shi, Q., Liu, S., Qian, L., Yu, Z., Wang, H., Lei, J., Gao, Z., Long, H., & Charles Xu, C. (2022). Volatile organic compounds (VOC) emissions control in iron ore sintering process: Recent progress and future development. *Chemical Engineering Journal, 448*, 137601. DOI: 10.1016/j.cej.2022.137601

Wang, Y., Zhao, L., Otto, A., Robinius, M., & Stolten, D. (2017). A Review of Post-combustion CO2 Capture Technologies from Coal-fired Power Plants. *Energy Procedia, 114*, 650–665. DOI: 10.1016/j.egypro.2017.03.1209

Winder, C. (2004). Toxicity of gases, vapours and particulates. *Occupational Toxicology*, 399–424. DOI: 10.1201/9781482289282-18

Wu, S. R., Chen, C. H., Chung, I. L., & Lee, H. T. (2011). Combustion of low-calorific waste liquids in high temperature air. *Fuel, 90*(8), 2639–2644. DOI: 10.1016/j.fuel.2011.04.001

Wuebbles, D. J., & Jain, A. K. (2001). Concerns about climate change and the role of fossil fuel use. *Fuel Processing Technology, 71*(1–3), 99–119. DOI: 10.1016/S0378-3820(01)00139-4

Xi, Z., Xi, K., Lu, L., & Zhang, M. (2023). Study on oxidation characteristics and conversion of sulfur-containing model compounds in coal. *Fuel, 331*, 125756. DOI: 10.1016/j.fuel.2022.125756

Xu, W., & Zhang, R. (2012). Theoretical Investigation of Interaction of Dicarboxylic Acids with Common Aerosol Nucleation Precursors. *The Journal of Physical Chemistry A, 116*(18), 4539–4550. DOI: 10.1021/jp301964u

Yang, Z., Wu, Y., Zhang, Z., Li, H., Li, X., Egorov, R. I., Strizhak, P. A., & Gao, X. (2019). Recent advances in co-thermochemical conversions of biomass with fossil fuels focusing on the synergistic effects. *Renewable & Sustainable Energy Reviews, 103*, 384–398. DOI: 10.1016/j.rser.2018.12.047

Yousefpour, A., Hojjati, M., & Immarigeon, J. P. (2004). Fusion bonding/welding of thermoplastic composites. *Journal ofThermoplastic Composite Materials, 17*(4), 303–341. DOI: 10.1177/0892705704045187

Yun, J., Zhu, C., Wang, Q., Hu, Q., & Yang, G. (2019). Catalytic conversions of atmospheric sulfur dioxide and formation of acid rain over mineral dusts: Molecular oxygen as the oxygen source. *Chemosphere, 217*, 18–25. DOI: 10.1016/j.chemosphere.2018.10.201

Zaman, S., Wang, M., Liu, H., Sun, F., Yu, Y., Shui, J., Chen, M., & Wang, H. (2022). Carbon-based catalyst supports for oxygen reduction in proton-exchange membrane fuel cells. *Trends in Chemistry, 4*(10), 886–906. DOI: 10.1016/j.trechm.2022.07.007

Zanurin, A., Johari, N. A., Alias, J., Ayu, H. M., Redzuan, N., & Izman, S. (2022). Research progress of sol-gel ceramic coating: A review. *Materials Today: Proceedings, 48*, 1849–1854. DOI: 10.1016/j.matpr.2021.09.203

Zare, N., Karimi, F., Altuner, E. E., Mashkoor, H., Tiri, R. N. E., Gulbagca, F., Bijad, M., Cherif, A., & Sen, F. (2024). Synthesis of Bimetallic Nickel-Cobalt Nanoparticles Supported by Functional Multi-Walled Carbon Nanotubes by Chemical Methods: Characterization and Hydrogen Energy Generation. *Topics in Catalysis, 67*(9–12), 606–614. DOI: 10.1007/s11244-024-01907-x

Zhang, T., Tang, Q., Pu, C., & Yan, Y. (2022). Simulation on Evaporation and Motion of Atomized Droplets in Spray Dry Flue Gas Desulfurization Tower. *Journal of Thermal Science 2022 31:6, 31*(6), 2252–2263. DOI: 10.1007/s11630-022-1612-y

Zhang, Z., Bai, L., Li, Z., Qu, Y., & Jing, L. (2019). Review of strategies for the fabrication of heterojunctional nanocomposites as efficient visible-light catalysts by modulating excited electrons with appropriate thermodynamic energy. *Journal of Materials Chemistry. A, Materials for Energy and Sustainability, 7*(18), 10879–10897. DOI: 10.1039/C9TA02373A

Zhao, H., Guo, S., & Zhao, H. (2018). Impacts of GDP, Fossil Fuel Energy Consumption, Energy Consumption Intensity, and Economic Structure on SO2 Emissions: A Multi-Variate Panel Data Model Analysis on Selected Chinese Provinces. *Sustainability 2018, Vol. 10, Page 657, 10*(3), 657. DOI: 10.3390/su10030657

Zhao, M., Peng, H. J., Li, B. Q., & Huang, J. Q. (2024). Kinetic Promoters for Sulfur Cathodes in Lithium–Sulfur Batteries. *Accounts of Chemical Research*, acs.accounts.3c00698. Advance online publication. DOI: 10.1021/acs.accounts.3c00698

Zhou, X., Wang, T., Liu, H., Gao, X., Wang, C., & Wang, G. (2021a). Desulfurization through Photocatalytic Oxidation: A Critical Review. *ChemSusChem, 14*(2), 492–511. DOI: 10.1002/cssc.202002144

Zhou, X., Wang, T., Liu, H., Gao, X., Wang, C., & Wang, G. (2021b). Desulfurization through Photocatalytic Oxidation: A Critical Review. *ChemSusChem, 14*(2), 492–511. DOI: 10.1002/cssc.202002144

Zou, H., Du, H., Wang, Y., Zhao, L., Mao, G., Zuo, J., Liu, Y., Liu, X., & Huisingh, D. (2017). A review of the first twenty-three years of articles published in the Journal of Cleaner Production: With a focus on trends, themes, collaboration networks, low/no-fossil carbon transformations and the future. *Journal of Cleaner Production, 163*, 1–14. DOI: 10.1016/j.jclepro.2017.04.157

Chapter 2
Synthesis and Characterization of Strontium Ferrites Nanocomposites for Green Fuel Desulfurization

N. S. Mohan
Erode Sengunthar Engineering College, India

S. Malathi
Vellalar College for Women, India

V. Vijayalakshmi
https://orcid.org/0000-0002-7157-2917
Erode Sengunthar Engineering College, India

G. Myvizhi
https://orcid.org/0000-0003-2326-6146
Vellalar College for Women, India

ABSTRACT

$SrFe_2O_4$ nanoparticles were synthesized via a citrate–nitrate sol–gel assisted auto-combustion route and investigated as efficient heterogeneous catalysts for biodiesel production from mustard seed oil. Transesterification was carried out under varying conditions of catalyst loading, methanol-to-oil molar ratio, reaction time, and

DOI: 10.4018/979-8-3373-5966-3.ch002

temperature. The catalyst exhibited excellent activity, achieving a maximum fatty acid methyl ester (FAME) yield of 88% at optimized conditions of 100 mg catalyst loading, 3:1 methanol-to-oil ratio, 100 minutes of reaction, and 80 °C temperature. The catalytic mechanism involved Lewis acid sites (Sr^{2+}, Fe^{3+}) facilitating triglyceride activation and methanol adsorption, leading to efficient conversion of triglycerides into FAMEs. Gas chromatography (GC-FID) confirmed the composition of the obtained biodiesel, which complied with ASTM standards. Reusability studies demonstrated that the catalyst maintained 81% activity after five cycles and 69% after eight cycles, indicating high durability and stability. A comparative evaluation with reported ferrite and titanate-based nanocatalysts revealed that $SrFe_2O_4$ offers a competitive and cost-effective pathway for biodiesel synthesis under mild conditions.

1. INTRODUCTION

The global demand for renewable and eco-friendly fuels has increased in recent years due to the depletion of fossil fuel reserves and the urgent need to reduce greenhouse gas emissions (Ghamarpoor et al., 2024). Among the various renewable alternatives, biodiesel has gained significant attention because it is biodegradable, sulfur-free, and capable of reducing particulate and carbon monoxide emissions when compared with conventional diesel (Keerthana et al., 2021). Despite these benefits, biodiesel production still faces challenges such as limited catalytic efficiency, longer reaction times, and the need for effective desulfurization strategies to ensure improved combustion quality (Rashimi et al., 2017; da Silva Júnior, 2024). The search for stable, reusable, and environmentally benign catalysts is therefore a critical step in advancing biodiesel technologies. Ferrite-based nanomaterials, particularly strontium ferrites ($SrFe_2O_4$), have emerged as promising candidates for energy and environmental applications (Raza et al., 2024; Elanthamilan et al., 2023). Their unique combination of magnetic behavior, chemical stability, and redox-active Fe^{3+}/Fe^{2+} sites makes them highly effective in catalytic processes. At the nanoscale, $SrFe_2O_4$ exhibits enhanced surface area and porous structures, which enable efficient interaction with organic molecules. These properties make $SrFe_2O_4$ nanoparticles suitable for two important roles in biodiesel technology: (i) as heterogeneous catalysts in the transesterification of oils to fatty acid methyl esters (FAMEs), and (ii) as active materials for the oxidative desulfurization of sulfur-containing compounds, which improves the purity and performance of biodiesel (Subarmanian et al., 2021; Singh et al., 2022).

Recent studies on ferrite nanocatalysts have reported improved biodiesel yields, reduced methanol-to-oil ratios, and higher reusability when compared with conventional catalysts (Aziz et al., 2012; Eskandari et al., 2019; Fang et al., 2000; Hedayati et al., 2017; Gonçalves et al., 2021; Miramontes et al., 2017). In addition, their ability to remove thiophenic sulfur compounds highlights their potential to meet stringent fuel quality standards. Strontium ferrite, with its stable crystal structure and strong catalytic activity, offers an attractive platform for integrating biodiesel production with green fuel purification. In the present work, pure $SrFe_2O_4$ nanoparticles were synthesized via a citrate–nitrate sol–gel auto-combustion method (Chen at al., 2019; Bo et al., 2019; Grewal et al., 2022). The structural and surface features of the synthesized nanoparticles were correlated with their catalytic potential for biodiesel-related applications, particularly desulfurization and esterification processes (Wu, 2019). This study demonstrates the role of $SrFe_2O_4$ nanomaterials as cost-effective, environmentally friendly catalysts that can contribute to the advancement of sustainable energy systems. Furthermore, the magnetic nature of $SrFe_2O_4$ nanoparticles provides an additional advantage, as they can be easily separated from the reaction mixture using an external magnetic field, reducing the cost and complexity of catalyst recovery. Their high thermal stability also ensures consistent performance under varying reaction temperatures, making them suitable for large-scale biodiesel processing. Compared to conventional alkaline catalysts, $SrFe_2O_4$ avoids soap formation and catalyst leaching, which are common drawbacks in biodiesel synthesis. The environmentally friendly synthesis route and reusability of these nanoparticles further enhance their economic feasibility. Overall, $SrFe_2O_4$-based nanocatalysts offer a sustainable pathway to improving both the efficiency and quality of biodiesel production, aligning with global efforts toward cleaner energy solutions (Raza et al., 2024; Elanthamilan et al., 2023).

2. EXPERIMENTAL TECHNIQUES

2.1. Materials

Strontium nitrate ($Sr(NO_3)_2$, ≥99%), ferric nitrate nonahydrate ($Fe(NO_3)_3 \cdot 9H_2O$, ≥98%), citric acid (CA, $C_6H_8O_7$, analytical grade), ethylenediaminetetraacetic acid (EDTA, analytical grade), aqueous ammonia (NH_4OH, ~25%), absolute ethanol, and deionized (DI) water were all used as received without any further purification.

Synthesis of Pure $SrFe_2O_4$ Nanoparticles

The pure $SrFe_2O_4$ nanoparticles were prepared by citrate–nitrate sol–gel auto-combustion method. In a typical synthesis, stoichiometric amounts of strontium nitrate and ferric nitrate nonahydrate were dissolved separately in minimal deionized water to obtain 0.5 M precursor solutions. The two solutions were then combined under constant magnetic stirring to ensure homogeneity of the total cation concentration. Citric acid was added in a molar ratio of 2:1 with respect to the total metal ions to act as a chelating agent, and EDTA was introduced in a 1:1 molar ratio to further stabilize the complexed ions. The pH of the mixture was carefully adjusted to 6.8–7.2 by dropwise addition of aqueous ammonia to prevent local precipitation.

The resulting solution was heated at 80–85 °C under continuous stirring for 2–3 h until a viscous sol was formed, which upon further heating at the same temperature transformed into a uniform gel. This gel was aged at room temperature for 12 h and then dried at 110 °C for 10–12 h to produce a porous xerogel. The xerogel was gradually heated to 250–300 °C in air at a rate of 3–5 °C min^{-1} and held for 1–2 h to initiate controlled combustion of the organic matrix, resulting in a fluffy, porous precursor powder. The powder was lightly ground in an agate mortar to ensure uniformity.

The obtained precursor was calcined in an alumina crucible in air by heating at a rate of 5 °C min^{-1} to 850–950 °C, with an optimized condition of 900 °C for 5 h. The product was cooled naturally in the furnace and then ground into a fine powder. In cases where minor secondary phases were detected by XRD, calcination was repeated either at a slightly higher temperature (+25–50 °C) or for a longer dwell time of 1–2 h. To enhance dispersion, the calcined powders were ultrasonicated in ethanol for 20–30 min, dried at 60 °C, and stored in airtight vials.

Following calcination, the powders were washed thoroughly with deionized water three times and ethanol once to remove residual by-products, and then dried at 60–70 °C for 6 h. The removal of nitrate species was confirmed by the disappearance of the characteristic ~1384 cm^{-1} band in the FTIR spectra. Each batch produced approximately 1–2 g of $SrFe_2O_4$ with an overall yield greater than 80% relative to the starting metal precursors. To validate reproducibility, three independent batches were synthesized under identical conditions, and the variations in crystallite size and phase purity were within 5%. All steps involving nitrate salts and ammonia were performed inside a fume hood, and appropriate safety measures such as the use of gloves, protective eyewear, and heat-resistant tools were strictly followed during high-temperature processing (Priya et al., 2025).

3. RESULTS AND DISCUSSIONS

3.1. Characterization of Biodiesel Properties

The ester exchange reaction was conducted in a 100 mL conical flask, with the setup maintained under reflux. In this technique, 2 grams of mustard seed oil were mixed with methanol at molar ratios ranging from 1:1 to 1:5. To enhance the reaction efficiency, 0.2 grams of La@Pr@$SrFeO_4$ nanoparticles were introduced as the catalyst. The reaction duration was adjusted between 10 and 100 minutes, while the temperature was controlled within a range of 25°C to 120°C to maximize the conversion of triglycerides into fatty acid methyl esters (FAMEs). Once the reaction was completed, the mixture was cooled and filtered. To facilitate phase separation, hexane was added to the filtrate in a 1:1 ratio and mixed using a vortex mixer for 30 seconds. Subsequently, the mixture was left to stand, allowing the phases to separate. A 1 μL aliquot was carefully collected from the upper hexane layer and analyzed using gas chromatography coupled with a flame ionization detector (GC-FID). The analysis was performed on an Agilent 5970A system equipped with an HP5 capillary column (30 m in length, 0.3 mm inner diameter, and 0.25 μm film thickness). The yield of fatty acid methyl esters (FAMEs) was determined using the internal standard method, as outlined in Equation (2).

FAME% = [[Σ *Area of FAMEs / Area of IS*) × *weight of IS]/ Weight of WCO]* X 100 → (2)

Methyl decanoate (C10) was used as the internal standard (IS), and the total area of the peaks corresponding to the methyl esters is denoted by ΣFAMEs (Golab et al., 2024). .

3.1.1. Mechanism of Catalytic Esterification

In this catalytic process, wild mustard seed oil provided the triglycerides required for esterification. These triglycerides interacted with the Lewis acid centers (Sr^{2+} and Fe^{3+}) present on the nanocatalyst, while methanol was simultaneously adsorbed onto the catalyst's oxygen-rich surface. The reaction commenced when the CH_3O^- anion attacked the carbonyl group (C=O) of the triglycerides, forming a tetrahedral intermediate. This intermediate underwent rearrangement to produce fatty acid methyl esters (FAMEs) and di-glyceride ions. The di-glyceride ions then reacted with the protonated catalyst, regenerating the catalyst and releasing di-glyceride molecules. This cycle repeated twice, ultimately yielding glycerol and FAMEs (biodiesel) as the final products. The physical and chemical properties of the FAMEs were evaluated in accordance with ASTM standards, as summarized in Table 2. The findings revealed that the composition of the synthesized FAMEs closely resembled that of the original mustard oil (Sharifi et al., 2024).

3.2. Key Factors

3.2.1. Significance of Catalyst Loading on FAME Yield

The quantity of catalyst employed plays a crucial role in determining both the reaction kinetics and the overall biodiesel yield. As shown in Figure 13(a), increasing the catalyst loading resulted in a notable enhancement in biodiesel production, with a fatty acid methyl ester (FAME) yield of 88% achieved at a catalyst dosage of 100 mg. Beyond this level, the yield reached a plateau, suggesting that additional catalyst did not contribute to further enhancements. This phenomenon could be attributed to either an increase in the viscosity of the reaction mixture or the saturation of the catalyst's active sites. Conversely, higher catalyst loading provides more active sites, which can enhance the overall catalytic efficiency. Therefore, a catalyst dosage of 100 mg (5% w/w) was identified as the optimal amount for achieving the maximum yield (Mohan et al., 2025a).

3.2.2. Impact of the Molar Ratio of Methanol to Oil

The methanol-to-oil ratio plays a pivotal role in determining the efficiency of converting feedstock into esters. Increasing the molar ratio from 1:1 to 3:1 significantly enhanced the production of fatty acid methyl esters (FAMEs), as illustrated in Figure 13(b), achieving a maximum yield of 88% at a 3:1 ratio. However, further increases in methanol led to a slight decline in conversion efficiency, likely due to excess methanol obstructing access to the catalyst's active sites. Consequently, a 3:1 molar ratio was established as the optimal condition.

3.2.3. Influence of Reaction Duration

The reaction duration significantly influences the catalyst's performance and biodiesel production. As depicted in Figure 13(c), extending the reaction duration from 20 to 100 minutes led to a gradual increase in FAME production, with the highest yield achieved at 100 minutes. However, prolonging the reaction beyond this point resulted in only marginal improvements, likely due to the reversible nature of transesterification and potential reduction in the catalyst's active surface area over time. Therefore, the optimal reaction duration was determined to be 100 minutes.

3.2.4. Impact of Temperature

During the transesterification process, temperature significantly impacts both the reaction rate and the equilibrium conversion. As depicted in Figure 13(d), raising the temperature initially enhanced the formation of FAMEs, reaching an optimal yield within a certain temperature range. However, exceeding this optimal temperature range resulted in a slight decrease in conversion efficiency, likely due to the initiation of side reactions or potential catalyst deactivation. Therefore, maintaining an optimal temperature is essential for maximizing biodiesel yield. Consequently, a 3 molar ratio was established as the ideal condition for subsequent optimization experiments (Mohan et al., 2025b).

3.2.5 Catalyst Regeneration

Evaluating the durability and reusability of solid catalysts is essential for determining their overall effectiveness. In this study, the reusability of the Sm@Ag@$SrTiO_3$ nanoparticle catalyst was investigated. The catalyst was recovered via centrifugation and thoroughly washed with acetone and n-hexane. Despite a slight decrease in biodiesel yield over multiple cycles (as shown in Figure 13(e)), the catalyst demonstrated significant stability. For example, biodiesel yields of 88%, 84%, and 81% were achieved in the first, third, and fifth cycles, respectively. The catalyst retained 69% efficiency after nine cycles and 65% after ten cycles, indicating only minimal deactivation.

3.2.6 Comparative Study

As outlined in Table 2, the catalyst exhibited remarkable efficiency in biodiesel production under relatively mild conditions. Unlike conventional methods that often require harsher reaction conditions, this catalyst enabled shorter reaction times, reduced reagent consumption, and operated effectively at moderate temperatures. Compared to other synthetic catalysts, our system demonstrated greater cost efficiency in terms of production, purification, and operation. For instance, a one-hour transesterification of residual mustard oil with methanol yielded 88% biodiesel. Moreover, the catalyst displayed excellent reusability, maintaining 81% efficiency after five cycles and 69% after eight cycles. The minimal decline in activity indicates that the catalyst's active sites remained largely unaffected under the tested conditions, highlighting its potential as a highly efficient and rapid solution for biodiesel production.

Figure 1. Influence of (a) catalyst loading (b) methanol-to-oil proportion and (c) reaction duration (d) reaction temperature and (e) catalyst reusability on the efficiency of $SrFe_2O_4$ NPs.

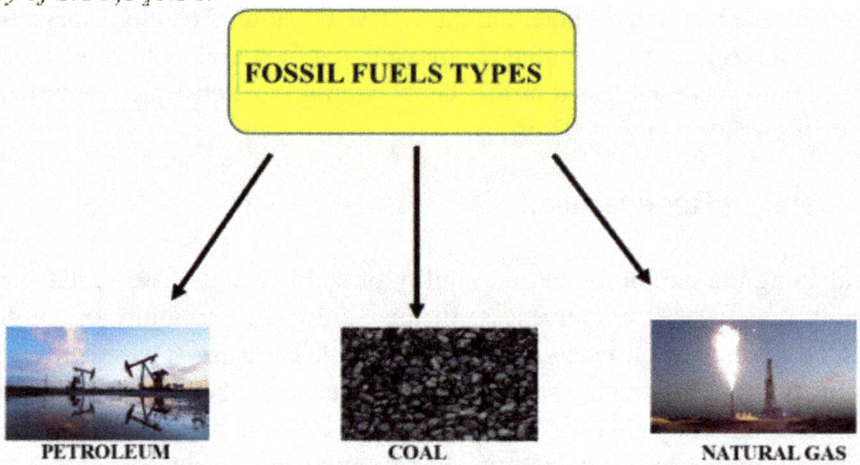

Table 1. Comparison of different heterogeneous catalysts utilized in biodiesel synthesis

Catalyst	Oil Source	Yield (%)	Reference
Sm and Ag co-doped $BaTiO_3$	Mustard oil	91%	(Mohan et al., 2025a)
Sm and Ag co-doped $SrTiO_3$	Mustard oil	85%	(Mohan et al., 2025a)
$SrFe_2O_4$	Mustard oil	88%	Present Work

At an optimal methanol-to-oil ratio of 3:1, a reaction time of 100 minutes, and a temperature of 80°C, the $SrFeO_4$ catalyst achieved a maximum biodiesel yield of 88%. It demonstrated excellent stability and reusability, retaining 81% efficiency after five cycles and 69% after eight cycles.

CONCLUSION

This study demonstrated that $SrFe_2O_4$ nanoparticles act as highly efficient and reusable catalysts for the transesterification of mustard seed oil to biodiesel. The catalyst achieved an optimal yield of 88% under relatively mild operating conditions, outperforming several conventional heterogeneous catalysts reported in literature.

Its excellent activity can be attributed to the presence of Lewis acid centers and the high surface area provided by the nanostructure, which enhanced the interaction between triglycerides and methanol. Moreover, the catalyst showed good stability, retaining significant activity over multiple cycles, thus offering a sustainable route for biodiesel production. The findings highlight the potential of $SrFe_2O_4$ nanocomposites as green, cost-effective, and durable catalysts for large-scale biodiesel synthesis, contributing toward cleaner energy technologies and reduced environmental impact.

REFERENCES

Aziz, A. A. A., Yong, K. S., Ibrahim, S., & Pichiah, S. (2012). Enhanced magnetic separation and photocatalytic activity of nitrogen doped titania photocatalyst supported on strontium ferrite. *Journal of Hazardous Materials, 199–200*, 1–9. DOI: 10.1016/j.jhazmat.2011.10.069

Bo, L., Hu, Y., Zhang, Z., & Tong, J. (2019). Efficient photocatalytic degradation of Rhodamine B catalyzed by $SrFe_2O_4$/g-C_3N_4 composite under visible light. *Polyhedron, 168*, 94–100. DOI: 10.1016/j.poly.2019.04.036

Chen, Y., Wu, Q., Bu, N., Wang, J., & Song, Y. (2019). Magnetic recyclable lanthanum–nitrogen co-doped titania/strontium ferrite/diatomite heterojunction composite for enhanced visible-light-driven photocatalytic activity and recyclability. *Chemical Engineering Journal, 373*, 192–202. DOI: 10.1016/j.cej.2019.05.047

da Silva Júnior, A. H., de Oliveira, C. R. S., Pellenz, L., Moraes, P. A. D., Marques, W. P., Mazur, L. P., Costa, T. G., Horn, A.Jr, Guelli Ulson de Souza, S. M. A., de Souza, A. A. U., da Silva, L., & da Silva, A. (2024). Perovskite-type strontium ferrite-based catalyst: Characterization and antibiotic degradation approach. *Process Safety and Environmental Protection, 187*, 1403–1421. DOI: 10.1016/j.psep.2024.05.048

Elanthamilan, E., Elizabeth, I. B., Wang, S.-F., & Lydia, I. S. (2023). Strontium hexaferrite microspheres: Synthesis, characterization and visible-light-driven photocatalytic activity towards the degradation of methylene blue dye. *Optical Materials, 137*, 113565. DOI: 10.1016/j.optmat.2023.113565

Elanthamilan, E., Elizabeth, I. B., Wang, S.-F., & Lydia, I. S. (2023). Strontium hexaferrite microspheres: Synthesis, characterization and visible-light-driven photocatalytic activity towards the degradation of methylene blue dye. *Optical Materials, 137*, 113565. DOI: 10.1016/j.optmat.2023.113565

Eskandari, N., Nabiyouni, G., Masoumi, S., & Ghanbari, D. (2019). Preparation of a new magnetic and photocatalyst $CoFe_2O_4$–$SrTiO_3$ perovskite nanocomposite for photodegradation of toxic dyes under short-time visible irradiation. *Composites. Part B, Engineering, 176*, 107343. DOI: 10.1016/j.compositesb.2019.107343

Fang, J., Wang, J., Gan, L. M., Ng, S. C., Ding, J., & Liu, X. (2000). Fine strontium ferrite powders from an ethanol-based microemulsion. *Journal of the American Ceramic Society, 83*(5), 1049–1055. DOI: 10.1111/j.1151-2916.2000.tb01329.x

Ghamarpoor, R., Fallah, A., & Eghbali, T. (2024). Design of bifunctional sandwich-like Co@Si/Ox-MXene nanocomposite to increase the supercapacitor properties and removal of pollutants from wastewater. *Journal of Alloys and Compounds*, *983*, 173920. DOI: 10.1016/j.jallcom.2024.173920

Golab, E. G., Ghamarpoor, R., Kondori, F. J., Hosseini, S., & Al-Saedi, H. N. (2024). Synthesis of hydrophobic polymeric surfactant (polyacrylamide/zwitterionic) and its effect on enhanced oil recovery (EOR). *Chemical Physics Impact*, *9*, 100756. DOI: 10.1016/j.chphi.2024.100756

Gonçalves, M. A., Mares, E. K. L., Zamian, J. R., da Rocha Filho, G. N., & da Conceição, L. R. V. (2021). Statistical optimization of biodiesel production from waste cooking oil using magnetic acid heterogeneous catalyst $MoO_3/SrFe_2O_4$. *Fuel*, *304*, 121417.

Grewal, J. K., Kaur, M., Sharma, R. K., Oliviera, A. C., Garg, V. K., & Sharma, V. K. (2022). Structural and photocatalytic studies on oxygen hyperstoichiometric titanium-substituted strontium ferrite nanoparticles. *Magnetochemistry (Basel, Switzerland)*, *8*(12), 120. DOI: 10.3390/magnetochemistry8100120

Hedayati, K., Behesht-Ara, Z., & Ghanbari, D. (2017). Preparation and characterization of various morphologies of $SrFe_{12}O_{19}$ nanostructures: Investigation of magnetization and coercivity. *Journal of Materials Science Materials in Electronics*, *28*(1), 1–9. DOI: 10.1007/s10854-016-5486-1

Keerthana, S. P., Yuvakkumar, R., Kumar, P. S., Ravi, G., & Velauthapillai, D. (2021). Rare earth metal (Sm) doped zinc ferrite ($ZnFe_2O_4$) for improved photocatalytic elimination of toxic dye from aquatic system. *Environmental Research*, *197*, 111047. DOI: 10.1016/j.envres.2021.111047

Miramontes, J. A. J., Arvizu, J. L. D., Gutierrez, S., Zaragoza, M. J. M., Ortiz, L., & Martinez, C. (2017). Synthesis, characterization and photocatalytic evaluation of strontium ferrites towards H_2 production by water splitting under visible light irradiation. *International Journal of Hydrogen Energy*, *42*(51), 30257–30266. DOI: 10.1016/j.ijhydene.2017.09.162

Mohan, N. S., Pandian, M., Vijayalakshmi, V., & Arulraj, A. (2025a). Improving catalytic performance of $BaTiO_3$ perovskite nanoparticles through Sm and Sm@Ag deposition for environmental sustainability. *Materials Science and Engineering B*, *311*, 117790. DOI: 10.1016/j.mseb.2024.117790

Mohan, N. S., Pandian, M., Vijayalakshmi, V., & Arulraj, A. (2025b). Enhancing catalytic efficiency of Sm and Sm@Ag doped strontium titanate for environmental and renewable energy applications. *Ceramics International, 51*(11), 13803–13814. DOI: 10.1016/j.ceramint.2025.01.215

Priya, V. N., Rajkumar, M., Rajendran, V., Mobika, J., Sibi, S. P. L., Veena, B., Vijayalakshmi, V., & Ahila, P. (2025). Sustainable selenium ions adsorption of cyclodextrin- and cellulose-functionalized layered double hydroxide/reduced graphene oxide nanocomposites. *Journal of Water Process Engineering, 69*, 106580. DOI: 10.1016/j.jwpe.2024.106580

Raza, Q., Bibi, I., Majid, F., Dahshan, A., Kamal, S., Jilani, K., Fatima, G., Taj, B., Alwadai, N., & Iqbal, M. (2024). Dielectric, ferroelectric, optical and photocatalytic properties of Al and Ag-doped strontium hexaferrite ($SrFe_{12}O_{19}$) synthesized via facile micro-emulsion approach. *Journal of Alloys and Compounds, 985*, 174098. DOI: 10.1016/j.jallcom.2024.174098

Sharifi, S., Javadpour, J., Rezaie, H., Jamshidi, M., & Ghamarpoor, R. (2024). Developing a deposited calcium-phosphate layer on zirconia surface by chemical grafting of L-serine molecules. *Journal of Materials Research and Technology, 28*, 66–77. DOI: 10.1016/j.jmrt.2023.11.285

Singh, G., Kaur, M., Garg, V. K., & Oliviera, A. C. (2022). Oxygen hyper-stoichiometric trimetallic titanium-doped magnesium ferrite: Structural and photocatalytic studies. *Ceramics International, 48*(17), 24476–24484. DOI: 10.1016/j.ceramint.2022.05.067

Subarmanian, K., Yuvakkumar, R., Ravi, G., Sehemi, A. A., & Velanthapillai, D. (2021). Synthesis of pure and lanthanum-doped barium ferrite nanoparticles for efficient removal of toxic pollutants. *Journal of Hazardous Materials, 424*, 127604.

Wu, Q. (2019). The fabrication of magnetic recyclable nitrogen-modified titanium dioxide/strontium ferrite/diatomite heterojunction nanocomposite for enhanced visible-light-driven photodegradation of tetracycline. *International Journal of Hydrogen Energy, 44*(16), 8261–8272. DOI: 10.1016/j.ijhydene.2019.01.145

Chapter 3
Strategic Heteroatom Doping for Electronic Tuning and Redox Optimization in Functional Materials

J. Mangaiyarkkarasi
https://orcid.org/0000-0003-1431-9584
Nadar Mahajana Sangam S. Vellaichamy Nadar College, India

J. Shanthalakshmi Revathy
https://orcid.org/0000-0003-1724-7117
University College of Engineering Ramanathapuram, India

ABSTRACT

The increasing need to use high-performance materials in energy storage, energy catalysis, electronics and sensing has heightened efforts to seek ways of precisely controlling the electronic and chemical characteristics of these materials. The deliberate introduction of foreign atoms into a bulk lattice, heteroatom doping, has proven to be a potent and scalable method of controlling the band structure, charge dynamics, and reactivity at the surface. This chapter will discuss strategic boron (B), nitrogen (N), and phosphorus (P) doping in functional materials with a particular focus on their specific contributions to the redox behavior and electronic structure. B represents the introduction of electron-deficient sites to increase the oxidation activity, N increases the electron density to promote conductivity and

DOI: 10.4018/979-8-3373-5966-3.ch003

reduce selectivity, and P brings about lattice distortion, creating more active sites and ion storage capacity. B-N and N-P combinations exhibit synergies whose effects are greater than those of single dopants.

1. INTRODUCTION

The search of high-performance functional materials has been more vigorous over the last decades, feverishly preoccupied with the pressing need to transform energy effectively, store energy sustainably, and develop catalytic architectures of the future to clean up the environment. The ability to manage and regulate the inherent nature of materials at their atomic and electronic level is at the core of this development. Doping can modify the local electronic environment by replacing or adding atoms to a host lattice to allow the tailoring of local conductivity, catalytic activity and stableness precisely. The doping of non-metallic heteroatoms (boron, N, and phosphorus) in particular has been of great interest (Shinde et al., 2025). Their relatively low atomic sizes, variety of valences state, and different electronegativities enable them to introduce controlled perturbations of the electronic configuration of the host material. Boron has its effect on the concentration of holes, with its deficiency of electrons, whereas nitrogen, more electronegative, has an effect on the concentration of electrons in active locations and better electron transport, and phosphorus, which has a larger atomic radius and lower electronegativity than nitrogen can expand the lattice, lowering the energy of bandgaps, and increasing the overall conductivity. The strategic placement of these heteroatoms does not simply affect the bulk electronic structure it also affects surface chemistry and interfacial phenomena which are important in redox reactions in photocatalysis, electrocatalysis, and energy storage. An example is that nitrogen doping of graphitic carbon materials can generate high-density active sites to undergo oxygen reduction reactions and boron doping of metal oxides can enhance charge separation after photogeneration during solar-driven water splitting. Phosphorus doping has been demonstrated to stabilize transition metal catalysts, to inhibit dissolution, and to enhance the life of the cycle in an electrochemical energy device (E. Zhang et al., 2021). More recent developments in synthetic methods, such as chemical vapor deposition and hydrothermal methods, as well as atomic layer deposition and plasma-assisted doping, have allowed control of both dopants concentration, distribution, and bondages to be controlled with accuracy. Such techniques as X-ray photoelectron spectroscopy (XPS), electron paramagnetic resonance (EPR) and state-of-the-art synchrotron techniques have given new insight on a hitherto unseen level of perception of the coordination and electronic states induced by doping (Song et al., 2022). These understandings are essential in creating effective structure property performance correlations, which allow logic and design, as opposed to trial and error methods.

The chapter is dedicated to strategic doping of B, N, and P in functional materials, with a special emphasis on how these heteroatoms allow to tune the electrons and optimize redox. It will start with concepts of the basic principles of heteroatom doping, delve into the separate and combined action of each dopant, describe more complex synthesis strategies, and describe the latest characterization methods. Lastly, the chapter will introduce various applications of energy conversion and storage to environmental catalysis that show how deliberate incorporation of heteroatoms can turn functional materials into next generation technology with high efficiency.

2. LITERATURE REVIEW

Doping of carbonaceous systems with nonmetals, doping of metal oxide, sulfides and other functional materials has emerged as an important strategy to tune electronic structure, create active sites and redox behavior. All observations and surveys point to the light nonmetals (B, N, P, S, O) altering charge density, introducing defect states, and changing the adsorption energetics of the reactants as one of the most frequently used tools to enhance electrocatalysis, photocatalysis and energy-storage behaviour, by doping.

Nitrogen (N) doping. The dopant, which has been studied most comprehensively in carbon materials, is N. The various N bonding patterns (pyridinic, pyrrolic, graphitic/quaternary, oxidized N) have different electronic effects: pyridinic N can normally increase adsorption and serve as redox-active site, and graphitic N can increase conductivity and electron transport (Mudoi & Saikia, 2025). Many experimental and theoretical studies have been associated with certain N species with high oxygen-reduction (ORR) selectivity/activity and with an increase of CO_2 capture or capacitive behavior in energy devices. Recent reviews are summarizing synthetic approaches and performance profiles in applications, and computational studies have mapped the locations of which N sites are most catalytically active.

Boron (B) doping. Boron being an electron deficient element relative to carbon is a natural p-type character and acceptor element upon incorporation into graphitic structure or oxides. B-doping can increase the amount of concentration of holes in semiconductors, edges of tuned bands in the photocatalysts and charge separation. Smaller than N in number, B-doped carbons and oxides have also shown hope in sensor, ORR (complementary to N) and in modifying electronic properties of host lattices (Kober et al., 2020).

Phosphorus (P) doping. Being more massive and less electronegative than carbon and nitrogen, phosphorus causes lattice strain and mid-gap states that may reduce bandgaps and enhance electronic conductivity. Li- and Na-ion batteries. Most recent experimental studies present P-doped hard carbons and P-doped silicon nanoparticles to provide better battery performance due to defect engineering and altered surface chemistry (M & G, 2023).

Synergy from co-doping (B–N, N–P, B–P, B–N–P). The non-additive effects of co-doping are frequently complementary electronic interactions (e.g., electron-rich N and electron-poor B) generate polarized bonds, increase the separation of charge, and provide neighboring sites with different chemical affinities that can undergo multi-electron redox reactions(K. Kim et al., 2020). A number of systematic studies and reviews show that B–N co-doped carbons, and N-P co-doped porous carbons are better at ORR, HER, and battery as electrode materials, due to synergistic tuning of DOS, adsorption energies and conductivity. Computational and spectroscopic data has shown that activity is determined by certain spatial positions of co-dopants (R. Zhang et al., 2022).

Synthesis advances. Chemical vapor deposition (CVD), hydrothermal/solvothermal, thermal annealing of dopant precursors, plasma and ion implantation, atomic-layer deposition and templated pyrolysis are all examples of an extensive repertoire of accurate doping techniques. Developments in precursor (e.g. boric acid, phytic acid, urea, organophosphorus compounds) design, and one-pot reactions have increased homogeneity of dopant, scalable processing, and control of the bonding state.

Characterization and mechanistic insights. The redistribution of charges of active sites under dopant bonding conditions have been analyzed by advanced characterization (XPS, EELS, XANES/EXAFS, EPR, synchrotron-based spectroscopies, high-resolution TEM) and DFT. Such forms of correlative analysis have had a central role in changing heuristic doping to rational design through correlating the dopant type/location against measurable changes of activity and electronic structure (Q. Wang & Frenkel, 2017).

3. FUNDAMENTALS OF HETEROATOM DOPING: MECHANISMS AND ELECTRONIC IMPLICATIONS

3.1 Electronic Structure Modulation through Dopant Incorporation

Heteroatomic doping is a powerful technique of electronic structure engineering of functional materials, where foreign atoms are inserted into the host lattice. The reaction changes the density of states (DOS), the Fermi level, and the energy of the bandgap, thus, allowing the conductivity, catalytic properties, and optical absorption to be precisely controlled. The extent and character of modulation is determined by the valence electron arrangement, electronegativity as well as the atomic radius of the dopant in comparison to the host atoms. As an example, boron (B), which has one valence less electron as carbon, is usually an electron acceptor when added to carbon structures (You et al., 2022). Doping sublattice symmetry in conductive structures such as graphene causes distribution changes of pi-electrons, which in turn, alters electrical conductivity and surface reactivity (Rabbani et al., 2025).

3.2 Defect Formation, Charge Redistribution, and Lattice Distortion

When heteroatoms are incorporated into a host material, it is bound to change the arrangement of atoms in the material and in most cases, results in the creation of structural defects such as vacancies, dislocations, and interstitial sites. Such defects are important in defining the electronic and catalytic properties of the material. As an example, since phosphorus (P) is larger than carbon, it will locally increase the lattice size, whereas boron (B) is electron-deficient and can produce compressive strain (W. Zhang et al., 2016). These distortions may affect phonon scattering, carrier mobility and mechanical stability. Charge redistribution is usually found to accompany defect formation. The dopant atom contains a different number of valence electrons than the host, and therefore alters the local density of electrons, creating some electron-rich and electron-deficient character regions. As an example, when nitrogen (N) dopes carbon structures this adds lone-pair electrons, creating electron-rich sites, whereas boron doping forms electron-deficient sites, which are taken up by Lewis acids. This redistribution alters the electrostatic potential surface, which affects the adsorption energy of reactants.

The lattice distortions may be used in energy storage to aid the diffusion of the ion by increasing the interlayer spacings or forming open channels(D. Liu et al., 2022). Thus defect formation, redistribution of charge, and lattice distortion are not incidental to doping, but instead functionality which can be exploited when carefully designed to maximize conductivity, catalytic reactions, and structural stability in specific uses in energy and environmental technologies.

3.3 Impact on Redox Properties and Charge Carrier Dynamics

The redox properties and charge carrier dynamics of functional materials are greatly affected by heteroatom doping, which changes the electronic structure, and surface chemistry of the materials. Introduction of dopants changes the location of the valence band maximum (VBM) and the conduction band minimum (CBM) and relocates the redox potentials in relation to the requirements of particular reaction. As an example, the VBM of photocatalysts is commonly increased by boron (B) doping, which increases oxidative capacity, and the CBM of photocatalysts is commonly reduced by nitrogen (N) doping, which increases reductive capacity of processes like CO_2 reduction. The intermediate electronegativity of phosphorus (P) doping is capable of simultaneously tuning both edges of the band since oxidation and reduction kinetics are balanced. On the charge transport aspect, the heteroatoms affect the carrier lifetime and generation (Bajjou et al., 2018). Doping may generate local states of energy that are stepping stones in electron or hole transfer, and decrease electron-hole recombination.

Boron Doping: Modulation of Band Structure and Charge Carrier Mobility

Boron (B) doping has been used as a flexible method to tune the electronic structure and dynamics of charge carriers of a vast array of functional materials such as carbon-based frameworks, metal oxides, and semiconductors. Due to its electron deficit boron has one less valence electron than carbon making it a p-type dopant that adds acceptor states, which conduce holes. This depletion of electrons essentially alters the density of states (DOS) and shifts the fermi level to the valence band which can narrow the bandgap or provide mid-gap states that increase light absorption and carrier use. Boron replacement distorts the ideal π-conjugated lattice in carbon-based materials, rupturing sublattice symmetry and distorting the π-distribution of electrons (Junaid et al., 2020). Boron is not an exclusive bulk conductor in catalytic systems: its electron-deficient character enhances hole transfer to reactants, which facilitates oxidative reaction pathways (Yao et al., 2023). When boron doping is co-purified with other heteroatoms, e.g. nitrogen, synergistic effects are frequently seen. B-N co-doping may form neighboring electron-deficient and electron-rich sites, which promote the spatial charge separation and the surface redox kinetics. This works especially well in electrocatalytic oxygen reduction, hydrogen evolution, and CO_2 electroreduction reactions. All in all, the use of boron doping is a multi-purpose tuning knob to modify the band structure, positioning of the Fermi level, and balancing the charge carrier mobility and catalyst site reactivity (Yadav et al., 2019).

4. NITROGEN DOPING: SURFACE CHEMISTRY AND REDOX ACTIVE SITE ENGINEERING

Among the best known methods to modify the surface chemistry and redox behavior of functional materials, especially carbon frameworks, metal oxides, and sulfur-based semiconductor photocatalysts is nitrogen (N) doping. The high electronegativity of nitrogen with a Pauling scale of 3.04 and the multiple bonding arrangement make it have a far-reaching effect on the electronic structure, catalytic activity, and surface reactivity. In contrast to boron, which is electron deficient, nitrogen is electron-rich, and frequently serves as an n-type dopant by providing electrons or introducing localized electron density to active sites (X. Wang et al., 2022). This characteristic allows nitrogen doping to be of particular importance in reactions that need a high ability to transfer electrons.

Pyridinic N, two carbon atoms located at the perimeter of a plane of graphene, donates a p-electron to a π-system in which the aromatic system contains only one electron in the sp^2 orbital. This lone pair performs the role of a Lewis base, which allows the adsorption of molecules such as O_2, CO_2, and protons and their activation.

The surface electronic density and local binding energies are changed by these nitrogen arrangements which have a direct effect on redox active site engineering. As an illustration, carbon catalysts with pyridinic N sites have been commonly referred as the most active sites in the 4-electron pathway in ORR, whereas graphitic N sites facilitate the potential in the onset and overall conductivity, which enables the efficient flow of electrons (J. Wang et al., 2025).

5. PHOSPHORUS DOPING: ENHANCING CONDUCTIVITY AND CATALYTIC STABILITY

Phosphorus (P) doping has become an effective and diverse method to achieve simultaneous high electrical conductivity and catalytic stability of functional materials, especially in energy conversion, storage and catalysis of the environment. Phosphorus being group 15 element, it has 5 valence electrons, but it is less electronegative (2.19) than nitrogen and has higher atomic radius than it and boron. These properties provide phosphorus with a special property to tune the local electronics environment, cause lattice expansion, and create defect sites enhancing the transfer of electrons and long-term stability of catalysts in severe operating conditions (Chen et al., 2023). This is especially beneficial in reactions such as water splitting, CO_2 electroreduction and nitrogen reduction, in which fine regulation of the adsorption/desorption processes defines overall performance. In semiconductors, such as TiO_2 and silicon, phosphorus is used as an n-type dopant, and it introduces shallow donor states that are beneficial to carrier density and electrical conductivity (Zakari et al., 2024). It is also important to control the amount of P content and bonding configuration, as an overload of phosphorus may cause amorphization or lower conductivity because of a high level of lattice disturbance (Long et al., 2020). The conclusion of the Single vs Multi-Heteroatom (B, N, P) Doping Effects is provided in Table 1.

Table 1. Summary of Single vs Multi-Heteroatom (B, N, P) Doping Effects

Doping type	Structural effects	Electronic effects	Catalytic property enhancements	Remarks
Single B doping	Slight lattice distortion due to smaller B atom; creates defects/vacancies	Introduces electron deficiency (p-type), shifts Fermi level, narrows bandgap in some oxides/carbons	Improves oxidation catalysis, enhances Lewis acidity, increases active sites	Beneficial for OER/ORR catalysts; stability sometimes limited by weak B–host bonding
Single N doping	Substitution in lattice or pyridinic/graphitic edge in carbons; stabilizes defect sites	Provides lone-pair electrons (n-type), increases conductivity; modulates band structure	Strongly enhances ORR activity, increases adsorption of O_2/intermediates; improves conductivity	Widely used in carbon-based catalysts; site configuration (pyridinic vs graphitic) critical
Single P doping	Expands lattice due to larger radius; induces strain; stabilizes amorphous phases	Induces electron donation (n-type), reduces work function; tunes charge carrier density	Enhances HER activity, improves electron transfer kinetics, stabilizes catalytic active phases	Often synergistic with metals (e.g., phosphides); can reduce stability in oxidative conditions
Dual B–N doping	Creates B–N bond domains in carbons; balances lattice strain	Synergistic tuning: B (electron-withdrawing) + N (electron-donating) → polarized charge centers	Increases density of active sites, enhances ORR selectivity and stability	Well-known in graphene and carbon catalysts; mimics catalytic behavior of noble metals
Dual N–P doping	P induces lattice expansion while N stabilizes local sites	Modulates charge density; strong electronic coupling increases carrier mobility	Boosts HER/OER performance, improves selectivity in CO_2 reduction; enhances conductivity	Effective for bifunctional electrocatalysts; good durability compared to single dopants
Dual B–P doping	Induces more lattice distortion than single doping; creates strain and defects	Charge polarization: B attracts electrons, P donates → strong local dipoles	Improves photocatalytic activity (H_2 evolution), enhances adsorption of reactants	Less common but emerging for visible-light photocatalysts
Tri-doping (B–N–P)	Complex lattice modifications with rich defect chemistry; stabilizes porous structures	Strong synergistic modulation of bandgap, carrier density, and Fermi level; enables broad electronic tuning	Maximizes active sites, improves bifunctional ORR/OER or HER/CO_2RR activity; enhances stability	Offers multifunctionality, tunable catalytic performance, but synthesis is challenging

6. SYNERGISTIC EFFECTS OF MULTI-HETEROATOM (B, N, P) CO-DOPING STRATEGIES

6.1 Complementary Electronic Effects and Charge Polarization Mechanisms

In the case of multi-heteroatomic co-doping of boron (B), nitrogen (N), and phosphorus (P), synergistic coverage of the electronic structure is achieved due to the complementary character of their intrinsic electronegativities, sizes and bonding tendencies. Boron is less electronegative (2.04 on the Pauling scale), and serves as an electron acceptor, producing electron deficient sites, which are able to transport holes. Nitrogen is additionally a more electronegative (3.04) electron donor that adds limited lone-pair electrons into the π coupled network, and increases the electron density around the fermi level. Phosphorus, a medium electronegative (2.19) element with a larger covalent radius can act as an electron donor as well as a structural modulator, increasing interlayer spacing and enhancing charge conduction. These dopants are then incorporated together to form an overall spatially diverse electronic potential map, which facilitates a more effective charge separation and directional carrier transport (Asif & Kumar, 2022). The polarization of charges in B, N, and P co-doped systems bonds is due to the asymmetry of dopant induced electronic states. These alterations do not only increase the absorption of visible light by the photocatalysts, but they also increase the conductivity and the turnover frequency of the electrocatalysts (Bo et al., 2023). Also, reaction intermediates can also be stabilized by the polarization fields produced by the co-doped network over long-range, which reduces the activation barrier of important reactions, including reactions with water, hydrogen evolution, or CO_2 reduction.

6.2 Enhanced Catalytic Activity through Spatially Distributed Active Sites

The formation of spatially distributed active sites with different electronic and chemical properties is one of the main benefits of multi-heteroatomic (B, N, P) co-doped materials. Individual dopants confer widely different functionalities: boron produces electron-deficient centers extremely efficient in adsorbing and activating electron-rich species, nitrogen offers electron-rich centers capable of binding electrophilic intermediates, and phosphorus offers tunable electronics states with a comparatively large atomic size capable of distorting local bonding geometries. Incorporating these heteroatoms in a non-uniform manner on the host lattice to form a network of active sites with different affinities to reaction intermediates, multi-step catalytic processes are facilitated more effectively (Sfeir et al., 2023).

Spatial allocation is a very important aspect in catalytic synergy. Too close, the active sites can compete; too distant, the charge transfer between them will not be efficient. The co-doping approaches allow to achieve an optimal arrangement of B, N and P sites so that the individual reactivity is maximized and the inter-site electronic communication is optimized. The structure enables cascade catalysis, a reaction intermediate formed at one location is then very quickly transferred to another location electronically and chemically compatible with the subsequent phase of the reaction sequence.

The difference in the local electronics environments also makes it less likely that active sites may be passivated. To illustrate, in electrocatalytic oxygen reduction reactions (ORR) N-doped sites can strongly adsorb oxygen intermediates, and neighboring sites are B or P sites that can fine-tune adsorption energies to avoid over-binding and thus increase turnover frequency (H. Zhang et al., 2024). Likewise, in photocatalysis, spatially discrete domains of dopants may serve as selective sites of oxidation or reduction with the assistance of the internal electric fields produced by polarization of charges. This geometry inherently pushes photogenerated electrons and holes to various active sites, minimizing recombination, and maximizing quantum efficiency.

6.3 Tailoring Band Structure and Stability via Co-Doping Configurations

Boron (B), nitrogen (N) and phosphorus (P) co-doping represents a potent approach to very finely tune the band structure of functional materials, optimizing its optoelectronic characteristics as well as structural stability. The dopants affect the host lattice in a special manner B tends to decrease the conduction band edge by deficiency of electrons, N tends to increase the conduction band edge through lone-pair electron states, P tends to decrease and increase the conduction and valence band edges respectively because of its intermediate electronegativity and high atomic radius. The interaction of these dopants in the form of definite spatial arrangements leads to alterations in the band gap or a selective shift in the band gap, and the material energy levels match the redox potentials needed to achieve a desired reaction like photocatalytic water splitting, CO_2 -reduction, or electrocatalysis (Poopanya et al., 2025).

The arrangement of dopants be it substitutional, interstitial or clustered is a conclusive factor in determining the resultant density of states (DOS) around the Fermi level. Replacement of B and N in a carbon or semiconductor lattice can often form complementary p and n domains, which give rise to internal heterojunction-like effects. Positioning P close to either B or N, it alters the orbital overlap and thus the effective mass of charge carriers changes, increasing their mobility. With this fine-tuning, optimality of light absorption in the visible region is achieved, recombination is minimized and catalytic turnover frequency is enhanced. Also, the presence of several heteroatoms may enhance chemical robustness through the regulation of surface electronic density, which dictates corrosion, oxidation, or dissolution resistance (Z. Zhang et al., 2024).

7. ADVANCED SYNTHESIS ROUTES FOR B, N, AND P DOPED FUNCTIONAL MATERIALS

7.1 Gas-Phase and Plasma-Assisted Doping Techniques

Gas-phase, as well as plasma-assisted doping methods, have become a very promising approach to the introduction of boron (B), nitrogen (N), and phosphorus (P) into functional materials, allowing the concentration, distribution and bonding environment of the dopant to be controlled with a high degree of accuracy. The process of such methods works under comparatively mild conditions of traditional high-temperature solid-state reactions, allowing the homogenous doping of the material and maintaining the integrity of the host material (Thakur et al., 2022).

In gas-phase doping, the substrate or precursor material is exposed to volatile species of dopant in the vapor phase whereby either the dopant atoms diffuse into the lattice or the atomic species chemically attaches to the surface. Under reactive conditions, typical precursors to B doping are diborane (B_2H_6) or boron trichloride (BCl_3) under reactive conditions; N, ammonia (NH_3) or nitrogen gas (N_2) and P, phosphine (PH_3) or organophosphorus compounds. An increase in temperature can be used as a catalyst to increase the reaction kinetics or catalytically activate the reaction. Gas-phase techniques are effective at conformal dopant coatings and obtaining depth-controlled diffusion, which are important to thin films, porous structures and nanoscale catalysts. Plasma-assisted doping is based on the fact that ionized species formed in a plasma discharge are highly reactive, and thus, the dopant incorporation is accelerated.

The use of defects and dopant incorporation synergy has two primary benefits:

(1) defects can be dopant activated, the plasma environment favors favorable bonding structures; and
(2) uniform fabrication, as the highly mobile plasma species can enter micro- and mesoporous structures.

Moreover, with the proper control of plasma power, exposure time, and gas composition, the researcher can carefully tune the chemistry of the dopant, enabling specific catalytic or electronic functionalities (Deng et al., 2023).

7.2 Solution-Based and Hydrothermal/ Solvothermal Methods

Solution-based and hydrothermal/solvothermal synthesis paths are popular in incorporation of boron (B), nitrogen (N), and phosphorus (P) in functional materials due to their ease, versatility and ability to obtain homogeneous distribution of dopants at relatively lower processing temperatures. These methods involve liquid-phase feeds, allowing close mixing of dopants and host material ingredients prior to crystallization, and increased dopant incorporation efficiency and structural uniformity (de Conti et al., 2023).

Solution-based doping usually starts with the solution of dopant precursor-boric acid or boron oxide to B, urea or ammonium salts to N and phosphoric or organophosphorus compounds to P in an aqueous or organic solvent containing the precursor of the desired material. Stoichiometry is easily controlled through the uniform distribution of dopant species, at the molecular or ionic level. The doped precursor solution may then be subjected to co-precipitation, sol-gel formation or direct drying to obtain dopant-containing intermediates, followed by calcination or annealing to obtain the desired phase and bonding geometry. The method is specifically adapted to the creation of doped nanoparticles, thin films, and porous structures.

Solvothermal and hydrothermal processes are special solution-based processes that are performed in a closed autoclave under high temperature and pressure. Hydrothermal system is a synthesis that uses water as the reaction medium whereas solvothermal synthesis uses high boiling organic solvents. These environments lead to greatly improved solubility of precursors, to low-thermal-budgets crystallization, and to control of particle size, morphology and distribution of dopants. The precursors of B, N, and P are dissolved together with the host lattice precursors, and at hydrothermal/solvothermal conditions, the dopant atoms are added to the growing crystal lattice or attached to the surfaces of particles (Nakashima et al., 2019). These techniques have a number of benefits:

(1) superior regulation of dopant homogeneity because chemical mixing occurs at the molecular level
(2) versatility in the adjustment of reaction parameters such as temperature, pH, and solvent polarity to affect doping performance and also material properties
(3) compatibility with co-doping techniques, so that B, N and P can be incorporated in a single synthesis reaction.

Due to their versatility in terms of scalability, low-energy needs, and high-crystallinity, defect-engineered materials, solution-based and hydrothermal/solvothermal processes continue to be the necessary techniques in the creation of next-generation B, N, and P co-doped catalysts, energy materials, and sensing platforms.

7.3 Solid-State, Pyrolytic, and Template-Assisted Approaches

Solid-state, pyrolytic, and template-assisted synthesis routes can offer versatile and scalable routes to the preparation of boron (B), nitrogen (N), and phosphorus (P) doped functional materials with high thermal stability, tunable morphologies, and electronic properties. They are especially beneficial in those applications that need bulk materials, high crystallinity, or architectured structures with a distinct porosity (Gangwar et al., 2024).

In solid-state doping, the precursor host materials are combined with dopant-containing compounds, e.g. boron oxide, urea or ammonium phosphate, without a solvent and then treated at high temperature. It is common that mechanical processes, such as ball milling are employed to provide intimate contact between reactants to improve the diffusion of dopants during the following calcination process.

Pyrolytic doping is anchored on the thermal decomposition of dopant-high precursors in the formation of the desired material or precursors. Pyrolytic techniques are normally applied in carbon-based systems and can generate B-, N-, or P-doped carbons using inert or reducing atmospheres by heating organic frameworks, polymers, or biomass impregnated by dopant sources. As an example, pyrolysis of melamine that is N source and boric acid that is B source and phosphoric acid that is P source can give co-doped porous carbons with high surface areas and heteroatom-enriched active sites. The pyrolytic condition does not only contain dopants, but also forms defect-rich surfaces which have increased catalytic activity.

In template-assisted techniques, the size, shape, and structure of doped materials are regulated with the help of physical or chemical templates like mesoporous silica, anodic alumina, or polymeric scaffolds. Dopant precursors are introduced into the template framework and on heating, the dopant precursors are incorporated into a doped network and the target nanostructure achieved. Etching or calcination of the template leaves a very ordered porous structure with uniformly dispersed dopants. This structural accuracy guarantees large surface to volume ratios, mass transport efficiency and increased exposure of active sites, which are important in catalytic and electrochemical applications (Pavlenko et al., 2022). The combination of these tools is a potent system of designing sophisticated B, N, and P co-doped materials to be optimized in highly active catalysis, energy storage, and sensing technologies. As mentioned in Table 2, the Comparison of advanced synthesis routes to heteroatom doping, it is possible to discuss the operating conditions, the configurations of dopants that may be achieved through the methods, benefits and limitations.

Table 2. Comparison of advanced synthesis routes for heteroatom doping.

Method	Typical operating conditions	Achievable dopant configurations	Main advantages
Solid-state diffusion / thermal annealing	High temperature (400–1200°C depending on host), controlled atmosphere (vacuum, O_2, N_2, Ar), long dwell (minutes–hours)	**Substitutional** and **interstitial** dopants near surface; graded profiles; sometimes clusters if concentration high	Simple, no specialized equipment beyond furnace; good for bulk crystalline hosts and tuning depth via time/temp
Ion implantation	Room temp to modest heating; high-energy ion beam (keV–MeV); vacuum; beam currents controllable	Highly controlled **substitutional** (after anneal) or **interstitial**, very precise depth/fluence profiles; can create defects	Extremely precise dose and depth control; compatible with semiconductor processing
Chemical Vapor Deposition (CVD)	400–1200°C (substrate-dependent), low pressure or atmospheric, gas precursors carrying dopant species	**Substitutional** incorporation in thin films and 2D materials; surface functionalized layers; dopant uniform in films	Excellent for uniform, conformal doping of films; scalable for wafers and coatings; good for covalent hosts (graphene, Si, GaN)
Atomic Layer Deposition (ALD)	Low-to-moderate temp (100–350°C typically), sequential self-limiting surface reactions, vacuum	**Surface/sub-monolayer** and controlled incorporation into thin films; enables atomic-scale dopant placement	Atomic-level thickness/dose control and excellent conformality on high-aspect features
Molecular Beam Epitaxy (MBE)	Ultra-high vacuum, substrate heating (200–800°C), elemental beams (effusion cells)	**Substitutional** dopants in epitaxial films with atomic-layer precision; delta-doping	Ultimate control for epitaxial, low-defect films; excellent for research-grade heterostructures
Solvothermal / hydrothermal synthesis	100–300°C, autogenous pressure (sealed autoclave), solution precursors, controlled pH	**Substitutional** or **surface-incorporated** dopants in nanoparticles/crystals; often homogeneous bulk doping in nanomaterials	Low-to-moderate temperatures preserve nanostructure; scalable for powders; good for complex oxides, sulfides
Co-precipitation / sol–gel	Ambient to 600°C (drying/calcination), solution chemistry control (chelators, pH), subsequent thermal treatment	**Bulk substitutional** doping in oxide powders, or dopants trapped in glassy matrices; can yield very homogeneous distribution	Good for homogeneous mixing at atomic level; low equipment cost; adaptable to many oxides and ceramics

continued on following page

Table 2. Continued

Method	Typical operating conditions	Achievable dopant configurations	Main advantages
Plasma-assisted methods (PECVD, plasma doping)	Low-to-moderate substrate temp (room to ~400°C), low pressure plasma, reactive species	**Surface functionalization**, shallow **substitutional/ interstitial** doping and defect engineering	Low thermal budget; can functionalize surfaces and introduce heteroatoms without high-T; fast
Electrochemical doping / intercalation	Ambient to moderate temp; liquid electrolyte, applied potentials; controlled charge transfer	**Interstitial/intercalated** dopants (ions inserted into layered hosts), surface charge doping	Precise stoichiometry control via coulometry; reversible doping (useful for batteries, sensors); low T
Mechanochemical (ball-milling)	Ambient to slightly elevated temps generated by milling; high-energy grinding; solid precursors	**Surface/substitutional** doping (often non-equilibrium), alloying, defect generation, amorphous phases	Solvent-free, scalable, straightforward; can activate solids and introduce dopants that are otherwise hard to incorporate
Flame spray / aerosol / spray pyrolysis	Very high local temp in flame, short residence times, precursor aerosols	**Rapid nucleation** doped nanoparticles, often with dopants incorporated or as surface species	High-throughput production of powders; continuous processing; good for oxide nanopowders
Vapor-phase transport / thermal evaporation with dopant source	Moderate-to-high temp, carrier gas flow, vacuum or low pressure	**Surface/substitutional** in transported crystals/films; gradient doping possible	Good for growing single crystals or depositing doped films; can create gradient/ segregated dopant profiles

8. CHARACTERIZATION TECHNIQUES FOR PROBING ELECTRONIC AND REDOX ALTERATIONS

8.1 X-ray Photoelectron Spectroscopy (XPS)

The strongest surface-sensitive methods of characterizing B, N and P doped functional materials are X-ray Photoelectron Spectroscopy (XPS) which provides quantitative and chemical information about elements in the few nanometers of the surface of the material. XPS measurement The binding energy of electrons in certain atomic orbitals is directly proportional to the kinetic energy of photoelectrons emitted by irradiating the sample with monochromatic X-rays. This allows establishing the elements involved, their oxidation states and even the chemical bonding of these elements (J. Liu et al., 2018).

In the case of B, N and P co-doped systems, XPS is capable of distinguishing amongst several bonding conditions. As an example, B -1s spectra can be used to detect the existence of BC_2O, BC_3, or B-N bonds, whereas N -1s spectra can be used to differentiate between pyridinic-N, pyrrolic-N, graphitic-N, and oxidized-N bonds in carbon-based hosts. On the same note, P 2p spectra are capable of detecting P–C, P-O and P-N bonding environments. This information is instrumental in being able to correlate particular dopant bonding states with useful performance in catalysis, sensing, or energy storage. As an example, an increase in C 1s binding energy of N-doped carbon materials can indicate electron withdrawal by electronegative nitrogen atoms whereas a decrease in B 1s may indicate electron donation by the host lattice (Loy et al., 2024).

XPS is essential in the study of the structure function relationships in B, N and P co-doped materials through the high resolution spectral deconvolution of dopant induced electronic and chemical changes, and plays a critical role in the rational design of complex functional systems.

8.2 Electron Paramagnetic Resonance (EPR) Spectroscopy

Electron Paramagnetic Resonance (EPR), also known as Electron Spin Resonance (ESR) is a very sensitive spectroscopic method of detecting and analyzing unpaired electrons within materials. When applied to B, N, P co-doped functional materials, EPR is an important tool in defect location, investigations of the electronic structure, and the redox-active species that directly affect catalytic and electronic activity. EPR is based on the idea that a sample can be irradiated with an unvarying magnetic field and exposed to a microwave radiation. This energy is absorbed by unpaired electrons when the strength of the magnetic field meets resonance condition and generates a typical EPR signal. The g-factor is a dimensionless resonance parameter based upon the resonance condition that gives details of the local electronic environment and symmetry of the unpaired site of the electron. In the case of doped systems, the changes in the g-factor may indicate the occurrence of defects formed by heteroatoms or the effect of redistribution of charges (Mourdikoudis et al., 2018).

Phosphorus doping can also cause the introduction of donor-like states, and co-doping can cause complex defect structures that are difficult to characterize using traditional structural characterization and can easily be characterized using EPR. A strong redox-active species identification tool that can be used by EPR is also in operando or in situ research. Reaction intermediates such as superoxide radicals (O_2^-), hydroxyl radicals (OH), or organic radicals may be produced directly in catalytic reactions in which O_2 is reduced, water is split, or CO_2 is reduced. Their tracking may give details on reaction mechanisms, role of dopants in electron transfer, and stability of active sites during working conditions. Furthermore, EPR is very sensitive, which allows observing even very low concentrations of paramagnetic centers, and it is therefore invaluable in the study of dilute doping concentrations, or low concentrations of active sites that are sparsely distributed.

8.3 X-ray Absorption Near Edge Structure (XANES) and Extended X-ray Absorption Fine Structure (EXAFS)

X-ray Absorption Spectroscopy (XAS) comprises two of the most potent, namely, X-ray Absorption Near Edge Structure (XANES) and Extended X-ray Absorption Fine Structure (EXAFS) methods, which furnishes a precise data on the local chemistry of dopants in B, N and P co-doped functional materials. The methods of these synchrotrons are especially useful since they examine short-range order about chosen atoms, hence suitable to investigate materials with high defect densities, amorphous phases or nanoscale dimensions whereby long-range crystallinity might not exist (Prasad et al., 2024).

XANES is focused on the near-edge portion of the absorption spectrum, which is usually located between the absorption edge (which is around 50 eV) and the x-ray portion of the absorption spectrum. Pre-edge features related to perfect symmetry and coordination environments can also be identified using XANES to let a researcher differentiate between tetrahedral, trigonal, or planar dopant atom coordination geometry (S. Y. Kim et al., 2023).

EXAFS is used to investigate the oscillatory structure in the absorption spectrum at energies as high as hundreds of eV above the absorption edge. In co-doped systems EXAFS is capable of distinguishing between isolated, clustered, and direct bonding of dopants to each other (e.g. B N, B P or NP bonds) and their distribution as compared to the host atoms. Combining XANES with EXAFS, complementary information is obtained regarding the local bonding and structural development under working conditions. In situ or operando measurements have the ability to trace the change in dopant oxidation states and coordination during catalytic reactions or at applied potential and this illuminates redox-active sites and structure stability. As an illustration, a reversible change in dopant valence states can be observed in XANES, on the contrary, during the reduction of oxygen in N, B and P co-doped electrocatalysts, whereas the change in bond length can be observed in EXAFS, indicating dynamic restructuring in active sites. XANES and EXAFS are essential in the disentangling of the structure-function relationships in the multi-faceted co-doped systems and allow the rational optimization of their electronic characteristics, stability, and catalytic activity due to element-specificity and the ability to analyze both crystalline and amorphous materials

8.4 Ultraviolet–Visible (UV–Vis) Diffuse Reflectance Spectroscopy (DRS)

The widely used optical characterization method to determine the electronic structure and light-harvesting characteristics of B, N, and P co-doped functional materials is called Ultraviolet Visible Diffuse Reflectance Spectrophotometer (UV-VIS DRS). In contrast to the older transmission UV-Vis spectroscopy, DRS is used to measure the reflected light of powder or opaque sources, thus being useful with catalysts, ceramics, and nanostructured solids in which transmission spectroscopy is not feasible. UV- Vis DRS in B, N and P co-doped systems direct information on changes in bandgaps caused by the addition of heteroatoms. The absorbance can be plotted versus photon energy to form Tauc plots by capturing the diffuse reflectance spectrum and performing the Kubelka-Munk transformation, to obtain some optical bandgap values. Co-doping can frequently reduce the bandgap, by localizing the states within the forbidden energy range or moving the conduction and valence bands.

To illustrate, nitrogen doping is more inclined to increase the valence band by lone-pair p-orbital contributions whereas boron doping may decrease conduction band and phosphorus doping is capable of varying both edges depending on bonding configuration (Morozzi et al., 2021). Changes in behavior of light absorption behavior across UV, visible and near-infrared regions are also disclosed in the spectral profile acquired by DRS.

Co-doping usually improves visible light absorption because of synergistic electronic effects, which allow improved photocatalytic or photoelectrochemical functionality under sunlight irradiation. Specifically, mid-gap states created by dopants allow sub-bandgap photon absorption to occur which extends light-harvesting range of the material to greater lengths than the pristine host. UV Vis DRS is also sensitive to defect related absorption characteristics, commonly found in co-doped systems. The existence of structural disorder or dopant-induced defect states can be revealed by broad absorption tails in the visible region (Urbach tails), which affect the dynamics of charge carriers. Comparing pre-thermal treatment and post-thermal treatment spectra, researchers are able to evaluate the stability of optically dopant-induced changes (Aceto et al., 2022).

8.5 Raman spectroscopy

Raman spectroscopy is an approach of vibrational spectroscopy that is common in the study of structural, chemical, and defect properties of B, N, P co-doped functional materials. Raman spectroscopy can determine the phonon mode which is sensitive to the atomic bonding, crystallinity and local structural distortions caused by heteroatom doping by measuring inelastic scattering of monochromatic light.

Raman spectra of B, N, and P co-doped carbon-based materials typically include both the D band (~1350 cm -1), associated with defects and disorder, and G band (~1580 cm -1), associated with the in-plane stretching vibration of carbon atoms in sp 2 bonds. The ratio of intensities is a measurable value of the density of defects, which tends to rise with the addition of dopants as a result of distortion of the lattice and the substitute of heteroatoms.

In the presence of B, N, and P, co-doping may result in a more complicated defect profile by affecting the electron transport and catalytic activity (Bumbrah & Sharma, 2016). In addition to carbon systems, Raman spectroscopy can be used to investigate dopant-induced vibrational modes of metal oxides, nitrides, and phosphides. A change in the characteristic lattice vibrations or a new peak are also possible signs of successful dopant incorporation, and altered strength of the bond. For example, B doping in TiO_2 may shift Ti–O stretching modes, while N or P incorporation can introduce additional vibrational features related to M–N or M–P bonds. Raman analysis can provide a more detailed view of how B, N, P co-doping can modify the structure-property relationships of a material when used together with other techniques, including XPS, EPR, or UV Vis DRS, to help design high-performance catalytic, sensing, and energy-conversion devices.

8.6 Electrochemical Impedance Spectroscopy (EIS)

Electrochemical Impedance Spectroscopy (EIS) is a useful, non-destructive technique that probes the frequency-dependent response of an electrochemical system to a small-amplitude AC voltage. For B, N, and P co-doped functional materials, EIS provides quantitative insights into charge transfer resistance (R_ct), electrical conductivity, and redox reaction kinetics, making it a key tool for evaluating performance in electrocatalysis, energy storage, and sensing applications.

In an EIS experiment, the material typically configured as an electrode is subjected to an AC perturbation over a wide frequency range. The resulting current response is analyzed to extract impedance parameters, which are often represented in Nyquist plots or Bode plots (Mc Carthy et al., 2022).

EIS is also invaluable for probing redox kinetics in real time. In electrocatalytic oxygen reduction (ORR), hydrogen evolution (HER), or oxygen evolution (OER), a smaller semicircle in the Nyquist plot corresponds to faster reaction kinetics, while variations reflect changes in electrochemically active surface area. For energy storage systems like supercapacitors or batteries, EIS can monitor charge–discharge processes, revealing how co-doping influences ion transport pathways and electrode stability over cycling.

Additionally, EIS can be performed under different potentials (in situ EIS) to monitor how doping affects electronic conductivity and catalytic behavior under operational conditions. This dynamic analysis helps establish direct correlations between dopant-induced structural/electronic modifications and the resulting electrochemical performance, guiding the optimization of B, N, and P co-doped materials for high-efficiency, durable electrochemical devices.

9. APPLICATIONS IN ENERGY CONVERSION, STORAGE, AND ENVIRONMENTAL REMEDIATION

9.1. Energy Conversion Applications

Electrocatalysis for Water Splitting

There has emerged the appearance of B, N and P co-doped materials as very efficient electrocatalysts in the hydrogen evolution reaction (HER) and the oxygen evolution reaction (OER). The interplay between the dopants produces an electronic configuration change and complementary active sites which allow rapid charge transfer. Electron deficient boron is an electron donor that draws in electron rich intermediates, whereas nitrogen brings in more electron density, and phosphorus adjusts the orbital overlap. The mix will decrease the overpotentials, increase current densities, and stabilize the catalytic long-term (Xue et al., 2021).

Oxygen Reduction Reaction (ORR) in Fuel Cells

ORR is one of the most important performance limiting reactions in alkaline and acidic fuel cells. B, N, P co-doped carbon-based catalysts exhibit a spectacular activity in comparison with platinum-based materials and are more durable and tolerant to methanol. The electronic polarization induced by heteroatom distribution increases the adsorption strength for oxygen intermediates (O_2, OOH*, OH*), thereby accelerating reaction kinetics and improving energy efficiency.

Photocatalytic Hydrogen Production

With co-doping of these materials, the bandgap can be customized, leading to long wavelength light absorption, considerably improving solar to hydrogen energy transformation efficiency. Introduction of dopants inhibits electron -hole recombination by formation of defect states which serve as charge separation centers which cause prolonged rates of hydrogen evolution under model sunlight.

CO_2 Reduction Reaction (CO_2RR)

B, N and P co-doped catalysts selectively reduce CO_2 to produce valuable fuels and chemicals, including CO, CH_4 or methanol, electrochemically or photochemically. The high concentrations of heteroatoms create strong diversity of charge-polarized active sites, which enhance CO_2 adsorption and stabilize reaction intermediates, and enhance selectivity and efficiency.

9.2. Energy Storage Applications

Supercapacitors

B, N and P addition to carbon based electrodes enhances double-layer capacitance and pseudocapacitance. Nitrogen functions lead to faradaic redox reactions, boron improves wettability to allow electrolyte to penetrate, phosphorus raises the count of defects. This leads to an increase in specific capacitance, charge discharge rate, and cycling stability (Dessie et al., 2024).

Lithium-Ion Batteries (LIBs)

The co-doped electrodes B, N, and P have good lithium storage capacity and the rate ability. The dopants enhance the density of electrochemically active sites, Li^+ diffusion kinetics and structural integrity at each repeated cycling, therefore improving battery life.

Sodium-Ion Batteries (SIBs)

The bigger ion radius of sodium decreases the easy Na^+ insertion/extraction is not much. Interlayer spacing is expanded by P doping to make ion movement easier and N and B doping enhance electronic conductivity. All these effects allow the retention of high capacity and long-term performance.

Metal–Air Batteries

The air electrode in zinc air and Lithium air battery needs effective ORR and OER catalysts. B–N-P co-doped carbon materials have a high bifunctional catalytic ability allowing efficient charge-discharge cycling, high energy density, and long service life.

9.3. Environmental Remediation Applications

Photocatalytic Degradation of Organic Pollutants

Co-doped photocatalysts of B, N, and P are found to perform better in dye degradation, pharmaceutical, pesticide, and other organic pollutants under visible light (Li et al., 2021). The dopants minimize the bandgap and increase the reactive oxygen species (ROS) that quickly oxidizes the pollutants into biodegradable products.

Adsorption of Heavy Metals and Toxins

The doped (B–OH, N–H, and P–O) functional groups offer powerful coordination domains to the toxic heavy metals such as Pb^{2+}, Hg^{2+}, As^{3+}. Co-doped materials are characterized by high defect density and porosity thereby supporting rapid adsorption and hence useful in water purification systems.

Electrocatalytic Nitrate Reduction

B-N-P co-doped catalysts are highly efficient in the conversion of nitrate (NO_3^-) in wastewater to nitrogen gas or ammonia under mild electrochemical conditions. The binding energy of the reaction intermediates is regulated by the synergistic action of the dopants and selectivity toward desirable products is increased and undesired side reactions reduced.

Gas Sensors for Pollution Monitoring

Co-doped nanostructures have stronger gas adsorption capacity and increased electron transfer rates, making them capable of sensitive detection of a wide range of gases, including NO_2, CO, and NH_3, at low concentrations. The charge polarization induced by the heteroatom notably enhances sensor response speed, recovery time and stability in different environmental conditions.

10. FUTURE PERSPECTIVES

Multi-heteroatomic (B, N, P) co-doping approaches to functional materials remain in their infancy, but seem to hold promise. Although much progress has been achieved in the field of tuning electronic structures, improving catalytic activity, and multi-functional functioning, there are many opportunities and challenges characterizing the path of the further research. To start with, tight control of dopant coverage and bonding conditions are one of the frontiers. The existing synthesis methods, though adaptable, tend to produce random dopant configurations with performance varying with lot to lot. Second, discovery is likely to be faster because of computational-experimental integration. Machine learning and density functional theory (DFT) simulations can quickly sample possible co-doping combinations and determine the best combinations to target a desired application. Such predictions, which have been confirmed by high-resolution in-situ spectroscopic methods like the synchrotron-based XPS, XANES, and EPR, will assist in defining definitive structure-activity correlations and will be used in rational material design. Third, stability in actual working environments is an issue of concern. Most B N P co-doped materials have high initial performance, however, they may have surface reconstruction, dopant leaching, or phase change during extended electrochemical cycling or photocatalytic cycles. It will be important to develop strong passivation schemes, resistance coating, or self-healing doped structures so that the long-term operating reliability is guaranteed (Thakur et al., 2022). Fourth, the shift in the lab to an industrial production should cover the issue of scalability, cost-effectiveness, and environmental sustainability. The use of inexpensive precursors, solvents that are green, and synthesis methods that consume minimal energy will enable large-scale implementation of the renewable energy and environmental remediation technologies. Fifth, the co-doped materials have a multi-functionality that presents a promising prospect to integrated systems.

CONCLUSION

Strategic heteroatomic doping has become a paradigm shift in the development of the structural, electronic, and catalytic characteristics of functional materials. Through the selective introduction of dopants, including boron (B), nitrogen (N), and phosphorus (P), researchers have the ability to fine tune band structures, control surface chemistry and make defect-rich architectures that boost charge transfer and redox activity. These alterations can not only enhance intrinsic conductivity and density of active sites, they also allow the determination of adsorption energies, reaction intermediates, and catalytic selectivity by precise control. The synergistic nature of the multi-heteroatom co-doping is in excess of additive characteristics of the single-element doping. The synergies between the electron-donating and electron-withdrawing dopants cause complementary effects of charge polarization, provide stability to reactive intermediates, and activate multi-pathway catalysis. These properties are important in high-performance use in energy conversion, energy storage and environmental remediation where efficiency and durability is needed. Development of synthesis methods of gas-phase and plasma-assisted methods to solution-based, hydrothermal, and template-assisted methods have expanded the range of possible doping patterns. Combined with advanced characterization methods, including XPS, EPR, XANES/EXAFS, UV Vis DRS, and EIS, these methods enable researchers to investigate the exact role of dopants to determine the material performance at the atomic and electronic scale. In the future, the integration of computational modeling, in-situ spectroscopy, and scalable and sustainable synthesis approaches will hasten the process of rational design of heteroatomic-doped materials to possess application-relevant properties. This capability to combine electronic tuning and redox optimization with multifunctional platforms will not just widen the vistas of catalysis and energy technologies, but also provide access to new uses in sensing, electronics and biomedical devices. Finally, strategic heteroatom doping has proven a highly flexible, potent, and future-oriented design paradigm one that evolves functional materials into highly versatile high-performance systems that have the potential to respond to the urgent technological and environmental challenges.

REFERENCES

Aceto, M., Calà, E., Gulino, F., Gullo, F., Labate, M., Agostino, A., & Picollo, M. (2022). The Use of UV-Visible Diffuse Reflectance Spectrophotometry for a Fast, Preliminary Authentication of Gemstones. *Molecules (Basel, Switzerland)*, *27*(15), 4716. DOI: 10.3390/molecules27154716 PMID: 35897892

Asif, M., & Kumar, A. (2022). Resistive switching in emerging materials and their characteristics for neuromorphic computing. *Materials Today Electronics*, *1*, 100004. DOI: 10.1016/j.mtelec.2022.100004

Bajjou, O., Bakour, A., Khenfouch, M., Baitoul, M., Mothudi, B., Maaza, M., & Faulques, E. (2018). Charge Carrier Dynamics and pH Effect on Optical Properties of Anionic and Cationic Porphyrin–Graphene Oxide Composites. *Journal of Electronic Materials*, *47*(5), 2897–2904. DOI: 10.1007/s11664-018-6139-3

Bo, W., Sheng, H., & Wang, J. (2023). Orbital Polarization-Dependent Fragment Twist-Induced Intramolecular Electric-Field-Driven Charge Transfer. *Molecules (Basel, Switzerland)*, *28*(4), 1801. DOI: 10.3390/molecules28041801 PMID: 36838789

Bumbrah, G. S., & Sharma, R. M. (2016). Raman spectroscopy – Basic principle, instrumentation and selected applications for the characterization of drugs of abuse. *Egyptian Journal of Forensic Sciences*, *6*(3), 209–215. DOI: 10.1016/j.ejfs.2015.06.001

Chen, N., Zeng, Y., Li, T., Cui, P., Dionysiou, D. D., Wang, X., Liu, C., Fang, G., Ding, C., Zhao, Y., Gao, J., Wang, Y., & Zhou, D. (2023). Phosphorus doping significantly enhanced the catalytic performance of cobalt-single-atom catalyst for peroxymonosulfate activation and contaminants degradation. *Journal of Hazardous Materials*, *454*, 131480. DOI: 10.1016/j.jhazmat.2023.131480 PMID: 37146341

de Conti, M. C. M. D., Dey, S., Pottker, W. E., & La Porta, F. A. (2023). An overview into advantages and applications of conventional and unconventional hydro(solvo) thermal approaches for novel advanced materials design. *Materials Today Sustainability*, *23*, 100458. DOI: 10.1016/j.mtsust.2023.100458

Deng, C., Xu, L., Hu, K., Chen, X., Gao, R., Zhang, L., Wang, L., & Zhang, C. (2023). Research Advances on Nitrogen-Doped Carbon Materials in COx Hydrogenation. *Atmosphere (Basel)*, *14*(10), 1510. DOI: 10.3390/atmos14101510

Dessie, Y., Tilahun, E., & Wondimu, T. H. (2024). Functionalized carbon electrocatalysts in energy conversion and storage applications: A review. *Heliyon*, *10*(20), e39395. DOI: 10.1016/j.heliyon.2024.e39395 PMID: 39492918

Gangwar, N., Gangwar, C., & Sarkar, J. (2024). A review on template-assisted approaches & self assembly of nanomaterials at liquid/liquid interface. *Heliyon, 10*(17), e36810. DOI: 10.1016/j.heliyon.2024.e36810 PMID: 39263084

Junaid, M., Khir, M. H. M., Witjaksono, G., Tansu, N., Saheed, M. S. M., Kumar, P., Ullah, Z., Yar, A., & Usman, F. (2020). Boron-Doped Reduced Graphene Oxide with Tunable Bandgap and Enhanced Surface Plasmon Resonance. *Molecules (Basel, Switzerland), 25*(16), 3646. DOI: 10.3390/molecules25163646 PMID: 32796504

Kim, K., Wie, J., & Kim, J. (2020). Synergistic interaction of P and N co-doping EDTA with controllable active EDTA-cobalt sites as efficient electrocatalyst for oxygen reduction reaction. *Journal of Industrial and Engineering Chemistry, 83*, 252–259. DOI: 10.1016/j.jiec.2019.11.035

Kim, S. Y., Kumachang, C. C. F., & Doumon, N. Y. (2023). Characterization Tools to Probe Degradation Mechanisms in Organic and Perovskite Solar Cells. *Solar RRL, 7*(13), 2300155. Advance online publication. DOI: 10.1002/solr.202300155

Kober, T., Schiffer, H.-W., Densing, M., & Panos, E. (2020). Global energy perspectives to 2060 – WEC's World Energy Scenarios 2019. *Energy Strategy Reviews, 31*, 100523. DOI: 10.1016/j.esr.2020.100523

Li, P., Lu, J., Cui, H., Ruan, S., & Zeng, Y.-J. (2021). The development, application, and performance of black phosphorus in energy storage and conversion. *Materials Advances, 2*(8), 2483–2509. DOI: 10.1039/D0MA01016B

Liu, D., Zhang, Q., Chen, X., Zhu, P., Yan, F., Wang, X., Dai, H., Chen, J., Gong, G., Shang, C., Xie, L., & Zhai, X. (2022). The Lattice Distortion, Defect Evolution and Electrochemical Performance Improvement in Zn-VO2(B) Nanorods. *Nanomaterials (Basel, Switzerland), 12*(18), 3196. DOI: 10.3390/nano12183196 PMID: 36144984

Liu, J., Bunes, B. R., Zang, L., & Wang, C. (2018). Supported single-atom catalysts: Synthesis, characterization, properties, and applications. *Environmental Chemistry Letters, 16*(2), 477–505. DOI: 10.1007/s10311-017-0679-2

Long, B., Zou, Y., Li, Z., Ma, Z., Jiang, W., Zou, H., & Chen, H. (2020). Effect of Phosphorus Doping on Conductivity, Diffusion, and High Rate Capability in Silicon Anode for Lithium-Ion Batteries. *ACS Applied Energy Materials, 3*(6), 5572–5580. DOI: 10.1021/acsaem.0c00534

Loy, A. C. M., Ng, W. L., & Bhattacharya, S. (2024). Advanced characterization techniques for the development of Subatomic scale catalysts: One step closer to industrial scale fabrication. *Materials Today Catalysis, 4*, 100033. DOI: 10.1016/j.mtcata.2023.100033

M, A., & G, P. K. (2023). Hydrogen towards sustainable transition: A review of production, economic, environmental impact and scaling factors. *Results in Engineering, 20*, 101456. DOI: 10.1016/j.rineng.2023.101456

Mc Carthy, K., Gullapalli, H., Ryan, K. M., & Kennedy, T. (2022). Electrochemical impedance correlation analysis for the estimation of Li-ion battery state of charge, state of health and internal temperature. *Journal of Energy Storage*, *50*, 104608. DOI: 10.1016/j.est.2022.104608

Morozzi, P., Ballarin, B., Arcozzi, S., Brattich, E., Lucarelli, F., Nava, S., Gómez-Cascales, P. J., Orza, J. A. G., & Tositti, L. (2021). Ultraviolet–Visible Diffuse Reflectance Spectroscopy (UV–Vis DRS), a rapid and non-destructive analytical tool for the identification of Saharan dust events in particulate matter filters. *Atmospheric Environment*, *252*, 118297. DOI: 10.1016/j.atmosenv.2021.118297

Mourdikoudis, S., Pallares, R. M., & Thanh, N. T. K. (2018). Characterization techniques for nanoparticles: Comparison and complementarity upon studying nanoparticle properties. *Nanoscale*, *10*(27), 12871–12934. DOI: 10.1039/C8NR02278J PMID: 29926865

Mudoi, R., & Saikia, L. (2025). Heteroatom Doping, Defect Engineering, and Stability of Transition Metal Diselenides for Electrocatalytic Water Splitting. *Chemistry, an Asian Journal*, *20*(19), e00755. Advance online publication. DOI: 10.1002/asia.202500755 PMID: 40783792

Nakashima, K., Toshima, Y., Kobayashi, Y., Ishikawa, Y., & Kakihana, M. (2019). Solvothermal synthesis and morphology control of NaNbO 3 nanocubes using a reaction medium of water and/or methanol. *Journal of Asian Ceramic Societies*, *7*(4), 544–550. DOI: 10.1080/21870764.2019.1683955

Pavlenko, V., Khosravi, H. S., Żółtowska, S., Haruna, A. B., Zahid, M., Mansurov, Z., Supiyeva, Z., Galal, A., Ozoemena, K. I., Abbas, Q., & Jesionowski, T. (2022). A comprehensive review of template-assisted porous carbons: Modern preparation methods and advanced applications. *Materials Science and Engineering R Reports*, *149*, 100682. DOI: 10.1016/j.mser.2022.100682

Poopanya, P., Gadhavi, P. M., Talati, M., Sivalertporn, K., Som, N. N., & Mishra, A. K. (2025). Tailoring graphene for thermoelectric: Theoretical insights into position-dependent effects of BN Co-doping. *Computational Condensed Matter*, *44*, e01106. DOI: 10.1016/j.cocom.2025.e01106

Prasad, R. D., Prasad, R. S., Prasad, R. B., Prasad, S. R., Singha, S. B., Prasad, R. J., Sinha, P., Saxena, S., Vaidya, A. K. B., Teli, S. B., Saxena, U. R., Harale, A., Deshmukh, M. B., Padvi, M. N., & Navathe, G. J. (2024). *A Review on Modern Characterization Techniques for Analysis of Nanomaterials and Biomaterials*. ES Energy & Environment., DOI: 10.30919/esee1087

Rabbani, M. S., Wei, S., Anwar, M. I., Zafar, Z., Liu, Z.-Y., Wang, Y., & Yue, X.-Z. (2025). Al3+ Doping-Induced Electronic Structure Regulation in NiFeO for Accelerated Oxygen Evolution Kinetics and Superior Stability. *Journal of Alloys and Compounds*, *182831*, 182831. Advance online publication. DOI: 10.1016/j.jallcom.2025.182831

Sfeir, A., Teles, C. A., Ciotonea, C., Reddy, G. N. M., Marinova, M., Dhainaut, J., Löfberg, A., Dacquin, J.-P., Royer, S., & Laassiri, S. (2023). Enhancing ammonia catalytic production over spatially confined cobalt molybdenum nitride nanoparticles in SBA-15. *Applied Catalysis B: Environmental*, *325*, 122319. DOI: 10.1016/j.apcatb.2022.122319

Shinde, P. A., Shrestha, L. K., & Ariga, K. (2025). Heteroatom-functionalized carbon nanoarchitectonics: Unlocking the doping effects for supercapacitor electrode design. *Green Energy & Environment*, *10*(9), 1838–1862. Advance online publication. DOI: 10.1016/j.gee.2025.02.007

Song, C., Zhang, B., Hao, L., Min, J., Liu, N., Niu, R., Gong, J., & Tang, T. (2022). Converting poly(ethylene terephthalate) waste into N-doped porous carbon as CO2 adsorbent and solar steam generator. *Green Energy & Environment*, *7*(3), 411–422. DOI: 10.1016/j.gee.2020.10.002

Thakur, A. K., Kurtyka, K., Majumder, M., Yang, X., Ta, H. Q., Bachmatiuk, A., Liu, L., Trzebicka, B., & Rummeli, M. H. (2022). Recent Advances in Boron- and Nitrogen-Doped Carbon-Based Materials and Their Various Applications. *Advanced Materials Interfaces*, *9*(11), 2101964. Advance online publication. DOI: 10.1002/admi.202101964

Wang, J., Lou, F., Zhang, M., & Yuan, J. (2025). Recent progress in nitrogen-doped activated carbon: Microstructural regulation and enhanced gas adsorption. *Journal of Environmental Chemical Engineering*, *13*(5), 118107. DOI: 10.1016/j.jece.2025.118107

Wang, Q., & Frenkel, A. I. (2017). *Characterization of Model Nanocatalysts by X-ray Absorption Spectroscopy*., DOI: 10.1016/B978-0-12-805090-3.00005-X

Wang, X., Zhu, W., Wang, N., & Zuo, X. (2022). Nitrogen-doped oxidized activated carbon as efficient metal-free bifunctional material for bisphenol removal: A tuning of surface chemistry. *Journal of Environmental Chemical Engineering, 10*(6), 108509. DOI: 10.1016/j.jece.2022.108509

Xue, X.-X., Meng, H., Huang, Z., Feng, Y., & Qi, X. (2021). Black phosphorus-based materials for energy storage and electrocatalytic applications. *Journal of Physics: Energy, 3*(4), 042002. DOI: 10.1088/2515-7655/abff6b

Yadav, V. K., Mir, S. H., & Singh, J. K. (2019). A computational study of structural, electronic and carrier mobility of boron and phosphorus/nitrogen co-doped graphene. *Physica B, Condensed Matter, 571*, 291–295. DOI: 10.1016/j.physb.2019.07.017

Yao, J., Liu, D., Zhao, B., Zhou, Y., Li, R., Zhang, Y., & Peng, T. (2023). Insight into the effect of B-doping on electronic structure and photocatalytic H2 evolution performance of C3N5 nanosheets. *Journal of Photochemistry and Photobiology A Chemistry, 436*, 114359. DOI: 10.1016/j.jphotochem.2022.114359

You, X.-X., Gao, J., Duan, Y.-C., Geng, Y., Zhang, M., Zhao, L., & Su, Z.-M. (2022). A theoretical analysis on the electron and energy transfer between host and guest materials in phosphor–doped OLED. *Journal of Photochemistry and Photobiology A Chemistry, 432*, 114058. DOI: 10.1016/j.jphotochem.2022.114058

Zakari, R. S. B., Guerrero Peña, G. D. J., Pasanaje, A. H., Elsayed, M., Kuppireddy, S., Hassan Ali, M. I., Raj, A., & Elkadi, M. (2024). Effect of phosphorus on the physicochemical and electrochemical properties of Ni/Carbon electrodes for glycerol electrooxidation. *Case Studies in Chemical and Environmental Engineering, 10*, 100768. DOI: 10.1016/j.cscee.2024.100768

Zhang, E., Zhu, Q., Huang, J., Liu, J., Tan, G., Sun, C., Li, T., Liu, S., Li, Y., Wang, H., Wan, X., Wen, Z., Fan, F., Zhang, J., & Ariga, K. (2021). Visually resolving the direct Z-scheme heterojunction in CdS@ZnIn2S4 hollow cubes for photocatalytic evolution of H2 and H2O2 from pure water. *Applied Catalysis B: Environmental, 293*, 120213. DOI: 10.1016/j.apcatb.2021.120213

Zhang, H., Chen, L., Dong, F., Lu, Z., Lv, E., Dong, X., Li, H., Yuan, Z., Peng, X., Yang, S., Qiu, J., Guo, Z., & Wen, Z. (2024). Dynamic transformation of active sites in energy and environmental catalysis. *Energy & Environmental Science, 17*(18), 6435–6481. DOI: 10.1039/D4EE02365J

Zhang, R., Du, X., Li, S., Guan, J., Fang, Y., Li, X., Dai, Y., & Zhang, M. (2022). Application of heteroatom doping strategy in electrolyzed water catalytic materials. *Journal of Electroanalytical Chemistry (Lausanne, Switzerland), 921*, 116679. DOI: 10.1016/j.jelechem.2022.116679

Zhang, W., Lu, W.-C., Zhang, H.-X., Ho, K. M., & Wang, C. Z. (2016). Lattice distortion and electron charge redistribution induced by defects in graphene. *Carbon*, *110*, 330–335. DOI: 10.1016/j.carbon.2016.09.031

Zhang, Z., Wu, G.-X., Tan, Z.-L., Chen, H.-Z., Yu, Z., Yu, Z.-R., Huang, X.-B., Yang, W.-J., & Huang, Y.-D. (2024). Lattice-Engineering modulated band structure facilitates enhanced stability of high-voltage LiCoO2 at 4.6 V. *Chemical Engineering Journal*, *502*, 158053. DOI: 10.1016/j.cej.2024.158053

Chapter 4
Experimental Approaches to Fuel Desulphurization: Catalysts, Conditions, and Performance

Subhraraj Panda
https://orcid.org/0000-0003-4224-4663
Centurion University of Technology and Management, Bhubaneswar, India

Niharika Das
https://orcid.org/0009-0008-4524-5198
Centurion University of Technology and Management, Bhubaneswar, India

Rajalaxmi Panda
S.K.C.G. Autonomous College, Paralakhemundi, India

ABSTRACT

Desulphurization of fossil fuels has emerged as a critical area of research in energy science and environmental engineering due to increasingly stringent environmental regulations. Sulphur compounds in fuels are precursors to SOx emissions, which contribute to acid rain, smog formation, and adverse health effects. This chapter provides an in-depth exploration of experimental approaches to fuel desulphurization, with a primary focus on catalysts, process conditions, and performance outcomes. Emphasis is placed on hydrodesulphurization (HDS), oxidative desulphurization (ODS), adsorptive desulphurization (ADS), biodesulphurization (BDS), and emerging nanocomposite-based methods. Comparative analysis of catalyst design, experimental methodologies, and operating conditions is presented alongside discussions on

DOI: 10.4018/979-8-3373-5966-3.ch004

reaction mechanisms and performance indicators. The chapter concludes by highlighting the future perspectives of nanostructured catalysts in achieving ultra-deep desulphurization under mild operating conditions.

1. INTRODUCTION

Sulphur-containing compounds are inherent in petroleum and its fractions, which creates significant challenges in both refining and environmental management. Song 2003; Stanislaus et al., 2010 The structural complexity of these compounds-from thiols and sulphides to complex cyclic structures like thiophenes, benzothiophenes, and multi-ring dibenzothiophenes-translates into differences in reactivities and removal efficiencies. Low molecular weight thiols and simple sulphides can be relatively easily removed from low sulphur streams using conventional refining methods; however, the more aromatically thiophenic species exhibit far greater chemical stability because of extensive electron delocalization within their ring structures. DBT and, in particular, its alkyl-substituted derivatives, especially 4,6-DMDBT, are considered among the most refractory species because of steric hindrance of the sulphur atom, limiting access to catalytic sites even under rigorous HDS conditions. Marafi & Stanislaus, 2008 Oxidation of thiophenic compounds to sulfoxides and sulfones, usually formed in the ODS process, represents another way of removal. Due to the higher polarity and elevated boiling points of these oxidized species, their separation by extraction or adsorption is possible. Since these species are seldom present in 'as-received' fuels, their controlled formation during experimental processes is of great importance in understanding reaction mechanisms and testing catalytic activities Campos-Martin et al. (2010); Otsuki et al. (2000). The proper characterization of sulphur compounds before and after treatment is crucial for process optimization. Advanced analytical techniques, such as GC–SCD, GC–MS, XPS, FT-IR, and ultraviolet fluorescence-based total sulphur analysers (ASTM D5453), now allow for the accurate quantification, structural elucidation, and oxidation state assessment of sulphur species (Hsu & Robinson, 2019; Lee et al., 2009). These data provide input to catalyst design through its correlations of molecular structure with desulphurization efficiency.

The persistence of refractory sulphur species has major implications for technology selection and fosters the development of nanocomposite catalysts, hybrid oxidative systems, and low-temperature processes suitable for ultra-low sulphur fuel production. Basically, other main reasons for this interest are linked with the environmental and operational impact caused by the sulphur oxides; in fact, combustion converts fuel sulphur to SO_2 and SO_3, contributing to acid rain, particulate pollution, and material corrosion (EPA, 2023; WHO, 2021).

Global fuel standards, including the recent Euro VI and Bharat Stage VI, require sulphur levels below 10 ppm, making effective and sustainable technologies for desulphurization highly demanding (Speight, 2020; Song, 2003). The distribution, reactivity, and transformation routes of sulphur compounds are essential information in elaborating next-generation catalytic materials and refining approaches for the fulfilment of both environmental and industrial goals.

2. CLASSIFICATION OF SULPHUR COMPOUNDS

2.1 Inorganic Sulphur Compounds

Inorganic sulfur in crude oil includes H_2S, elemental sulfur, and metal sulfides such as FeS, Ni_3S_2, and CuS. H_2S usually develops during early distillation or thermal cracking and is eliminated through amine scrubbing or Claus processes. Elemental sulfur may exist in oxidized oils or be of biological origin. Metal sulfides are formed through the reaction of H_2S with metallic constituents of pipelines and equipment. While these compounds can easily be separated by using conventional methods, they pose operational hazards and result in corrosion and catalyst poisoning.

According to Speight 2014, and Hsu and Robinson 2006, inorganic sulfur in crude oil is mainly made up of hydrogen sulfide, H_2S, elemental sulphur, and metal sulphides like FeS, Ni_3S_2, and CuS. Speight (2014) points out that hydrogen sulfide is often released during initial distillation or thermal cracking and is thus usually eliminated by gas-treating operations, including amine scrubbing along with Claus sulphur recovery systems. Elemental sulphur can also be present in minor amounts, especially in oxidized crude oils or due to microbial activity in storage and processing facilities (Hsu & Robinson, 2006). Additionally, metal sulphides usually occur when H_2S reacts with metallic infrastructure components in pipelines or refinery equipment, according to Gary, Handwerk, and Kaiser 2007. Even though these inorganic species are generally more straightforward to eliminate through conventional approaches such as gas scrubbing or filtration, their presence remains problematic because they catalyze corrosion, poison catalysts, and introduce major operational hazards in refining systems (Speight, 2014; Gary et al., 2007).

2.2 Organic Sulphur Compounds

Organic sulphur is predominant in petroleum fractions and encompasses thiols, sulphides, disulphides, thiophenes, benzothiophenes, and dibenzothiophenes. Thiols, existing in light gasoline, are pretty reactive and eliminated by sweetening processes, many of which involve NaOH and catalysts, such as the Merox process.

Sulphides and disulphides are more stable and thus demand more intensive oxidative or hydrodesulphurization conditions. Compounds of the thiophenic series prevail in diesel and heavy gas oils. In these compounds, the aromatic nature of the ring and steric hindrance make them resistant to HDS. Among them, the alkyl-substituted DBTs, such as 4,6-DMDBT, can be considered refractory (Stanislaus et al. 2010). In oxidative desulphurization, these would form sulfoxides and sulfones, more polar products with higher boiling points, and thus could be separated by extraction or adsorption.

Organic sulphur compounds make up the major sulphur inventory in fossil fuels, especially in the middle and heavy distillates. They are covalently bound within the hydrocarbon matrix and are thermally and chemically stable, hence very difficult to sulphurise.

The most prevalent classes include:

(a) Thiols (Mercaptans, R–SH)

According to Speight, 2014; Hsu and Robinson, 2006, thiols (also named mercaptans) are the simplest class of organic sulphur compounds and predominantly occur in light gasoline fractions with a boiling point below 100 °C. They give a petroleum stream a strong, unpleasant odor and present pronounced corrosivity toward metals; for these reasons, they can be particularly problematic in handling and storage. Due to their relatively high chemical reactivity, thiols can be easily transformed into disulphides or other oxidized sulphur species through mild oxidative desulphurization approaches, as reported by Song, 2003. In modern petroleum refining, mercaptans are often removed or transformed into less odorous and less corrosive derivatives through sweetening processes, the most notable being the Merox process, using sodium hydroxide and proprietary catalysts to oxidize mercaptans into disulphides. Gary, Handwerk, & Kaiser, 2007

(b) Sulphides (R–S–R') and Disulphides (R–S–S–R')

The sulphides and disulphides are more stable than thiols and commonly appear in heavier petroleum fractions such as kerosene and diesel, according to Speight 2014 and Hsu and Robinson 2006. In addition, sulphides possess relatively low polarity and limited chemical reactivity, necessitating the utilization of stronger oxidative or hydrodesulphurization conditions for their effective removal during refining. According to Song 2003, disulphides are mainly formed through the oxidation of thiols and though generally less environmentally harmful they can revert to thiols rather easily under reducing conditions, which are common in many refining processes. These sulphur species often serve as intermediates along desulphurization pathways, acting as transitional compounds between a primary sulphur-containing molecule and its fully desulphurized end product Speight, 2014; Hsu & Robinson, 2006.

(c) Thiophenes (C_4H_4S) and Their Derivatives

As identified by Speight 2014 and Hsu and Robinson 2006, thiophenic compounds represent the major class of organic sulphur species in diesel and heavy gas oil fractions. Of these, thiophene, benzothiophene BT, dibenzothiophene DBT, and 4,6-dimethyldibenzothiophene 4,6-DMDBT, in particular, are extensively investigated due to their strong resistance to HDS processes. Song 2003 has noted that the aromatic ring system surrounding the sulphur atom promotes extended electron delocalization, which stabilizes the C-S bond and drastically reduces its reactivity relative to hydrogenation or bond cleavage. Additionally, steric hindrance developed by alkyl substituents, notably the 4- and 6-methyl groups in 4,6-DMDBT, prevents effective access of the sulphur atom to catalytic active sites, as discussed in detail by Song 2003, with support in refinery reaction studies reported by Gary, Handwerk, and Kaiser 2007. For all of these combined structural effects, thiophenic molecules are commonly described as refractory sulphur species, reflecting their persistence under even aggressive desulphurization conditions.

(d) Sulfoxides (R_2SO) and Sulfones (R_2SO_2)

According to Song (2003) and further evidenced through refinery chemistry analyses by Speight (2014), sulfoxides and sulfones are the oxidized derivatives of sulphides or thiophenic compounds, commonly formed in most ODS processes. These polar forms of sulphur compounds have much higher polarity and higher boiling points than their parent sulphur species; thus, they can easily be separated from fuel streams by extraction or adsorption methods (Hsu & Robinson, 2006). While sulfoxides and sulfones are not typically found in untreated petroleum fractions, Song (2003) identifies them as important intermediates in both experimental and industrial ODS routes, whereby the sulphur-containing molecules are stepwise oxidized into these more polar forms, which then become separable from the hydrocarbon matrix.

Analytical Identification of Sulphur Compounds

According to Speight (2014) and Hsu and Robinson (2006), experimental desulphurization studies rely heavily on the accurate characterization of sulphur species before and after treatment, placing immense demands for precision and selectivity on analytical techniques. GC–SCD has the advantage of compound-specific detection with very high sensitivity, often down to the parts-per-billion (ppb) level, which makes the technique particularly useful for tracing sulphur species at low concentration levels (Song, 2003). In terms of structure identification, the use of GC–MS gives detailed molecular information on both thiophenic and aliphatic sulphur compounds, enabling the refined characterization of desulphurization pathways (Speight, 2014). The XPS technique is applied conventionally to differentiate sulphur atoms by their states of oxidation: –2, 0, +4, and +6. This helps to identify the chemical changes undergone by sulphur during catalytic or oxidative desulphurization (Hsu & Robinson, 2006). Further support to the identification of sulfoxide and sulfone functional groups resulting from oxidative reaction has been provided by FT-IR, while the determination of the overall sulphur content in fuels accurately has been made possible by Total Sulphur Analyzers using the ultraviolet fluorescence methodology according to ASTM D5453 (Speight, 2014). These techniques collectively help in providing the data necessary for correlating molecular structure with desulphurization efficiency, thus guiding catalyst development and optimization of reaction conditions.

Reactivity of sulphur species generally follows:

Thiols > Sulphides > Disulphides > Thiophenes > Alkyl-substituted thiophenes

As noted by Song (2003) and described for refinery applications by Speight (2014), the order of reactivity of the sulphur compounds directly dictates the experimental parameters of temperature, pressure, and catalyst formulation that are required for effective desulphurization. The most refractory species, such as 4,6-DMDBT, resist extreme HDS conditions and are primarily responsible for the residual sulphur present in ultra-low sulphur diesel fuels. This has driven the development of nanocomposite catalysts and hybrid oxidative–catalytic systems capable of targeting sterically hindered refractory molecules (Song, 2003). From an environmental perspective, combustion of these sulphur-containing compounds results in the formation of SO_2, which is further oxidized to H_2SO_4, contributing to acid rain, particulate emissions, and corrosion in both engines and exhaust systems (Speight, 2014). Modern legislation, such as the Euro VI and Bharat Stage VI requirements for sulphur below 10 ppm, highlights the importance of sulphur speciation studies in the design of effective and compliant desulphurization methodologies (Hsu & Robinson, 2006). The molecular structure of sulphur compounds in fuels determines their reactivity and, therefore, their ease of removal. Thiols (mercaptans) are low molecular weight compounds that can be relatively easily removed by conventional refining processes. Sulphides and disulphides exhibit moderate reactivity and may provide intermediates during reactions of oxidation or hydrodesulphurization. Thiophenes and benzothiophenes, due to their aromatic nature, are resistant to removal because of the stability imparted by their conjugated ring structures. Dibenzothiophenes and their alkyl-substituted derivatives are highly refractory species of sulphur, since steric hindrance around the sulphur atom reduces accessibility to catalytic active sites, thereby making the desulphurization of such molecules quite challenging.

2.3 Environmental and Industrial Implications

2.3.1 Environmental Impact

Combustion of sulphur-containing fuels forms SO_2 and SO_3, which react with moisture to produce H_2SO_4-the major contributor to acid rain (EPA, 2023). Acid rain reduces the fertility of the soil, leaches nutrients, and disrupts aquatic ecosystems, causing the mortality of fish and loss of biodiversity. Infrastructure, particularly those made from limestone and marble, undergoes chemical erosion (WHO, 2021). Human exposure to SO_2 and sulfate aerosols leads to respiratory illnesses such as asthma, bronchitis, and COPD, while long-term exposure increases cardiovascular risks (EPA, 2023; WHO, 2021).

Impact on Environment and Industry

Most significantly, the presence of sulphur compounds in fossil fuels affects environmental sustainability and the operation of industry alike. Sulphur-containing molecules are oxidized during combustion into sulphur oxides, SO_X, that degrade air quality and place technological, economic, and regulatory burdens on refineries and end-users alike (Speight, 2020; Song, 2003). Understanding these impacts provides the rationale for stringent desulphurization targets and the ongoing evolution of catalytic technologies.

2.3.2 Environmental Impact

(a) Formation of Sulphur Oxides (SO_2 and SO_3)

When sulphur-containing fuels are burned, the sulphur is mostly oxidized into sulphur dioxide, SO_2, which is the main sulphur oxide emitted during combustion. Only a fraction of it gets further oxidized to sulphur trioxide, especially in the presence of metal oxides or catalytic surfaces. It then reacts with atmospheric moisture to form sulphuric acid, one of the most important contributors to acid rain.

(b) Acid Rain and Ecosystem Damage

Sulphur emissions from combustion contribute considerably to environmental quality. Acid deposition causes acidification of the soil, leaching of nutrients, and reduced agricultural productivity. In aquatic environments, the low pH leads to mortalities in fish and changes in biodiversity. Built heritage undergoes chemical weathering of limestone and marble through sulphuric acid attack, wherein there is surface erosion and structural weakening.

(c) Human Health Effects

Exposure to SO_2 and particulate sulfate aerosols can carry substantial health risks. Longer-term exposure has been linked with asthma, bronchitis, and COPD, especially among those populations that are more sensitive. Acid mists formed from sulphur oxides irritate the eyes, skin, and respiratory tract, while long-term exposure increases cardiovascular disease risk (EPA, 2023; WHO, 2021). WHO and the U.S. EPA recommend that ambient SO_2 concentrations not exceed 20 $\mu g/m^3$ (24-hour mean), although levels in industrial regions can far exceed this level in the absence of adequate controls on emissions. WHO, 2021; EPA, 2023

2.3.3 Industrial Impact

Sulphur compounds poison platinum-group metal catalysts, diminishing activity and lifespan. SO_2 and H_2S accelerate sulfidation corrosion in pipelines, heat exchangers, and combustion systems of various industries. In hydrotreating units, high sulphur requires higher hydrogen pressures, temperatures, and longer residence times, increasing energy consumption. Residual sulphur forms corrosive acids, deactivates emission control catalysts, and lowers fuel stability that impacts gasoline, diesel, and aviation fuels. In general, hydrotreating costs via consumption of hydrogen, catalyst replacement, and compliance with ultra-low sulphur fuel regulations are an important economic load.

(a) Catalyst Poisoning and Equipment Corrosion

Sulphur compounds are potent catalyst poisons during refining and petrochemical processes. During reforming and hydrogenation, PGM catalysts are readily susceptible to sulphur adsorption on active sites, inhibiting hydrogenation pathways and catalytic activities.

Sulphur also increases the rate at which equipment corrodes. Hydrogen sulfide arising from H_2S gives rise to hydrogen-induced cracking in steel pipelines, commonly referred to as HIC. In flue gas systems, the formation of highly corrosive sulphuric acid condensates arises from SO_2 and SO_3, while mercaptans in fuels give copper corrosion in storage and distribution networks.

(b) Operational Challenges in Refineries

Sulphur-rich feedstock impacts many refinery units: In catalytic cracking, sulphur promotes catalyst sulfidation and forms hydrogen sulfide (H_2S) in regenerator gas, which needs extra cleanup (Song, 2003); hydrotreating units will have to work at more elevated temperatures, pressures, and hydrogen consumption to remove refractory sulphur species, increasing energy and operational costs of the process (Stanislaus et al., 2010).

Sulphur also deteriorates lubricant stability and promotes oxidation, corrosive wear, and extra post-treatment or special additive packages.

(c) Impact on Fuel Quality and Performance

Residual sulphur impairs fuel quality. The combustion of sulphur compounds originates with sulphurous and sulphuric acids, which cause corrosion in parts of engines, exhaust systems, and fuel lines. Sulphur oxides further deactivate three-way catalytic converters, diminishing their effectiveness in controlling the emissions of CO, NO_X, and hydrocarbons.

Sulphur further reduces fuel stability by initiating oxidation and gum formation, leading to injector fouling and performance deterioration.

Examples include:

- Gasoline: Thiols and sulphides cause objectionable odors and corrode storage systems.
- Diesel: Thiophenic sulphur decreases cetane number and promotes particulate emissions. - Song, 2003
- Aviation fuel: Sulphur residues impede thermal stability at high temperatures, promoting deposits and clogging fuel systems (Hsu & Robinson, 2017).

(d) Economic Costs

Desulphurization has greatly increased the cost of running a refinery. Hydrotreating units alone account for 15–20% of total refinery hydrogen consumption. Catalyst poisoning requires regeneration or its replacement quite frequently, at considerable costs to refineries of millions per year. ULSF standards (<10–15 ppm) often require upgrading existing units or adoption of advanced methods such as ODS and ADS. So, desulphurization is not only required by environmental considerations but has also become a strategic industrial need for both competitiveness and efficiency in compliance with the law.

2.3.4 Socioeconomic and Policy Dimensions

Low-sulphur fuels reduce respiratory illnesses, healthcare costs, and improve urban air quality. WHO (2021), EPA (2023) Compliance with international standards, such as IMO 2020 for marine fuels, impels refinery modernization and trade compliance. IMO (2020) National energy security is enhanced through deep desulphurization-supporting industrial competitiveness, exports, and foreign investment. Speight (2020) Advanced catalytic and nanocomposite-based desulphurization strategies are considered essential for sustainable environment and socioeconomic outcomes. Song (2003); Stanislaus et al. (2010)

Socioeconomic and Policy Dimensions (with APA In-Text Citations)

Effective fuel desulphurization provides broad benefits, extending from protection of the environment to impacts on public health, international trade, and energy security for nations. Due to this fact, from the point of view of public health, application of low-sulphur fuels notably decreases respiratory diseases such as asthma, bronchitis, and COPD, strongly connected with sulphur oxides and particulate sulfate aerosols exposure (WHO, 2021; EPA, 2023). Such changes help reduce the health burden and improve urban air quality.

On a global scale in terms of trade, compliance with international sulphur regulations, such as the IMO 2020 mandate, requiring the sulphur content in marine fuels to $\leq$0.5 wt%, has become critical to international shipping operations and has necessitated modernization and compliance at refineries around the world.

The expansion of deeply desulphurized fuels production capacity, from a national energy security perspective, enhances industrial competitiveness, strengthens export potential, and stimulates foreign investment, thus contributing to economic resilience and long-term sustainable energy development.

Thus, the drive toward low-sulphur fuels reflects both environmental responsibility and socio-economic advancement, emphasizing the need for continued research into advanced catalytic and nanocomposite-based desulphurization technologies. Song, 2003; Stanislaus et al., 2010

Figure 1. Schematic illustration of sulphur oxide formation and its environmental consequences, showing emission from fuel combustion, atmospheric transformation to sulphuric acid, and effects on soil, water, and infrastructure. (Image generated by AI)

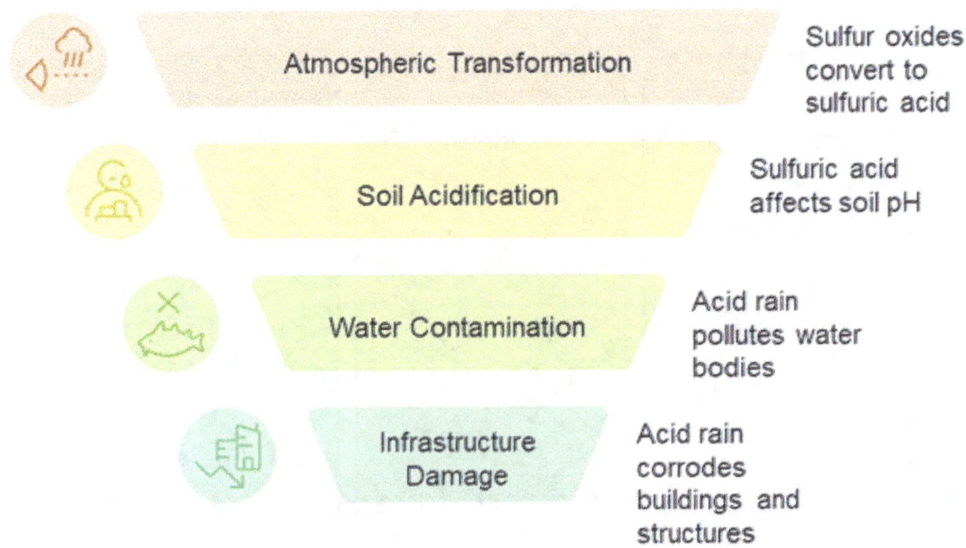

Sulphur oxide (SOx) emissions have several negative impacts across different sectors. Environmentally, they contribute to acid rain and air pollution, which can harm ecosystems, damage vegetation, and deteriorate buildings. In industrial settings, SOx can cause catalyst poisoning and corrosion in pipelines and engines, reducing efficiency and increasing maintenance costs. From a health perspective, exposure to sulphur oxides can lead to respiratory problems and cardiovascular issues, posing serious risks to human well-being.

3. EXPERIMENTAL APPROACHES TO FUEL DESULPHURIZATION

Sulphur removal from petroleum fuels is one of the most important refining steps with regard to environmental compliance and avoidance of operational problems resulting from sulphur compounds. Experimental methods of desulphurization broadly fall into HDS, ODS, ADS, and BDS. Capabilities of these techniques have been extended by advanced hybrid systems and nanocomposite catalysts that enable refractory sulphur compounds to be treated under milder conditions.

Figure 2. Mechanism of oxidative desulphurization showing oxidation of DBT to sulfone. (Image generated by AI)

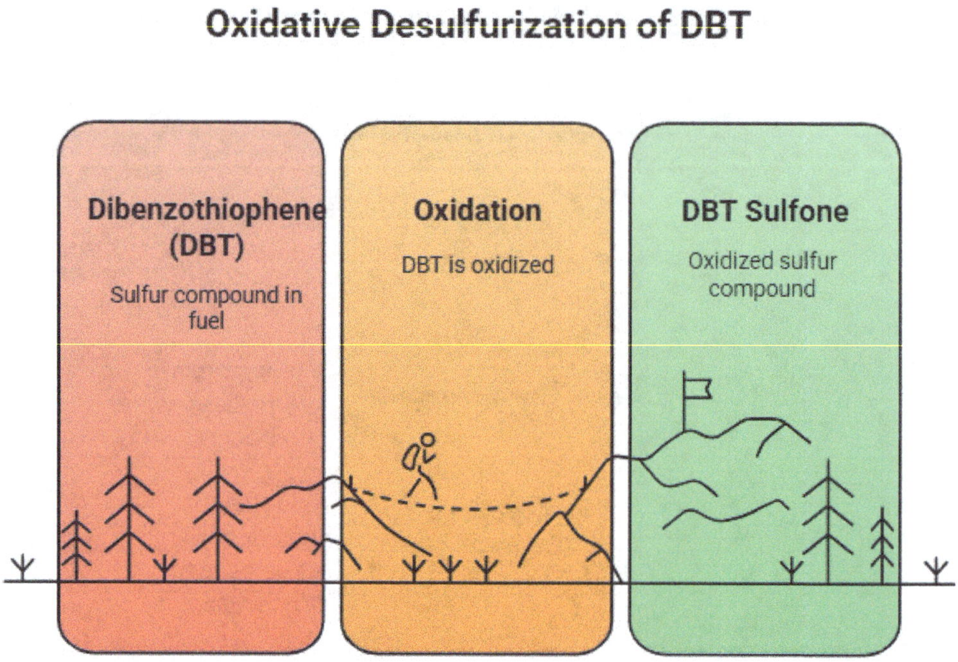

Figure 3. Adsorption process schematic illustrating π-complexation and pore diffusion.(Image generated by AI)

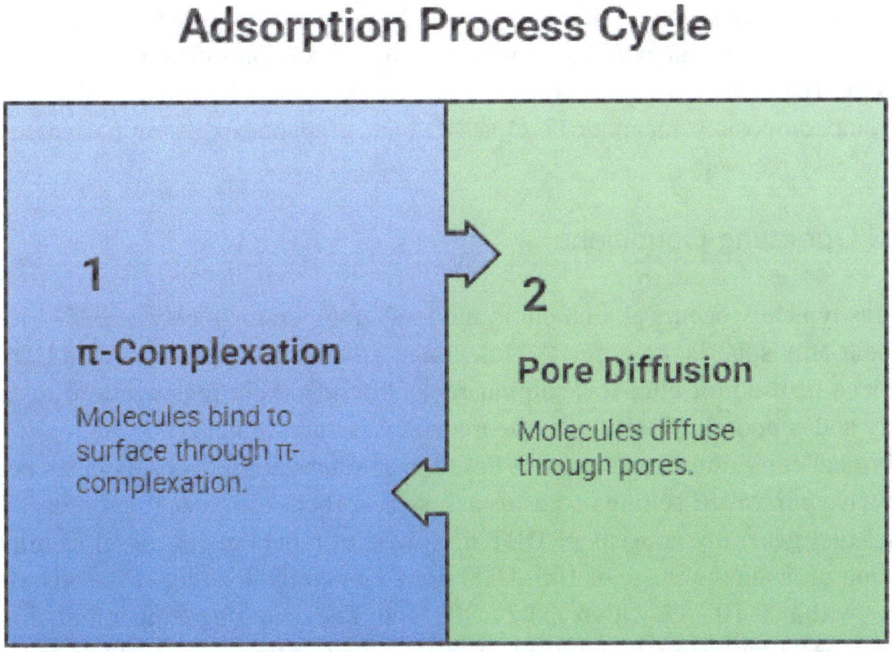

3.3 Adsorptive Desulphurization (ADS)

3.3.1 Principles

In most cases, ADS relies on the selective adsorption of sulphur compounds onto solid supports such as activated carbon, zeolites, MOFs, or mesoporous silica. Adsorption could happen via π-π interactions, metal-sulfur coordination, or hydrogen bonding, depending on the nature of the adsorbent and sulphur compound.

3.3.2 Adsorbent Design

High surface area and pore volume are crucial for effective adsorption. Functionalization with transition metal ions such as Cu^{2+}, Ag^{+}, or Ni^{2+} imparts selectivity toward thiophenic compounds. Nanocomposite adsorbents, in which MOFs are combined with graphene or silica, show improved kinetics and capacity Campos-Martin et al. (2010)

1. Activated Carbons: high surface area - effective for thiols and sulphides.
2. Metal-Exchanged Zeolites: Cu^+, Ag^+, or Ni^{2+} ions interact with sulphur π-electrons through π-complexation.
3. Mesoporous Materials: MCM-41, SBA-15 functionalized with transition metals.
4. MOFs have tunable pore sizes and active sites. Examples include Cu–BTC and MIL-101.
5. Nanocomposites: magnetic $Fe_3O_4@SiO_2$–Cu, graphene oxide-supported metals for easy recovery.

3.3.3 Operating Conditions

This typically occurs at ambient to moderate temperatures between 25–150 °C and near-atmospheric pressures. An integrated process in conjunction with HDS or ODS can be used for ultra-low sulphur fuels. For adsorbent regeneration, sustainability and economic viability are the main key factors. Song, 2003

Typical conditions in adsorption-based desulphurization experiments generally involve controlled settings so as to assess adsorbent efficiency and reuse. The model fuel generally consists of DBT dissolved in n-heptane or diesel at sulphur concentrations in the range of 100–1000 ppm. Generally, loading of the adsorbent is kept within 5–10 wt% with respect to the feed. The operating temperature during this process is within a mild range of 25–80 °C; contact times of 1–3 hours in batch modes or continuous flow through an adsorbent-packed column are employed. Subsequently, the adsorbed sulphur compounds can be desorbed using toluene or ethanol, while regeneration of the adsorbent through calcination or solvent washing restores the adsorption capacity for successive cycles.

3.3.4 Performance Metrics

Sulphur removal efficiency (η):

$$\eta\left(\%\right) = \frac{C_0 - C_t}{C_0} \times 100$$

Adsorption capacity (q):

$$q = \frac{\left(C_0 - C_t\right)V}{m}$$

where C_0 and C_t are initial and final sulphur concentrations, V is fuel volume, and m is adsorbent mass.

Nanocomposite adsorbents, due to synergistic effects of metal–support interactions and surface polarity, can achieve sulphur removals exceeding 90% at ambient temperature.

3.3.5 Advantages and Limitations

The approach of ADS has the advantage of avoiding hydrogen use and high pressures but suffers from adsorbent saturation and possible fouling by other fuel components. Higher sulphur removal efficiency could be achieved when ADS was combined with oxidative or hydrodesulphurization methods.

3.4 Biodesulphurization (BDS)

3.4.1 Principles

The principal of BDS is to utilize microorganisms such as Rhodococcus or Bacillus species for the selective removal of sulphur without breaking the hydrocarbon chains. These processes have the advantage of mild operating conditions and are also environmentally friendly.

Biodesulphurization uses microorganisms or enzymes selectively to cleave the C-S bonds of organosulphur compounds without hydrogenation. Among microbial pathways, the 4S pathway has been most studied for its desulphurization of DBT into 2-hydroxybiphenyl (2-HBP) and sulfate by leaving the carbon skeleton intact.

Reaction sequence:

The BDS of DBT proceeds via a well-defined 4S pathway consisting of a series of selective enzymatic reactions that remove sulphur without rupturing the carbon skeleton. First, DBT is oxidized to DBT sulfone by the enzyme DszC. Then, the enzyme DszA catalyzes the conversion of DBT sulfone to 2-hydroxybiphenyl-2-sulfinate (HPBS). Finally, HPBS is desulfinated by the enzyme DszB to form 2-hydroxybiphenyl (2-HBP) and sulfate ions (SO_4^{2-}) as end products. This pathway represents one mild and selective biological route of sulphur removal, leaving the hydrocarbon structure of the fuel intact.

3.4.2 Microbial Pathways

The most common pathway is the 4S pathway, in which the sulphur atoms undergo oxidation and subsequent cleavage leaving the fuel carbon skeleton intact. Examples of enzymes catalyzing such microbial reactions have been reported to be dibenzothiophene monooxygenase, among others (Stanislaus et al., 2010).

3.4.3 Advantages and Limitations

Results from BDS provide high selectivity and energy efficiency, but reaction rates are slow. The number of industrial applications involving fuel toxicity, oxygen supply, and scale-up is limited. Song 2003; Campos-Martin et al. 2010

3.5 Hybrid and Emerging Approaches

3.5.1 HDS-ODS Combinations

Hybrid processes, which couple HDS with ODS, resulted in an improved removal of refractory compounds. For instance, the pre-oxidation of DBT derivatives to sulfones increases their susceptibility to hydrodesulphurization, which reduces reaction severity (Otsuki et al., 2000; Lee et al., 2009).

3.5.2 Nanocomposite Catalysts

Nanostructured materials, like MoS_2–graphene, Co–Mo–TiO_2, and supported polyoxometalates, offer high active surface area, enhanced electron transfer, and improved accessibility to sterically hindered sulphur compounds (Campos-Martin et al., 2010; Marafi & Stanislaus, 2008).

3.5.3 Ionic Liquid-Based Desulphurization

Thus, ILs are considered as prospective media for selective extractive removal of sulphur species. Thiophenes can be oxidized and extracted by task-specific ILs, which thus act together with ODS or ADS processes (Song, 2003).

3.6 Reactor Design and Process Optimization

Experimental studies require care in the control of reactor parameters. For the studies related to HDS, fixed-bed, slurry, and trickle-bed reactors have been widely used, while for ODS and BDS experiments, batch and continuous stirred-tank reactors are preferred (Speight, 2020). These parameters include temperature, pressure, LHSV, oxidant concentration, and catalyst loading, varied systematically to find optimal conditions that achieve the highest efficiency in the removal of sulfur.

3.7 Analytical Techniques for Performance Evaluation

Accurate evaluation of desulphurization requires sensitive analytical methods.

Gas Chromatography–Sulphur Chemiluminescence Detection (GC–SCD): This is sensitive, compound-specific detection down to ppb levels. Hsu & Robinson, 2019

- GC–MS: Determination of the structure of thiophenic and aliphatic sulfur compounds (Lee et al., 2009).
- X-ray Photoelectron Spectroscopy: XPS distinguishes between sulphur oxidation states (-2, 0, +4, +6) (Campos-Martin et al., 2010).
- Fourier-Transform Infrared Spectroscopy: Detects sulfoxide and sulfone functional groups.
- Total Sulphur Analysis by ASTM D5453: UV fluorescence measures the total sulphur content to ensure the regulatory compliance of the fuel. (Stanislaus et al., 2010) These techniques enable the correlation of molecular structure with desulphurization efficiency, thereby allowing the systematic optimization of catalysts and reaction conditions.

4. CASE STUDIES, COMPARATIVE EVALUATION, AND INDUSTRIAL PERSPECTIVES IN FUEL DESULPHURIZATION

4.1 Detailed Case Studies of Catalyst Performance

4.1.1 Co–Mo and Ni–Mo Catalysts for HDS

Hydrodesulphurization remains the cornerstone of sulphur removal in industry. Co–Mo and Ni–Mo catalysts supported on γ-Al_2O_3 are commonly used in the petroleum refineries for the hydrogenolysis of sulphur containing compounds. Co–MoS_2 nanoparticles indeed showed 95% sulphur removal from middle distillates containing thiophenic compounds at reaction conditions of 350 °C and 100 bar hydrogen pressure (Marafi and Stanislaus, 2008). Ni–Mo catalysts behaved similarly except that they showed enhanced activity for the refractory dibenzothiophenes due to improved hydrogenation sites and sulfur vacancy formation.

Nanostructuring of the catalysts significantly increased the dispersion of the active site. MoS_2-based catalysts supported on mesoporous silica showed the best activity for 4,6-DMDBT with near-complete conversion under moderate HDS conditions (Campos-Martin et al., 2010). This enhanced reactivity was related to the expansion of interlayer spacing and increased edge sites that improved sulfur accessibility.

In general, sulphur removal in HDS proceeds by two major mechanistic pathways. The hydrogenation pathway includes the saturation of an aromatic ring prior to the cleavage of a carbon–sulphur bond, which is effective for bulky or refractory sulphur compounds. This is in contrast to the direct desulphurization pathway, where C–S bond cleavage occurs directly without prior hydrogenation, a pathway that is more efficient for simpler thiophenic species. The balance between these pathways depends on catalyst composition, reaction conditions, and the structure of sulphur-containing molecules.

4.1.2 Oxidative desulphurization catalysts

POMs, transition metal oxides, and supported metal complexes are generally used in ODS. According to Song (2003), $H_3PMo_{12}O_{40}$ supported on silica effectively oxidized DBT and 4,6-DMDBT to their corresponding sulfones using hydrogen peroxide at 60 °C and attained 90–95% sulphur removal. Supported vanadium-based catalysts on TiO_2 and ZrO_2 also showed fast oxidation of refractory sulphur compounds in diesel fractions and therefore showed possibilities of integration with mild hydrodesulphurization steps.

Generally, the mechanism of ODS involves a two-step oxidation of the sulphur compound to a sulfoxide followed by further oxidation to a sulfone. It involves electron transfer from the sulphur atom to the oxidant through a catalytic cycle, usually mediated by a metal-based or polyoxometalate catalyst. This successive oxidation increases the sulphur polarity, which allows easy extraction or adsorption separation of the resulting sulfones, hence making efficient sulphur removal possible under mild conditions.

4.1.3 Adsorptive Desulphurization Materials

Activated carbons, zeolites, and metal-organic frameworks have been explored for selective adsorption of sulphur compounds. In a recent study conducted by Hsu and Robinson (2019), it was found that Cu(II)-functionalized MOFs achieved between 80 to 90% removal of thiophenic compounds at room temperature and atmospheric pressure, thus further proving the potential of ADS for energy-efficient desulphurization. Similarly, functionalized mesoporous silicas further increased selectivity toward sterically hindered DBTs.

Adsorptive desulphurization relies on multiple key interaction mechanisms between sulphur compounds and the adsorbent surface. π-Complexation with metal sites allows sulphur-containing aromatic compounds, such as thiophenes, to form coordination bonds with transition metals like Cu^+ or Ag^+, enhancing selectivity. Acid–base interactions with functional groups on the adsorbent surface—such as hydroxyl or carboxyl moieties—facilitate binding through polarity differences or electron donation. In addition, size-selective molecular sieving is realized in porous adsorbents like zeolites and MOFs due to the dimensions of pores, which restrict access according to the size and shape of molecules and further enhance sulphur compounds' selectivity and separation efficiency.

4.1.4 Biodesulphurization Systems

Biodesulphurization through the 4S pathway ensures selective removal of sulphur without affecting the hydrocarbon structure. Laboratory experiments with Rhodococcus erythropolis have exhibited 60-70% sulphur removal from model DBT-containing fuels for 72 hours at ambient temperature (Stanislaus et al., 2010). The combination of BDS with mild ODS or ADS resulted in an increase in overall sulphur reduction and also opened up new vistas for sustainable low-energy desulphurization techniques.

4.2 Comparative Evaluation of HDS, ODS, ADS, and BDS

Table 1.

Method	Operating Conditions	Advantages	Limitations
HDS	300–400 °C, 30–130 bar H_2	Established industrial method; high efficiency for thiols/sulphides	Poor for refractory DBTs; high energy; catalyst deactivation
ODS	25–80 °C, atmospheric pressure, oxidant	Effective for refractory sulphur; mild conditions	Oxidant cost; solvent handling; catalyst recovery
ADS	25–150 °C, atmospheric pressure	Hydrogen-free; selective	Adsorbent saturation; fouling; regeneration required
BDS	Ambient temperature, aerobic conditions	Environmentally friendly; selective	Slow reaction; fuel toxicity limits; scale-up challenges

This comparative analysis brings out the fact that no single approach is sufficient for ultra-deep desulphurization, particularly in the case of 4,6-DMDBT and other sterically hindered compounds. Hybrid approaches that combine HDS with ODS, ADS, or BDS yield synergistic effects allowing lower energy requirements and higher overall sulphur removal.

Thermodynamic and Kinetic Correlations

Desulphurization reactions are either endothermic (HDS) or mildly exothermic (ODS, BDS), and their feasibility can be computed by the use of thermodynamic parameters:

$\Delta G = \Delta H - T\Delta S$

The kinetic-thermodynamic correlation gives information on the reaction control mechanism:

Reaction control in many desulphurization processes can often be categorized based on apparent activation energy, Ea. Kinetically controlled reactions have a high activation energy, typically greater than 60 $kJ \bullet mol^{-1}$, representing the reaction rate that is dominated by chemical transformation steps. On the other hand, diffusion-controlled processes usually have a low activation energy of less than 20 $kJ \bullet mol^{-1}$, and the overall rate is dominated by mass transfer limitations. Both ODS and ADS display mixed control regimes since chemical oxidation and diffusion through liquid or solid phases occur together and influence the overall reaction kinetics and efficiency.

4.7 Mechanistic Modelling and Computational Insights

DFT and MD are some of the computational chemistry techniques that have become indispensable in determining desulphurization mechanisms at a molecular level. DFT calculations provide valuable information on catalyst behavior, such as identifying active sites, calculating adsorption energies of sulphur compounds, and determining transition states in bond cleavage and oxidation reactions. The kMC method allows modeling of dynamic surface phenomena, such as the evolution of site coverage and deactivation of catalysts, hence providing a realistic picture of catalytic performance in real time. These computational methods bridge the atomic-level understanding with optimization of macroscopic processes in advanced desulphurization research. As an example, DFT studies on Ni–MoS_2 catalysts prove that sulphur compounds preferentially adsorb on CUS edge sites, confirming experimental HDS pathways.

4.8 Summary of Mechanistic and Kinetic Insights

Table 2.

Desulphurization Type	Mechanism	Rate Law	Kinetic Order	Typical Activation Energy (kJ/mol)	Key Control Parameter
HDS	DDS & HYD pathways	Langmuir–Hinshelwood	1	80–150	H_2 partial pressure
ODS	Oxidation → Extraction	Pseudo-first-order	1	25–60	Oxidant/Catalyst ratio
ADS	π-complexation, S–metal interaction	Pseudo-second-order	2	20–45	Adsorbent surface area
BDS	Enzymatic oxidation (4S Pathway)	Michaelis–Menten	Variable	15–30	Enzyme activity, cell density

Figure 4. Two-step mechanism of oxidative desulphurization. (Image generated by AI)

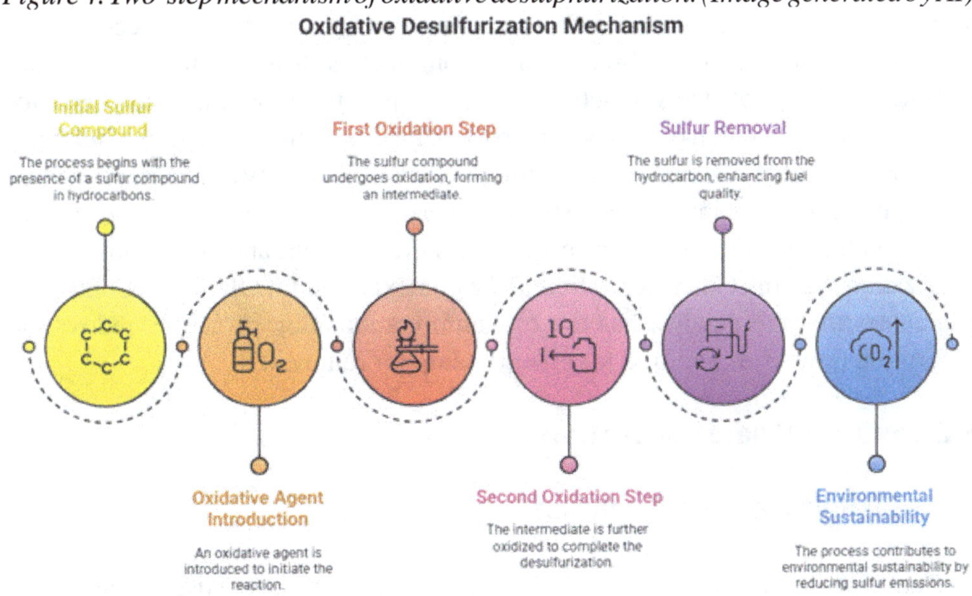

5 CATALYST CHARACTERIZATION AND PERFORMANCE EVALUATION

The performance of a desulphurization catalyst is closely related to its surface structure, morphology, active site distribution, and physicochemical properties. Precise characterization of the mentioned parameters is indispensable for catalytic behavior understanding, synthesis conditions optimization, and establishing relationships between structure and activity. This section details the experimental and analytical approaches used for the characterization and performance evaluation of desulphurization catalysts, covering both conventional and nanocomposite materials.

5.1 Importance of Catalyst Characterization

Characterization of the catalyst bridges synthesis with catalytic performance.

It helps in the following ways:

Characterization of desulphurization catalysts is indispensable in order to understand their structure-activity relationship and to optimize their performance. The first step is the identification of the active species, which is responsible for the desulphurisation and hence decides the catalytic selectivity and efficiency of the materials. The textural characteristics, such as surface area, porosity, and dispersion of the active metals, are quantified because these parameters affect the accessibility of reactant molecules and subsequent reaction kinetics. It is also important to determine the chemical states of active sites and supports so that the redox behavior and catalytic mechanisms could be understood. Additionally, evaluation of textural/structural stability under reaction conditions ensures durability and reusability over a long period. Finally, insights into structure-performance correlation are utilized to optimize reaction conditions for developing more efficient and sustainable desulphurization systems.Characterization in the context of HDS, ODS, ADS, and BDS guarantees that the catalyst has the optimum combination of acidic, redox, and adsorption properties required for effective sulphur removal.

5.2 Textural Characterization

(a) BET Surface Area Analysis

The Brunauer-Emmett-Teller BET method is the most widely used technique in determining the surface area and porosity of catalysts. A high surface area increases the number of available active sites for sulphur compound adsorption and reaction.

Principle:

Based on the nitrogen adsorption–desorption isotherms at liquid nitrogen temperature (77 K).

Key output parameters for textural characterization include surface area (m^2/g), which provides a measure of available active surface for adsorption or catalysis, commonly measured through BET analysis; total pore volume, cm^3/g, which gives the material's overall void space that controls capacity for adsorbate uptake; and pore size distribution, nm, providing the range and abundance of micropores, mesopores, and macropores, often obtained from BJH or DFT analysis of adsorption/desorption isotherms and strongly influencing mass transfer, selectivity, and accessibility to active sites.

Observations: Higher BET surface areas in the range of 200–350 m^2/g are related to better desulphurization activity among nanocomposite catalysts, like MoS_2–TiO_2 and Ni–Mo/γ-Al_2O_3.

(b) Pore Structure Analysis (BJH Method)

The BJH analysis gave pore size distribution and helped classify the materials as microporous (<2 nm), mesoporous (2–50 nm), or macroporous (>50 nm). Mesoporous supports facilitate the diffusion of the reactants and minimize mass transfer resistance, crucial for bulky sulphur molecules like DBT.

5.3 Structural Characterization

a) X-Ray Diffraction (XRD):

XRD is one of the most valuable techniques for characterizing structural properties in desulphurization catalysts, yielding crystalline phase information along with data on particle size and lattice defects. The diffraction pattern of a supported catalyst can give insight into material composition and the degree of crystallinity. Generally speaking, broad peaks in an XRD spectrum denote either a nanocrystalline or amorphous structure, reflecting small particle sizes and high dispersion of an active phase. Sharp and well-defined peaks correspond to highly crystalline oxides or sulfide phases, such as MoS_2 or other metal oxides, generally meaning larger crystal domains or well-ordered structures. These observations can help assess catalyst morphology and its correlation with catalytic performance.

Example: In Ni–MoS_2 catalysts, characteristic peaks at $2\theta \approx 14°$, 33°, and 58° confirm the (002), (100), and (110) planes of MoS_2. Peak broadening signifies enhanced dispersion of active species. Raman Spectroscopy:

Raman spectroscopy has been used to gain information about vibrational modes of metal–sulphur bonds and defect sites. The D and G bands at ≈1350 cm^{-1} and 1580 cm^{-1}, respectively, in carbon-supported catalysts evidence graphitic ordering, while Mo–S stretching bands at ~380–410 cm^{-1} confirm the formation of MoS_2 phases.

5.4 Morphological and Surface Characterization

(a) Scanning Electron Microscopy (SEM)

Scanning Electron Microscopy (SEM) is one of the key techniques utilized in the study of the surface morphology, particle agglomeration, and uniformity of catalyst coatings in desulphurization materials. With SEM, one can visually distinguish between smooth, dense surfaces, normally characteristic of compact structures, and the rough and porous surfaces that are an indication of high surface area, hence the potential accessibility of active sites. More importantly, SEM combined with Energy Dispersive X-ray Spectroscopy (EDS) allows elemental mapping of key components including Ni, Mo, S, O, among other support materials. This combined analysis ascertains that there is homogeneous dispersion of the active metals across the catalyst surface, which is very critical to ensure consistent catalytic performance and sulphur removal.

Figure 5. SEM–EDS micrograph showing uniform Ni–Mo dispersion on Al_2O_3.

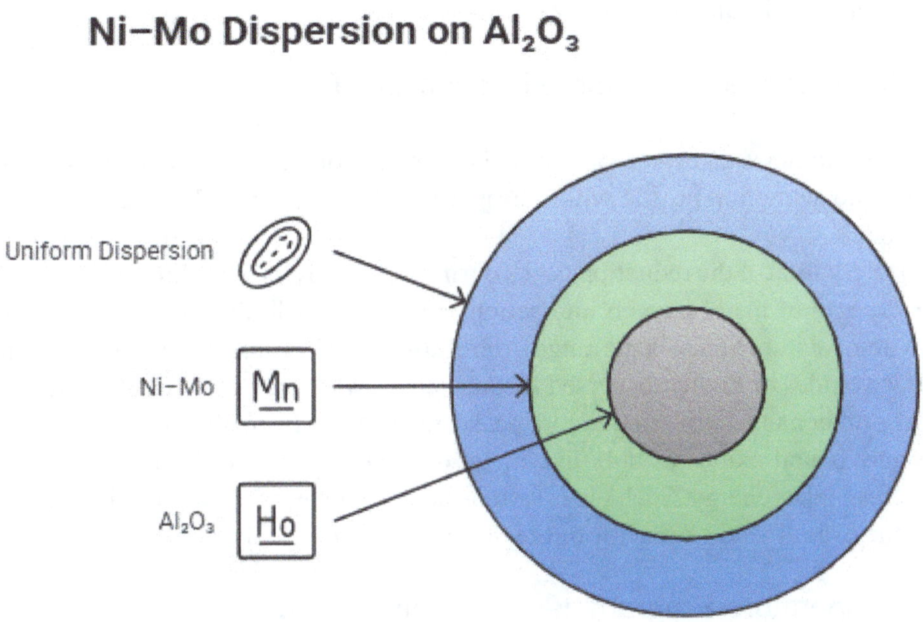

(b) Transmission Electron Microscopy (TEM)

TEM provides atomic-scale information on the particle size, crystal orientation, and the number of layers of MoS_2. The length of the slab and the stacking number of the MoS_2 nanocrystallites are related to the catalytic performance:

Shorter slabs (2–5 nm) → more edge sites → higher activity.

Higher stacking ≥5 layers results in lower accessibility and thus reduced activity.

5.5 Surface Chemistry and Electronic Properties

(a) X-Ray Photoelectron Spectroscopy (XPS)

XPS shows the oxidation states of metal species and surface composition. Binding energy values reflect the degree of sulfiding or oxidation.

Examples:

- Mo $3d_{5/2}$ peak at ~228.5 eV indicates Mo^{4+} (MoS_2).
- Ni $2p_{3/2}$ peak near 853.0 eV corresponds to Ni^{2+} in the Ni–Mo–S phases.

S 2p peak at about 162 eV indicates sulfide species.

Changes in the BE values indicate metal-support electronic interactions, which are crucial for desulphurization activity.

(b) Temperature-Programmed Reduction (TPR)

TPR examines catalyst reducibility and interaction between metal and support. The H_2 consumption profile will indicate the reduction of metal oxides into active sulphides.

The position of the reduction peaks during TPR analysis gives information about the strength of metal-oxygen interactions and the reducibility of catalyst species. Low-temperature peaks, in the range from 300 to 500 °C, correspond to weakly bound surface oxides or highly dispersed metal species, which are more easily reducible. On the other hand, high-temperature peaks, in the range from 600 to 800 °C, suggest strongly bound oxides or bulk-like metal-oxygen species, the reduction of which requires higher energy. Such observations help evaluate metal-support interactions and the overall redox behavior of the desulphurization catalysts.

(c) Temperature-Programmed Desorption (TPD)

Acid-base properties of the catalysts are determined by NH_3-TPD or CO_2-TPD:

In the analysis of NH_3-TPD, the temperature of ammonia desorption corresponds to the strength and distribution of the acid sites on the catalyst surface. The low-temperature desorption peak, between 100 and 250 °C, represents weak acid sites with physisorbed ammonia or loosely bound acid sites. On the contrary, strong acid sites are represented by high-temperature peaks, greater than 400 °C, corresponding to chemisorbed ammonia or strong Brønsted and Lewis acid sites. Knowing the acidity profile by using this technique is important for desulphurization catalysts since acid site strength and density are determining factors in adsorption, activation, and overall catalytic performance.

5.6 Optical and Redox Characterization

(a) UV–Visible Diffuse Reflectance Spectroscopy (UV–Vis DRS)

This technique characterizes electronic transitions and gives information about band gaps. In nanocomposites-for example, TiO_2-based catalysts-red-shifts in absorption edge indicate enhanced visible-light activity, which in turn benefits photocatalytic oxidative desulphurization (PODS).

(b) Fourier Transform Infrared Spectroscopy (FTIR)

FTIR identifies functional groups and surface species. Peaks near:

- 1050–1100 cm^{-1} → M–O–M vibrations
- 560–600 cm^{-1} → M–S stretching (for sulfided catalysts)
- 1450–1650 cm^{-1} → aromatic ring vibrations (from adsorbed DBT)

FTIR spectra before and after reaction help trace intermediate sulphur species and evaluate catalyst stability.

5.7 Thermal and Mechanical Stability

(a) Thermogravimetric analysis (TGA)

TGA measures the weight loss upon heating and delivers thermal stability data:

The weight loss in the course of TGA may provide useful information on the material thermal stability and composition. Mass losses under 200 °C usually correspond to the desorption of physically adsorbed water or residual solvent. The temperature range from 200 to 400 °C is commonly associated with the decomposition of organic residues or surface functional groups that often remain after synthesis or impregnation steps. A weight loss around 600 °C indicates the structural breakdown of the support or the decomposition of thermally stable components, hence marking the limit of the stability of this material in heating conditions.

Catalysts with >90% retention in weight up to 800 °C are termed thermally robust.

(b) Mechanical Strength Tests

For fixed-bed reactors, crushing strength and attrition resistance are tested to ensure the catalysts are durable in high-pressure conditions.

5.8 Performance Evaluation of Catalysts

Catalytic activity, with regards to sulphur conversion and selectivity, is quantitatively determined under controlled experimental conditions.

(a) Batch Reactor Studies

Batch reactors are used for fundamental kinetic studies. Parameters such as temperature, pressure, hydrogen flow rate, and catalyst loading are varied systematically.

Desulphurization efficiency (η) is calculated as:

$$\eta = \frac{C_0 - C_t}{C_0} \times 100\%$$

where C_0 and C_t are initial and final sulphur concentrations, respectively.

(b) Continuous Flow Reactor (CFR) Studies

CFR experiments provide long-term performance data, simulating industrial hydroprocessing conditions. Catalyst deactivation, coke formation, and metal sintering are monitored.

5.9 Evaluation Parameters

Table 3.

Parameter	Definition	Typical Range / Ideal Value
Sulphur Conversion (%)	Percentage removal of sulphur compounds	85–99%
Selectivity (%)	Ratio of desired desulphurized products to by-products	>90%
Turnover Frequency (TOF)	Moles of reactant converted per mole of active site per unit time	10^{-3}–10^{-1} s^{-1}
Catalyst Stability	% activity retained after multiple cycles	>80%
Regeneration Efficiency	% recovery of initial activity after regeneration	>90%

5.10 Catalyst Regeneration and Reusability

Repeated use of catalysts results in deactivation due to:

The deactivation of catalysts in desulphurization processes may be due to several mechanisms that reduce activity and stability over time. Carbon deposition (coke formation) results in blockage of active sites and pores, thereby decreasing accessibility of reactants to catalytic surfaces. Sintering of active sites promotes particle growth and loss of metal dispersion, hence reducing the number of available active centres. Moreover, oxidation of sulphides-especially in catalysts that are normally based on sulfides, like MoS_2-can promote changes in the composition of the active phase and decrease catalytic effectiveness. Minimizing these deactivation routes is key to achieving better longevity of the catalysts and overall process efficiency.

Regeneration techniques include:

Catalyst regeneration is an important step to maintain activity and extend the catalyst lifespan in the desulphurization process. The oxidative regeneration involves the burning-off of deposited coke in air at 400–500 °C and restores surface area and active site accessibility. Following this, a sulfidation treatment is carried out to reconvert the oxidized metal species into active sulfide phases. This ensures that catalytic performance will be optimum for sulfide-based catalysts. For adsorptive catalysts, the adsorbed sulphur compounds are effectively removed by washing with appropriate solvents such as toluene and ethanol, or any other polar solvent, without affecting the structural integrity of the material; thereby, its adsorption capacity is restored.Regenerable catalysts reduce operation costs and environmental wastes, satisfying major objectives in sustainable fuel processing.

5.11 Industrial Implementation and Scaling

5.11.1 HDS Plants

The industrial HDS units are designed as fixed-bed or trickle-bed reactors, where hydrogen flow, temperature, and liquid space velocity are controlled precisely. In the commercial plants, scaling of catalysts optimized in the laboratory needs to take into consideration heat transfer, mass transport, and catalyst attrition. Poisoning of catalysts by heavy metals is one of the major operational challenges, apart from feed variability.

5.11.2 ODS Integration

ODS normally appears as a finish step after HDS. Diesel fractions, pre-treated by HDS were oxidized with hydrogen peroxide over polyoxometalate catalysts followed by extraction of sulfones. Demonstration on a pilot-scale basis reported sulphur reductions from 500 ppm to <10 ppm (Lee et al., 2009). While process integration is concerned with optimization of oxidant consumption, solvent recovery, and continuous regeneration of catalysts.

5.11.3 ADS in Industry

Continuous flow ADS has been demonstrated successfully on pilot scales with Cu-MOF and activated carbon columns. These continuous flow adsorption systems permit regeneration by means of either thermal or chemical treatments, reducing operation costs. When integrated into HDS, it extends catalyst life and maintains ultra-low sulphur without needing extreme reaction conditions.

5.11.4 BDS Prospects

Though BDS is still largely at laboratory and pilot scale, advances in bioreactor design, oxygenation, and microbial tolerance are addressing issues of scalability. Encapsulation of microbial cells and co-culturing with mild oxidative systems have given increased reaction rates for the development of BDS as a viable complementary technology in the near future (Song, 2003; Stanislaus et al., 2010).

5.12 Economic and Environmental Considerations

5.12.1 Energy and Hydrogen Costs

These HDS units are hydrogen intensive and account for 15–20% of the refinery hydrogen requirements. Energy consumption and hydrogen usage can therefore be minimized by optimizing process severity through hybrid techniques or using nanocomposite catalysts.

5.12.2 Catalyst Cost and Longevity

Catalyst replacement and regeneration account for considerable operational expenses. Nanostructured catalysts with higher dispersion and longer lifetime reduce the expenses while maintaining high efficiency of desulphurization (Campos-Martin et al., 2010).

5.12.3 Environmental Impacts

Sulphur oxides formed during combustion lead to acid rain, particulate pollution, and respiratory diseases. Deep desulphurization reduces emissions, thus playing a vital role in meeting regulations such as Euro VI, Bharat Stage VI, and IMO 2020.

5.12.4 Waste Management

ODS and ADS yield oxidized sulphur-containing byproducts that demand adequate disposal or recovery. Ideally, industrial practice should be sustainability-based, with solvent recycling, oxidant recovery, and solid adsorbent regeneration.

5.12.5 Emerging Research Trends and Future Directions

1. Nanocomposite Catalysts: The integration of metal sulfides with carbon-based supports, represented by graphene and carbon nanotubes, improves reactivity toward sterically hindered sulphur compounds (Campos-Martin et al., 2010).
2. Task-Specific Ionic Liquids: ILs are investigated for selective extraction of thiophenic compounds, often in combination with oxidation steps by Song (2003).
3. Hybrid Catalytic Systems: Combining HDS, ODS, ADS, and BDS in sequence or tandem processes can achieve ultra-low sulphur fuels (<10 ppm) while minimizing energy consumption.
4. Computational Catalyst Design: Molecular modeling and DFT guide the design of catalysts for targeted removal of refractory compounds (Marafi & Stanislaus, 2008).
5. Sustainable Biodesulphurization: Genetic engineering of microbial strains and immobilization techniques strive to improve reaction rates and fuel tolerance for large-scale applications.
6. Integration with Renewable Feedstocks: Desulphurization technologies are being adapted to biofuels and hybrid fossil–biofuel systems supporting low-carbon energy initiatives.

Case Studies and Pilot-Scale Performance

Case 1: Ni–Mo/Al_2O_3 catalyst in diesel HDS pilot plant (350 °C, 80 bar) → 98.5% sulphur removal.

Case 2: MoO_3–TiO_2 ODS reactor using H_2O_2 (60 °C, 1 bar) → 95% DBT oxidation in 40 min. Case 3: Cu–Y zeolite ADS unit → 90% sulphur reduction, regenerable up to 5 cycles. Case 4: Immobilized Rhodococcus bioreactor → 82% sulphur removal at 35 °C, pH 7.

Sensitivity and Response Analysis

Sensitivity analysis quantifies the influence of each variable on overall sulphur removal. Typically:

$$S_{\text{efficiency}} = f\left(T, C, t, P, \frac{O}{S}\right)$$

Sensitivity analysis of desulphurization processes provides a clear overview of the relative influence of important operating parameters on total sulphur removal efficiency. The sensitivity is in ranked order from most to least influential as follows:

1 Temperature - has the largest influence since it directly affects reaction kinetics, catalyst activity, and equilibrium conversion.

2 Catalyst loading: it determines the amount of available active sites and hence greatly influences the rates of reaction.
3 Reaction time – controls the degree of conversion, especially for the slower steps of oxidation or adsorption.
4 Hydrogen or oxidant pressure: affects the availability of reactants and mass transfer but is less critical than temperature and loading.
5 Intensity of mixing - homogenizes the distribution of reagents, but affects primarily external mass transfer, not intrinsic rates of reaction.

Therefore, such a ranked sensitivity analysis provides invaluable information for real-time process control, whereby operators can target adjustments to parameters that provide the highest payoff on maximizing desulphurization performance and operational efficiency.

CONCLUSION

The experimental methods for fuel desulphurization have undergone great changes in the past few decades, thanks to the twin imperatives of environmental protection and energy efficiency. Because global regulations require ultra-low sulphur fuels, innovative desulphurization technologies are core features of current refining strategies. This chapter synthesized principles, methods, and experimental insights surrounding catalytic and non-catalytic desulphurization with an emphasis on process parameters and catalyst design that determine overall performance and sustainability.

HDS remains the industrial benchmark for large-scale sulphur removal due to its proven efficiency in treating aliphatic and low-molecular-weight aromatic sulphur compounds. However, it suffers from a number of disadvantages that raise the need for complementary or alternative approaches: it cannot effectively remove refractory sulphur species such as 4,6-DMDBT; it consumes high amounts of hydrogen; and it requires elevated operating temperatures and pressures. In this respect, ODS, ADS, and BDS have gained increasing attention under mild operating conditions, each based on specific mechanisms that target sulphur compounds not efficiently removable by conventional hydroprocessing.

Catalyst design is still the most important factor in effective desulphurization. The use of nanostructured materials, bimetallic catalysts, ionic liquids, and supported metal oxides has been extended to broaden the performance envelope by allowing the selective activation of C-S bonds under environmentally benign conditions. Characterization techniques such as XRD, BET, SEM, XPS, and FTIR are necessary to understand active sites, surface morphology, and catalyst stability for assuring reproducibility and insight into deactivation pathways. Kinetic modeling has further helped process optimization, especially in elucidating reaction order, rate constants, and diffusion limitations, which guide reactor design and scale-up strategies.

The experimental and case analyses establish the potential for synergistic benefits from hybrid approaches such as ODS followed by ADS or catalytic hydrogenation. These combinations result in deeper removal of sulphur at a lower utilization of hydrogen and, hence, operational costs. However, a number of challenges persist, including catalyst deactivation through sulphur poisoning, mass-transfer limitations, and the difficulty of regenerating adsorbents or biological systems for repeated use.

The future of desulphurization research will increasingly focus on:

- Nanocomposite catalysts with tunable surface functionalities.
- Photocatalytic and electrochemical desulphurization which utilize renewable energy inputs.
- Green solvents and ionic liquids for minimizing environmental impact.
- Integration of AI and machine learning for predictive modeling and process control.

In fact, these directions will improve not only efficiency and selectivity but also contribute to the development of sustainability and circular economy principles in the fuel industry.

REFERENCES

ASTM International. (2019). *ASTM D5453–19: Standard test method for determination of total sulfur in light hydrocarbons, motor fuels, and oils by ultraviolet fluorescence.*

Campos-Martin, J. M., Capel-Sanchez, M. C., Perez, A., & Fierro, J. L. G. (2010). Oxidative desulfurization of diesel fuels. *Catalysis Today*, *158*(1–2), 62–68. DOI: 10.1016/j.cattod.2010.05.005

Campos-Martin, J. M., Capel-Sanchez, M. C., Perez-Presas, P., & Fierro, J. L. G. (2010). Oxidative processes of desulphurization of liquid fuels. *Journal of Chemical Technology and Biotechnology*, *85*(7), 879–890. DOI: 10.1002/jctb.2394

Environmental Protection Agency. (2023). *Integrated science assessment for particulate matter*. U.S. Environmental Protection Agency. https://www.epa.gov/

EPA. (2023). *National ambient air quality standards for sulfur oxides*. U.S. Environmental Protection Agency. https://www.epa.gov/sox-pollution

(Euro VI sulphur limits reference)

European Commission. (2009). *Directive 2009/30/EC amending Directive 98/70/EC relating to the quality of petrol and diesel fuels*. Official Journal of the European Union.

Gao, H., Guo, C., & Jiang, X. (2020). Nanocomposite catalysts for oxidative desulphurization: Recent advances and perspectives. *Journal of Energy Chemistry*, *46*, 1–14. DOI: 10.1016/j.jechem.2019.11.019

Gary, J. H., & Handwerk, G. E. (2019). *Petroleum refining: Technology and economics* (5th ed.). CRC Press.

Government of India. (2016). *Bharat Stage (BS) VI Emission Standards—Automotive Industry Standards (AIS)*. Ministry of Road Transport & Highways.

Hsu, C. S., & Robinson, P. (2019). Characterization techniques for sulfur compounds in petroleum fuels. *Energy & Fuels*, *33*(4), 3030–3045. DOI: 10.1021/acs.energyfuels.9b00345

Hsu, C. S., & Robinson, P. R. (2006). *Practical advances in petroleum processing*. Springer.

Hsu, C. S., & Robinson, P. R. (Eds.). (2017). *Petroleum science and technology* (2nd ed.). Springer.

IMO. (2020). *IMO 2020 global sulphur cap*. International Maritime Organization. https://www.imo.org/en/MediaCentre/HotTopics/Pages/Sulphur-2020.aspx

International Maritime Organization. (2020). *IMO 2020 – Regulation 14: Sulphur oxides (SOx) and particulate matter*. International Maritime Organization. https://www.imo.org/

Lee, W. S., Choi, J. H., & Kim, D. H. (2009). Analytical approaches for characterization of sulfur compounds in fuels. *Journal of Analytical Chemistry*, *64*(5), 423–435. DOI: 10.1021/ac900987t

Li, C., Zhao, D., & Shi, W. (2018). Metal–organic frameworks for the adsorptive desulphurization of fuels: A review. *Energy & Fuels*, *32*(12), 12439–12460. DOI: 10.1021/acs.energyfuels.8b03092

Ma, X., Sakanishi, K., & Isoda, T. (2002). A review of selective oxidative desulphurization of liquid fuels with hydrogen peroxide catalyzed by polyoxometalates. *Applied Catalysis A, General*, *235*(1–2), 79–92. DOI: 10.1016/S0926-860X(02)00240-0

Marafi, M., & Stanislaus, A. (2008). Hydrodesulfurization of refractory sulfur compounds in diesel fuels. *Catalysis Reviews*, *50*(1), 1–33. DOI: 10.1080/01614940701777791

Otsuki, J., Ishihara, A., & Nakayama, T. (2000). Oxidative desulfurization of diesel fuel using hydrogen peroxide. *Energy & Fuels*, *14*(2), 279–284. DOI: 10.1021/ef990125p

Otsuki, S., Nonaka, T., Takashima, N., Qian, W., Ishihara, A., Imai, T., & Kabe, T. (2000). Oxidative desulphurization of light gas oil and vacuum gas oil by oxidation and solvent extraction. *Energy & Fuels*, *14*(6), 1232–1239. DOI: 10.1021/ef000096e

Song, C. (2003). An overview of new approaches to deep desulphurization for ultra-clean gasoline, diesel fuel and jet fuel. *Catalysis Today*, *86*(1–4), 211–263. DOI: 10.1016/S0920-5861(03)00412-7

Speight, J. G. (2014). *Fuel chemistry and technology* (3rd ed.). CRC Press.

Speight, J. G. (2020). *Fuel science and technology handbook* (2nd ed.). CRC Press.

Speight, J. G. (2020). *The chemistry and technology of petroleum* (6th ed.). CRC Press.

Stanislaus, A., Marafi, A., & Rana, M. S. (2010). Recent advances in the science and technology of ultra-low sulfur diesel (ULSD) production. *Catalysis Today*, *153*(1–2), 1–68. DOI: 10.1016/j.cattod.2010.05.011

United States Environmental Protection Agency (EPA). (2020). *Regulations for Fuels and Fuel Additives: Gasoline and Diesel Fuel Standards*. Retrieved from https://www.epa.gov

WHO. (2021). *Air quality guidelines: Sulfur dioxide (SO_2)*. World Health Organization. https://www.who.int/publications/i/item/9789240034228

World Health Organization. (2021). *WHO global air quality guidelines: Particulate matter, ozone, nitrogen dioxide, sulfur dioxide and carbon monoxide*. World Health Organization. https://www.who.int

Chapter 5
Plant-Derived Nanocomposites for Sustainable Desulfurization With a Focus on Materials, Applications, and Environmental Impact

Sidra Ahmad
https://orcid.org/0009-0004-4831-5906
University of Agriculture, Peshawar, Pakistan

Nayab Ahmad
University of Agriculture, Peshawar, Pakistan

Tuba Zafar
https://orcid.org/0009-0001-9158-5053
The University of Agriculture, Peshawar, Pakistan

Muhammad Haseeb Javaid
College of Agriculture and Biotechnology, Zhejiang University, Hangzhou, China

Micheal Ayodeji Sokan-Adeaga
https://orcid.org/0000-0001-8550-1960
Department of Community Health and Primary Care, Faculty of Clinical Sciences, College of Medicine, University of Lagos, Lagos, Nigeria

Adewale Allen Sokan-Adeaga
https://orcid.org/0000-0002-0885-1526
Public Health Program, College of Health Sciences, Bowen University, Iwo, Nigeria

Sezai Ercisli
Department of Horticulture, Faculty of Agriculture, Ataturk University, Erzurum, Turkey & HGF Agro, Ata Teknokent, Erzurum, Turkey

DOI: 10.4018/979-8-3373-5966-3.ch005

ABSTRACT

The removal of sulfur from fossil fuels is crucial for achieving clean energy and complying with stringent environmental regulations. Although traditional desulfurisation methods are effective, they are often energy-intensive, costly, and inefficient for addressing recalcitrant sulfur compounds. Plant-based nanocomposites present a promising alternative owing to their tunable structures, high porosity, and unique surface properties that facilitate sulfur removal. This review examines the utilisation of agricultural and forestry wastes, such as rice husks, coconut shells, sugarcane bagasse, and fruit peels, in the synthesis of these materials. Eco-friendly techniques, including pyrolysis, hydrothermal carbonisation, and steam activation, can be used to produce materials with high surface areas and advantageous properties. Additionally, it discusses the necessity of balancing the hydrophobic and hydrophilic properties for effective application with hydrocarbons.

INTRODUCTION

For a long time, sulfur in fossil fuels has plagued the energy and environmental sectors. When burned, they produce sulfur oxides, pollutants that are the main cause of acid rain, which destroys trees, soils, and aquatic life, and is also a major contributor to respiratory diseases affecting humans. Although progress may be made in emission control technology regarding the removal of sulfur from exhaust technologies, the most preferable solution to avoid such adverse implications is the removal of sulfur at the fuel treatment level (Sedyaaw et al., 2024). However, desulfurisation techniques, such as hydrodesulfurization, require high temperature, high pressure, and expensive catalysts, use non-renewable resources, and are unsustainable. Thus, the need for highly innovative and environmentally friendly alternatives to synthetic polymers of mineral oil origin is becoming increasingly urgent, and the global trend in the construction industry's dedication to green energy and environmental protection is increasing (Dembaremba et al., 2022).

The transition to green alternatives is clearly related to focusing on materials and processes based on the circular economy approach. In contrast to finite or costly feedstocks, the utilisation of renewable and naturally abundant feedstocks is becoming increasingly popular. Among the potential alternative feedstocks, plant-based materials, particularly agriculture and forestry residues, are the most widely studied (Le et al., 2024). As sources of carbon, these resources are environmentally sound and renewable materials with intrinsic carbon content that can be easily developed into functional materials with interesting physicochemical properties(He et al., 2023). The utilisation of nanocomposites based on these adsorbents for desulfurisation is not only a way to solve the problem of fuel oil purification, but also an opportunity to contribute to waste control, resource economy, and environment conservation (Jha et al., 2025).

Unlike conventional inorganic sorbents, the above-mentioned properties of plant-material-based sorbents can be enhanced by mild and environmentally friendly modifications to improve the selectivity and sorption of sulfur species, such as thiophens, benzothiophens, and dibenzothiophens. Furthermore, in nanocomposites, these bio-based materials can be instantly blended with metals, metal oxides, or other benign additives, resulting in a trade-off between cost, performance, and the environment (Saleh et al., 2017).

Moreover, plant-based desulfurisation nanocomposites are attractive because they are intentionally designed for sustainable energy construction worldwide. Legislation (such as the increasing regulation of sulfur emissions globally) has already driven the search for cleaner burning fuels (Qiu et al., 2024). Concurrently, nanotechnology has the potential to allow fine tuning of materials even down to the molecular level for the design of adsorbents which will selectively go after individual sulfur molecules (Mabena et al., 2024). Therefore, plant-derived nanocomposites are likely to play a role at the junction of two evolving concepts: nanomanufacturing and renewability. These results also indicate the possibility of using common biomass as a high-value-added functional material for promising energy purification applications (Yang et al., 2024).

Plant-derived DoP materials are technically feasible and have high socioeconomic and environmental significance. In addition, available waste from agriculture and forestry sources is not being exploited which is unsustainable with respect to waste disposal and open burning (Popa et al., 2021). Value-added utilisation of these residues, which can be converted into high-efficiency desulfurisation materials, not only reduces the pressure of waste but also creates an economy for rural agricultural-based regions. Furthermore, as they are produced at a low energy cost, in comparison to conventional catalytic or synthetic sorbents, their global environmental impact, to a great extent, is minimised. This makes plant-based nanocomposites both efficient and environmentally friendly for industrial applications (Oyedeji et al., 2024).

Simultaneously, the scientific and engineering communities must work on scale-up issues close to the industry-scale production of these materials. Nevertheless, the reproducibility, stability, and long-term stability of plant-based nanomaterials under realistic fuel-reforming conditions have not been well established (Hachhach et al., 2025). Feed stock diversity of biomass (species, growing conditions, and post-harvesting treatment) makes the assurance of the reliability of the materials difficult. However, there is a continued exploratory effort towards the establishment of standard preparation methods, surface functionalization, and composite structures which could help to resolve these problems (Mignogna et al., 2024).

The general direction, so far, for this tale of plant-based nanocomposites for desulfurisation is environmental demand, technology innovation, and sustainable materials all in one. For a world in which the energy industry is torn between a need to lower greenhouse gases and one to supply increasingly more power, solutions that work and are made to last, such as this one, are going to be even more essential (Segovia et al., 2025). In the process, they are shifting the way we think about unwanted agricultural tillage and plant-based matter and changing it from something to be disposed of into something we can tap as a source to create nanomaterials — and they are doing it more greenly, more circularly, to manage our resources — to make cleaner, more efficient fuel, and generate less waste (Nezafat et al., 2024).

In this article, recent advances in plant-based nanocomposites for sustainable desulfurisation are reviewed, with an emphasis on biomass feedstock for the materials. It further focuses on the conversion reactions, property enhancement pathways, and underlying material chemistries related to sulfur adsorption. The following sections concentrate on the role of such materials in hydrocarbon cleaning in terms of their effects on the environment and economy and the limitations for industrial use. Drawing on the principles of materials science, nanotechnology, and environmental engineering, we provide an overview of the future perceptions of plant-based nanocomposites for the next era of clean and sustainable energy (Azeez et al., 2023).

2. PLANT-BASED RESOURCES FOR NANOMATERIAL DEVELOPMENT

This new type of plant-mediated nanocomposite with desulfurisation is attributed to abundant and diverse biomass resources. Millions of tons of waste generated in the agricultural and forestry sectors annually are still underused and mismanaged through methods such as open burning or landfills. Commercially useful raw materials waste into thin air by destroying the environment, causing pollution and greenhouse gas (GHG) generation. By converting plant-based waste to nanomaterials, researchers can capitalise on renewable, cost-effective, and environmentally friendly feedstock to meet the global needs for waste valorization and circular economy (Bhatia et al., 2024).

Common Agricultural and Plant Wastes

Agricultural waste is a huge biomass resource for nanomaterial synthesis. Examples include rice husks, wheat straw, sugarcane bagasse, corn stover, coconut shells, and palm residue. Most of these wastes consist primarily of lignocellulosic substrates, including cellulose, hemicellulose, and lignin (Gupta et al., 2022).These polymers act as carbon-rich precursors and structurally stable chemicals. For example, rice husks include large amounts of silica and carbon and are interesting as carbon-based nanomaterials with a hybrid structure (Mubarik et al., 2021). Coconut shells and nutshells with dense lignocellulosic architectures can afford high carbon with a reasonable carbon yield from pyrolysis and yield tough activated carbon materials (TONGDANG et al., 2023).

Forestry residues and wood industry waste are also important resources. Sawdust, bark, and wood chips are produced in large quantities and can be discarded, although they have high lignocellulose content. In many senses, they are good candidates for the thermochemical production of biochar or activated carbon, with a high degree of porosity, as they are promising for adsorption applications (Thakur et al., 2024). Natural materials such as fruit pericarps, seeds, and shells of agro-industrial liquids such as orange peel, banana peel, mango seed, and groundnut shells have likewise been investigated as low-cost renewable feedstocks for nanomaterials. The relatively heterogeneous chemical compositions of the extracts, which can consist of natural oils, tannins, and polyphenols, provide functional groups for surface modification and adsorption (Basri et al., 2021).

Increasing attention towards aquatic and marine biomass has broadened the horizon for plant-based resources. Although algae and seaweed are not terrestrial plants, they do possess many compositional similarities to lignocellulosic biomass, and their combined use as sustainable feedstock is becoming popular. The high growth rate and carbon content of these species make them attractive options, especially in regions with limited access to agricultural residues. Taken together, the variety of biomass feedstocks that can be used highlights the versatility of plant-based materials in developing nanocomposites, depending on local availability and resource type (Amulya et al., 2023).

Availability and Environmental Benefits

Abundant plant sources are the most compelling sources of nanomaterials from plant residues. Every year, millions of tons of crop residue are produced worldwide. For instance, rice farming generates more than 700 million tons of rice straw and husks annually, a considerable portion of which is discarded. In other countries, sugarcane is grown as well, and a large amount of bagasse that results from thousands of tons of sugarcane being harvested is burnt in an extremely inefficient way to generate power or else wasted. The search for these waste streams as suitable raw materials for advanced functional materials is likely to not only discover an excellent way to manage waste products, but also be a new step forward in sustaining the development of nanotechnology applications (Nath et al., 2023).

Lastly, value-added utilisation of plant residue has an environmental aspect as well. These types of procedures are a source of carbon footprint, and this could be avoided through the generation of desulfurisation nanomaterials from biomass feedstock, with the possibility of eliminating synthetic sorbents and energy-consuming catalysts. Furthermore, the plant biomass feedstock can be used to redirect waste from polluting end-of-life pathways. Stubble burning is a key cause of particulate matter (PM) pollution, smog, and increasing cases of respiratory diseases in many parts of the country. By converting these residues to sustainable value-added products, potential emissions are avoided, and clean air programs are enabled. Moreover, biomass use also offers an opportunity for carbon sequestration because the carbon of plants, in other words, taken from it, is stored in the functional material as stored carbon and is no longer released into the atmosphere (Singh et al., 2024).

Plant-based materials have decentralisation benefits from a resource point of view. Unlike common sorbent materials that rely on mining or specialised chemical synthesis, biomass feedstocks can be found locally, minimising transportation costs and emissions. This regional supply chain is beneficial not only for sustainability but also provides the community with an economic buffer, especially for rural farming areas (Viell et al., 2025).

Figure 1. Generalised drawing of plant-based nanocomposites for sustainable desulfurisation. The figure reflects the interdependence of the relationships between plant-derived materials, their use in the process of removing sulphur from fuels, and their overall impact on the environment. Today, a cycle-like approach to reducing pollution and green energy is achieved by combining renewable biomass sources with nanotechnology.

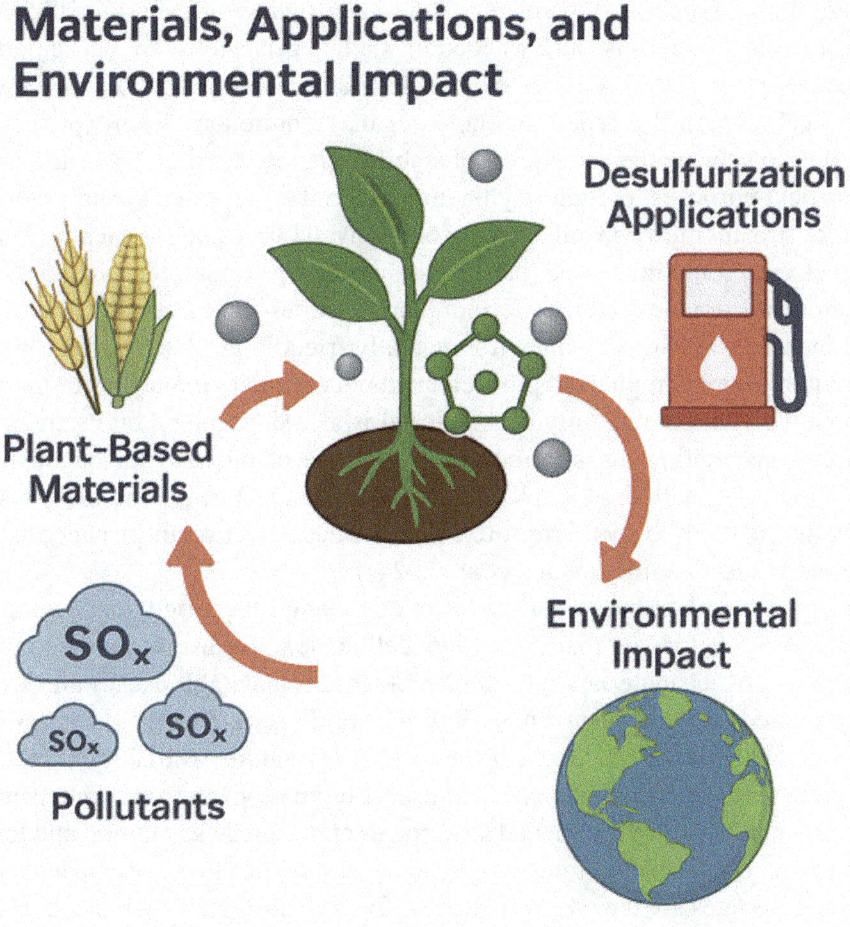

Selection Criteria for Effective Use

Plant-based residues are abundantly available, with a large variety of options. The selection of suitable feedstocks for nanocomposite development and desulfurisation applications must be carefully made. Many factors can influence the performance of biomass feedstock (Shah et al., 2023).

The response of fibre yield and quality was dependent on the cellulose, hemicellulose, lignin, and other mineral content ratios. A high carbon content containing functional groups that can be controlled is attractive. Isotropic materials containing high mineral content, such as silica-rich rice hulls, present composite possibilities with additional processing to enhance the superstructure (Dey et al., 2025). Hard and brittle plant materials, such as coconut shells or nutshells, are particularly advantageous because they produce an activated material with a high surface area and mechanical strength. Softer deposits, however, may require some form of pretreatment or mixing to achieve the same level of stability (Ajien et al., 2023). Furthermore, for practical purposes, particularly on an industrial scale, reliable and predictable feedstock streams must be available. Seasonally available or geographically restricted feedstocks are sometimes used, but may be inhibiting if not underpinned by good collection and storage systems. Certain plant residues can be converted into functional materials by single and environmentally friendly processes, whereas others may require more complicated pre-treatments to eliminate impurities or unwanted compounds. The sustainability of a material is also determined by its energy and resource consumption during processing. The use of biomass for nanomaterials must not conflict with food and land use. Attention must be paid to these residues and wastes, and not to food crops, to achieve a net reduction in demand that does not threaten sustainability (Wang et al., 2024).

In summary, plant-based materials are abundant, diversified, and ecologically friendly resources for nanomaterials. Lignocellulosic by nature, they provide a strong view of carbon-rich materials, and their abundance is inevitable as they are designed to be upscaled into applied settings. With the appropriate choice of biomass-based nanocomposites and depending on their chemical nature, availability, and structural properties, scientists can produce useful biomass-based nanocomposites for desulfurisation. The environmental advantages of utilising agricultural and forestry residues are related and therefore increase the appeal of the process; hence, plant-powered resources are the centre of interest for both sustainable nanotechnological and clean fuels (Sulis et al., 2025).

CONVERSION OF PLANT WASTES INTO USABLE MATERIALS

The transformation of vegetal trash to functionalized nanomaterials in the area of desulfurisation has emerged as a significant methodology in linking uncooked biomass supplies and actual purification of fuels. Raw plant residuals typically lack proper physicochemical characteristics that can support sulphur adsorption and catalytic reactions (Lim et al., 2021). Such low-value wastes can be intentionally transformed into high-performance materials, including activated carbon, biochar, nanocellulose, and hybrid composites, through specific conversion pathways. Innovations in the sector have concentrated primarily on environmentally friendly treatments, material finishes, and extraction and disposal low-impact solutions, so that the entire process reflects sustainability and a circular use of resources (Lehmann et al., 2025).

Eco-Friendly Preparation Methods

The pretreatment step is usually the first step in the development of biomass into functional materials that make plant waste usable for the following modifications. Eco-friendly alternatives have been developed to eliminate the use of chemicals and electricity waste. For instance, surface impurities and agrochemical residues present in biomass can be removed by washing and drying. The later stages of grinding and milling decrease the particle size to provide additional surface area and adjust the feedstock for thermal or aqueous treatment (Trivedi et al., 2025). Pyrolysis is one of the most commonly used techniques involving the thermal degradation of biomass in the presence of restricted amounts of oxygen to create biochar. Controlled area pyrolysis, on the other hand, results in the preparation of carbon-rich materials with adjustable surface properties (Zaman et al., 2025). Pyrolysis at low temperatures may not remove all functional groups of the biomass, which may be beneficial for surface reactivity, whereas higher temperatures would increase the porosity and surface area. Recent pretreatment processes, such as microwave-assisted pyrolysis, have also resulted in the efficient use of energy and a short reaction time, as opposed to traditional heating processes (Zhang et al., 2025).

Alternatively, biomass can be operated through the hydrothermal carbonisation (HTC) process. This method uses subcritical water at ambient temperature to transform biomass to hydro car which is a structurally different carbonaceous substance. The HTC imitates the formation process of natural coal, which takes several hours rather than millions of years (Ighalo et al., 2025). Specifically, wet biomass can be treated without drying, which is a process that consumes energy and is promising for agricultural and food processing residues, a process to which HTC can be applied in large amounts. In addition, novel alternatives to deconstructing lignocellulosic architecture should be considered based on enzymatic and microbial processes to facilitate benign technologies. Certain enzymes can selectively depolymerise hemicellulose or lignin and extract cellulose nanofibres used as precursors of nano- and microscale cellulose-based products. The fact that this biosynthesis is solvent- and high-temperature-free is appealing; however, the scale-up of the processes is limited by the same problem (Abolore et al., 2024).

Therapeutic: Natural Therapies, Heat, or Hot Water

Thermal effects are the primary pillars in the establishment of material characteristics in relation to temperature, gaseous phase, and activator effects. Maximising the protocols to enhance the adsorption process, in addition to pyrolysis and HTC, would enhance the adsorption properties (Mishra et al., 2024). Physical: porosity is formed owing to the impact of hot flue gases, steam, and carbon dioxide on the raw material. This approach prevents the use of corrosive reagents and achieves a high specific surface area, which is beneficial for a suitable desulfurisation performance (Hanif et al., 2020). Chemical activation is also conductible in an eco-friendly manner with naturally occurring activation. For example, plant acids, salts, and even activating compounds in waste have been studied as more environmentally friendly alternatives to the use of conventional reagents, including potassium hydroxide and zinc chloride which are difficult to treat. Concurrent thermal and chemical activation typically results in the cooperative promotion of pore generation and surface chemistry (Sevilla et al., 2021). The conversion of biomass is also necessary for it to be water-based. In addition, critical hydrolysis weakens hemicellulose by breaking it into sugars and removing it, and the cellulose of fibres with a high proportion is available for users. Regarding chemical pretreatment, alkaline pretreatment using natural and environmentally friendly mild bases, such as sodium bicarbonate, has been applied to partially delignify biomass while improving its porosity (Kumari et al. Steam explosion is a water-based method where cellulose fibres are exposed to super-heated steam and then the equipment decompressed in an explosive manner, producing micro- and nanofibrillation of the lignocellulosic structure that can be re-assembled into useful materials (Kumari et al., 2019).

Key Properties Achieved Through Simple Processing

In addition, the green treatment of plant waste offers excellent desulfurisation performance. These are mostly high surface areas and porosities. The fuel sulphur compounds are either large aromatic molecules such as dibenzothiophens, the size of which will require a permeable sorbent to allow diffusion and adsorption (Azeez et al., 2023). Independent of the thermal activation and controlled carbonisation processes, their synergistic effect can produce a hierarchical porous structure of MIROPs as the site of molecular adsorption and MEFPs as the site of diffusion. Surface chemistry, which also occurs during treatment, is a necessary aspect to be mentioned. Certain processing conditions can leave functional groups such as hydroxyl, carboxyl, and carbonyl groups on biomass carbons. Such groups encourage the material to possess greater angular forces to facilitate the adsorption of polarisable sulphur species. Furthermore, the degree of carbonisation is tenable, and one can trade between a high surface area and the presence or absence of functional groups, depending on the specific application (Zhang et al., 2020).

Thermal treatment can enhance the structural stability of biomass-based materials and ensure that they do not collapse after several adsorption and regeneration cycles. Mechanical strength is important for industrial fuel purification systems and increases when biomass is carbonised under controlled conditions to avoid over-fragmentation (Zhu et al., 2024).

Improving Material Properties with Natural and Added Components

Natural plant-derived sources have been modelled as potential precursors to the production of functional nanocomposites in recent years, as the desulfurisation activity elicited by them may be constrained by some of their characteristics (low specific surface area, disfavoured pore structure, or low adsorption selectivity of sulphur compounds) (Sadrara et al., 2025). To eliminate these deficiencies, the research community has attempted to improve the physicochemical attributes of PBMs through various mild treatments, the addition of natural additives, and/or the use of safe, complementary materials. These methods are intended to enhance the performance and green and sustainable nature of the entire process. Nanomaterials that can be developed to achieve increased sulphur removal capacities cheaply and environmentally friendly are plant-derived nanomaterials with altered surface structures, more active sites, or combinations thereof (Roy et al., 2021).

Natural Products or Healing Process Additions

These functional groups, viz., hydroxyl, carboxyl, and carbonyl groups, are significant for the correct construction of a hydrogen bond or dipole -dipole interaction with a sulfur-containing species to enhance adsorption. Activation of the alkali by innocuous agents (sodium bicarbonate and calcium carbonate) may also increase porosity and allow an increase in the number of basic sites on the carbon surface (Chen et al., 2024).

A mild additional change after the treatment is the thermal requirement. Additional oxygen groups on the surface which do not damage the structure, are added at moderately higher temperatures than room temperature using I 2 controlled oxidation. The microporosity and surface area - significant characteristics of an effective desulfuriser–will be enhanced by steam activation or, say, low-energy input carbon dioxide activation as well. More importantly, these feedstocks are feedstock-specific in terms of their received and mineral composition (Hou et al., 2022).

Currently, there is a trend to use natural components sourced directly from plants. For example, tannins and polyphenols from agricultural residues can be used to functionalize carbon surfaces to create additional adsorption sites. Biomass-derived polysaccharides, including starch or chitosan (from chitin as a biomass), can be employed to coat biomass-derived carbons, and the amount of sulphur immobilised in them is increased by introduction during their synthesis of amine groups or OH-groups. These bio-alternatives are recommended to have the medication of paint easily reinable and their preparation amenable to green and the general principle of green chemistry (Behrooznia et al., 2024).

Combining Plant-Based Materials with Other Safe Elements

Nanomaterials and plant-derived carbons have gained a good background in the previous decade; however, their potential can be enhanced through combination with other safe and environmentally friendly constituents, and nanocomposites can be achieved. Biomass-based carbon synergistic effects have been exploited and additional components have been added to improve the adsorption capacity, selectivity, and regeneration of composites (Kanagaraj et al., 2025). Metal oxides are among the most widely used options for binary combinations. The active centres on carbon can be coordinated by growthwths of oxides such as iron oxide, zinc oxide, and titanium dioxide with sulphur. These oxides have a high SO 2 affinity through adsorption on the surface or catalytic oxidation (Sadegh et al., 2019). Importantly, most of these metals are found in high concentrations, are non-toxic, and are cheap; hence, they can be utilised to perform sustainable activities. These metal oxides are stabilised by the biomass-derived carbon matrix, and agglomeration is also inhibited by the biomass-derived carbon matrix, thereby resulting in increased dispersion and accessibility of the active sites. One approach involves the incorporation of indigenous minerals existing in the plant matrix into the composite (Yang et al., 2023). Silica, for example, is abundant in rice husks and can be retained and utilised as a reinforcing phase in the carbon matrix. Silica promotes structural integrity and provides adsorption sites for the formation of natural composites. Similar beneficial effects have been observed when biomass residuals with calcium, potassium, or magnesium were prepared to maintain the natural mineral content of the feedstock (Taiye et al., 2024). Biopolymer coatings and hybridisation are also promising avenues to improve properties. Plant-based carbon can also be functionalized with natural polymers, such as lignin, starch, or cellulose derivatives, to prepare hybrid sorbents in specific chemical environments. These composites have the advantage of the inherent stability of carbon and the synthetic flexibility of polymers and are more suitable for the interaction with various sulphur species (Yaashikaa et al., 2022). Finally, plant-based carbon-supported green-synthesised nanoparticles have become popular. Reduction capabilities-containing plant extracts can lead to the green synthesis of metal nanoparticles that can be deposited on biomass-based supports. For instance, seed- or leaf-extract-fabricated silver or iron nanoparticles can amplify the catalytic or adsorptive potential of plant-based materials without the need for synthetic reducing agents. Not only does this approach improve performance, but it also sustains the general utility of the system (Dhaka et al., 2023).

To enhance the performance of nanomaterials in desulfurisation, nanomaterials obtained from plants must be prepared through the introduction of green reagents, low-cost production, and environmentally friendly composites. They can reduce the inherent disadvantages of raw biomass-derived carbons and meet the requirements of environmental friendliness through the manipulation of surface structures and the introduction of synergistic ingredients. Based on this plan, it can be concluded that plant-based materials are suitable and cost-effective for high-performance adsorption and catalysis. The demand for cleaner fuel is soaring, and efficient processing of plant-based nanocomposites should facilitate scalable, green desulfurisation technology (Lahiri et al., 2021).

LEARNING MATERIAL FEATURES AND BEHAVIOUR

The ability of nanocomposites comprising plant-derived ingredients to desulfurise primarily relies upon the nature of the material used to prepare the nanocomposite and the impact that this nature has on the reaction of the nanocomposite with sulfur-containing substances in the fuel streams. Although the starting structural characteristics are those imparted by the raw material used and the method used for processing, the overall efficacy of desulfurisation also depends on the efficient adjustment of the surface area, porosity, and texture of the material for adsorption or catalysis. Knowledge of these properties is thus essential for the characterization of existing materials and to inform the rational design of novel bio-based nanocomposites (Mansouri et al., 2025)

Structure, Texture, and Surface Characteristics

The skeletal carbon structure of plant-based materials is mainly derived from the decomposition of lignocellulosic biomass. The general structure of the final carbon material varies between amorphous and partly graphitic and depends on the carbonisation temperature and method of processing. Amorphous carbons usually have a higher functional group content, whereas graphitic parts exhibit elevated electrical conductivity and stability. Both aspects are likely to contribute to sulphur removal, especially in the case of hybrid methods, which combine adsorption with catalytic conversion (Herold et al., 2021).

Just as texture is crucial, the arrangement of pores and particles at the microscopic scale of a material is also crucial. Shell and Coconut shell give origin to hard carbons of lowest Imodium extremely developed porosity, while the aqua cultural residues (e.g. straw ortho: Bagasse) yield a way of producing fibrous light weight carbon (Jasni et al., 2018). The degree of fragmentation, fibre alignment, and surface roughness not only define the available surface area but also the integrity of the sorbent after repeated use. Therefore, understanding the surface chemistry of plant-based nanocomposites is essential. Here, the biomass has oxygenated groups (e.g. hydroxyls, carboxyl, and phenolic groups) in its structure. These groups may be retained, modified, or deleted as required during processing (Karthik et al., 2024). Their presence is crucial in the interaction with sulfur-bearing systems, not only in terms of hydrogen bonding and dipole interactions, but also for electron-donating or electron-withdrawing effects. High-temperature carbonisation, on the contrary, can lower the relative population of these groups and produce a more hydrophobic surface that could foster compatibility with non-polar substrates but reduce sites for targeted chemical functionalization. Thus, optimising the balance between hydrophilicity and hydrophobicity is a key design issue (Gonçalves et al., 2010).

Role of Porosity and Surface Area

The porosity and surface area, the main properties of biomass-based adsorbents, have a direct impact on the capability of absorbing and storing sulphur species. In particular, micropores (pore diameter < 2 nm) are very important because they are similar in size to sulfur-containing species, such as thiophene and dibenzothiophene (Ozcan et al., 2024). A high micropore content allows it to accommodate more molecules, thereby further increasing the adsorption capacity. Mesopores (2–50 nm) serve as adjuncts by enhancing molecular diffusion in the adsorbent. Under low microporosity, the bulky sulphur species can merely be located at the external surface, causing a decrease in the activity. A hierarchical pore structure with micro-and microporosity for adsorption and diffusion is the most favourable one for desulfurisation. Hierarchical control: Plant-based materials, if properly activated, are instinctively predisposed to this type of hierarchical control because of the natural variability of lignocellulosic structures (Studart et al., 2015).

Another important parameter is the surface area. Biomass-based activated carbons can usually achieve surface areas of over 1000 m^2g^{-1}, depending on the activation method. In addition to enhancing the number of available adsorption sites, a high surface area allows for more precise control of surface functionalities (Malini et al., 2023). Nevertheless, the ultrahigh SSA obtained through extreme activation may be detrimental to structural integrity (i.e. an inverse relationship between porosity level and mechanical stability). The type and arrangement of pores also influence the performance. A continuous pore network allows fuel molecules to quickly access the interior adsorption sites; however, in the case of isolated pores, the cores of these pores may be unused. Therefore, the optimal material is a combination of high micropore density and interconnected mesopores supported by a mechanically stable carbon scaffold (Khamkeaw et al., 2020).

How Features Influence Interaction with Sulphur Compounds

Physical absorption and chemical combination (in some cases, catalytic effect) play a role in plant-derived nanocomposite-captured sulphur species. Each material property plays a unique role in these processes. Hydroxyl and carboxyl groups can be beneficial for adsorption owing to their weak interactions with sulphur functional groups. Such interactions are especially strong with polarisable thiophene derivatives that can partake in hydrogen bonding. Nitrogen heteroatoms inserted either by doping or with natural additives can enhance affinity by providing basic sites that strongly interact with acidic sulphur compounds. Excluding this, nitrogen and oxygen groups may induce local charge distributions on the carbon surface, and accordingly, the binding of the compounds can become stronger (Dong et al., 2024).

Fuel sulphur compounds range from smaller thiophenes to bulky polyaromatic dibenzothiophenes. Small molecules may be trapped in the micropores, whereas larger substances may gain access to the mesopores. It follows that materials with a mixed pore size distribution are more suitable for thorough removal of sulphur (Moosavi et al., 2012). The confinement of micropores can increase the adsorption force by limiting the distance between the adsorbent wall and the sulphur compound, resulting in stronger van der Waals and π–π interactions. Graphitic layers of vegetable carbon allow π–π stacking interactions with organic sulfur. Being aromatic, dibenzothiophene closely associates with graphitic surfaces through overlapping, delocalised electrons. These are not covalent, but strong interactions enhance the adsorption selectivity for aromatic sulphur over aliphatic hydrocarbons (Nahian et al., 2024). Because hydrocarbon fuels are mostly nonpolar, adsorbents must have adequate hydrophobic properties to be compatible with the fuel matrix. Hydrophobic sites enhance wetting by fuel molecules, including sulphur compounds, to allow for their interaction with active sites. Meanwhile, polar groups are localised for the selective adsorption of sulphur molecules. The dual affinity for selective desulfurisation is due to the synergistic effect of the hydrophobic carbon backbones contacting a polar functional group (Guo et al., 2022).

After processing, plant sources may contain minerals such as silica, calcium, and potassium. These components may influence sulphur removal by adding more adsorption sites or by being used as catalyst modifiers. For instance, silica-rich carbons prepared from rice husks may enhance adsorption through physical trapping, while calcium-based residuals may chemically combine with sulphur functionalities. Whether we will retain or remove them in the NP's is another parameter which can be adjusted. The characteristics of a material also affect its regeneration and reusability as an adsorbent. Heavier bound sulphur compounds may be less easily regenerated under conditions which would not otherwise harm the material. In contrast, optimum moderate and strong adsorption sites could make not only the sulphur capture performance but also the recyclability satisfactory. Hence, porous stability and surface robustness under thermal or solvent regeneration are essential features for practical applications (Hanif et al., 2021).

Role of Plant-Based Materials in Fuel Purification

The primary goal of making plant-based nanocomposites is to use them in fuel cleaning, particularly to remove sulphur compounds (Rajendran et al., 2025). Desulfurisation is important because the sulphur oxides produced during burning are the primary cause of acid rain, airborne particulates, and smog, a human medical condition. Furthermore, increasingly stringent environmental legislation across the globe requires transport fuels to contain significantly reduced sulphur compared to previous levels, adding pressure on the refining industry to find economically viable, sustainable alternatives. Standard HDS is still the prevailing technology used in the industry, but its restrictions on refractory sulphur compound removal, as well as its high processing cost, necessitate alternatives compared with such technologies. Plentiful, inexpensive, and adjustable surface chemistry, plant-based nanomaterials offer a promising approach for green fuel purification technology. Sulphur compounds in fuels are a complex group (Ng et al., 2025). Among these, thiophenes, polythiophene, dibenzothiophene (DBT), and their alkylated homologues are the most prevalent. However, DBT and its homologues are difficult to convert because of their steric hindrance and strong aromaticity, and they are less reactive under HDS conditions. Plant-based NMs respond to this challenge mainly by adsorption-based mechanisms and some by catalytic or hybrid processes. Plant carbon-based materials (carbon, biochar, and nanocellulose composites) are effective adsorbents owing to their porous structures, high surface areas, and tenable surface functionalities. Adsorption-based desulfurisation is based on the physical and chemical contact between sulphur compounds and the surface of the adsorbent. The micropore structures enable the capture of small sulphur molecules and the mesopore structures enable the diffusion and adsorption of bulky DBT derivatives. Surface groups, such as hydroxyl, carboxyl, and nitrogen groups, may form dipole interactions, hydrogen bonding, or acid–base interactions with sulphur compounds, making the adsorption more selective (Saha et al., 2021).

Graphitic regions in plant-based carbons facilitate π–π stacking interactions with aromatic sulphur molecules such as DBT. Although non-covalent, they are strong enough to widen the selectivity between sulphur and hydrocarbons (Nahian et al., 2024). Finally, the controlled polarity of plant-derived surfaces provides a beneficial synergy: hydrophobicity allows for compatibility with non-polar fuels, while the localised polar groups are used as a site for selective binding of sulphur species. Plant-based materials can also participate in catalytic desulfurisation by the addition of minerals composed of natural or harmless metal oxides. For instance, the incorporation of iron or zinc oxides into biomass-based carbon meso phases promotes the creation of active sites capable of both adsorbing and mildly oxidising sulphur species. This EAT+SAT effect contributes to the improvement in desulfurisation efficiency and the regeneration cycle of adsorbents (Hanif et al., 2020).

Key Advantages over Traditional Methods

Compared to classic desulfurisation techniques, particularly when compared to hydrodesulfurization, plant-produced nanomaterials have some distinctive strengths. HDS is conducted at high pressures of hydrogen gas and at high temperatures (300–400 °C), which are energy- and cost-intensive. In contrast, plant-based adsorbents operate under ambient/mild conditions and drastically reduce energy consumption. This makes them particularly attractive to decentralised or small refinery operators (Ahmad et al., 2017).

Despite the weak interactions of 4, 6-dimethyldibenzothiophene like molecules with HDS, adsorption-based methods using plant-based nanomaterials are more effective for adsorbing the target molecule because of the pore confinement effect and strong surface interactions. Feedstocks of agricultural and forestry residues were adopted to reduce the cost of raw materials. However, owing to the disposal problem of the residues, their vaporisation into a desulfurisation material could not only effectively avoid the issues in disposal but also realise the resource of the spent adsorbent. Process options such as pyrolysis and hydrothermal carbonisation are inexpensive alternatives to advanced catalyst synthesis, further enhancing cost efficiency (Areti et al., 2025).

Procedures and Comparisons Based on Research for Simple Tests

Nanomaterials derived from this plant have been proven in numerous laboratory studies as successful when used in sulphur cleaning, primarily through straightforward adsorption studies. Specifically, the Coconut shell has been reported to exhibit high adsorption performance for DBT because of its well-formed microporous structure and graphitic surface areas (Dordević et al., 2025). Carbon that is silica-modified and oxygenated group-modified rice husk-based carbons have also exhibited better affinity with theophanic sulphur compounds. Other studies have documented the excellence of plant-based carbons as opposed to commercial activated carbons for the cases they are adjusted to using environmentally sensitive activation methods (Jha et al., 2021). An example is steam or carbon dioxide activated materials which are more effective than chemically activated carbons that have aggressive agents. This explains why green processes are useful for the preparation of high-performance sorbent materials (Serafin et al., 2023).

Finally, natural materials are very promising in relation to fuel cleaning owing to their friendship to the environment, ability to tune their structure, and abundance. Sulfur that has been extracted by various processes, such as physical adsorption, chemical reaction, and catalysis promotion, has greater alternatives to the traditional desulfurisation process. Energy conservation, cost efficiency, environmental friendliness, and regeneration using refractory sulphur compounds have potential benefits that can bring a great deal to the desulfurisation process of fuels in the future. They have great laboratory support, have been compared, and are being extended to other industrial applications (Liu et al., 2024).

Environmental and Economic Goods

The case described above of plant-based nanocomposites in desulfurisation is not exclusive to the direct application of these materials in fuel cleaning. These characteristics render these alternative appropriate sustainable as alternative technologies in traditional desulfurisation, in addition to their overwhelming environmental and economic advantages (Norizan et al., 2022). Through this revaluation of the agro- and forestry residues, the process of resource saving can be seen as a possibility, and therefore, the possibility of processing these materials in a cost-effective manner, they would also fit in the strategy towards global goals of sustainability, and lastly, it would avail an opportunity to our economy. To assess the feasibility and acceptability of a shift towards plant-based nanomaterials in large-scale fuel processing processes, it is essential to understand the larger implications of the problem and its effects on the sector (Zhang et al., 2025).

Waste Reduction by Value-Added Used

Farming and timber harvesting produce millions of tons of biomass landfillsill every year. In most instances, these remaining residues are thrown away or burnt as a by-product or allowed to decompose over time, resulting in severe environmental issues such as air pollution, greenhouse gases, and soil- and river-polluting nutrient runoffs. Consider an example of rice production that annually exudes hundreds of million tons of husks and straw, many millions of which fall into dumps. Furthermore, sugarcane products such as bagasse and coconut and palm products such as shells/husks exist in bulk numbers (Riseh et al., 2024).

The use of waste in the production of useful nanocomposites transforms the waste disposal problem into value-additivity. Rather than adding to the issue, these wastes are used as feedstocks by new generations of improved materials that eventually address one of the most serious problems in energy production and storage: the production of a clean source of fuel (Dave et al., 2024). This vaporisation action characterizes the cyclic economic manner of reusing waste in productive streams because it constitutes a closed circuit system. In addition, the use of plant residues will decrease the pressure on landfill facilities and abolish untidy waste management practices, such as open burning. The transformation of crop residues into desulfurisation materials could lower the air pollution and health burden at the local level in areas where crop residues are used. It not only does not harm the environment but also leads to better social health of the people, and thus, there is also a social benefit to this technical innovation (Singh et al., 2025).

LOW-COST FEEDSTOCKS

Raw materials – agro and forestry residues–are low cost and, in some cases, free of cost. In fact, in several cases, the residues are paid for disposal, and transformation into nanocomposites means a cost was changed into an economical property (Salazar Sandoval et al., 2024).

Simple Processing Requirements

Although expensive metals such as cobalt, nickel, and molybdenum are used as high-performance catalysts for HDS, biomass-based sorbents can be fabricated using relatively easy techniques (Akhtar et al., 2025). Pyrolysis or steam activation processes require lower initial capital investment and operating costs; hence, they are suitable for deployment in small or decentralised fuel purification plants. Because plant-based adsorbents function under moderate conditions, no high-pressure hydrogen or harsh heat is utilised, reducing costs significantly. This allows producers of 1 fuel in less developed areas to meet sulphur reduction limits without incurring unreasonable costs. The reusability of such plant-derived sorbents is also stronger where mild heating or solvent washing results in the regeneration of such sorbents.

One batch of adsorbent can be reused many times which decreases the replacement frequency and reduces the long-term material cost (Bunker et al., 2023).

In short, the environmental and economic benefits associated with plant-fabricated nanocomposites outweigh their technical contribution to desulfurisation. They transform agricultural pollutants into value-added products, reduce waste and pollution, and lower energy/resource inputs compared to traditional chemical processes (Naseer et al., 2024). They provide economic benefits such as low-cost feedstock production, regenerator operation cost savings, and semi-social benefits (such as provision of income to rural people and employment generation). All these advantages render plant-based nanomaterials an attractive and versatile approach to green fuel polis-Hing, offering opportunities wherein environmental relevance has economic implications (Singh et al., 2023).

Challenges and Future Opportunities

Plant-derived nano nanocomposites have enormous potential for desulfurisation, but demonstrating this potential in the laboratory is challenging compared to the industrial scale. Their beneficial environmental and economic impacts have been observed, although the current purification facilities have current issues with material stability, long-term stability, scalability, and integration (Jha et al., 2025). This will also create space for new, creative, collaborative, and policy-knowledgeable support that would be stimulated to take over time. The identification of stagnations, bottlenecks, and knowledge on probable RD&D future prospects is inevitable in guiding R&D towards practical application (Georgiou et al., 2025).

Stability and durability

The heterogeneity of biomass feedstocks is one of the problems facing plant-based nanomaterials. The residue of agriculture is large since the type of plants grown, the mode of cultivation, climate, and harvesting pattern are different (Bhushan et al., 2024). Naturally, other crop residues, such as rice husks or sugarcane bagasse, vary on both local and seasonal levels in both composition and structure. The variation in the profile was not conducive to attain consistent nanocomposite properties that would be obligatory for industry implications. Moreover, when bio-materials are plant-based, long-lasting solid strength and resistance to chemicals is one of the issues. The pore structure and surface functionality are easily destroyed during absorption-regeneration cycles, despite the promising adsorption capacity that can be attained in laboratory experiments (Barbhuiya et al., 2025). Sulphur species that are of high aroma, such as dibenzothiophene, may be strongly adsorbed on the surface of an adsorbent because they are very difficult to regenerate and slow to deactivate the media. To be economically attractive, these materials should not lose much of their catalytic activity when subjected to repeated cycling. Such aspects as standardisation of feedstock pre-processing, control of a variety of activation techniques, stabilisation by mild doping, and so forth appear to be good opportunities to alleviate problems of reproducibility and stability. Nevertheless, further studies are needed to identify a procedure that can be scaled and reproduced at a low cost (Mane et al., 2021).

Laboratory to Factory transfer

One point that needs to be explained is that the research on plant-based nanocomposites used in desulfurisation reactions is mostly of a laboratory nature, using model fuels. Controlled experiments under such conditions are informative about the mechanistic aspects of the problem; however, in many cases, they are unable to capture the complexity of real fuels which are mixtures of many different hydrocarbons and a range of sulphur species at varying concentrations. Fuel impurities can significantly affect the behaviour of industrial materials owing to pressure and temperature fluctuations (Kadhum et al., 2022).

Research Potential and Trends

However, there is no easy research on the process of desulfurisation with plant-based nanomaterials; however, the literature in this field is growing and a number of research directions are rather promising (Ghotekar et al., 2020). Recently, attention has been devoted to improving the selectivity of adsorption by doping biomass-derived carbons with heteroatoms (N, S, or P). Biogenic precursors are more sustainable in making these modifications because of the common use of sulphur functionality. In the past, there has been increasing interest in the use of plant-based carbons along with green metal oxides, photocatalysts, and ionic liquids (Mihai et al., 2025). They enjoy the benefits of synergetic effects, thus realising the concomitant adsorption and catalytic transformation of sulphur compounds. This multimodal behaviour may provide an answer to some of the limitations of adsorptive methods of special research interest, such as nanocellulose nanofibres and nanosheets manufactured using plant biomass as templates (Krishani et al., 2025).

The remaining challenges that plant nanocomposites can address through technological development and cooperation are desulfurisation opportunities. Pilot-scale validation and life-cycle sustainability analysis will also have to be established. The influence of hybrid composites and nano-biomass materials will be promoted in the future, and industrialisation may be encouraged by this policy (Sankar et al., 2022). To conclude, plant-derived nanomaterials will be found in the future not only for practical and reliable service but also for high desulfurisation efficiency, cost effectiveness, environmental friendliness, and high quantities, to the global scenario. Thus, such materials are already in the pipeline to establish a precedent on the issue of green purification of fuel and demonstrate possibilities of reaching the environmental, technological, and economic balance which can guide the solutions for future economic and environmental issues (Balusamy et al., 2023).

CONCLUSION

One of the key points for promoting clean and sustainable energy systems is still the desulfurisation of fossil fuels. Sulphur oxides produced by burning are air pollutants responsible for acid rain and are also known to impact the health of humans; therefore, alcohol needs to have a low sulphur content to comply with environmental standards. Although conventional hydrodesulfurization is a well-known process, it has problems such as the necessity for high-energy requirement, expensive catalysts, inability to desulfurise some very recalcitrant sulphur compounds, and other limitations so that there should be some way of removing the same which is non-polluting and cost effective. In summary, plant-based nanocomposites are transforming as a global cure for fuel depuration and can be visualised as a model of ethically admissible, economically feasible, and technologically aspiring practices that would offer purifying life to generations ahead. In this contribution, accompanied by the waste value nanocomposite generation, we are able to solve the key problem of sulphur rejection and at the same time reveal a more underlying transition trajectory from the traditional energy to the green energy concept and design of energy materials. If further developed and evolved, they could potentially unlock cleaner fuels and environments, a future of circles that are peril-proof for energy.

AI DISCLOSURE

During the preparation of this work, the author(s) used *Paperpal* solely for grammar and language refinement. After using this tool, the author(s) carefully reviewed and edited the content as needed and take full responsibility for the integrity and accuracy of the scientific content presented in this chapter.

REFERENCES

Abolore, R. S., Jaiswal, S., & Jaiswal, A. K. (2024). Green and sustainable pretreatment methods for cellulose extraction from lignocellulosic biomass and its applications: A review. *Carbohydrate Polymer Technologies and Applications*, *7*, 100396. DOI: 10.1016/j.carpta.2023.100396

Ahmad, W., & Ahmad, I. (2017). Mechanisms of desulfurization by nanomaterials. In *Nanotechnology in Oil and Gas Industries: Principles and Applications* (pp. 211-243). Springer.

Ajien, A., Idris, J., Md Sofwan, N., Husen, R., & Seli, H. (2023). Coconut shell and husk biochar: A review of production and activation technology, economic, financial aspect and application. *Waste Management & Research*, *41*(1), 37–51. DOI: 10.1177/0734242X221127167 PMID: 36346183

Akhtar, M. S., Naseem, M. T., Ali, S., & Zaman, W. (2025). Metal-Based Catalysts in Biomass Transformation: From Plant Feedstocks to Renewable Fuels and Chemicals. *Catalysts*, *15*(1), 40. DOI: 10.3390/catal15010040

Amulya, K., Morris, S., & Lens, P. N. (2023). Aquatic biomass as sustainable feedstock for biorefineries. *Biofuels, Bioproducts & Biorefining*, *17*(4), 1012–1029. DOI: 10.1002/bbb.2471

Areti, H. A., Hamda, A. S., Abo, L. D., Jabesa, A., & Muleta, M. D. (2025). Biowastes as sustainable catalysts for water treatment: A comprehensive overview. *Environmental Advances*, *19*, 100610. DOI: 10.1016/j.envadv.2025.100610

Azeez, M. O., & Ganiyu, S. A. (2023). Review of biomass derived-activated carbon for production of clean fuels by adsorptive desulfurization: Insights into processes, modifications, properties, and performances. *Arabian Journal of Chemistry*, *16*(10), 105182. DOI: 10.1016/j.arabjc.2023.105182

Balusamy, S. R., Joshi, A. S., Perumalsamy, H., Mijakovic, I., & Singh, P. (2023). Advancing sustainable agriculture: A critical review of smart and eco-friendly nanomaterial applications. *Journal of Nanobiotechnology*, *21*(1), 372. DOI: 10.1186/s12951-023-02135-3 PMID: 37821961

Barbhuiya, S., Das, B. B., Kapoor, K., Das, A., & Katare, V. (2025). Mechanical performance of bio-based materials in structural applications: A comprehensive review. *Structures (Oxford)*, *75*, 108726. DOI: 10.1016/j.istruc.2025.108726

Behrooznia, Z., & Nourmohammadi, J. (2024). Polysaccharide-based materials as an eco-friendly alternative in biomedical, environmental, and food packaging. *Giant (Oxford, England)*, *19*, 100301. DOI: 10.1016/j.giant.2024.100301

Bhatia, T., & Sindhu, S. S. (2024). Sustainable management of organic agricultural wastes: Contributions in nutrients availability, pollution mitigation and crop production. *Discover Agriculture*, *2*(1), 130. DOI: 10.1007/s44279-024-00147-7

Bhushan, D., Shoran, S., Kumar, R., & Gupta, R. (2024). Plant biomass-based nanoparticles for remediation of contaminants from water ecosystems: Recent trends, challenges, and future perspectives. *Chemosphere*, *365*, 143340. DOI: 10.1016/j.chemosphere.2024.143340 PMID: 39278321

Bunker, B., Dvorak, B., & Aly Hassan, A. (2023). Thermal regeneration of activated carbon used as an adsorbent for hydrogen sulfide (H2S). *Sustainability (Basel)*, *15*(8), 6435. DOI: 10.3390/su15086435

Chen, J., Ma, X., Liang, M., Guo, Z., Cai, Y., Zhu, C., Wang, Z., Wang, S., Xu, J., & Ying, H. (2024). *Physical–chemical–biological pretreatment for biomass degradation and industrial applications: A review*. Waste. DOI: 10.3390/waste2040024

Dave, V., Sharma, P., Kuila, A., & Sahay, S. S. (2024). Revolutionizing environmental sustainability: Exploring the transformative potential of nanocomposites/nanomaterials from recycled waste. *Environmental Science and Pollution Research International*, *31*(29), 41621–41623. DOI: 10.1007/s11356-024-33435-2 PMID: 38671270

Dembaremba, T. O., Majodina, S., Walmsley, R. S., Ogunlaja, A. S., & Tshentu, Z. R. (2022). Perspectives on strategies for improving ultra-deep desulfurization of liquid fuels through hydrotreatment: Catalyst improvement and feedstock pretreatment. *Frontiers in Chemistry*, *10*, 807225. DOI: 10.3389/fchem.2022.807225 PMID: 35936099

Dey, N., Bhardwaj, S., & Maji, P. K. (2025). Recent breakthroughs in the valorization of lignocellulosic biomass for advancements in the construction industry: A Review. *RSC Sustainability, 3*(8), 3307–3357. DOI: 10.1039/D5SU00142K

Dhaka, A., Mali, S. C., Sharma, S., & Trivedi, R. (2023). A review on biological synthesis of silver nanoparticles and their potential applications. *Results in Chemistry, 6*, 101108. DOI: 10.1016/j.rechem.2023.101108

Dong, W., Xing, J., Chen, Q., Huang, Y., Wu, M., Yi, P., Pan, B., & Xing, B. (2024). Hydrogen bonds between the oxygen-containing functional groups of biochar and organic contaminants significantly enhance sorption affinity. *Chemical Engineering Journal, 499*, 156654. DOI: 10.1016/j.cej.2024.156654

Đorđević, A. M., Milikić, J., Milanković, V., Bogdanović, D. B., Radinović, K., Kaninski, M. M., Relić, D., Stanković, D., & Šljukić, B. (2025). Efficient Removal of Nitrobenzene and Its Compounds by Coconut Shell-Derived Activated Carbon. *Processes (Basel, Switzerland), 13*(7), 2072. DOI: 10.3390/pr13072072

Georgiou, O., Maksymenko, M., & Russo, S. (2025). Deep Insights and Advances in R&D Management. In *R&D Management and Technology Commercialization: Practical Concepts, Innovation, and Case Studies* (pp. 39-66). Springer. DOI: 10.1007/978-3-031-86789-7_5

Ghotekar, S., Pagar, T., Pansambal, S., & Oza, R. (2020). A review on green synthesis of sulfur nanoparticles via plant extract, characterization and its applications. *Advanced Journal of Chemistry-Section B, 2*(3), 128–143.

Gonçalves, M., Molina-Sabio, M., & Rodriguez-Reinoso, F. (2010). Modification of activated carbon hydrophobicity by pyrolysis of propene. *Journal of Analytical and Applied Pyrolysis, 89*(1), 17–21. DOI: 10.1016/j.jaap.2010.04.009

Guo, Y., Xie, W., Li, H., Li, J., Hu, J., & Liu, H. (2022). Construction of hydrophobic channels on Cu (I)-MOF surface to improve selective adsorption desulfurization performance in presence of water. *Separation and Purification Technology, 285*, 120287. DOI: 10.1016/j.seppur.2021.120287

Gupta, N., Mahur, B. K., Izrayeel, A. M. D., Ahuja, A., & Rastogi, V. K. (2022). Biomass conversion of agricultural waste residues for different applications: A comprehensive review. *Environmental Science and Pollution Research International, 29*(49), 73622–73647. DOI: 10.1007/s11356-022-22802-6 PMID: 36071366

Hachhach, M., Bayou, S., El Kasmi, A., Saidi, M. Z., Akram, H., Hanafi, M., Achak, O., El Moujahid, C., & Chafik, T. (2025). Towards Sustainable Scaling-Up of Nanomaterials Fabrication: Current Situation, Challenges, and Future Perspectives. *Eng*, *6*(7), 149. DOI: 10.3390/eng6070149

Hanif, M. A., Ibrahim, N., & Abdul Jalil, A. (2020). Sulfur dioxide removal: An overview of regenerative flue gas desulfurization and factors affecting desulfurization capacity and sorbent regeneration. *Environmental Science and Pollution Research International*, *27*(22), 27515–27540. DOI: 10.1007/s11356-020-09191-4 PMID: 32415453

Hanif, M. A., Ibrahim, N., & Abdul Jalil, A. (2020). Sulfur dioxide removal: An overview of regenerative flue gas desulfurization and factors affecting desulfurization capacity and sorbent regeneration. *Environmental Science and Pollution Research International*, *27*(22), 27515–27540. DOI: 10.1007/s11356-020-09191-4 PMID: 32415453

Hanif, M. A., Ibrahim, N., Isa, K. M., Abdullah, T. A. T., & Jalil, A. A. (2021). Sulfur dioxide removal by mesoporous silica KCC-1 modified with low-coverage metal nitrates. *Materials Today: Proceedings*, *47*, 1323–1328. DOI: 10.1016/j.matpr.2021.02.807

He, H., Zhang, R., Zhang, P., Wang, P., Chen, N., Qian, B., Zhang, L., Yu, J., & Dai, B. (2023). Functional carbon from nature: Biomass-derived carbon materials and the recent progress of their applications. *Advancement of Science*, *10*(16), 2205557. DOI: 10.1002/advs.202205557 PMID: 36988448

Herold, F., Prosch, S., Oefner, N., Brunnengräber, K., Leubner, O., Hermans, Y., Hofmann, K., Drochner, A., Hofmann, J. P., Qi, W., & Etzold, B. J. M. (2021). Nanoscale hybrid amorphous/graphitic carbon as key towards next-generation carbon-based oxidative dehydrogenation catalysts. *Angewandte Chemie International Edition*, *60*(11), 5898–5906. DOI: 10.1002/anie.202014862 PMID: 33497000

Hou, J., Yu, J., Li, W., He, X., & Li, X. (2022). The effects of chemical oxidation and high-temperature reduction on surface functional groups and the adsorption performance of biochar for sulfamethoxazole adsorption. *Agronomy (Basel)*, *12*(2), 510. DOI: 10.3390/agronomy12020510

Ighalo, J. O., Akaeme, F. C., Georgin, J., de Oliveira, J. S., & Franco, D. S. (2025). Biomass hydrochar: A critical review of process chemistry, synthesis methodology, and applications. *Sustainability (Basel)*, *17*(4), 1660. DOI: 10.3390/su17041660

Jasni, M. Z., Abdullah, N. H., Abdullah, S., Razab, M. A., Noor, A. M., Mohamed, M., Yusuf, N. N., Amin, M. M., Rasat, M. M., & Amin, M. M. (2018). Preparation and characterization of activated carbon from cocos nucifera l.(coconut) shell and sugarcane bagasse. *Int. J. Cur. Res. Eng. Sci. Tech, 1*(Spl-1), 416–421. DOI: 10.30967/ijcrset.1.S1.2018.416-421

Jha, D., Gomey, A. K., Kumari, G., Maheswari, P., Haider, M. B., Kumar, R., Balathanigaimani, M., & Sahu, J. (2025). A review on the role of nanocomposites for desulfurization of liquid transportation fuels. *Journal of Environmental Management, 375*, 124255. DOI: 10.1016/j.jenvman.2025.124255 PMID: 39879928

Jha, D., Gomey, A. K., Kumari, G., Maheswari, P., Haider, M. B., Kumar, R., Balathanigaimani, M., & Sahu, J. (2025). A review on the role of nanocomposites for desulfurization of liquid transportation fuels. *Journal of Environmental Management, 375*, 124255. DOI: 10.1016/j.jenvman.2025.124255 PMID: 39879928

Jha, M. K., Joshi, S., Sharma, R. K., Kim, A. A., Pant, B., Park, M., & Pant, H. R. (2021). Surface modified activated carbons: Sustainable bio-based materials for environmental remediation. *Nanomaterials (Basel, Switzerland), 11*(11), 3140. DOI: 10.3390/nano11113140 PMID: 34835907

Kadhum, A. T., & Albayati, T. M. (2022). Desulfurization techniques process and future challenges for commercial of crude oil products. *AIP Conference Proceedings*, 030039. DOI: 10.1063/5.0092049

Kanagaraj, K., Nangan, S., & Thirumalaivasan, N. (2025). Applications of carbon-based nanomaterials derived from plant extracts in composite materials. In *Nanostructured Carbon Materials from Plant Extracts* (pp. 351–380). Elsevier. DOI: 10.1016/B978-0-323-95126-5.00007-X

Karthik, A., Bhuvaneshwaran, M., Senthil Kumar, M., Palanisamy, S., Palaniappan, M., & Ayrilmis, N. (2024). A review on surface modification of plant fibers for enhancing properties of biocomposites. *ChemistrySelect, 9*(21), e202400650. DOI: 10.1002/slct.202400650

Khamkeaw, A., Asavamongkolkul, T., Perngyai, T., Jongsomjit, B., & Phisalaphong, M. (2020). Interconnected micro, meso, and macro porous activated carbon from bacterial nanocellulose for superior adsorption properties and effective catalytic performance. *Molecules (Basel, Switzerland), 25*(18), 4063. DOI: 10.3390/molecules25184063 PMID: 32899569

Krishani, M., Sambudi, N. S., & Suhaimi, H. (2025). Green Catalysts in the Synthesis of Biomaterials for Biomedical Applications. In *Sustainable Green Biomaterials As Drug Delivery Systems* (pp. 217–252). Springer. DOI: 10.1007/978-3-031-79062-1_9

Kumari, P., Seth, R., Meena, A., & Sharma, D. (2023). Enzymatic synthesis of cellulose nanocrystals from lemongrass and its application in improving anti-cancer drug release, uptake and efficacy. *Industrial Crops and Products, 192*, 115933. DOI: 10.1016/j.indcrop.2022.115933

Lahiri, D., Nag, M., Ghosh, S., & Ray, R. R. (2021). Green synthesis of nanoparticles and their applications in the area of bioenergy and biofuel production. In *Nanomaterials* (pp. 195–219). Elsevier. DOI: 10.1016/B978-0-12-822401-4.00017-9

Le, D. L., Salomone, R., & Nguyen, Q. T. (2024). Sustainability assessment methods for circular bio-based building materials: A literature review. *Journal of Environmental Management, 352*, 120137. DOI: 10.1016/j.jenvman.2024.120137 PMID: 38266527

Lehmann, J., Barrios, E., Devault, M., Li, L., Nelson, R., Six, J., & Trimmer, J. (2025). Biochar in the circular bionutrient economy. *Proceedings of the National Academy of Sciences of the United States of America, 122*(33), e2503668122. DOI: 10.1073/pnas.2503668122 PMID: 40789023

Lim, X. B., & Ong, W.-J. (2021). A current overview of the oxidative desulfurization of fuels utilizing heat and solar light: From materials design to catalysis for clean energy. *Nanoscale Horizons, 6*(8), 588–633. DOI: 10.1039/D1NH00127B PMID: 34018529

Liu, Y., Biswas, B., Hassan, M., & Naidu, R. (2024). Green adsorbents for environmental remediation: Synthesis methods, ecotoxicity, and reusability prospects. *Processes (Basel, Switzerland), 12*(6), 1195. DOI: 10.3390/pr12061195

Mabena, K. G., Nomngongo, P. N., & Mketo, N. (2024). Selective and precise solid phase extraction based on a magnetic cellulose gold nanocomposite followed by ICP-OES analysis for monitoring of total sulfur content in liquid fuels. *RSC Advances, 14*(38), 27990–27998. DOI: 10.1039/D4RA04427D PMID: 39224633

Malini, K., Selvakumar, D., & Kumar, N. (2023). Activated carbon from biomass: Preparation, factors improving basicity and surface properties for enhanced CO2 capture capacity–A review. *Journal of CO2 Utilization, 67*, 102318.

Mane, M. B. (2021). Studies in newer methodologies and materials/material modifications for environmental, pollution control applications specifically for water treatment and desulfurization.

Mansouri, M., Shayanmehr, M., & Ghaemi, A. (2025). Exploring the adsorption desulfurization efficiency using RSM and ANN methodologies. *Scientific Reports, 15*(1), 20869. DOI: 10.1038/s41598-025-05688-5 PMID: 40595999

Mignogna, D., Szabó, M., Ceci, P., & Avino, P. (2024). Biomass energy and biofuels: Perspective, potentials, and challenges in the energy transition. *Sustainability (Basel), 16*(16), 7036. DOI: 10.3390/su16167036

Mihai, S., Bondarev, A., & Necula, M. (2025). The potential of biogenic materials as sustainable and environmentally benign alternatives to conventional adsorbents for dyes removal: A review. *Processes (Basel, Switzerland), 13*(2), 589. DOI: 10.3390/pr13020589

Mishra, R. K., Singh, B., & Acharya, B. (2024). A comprehensive review on activated carbon from pyrolysis of lignocellulosic biomass: An application for energy and the environment. *Carbon Resources Conversion, 7*(4), 100228. DOI: 10.1016/j.crcon.2024.100228

Mohd Basri, M. S., Abdul Karim Shah, N. N., Sulaiman, A., Mohamed Amin Tawakkal, I. S., Mohd Nor, M. Z., Ariffin, S. H., Abdul Ghani, N. H., & Mohd Salleh, F. S. (2021). Progress in the valorization of fruit and vegetable wastes: Active packaging, biocomposites, by-products, and innovative technologies used for bioactive compound extraction. *Polymers, 13*(20), 3503. DOI: 10.3390/polym13203503 PMID: 34685262

Moosavi, E. S., Dastgheib, S. A., & Karimzadeh, R. (2012). Adsorption of thiophenic compounds from model diesel fuel using copper and nickel impregnated activated carbons. *Energies, 5*(10), 4233–4250. DOI: 10.3390/en5104233

Mubarik, S., Qureshi, N., Sattar, Z., Shaheen, A., Kalsoom, A., Imran, M., & Hanif, F. (2021). Synthetic approach to rice waste-derived carbon-based nanomaterials and their applications. *Nanomanufacturing, 1*(3), 109–159. DOI: 10.3390/nanomanufacturing1030010

Nahian, M. S., Pritom, R., & Islam, M. M. (2024). Mechanistic Insights into the Evolution of Graphitic Carbon from Sulfur Containing Polar Aromatic Feedstock. *ACS Applied Materials & Interfaces, 16*(41), 55478–55489. DOI: 10.1021/acsami.4c12324 PMID: 39356548

Naseer, A. (2024). Role of nanocomposites and nano adsorbents for heavy metals removal and dyes. An overview. *Desalination and Water Treatment, 320*, 100662. DOI: 10.1016/j.dwt.2024.100662

Nath, P. C., Ojha, A., Debnath, S., Sharma, M., Sridhar, K., Nayak, P. K., & Inbaraj, B. S. (2023). Biogeneration of valuable nanomaterials from agro-wastes: A comprehensive review. *Agronomy (Basel), 13*(2), 561. DOI: 10.3390/agronomy13020561

Nezafat, Z., Dong, Y., Nasrollahzadeh, M., Shafiei, N., Gharoubi, H., & Javanshir, S. (2024). Recent progresses in energy conversion and storage of agricultural waste-derived (carbon/nano) materials: A review. *Green Chemistry*, *26*(21), 10687–10717. DOI: 10.1039/D3GC04332K

Ng, S. J. C., & Choi, A. E. S. (2025). Advancements in New Desulfurization Technology for Industry Scale Applications. *Results in Engineering*, *27*, 106577. DOI: 10.1016/j.rineng.2025.106577

Norizan, M. N., Shazleen, S. S., Alias, A. H., Sabaruddin, F. A., Asyraf, M. R. M., Zainudin, E. S., Abdullah, N., Samsudin, M. S., Kamarudin, S. H., & Norrrahim, M. N. F. (2022). Nanocellulose-based nanocomposites for sustainable applications: A review. *Nanomaterials (Basel, Switzerland)*, *12*(19), 3483. DOI: 10.3390/nano12193483 PMID: 36234612

Oyedeji, S., Patel, N., Krishnamurthy, R., & Fatoba, P. O. (2024). Agricultural wastes to value-added products: economic and environmental perspectives for waste conversion. *Biowaste to Value-added Products: Economics and Technologies*, 215-248.

Ozcan, D. O., Hendekci, M. C., & Ovez, B. (2024). Enhancing the adsorption capacity of organic and inorganic pollutants onto impregnated olive stone derived activated carbon. *Heliyon*, *10*(12), e32792. DOI: 10.1016/j.heliyon.2024.e32792 PMID: 38975120

Qiu, X., Wang, B., Wang, R., & Kozhevnikov, I. V. (2024). New adsorption materials for deep desulfurization of fuel oil. *Materials (Basel)*, *17*(8), 1803. DOI: 10.3390/ma17081803 PMID: 38673161

Rajendran, S., Al-Samydai, A., Palani, G., Trilaksana, H., Sathish, T., Giri, J., Saravanan, R., Lalvani, J. I. J., & Nasri, F. (2025). Replacement of Petroleum Based Products With Plant-Based Materials, Green and Sustainable Energy—A Review. *Engineering Reports*, *7*(4), e70108. DOI: 10.1002/eng2.70108

Riseh, R. S., Vazvani, M. G., Hassanisaadi, M., & Thakur, V. K. (2024). Agricultural wastes: A practical and potential source for the isolation and preparation of cellulose and application in agriculture and different industries. *Industrial Crops and Products*, *208*, 117904. DOI: 10.1016/j.indcrop.2023.117904

Roy, A., Sharma, A., Yadav, S., Jule, L. T., & Krishnaraj, R. (2021). Nanomaterials for remediation of environmental pollutants. *Bioinorganic Chemistry and Applications*, *2021*(1), 1764647. PMID: 34992641

Sadegh-Vaziri, R., & Babler, M. U. (2019). Removal of hydrogen sulfide with metal oxides in packed bed reactors—A review from a modeling perspective with practical implications. *Applied Sciences (Basel, Switzerland), 9*(24), 5316. DOI: 10.3390/app9245316

Sadrara, M., & Khorrami, M. K. (2023). Designing an efficient organic–inorganic hybrid nanocomposite for simultaneous oxidative/adsorptive desulfurization of model and real fuel oils. *Scientific Reports, 13*(1), 15134. DOI: 10.1038/s41598-023-42392-8 PMID: 37704702

Saha, B., Vedachalam, S., & Dalai, A. K. (2021). Review on recent advances in adsorptive desulfurization. *Fuel Processing Technology, 214*, 106685. DOI: 10.1016/j.fuproc.2020.106685

Salazar Sandoval, S., Amenábar, A., Toledo, I., Silva, N., & Contreras, P. (2024). Advances in the sustainable development of biobased materials using plant and animal waste as raw materials: A review. *Sustainability (Basel), 16*(3), 1073. DOI: 10.3390/su16031073

Saleh, T. A., Sulaiman, K. O., Al-Hammadi, S. A., Dafalla, H., & Danmaliki, G. I. (2017). Adsorptive desulfurization of thiophene, benzothiophene and dibenzothiophene over activated carbon manganese oxide nanocomposite: With column system evaluation. *Journal of Cleaner Production, 154*, 401–412. DOI: 10.1016/j.jclepro.2017.03.169

Sankar, M. K., Parvez, M., Rinawa, M. L., Chaudhary, V., Gupta, S., & Gupta, P. (2022). The aspect of green nanocomposites in green technology and sustainable development: state of the art and new challenges. *Nanomaterials for Innovative Energy Systems and Devices*, 583-599.

Sedyaaw, P., Pandey, A., Kawade, S. S., Ade, P. J., Nimbarte, P. V., Bhaladhare, D. R., & Kamble, P. A. (2024). A review on acid rain it's causes, effects and management measures. [IJCRT]. *International Journal of Creative Research Thoughts, 12*(4), q959–q968.

Segovia-Hernández, J. G., Hernández, S., Cossío-Vargas, E., Juarez-García, M., & Sánchez-Ramírez, E. (2025). Green hydrogen production for sustainable development: A critical examination of barriers and strategic opportunities. *RSC Sustainability, 3*(1), 134–157. DOI: 10.1039/D4SU00630E

Serafin, J., Dziejarski, B., & Sreńscek-Nazzal, J. (2023). An innovative and environmentally friendly bioorganic synthesis of activated carbon based on olive stones and its potential application for CO2 capture. *Sustainable Materials and Technologies, 38*, e00717. DOI: 10.1016/j.susmat.2023.e00717

Sevilla, M., Díez, N., & Fuertes, A. B. (2021). More sustainable chemical activation strategies for the production of porous carbons. *ChemSusChem*, *14*(1), 94–117. DOI: 10.1002/cssc.202001838 PMID: 33047490

Shah, M. A., Hayder, G., Kumar, R., Kumar, V., Ahamad, T., Kalam, M. A., Soudagar, M. E. M., Mohamed Shamshuddin, S. Z., & Mubarak, N. M. (2023). Development of sustainable biomass residues for biofuels applications. *Scientific Reports*, *13*(1), 14248. DOI: 10.1038/s41598-023-41446-1 PMID: 37648719

Singh, H., Desimone, M. F., Pandya, S., Jasani, S., George, N., Adnan, M., Aldarhami, A., Bazaid, A. S., & Alderhami, S. A. (2023). Revisiting the green synthesis of nanoparticles: Uncovering influences of plant extracts as reducing agents for enhanced synthesis efficiency and its biomedical applications. *International Journal of Nanomedicine*, *18*, 4727–4750. DOI: 10.2147/IJN.S419369 PMID: 37621852

Singh, P., & Singh, J. S. (2025). Environmental Consequences of Crop Residue Burning and Its Threats to Urbanization.

Singh, Y., Sharma, S., Kumar, U., Sihag, P., Balyan, P., Singh, K. P., & Dhankher, O. P. (2024). Strategies for economic utilization of rice straw residues into value-added by-products and prevention of environmental pollution. *The Science of the Total Environment*, *906*, 167714. DOI: 10.1016/j.scitotenv.2023.167714 PMID: 37832665

Studart, A. R., Erb, R. M., & Libanori, R. (2015). Bioinspired hierarchical composites. In *Hybrid and Hierarchical composite materials* (pp. 287–318). Springer. DOI: 10.1007/978-3-319-12868-9_8

Sulis, D. B., Lavoine, N., Sederoff, H., Jiang, X., Marques, B. M., Lan, K., Cofre-Vega, C., Barrangou, R., & Wang, J. P. (2025). Advances in lignocellulosic feedstocks for bioenergy and bioproducts. *Nature Communications*, *16*(1), 1244. DOI: 10.1038/s41467-025-56472-y PMID: 39893176

Taiye, M. A., Hafida, W., Kong, F., & Zhou, C. (2024). A review of the use of rice husk silica as a sustainable alternative to traditional silica sources in various applications. *Environmental Progress & Sustainable Energy*, *43*(6), e14451. DOI: 10.1002/ep.14451

Thakur, A., Kumar, A., & Somya, A. (2024). Forestry and Agricultural Residues-Based Wastes. *Biomass Wastes for Sustainable Industrial Applications: A Waste-to-Wealth Approach*, 95.

TONGDANG. I., & Kiatkittipong, W. (2023). *THE CATALYTIC PYROLYSIS OF COCONUT SHELL OVER Ni-DOPED ACTIVATED CARBON CATALYST FOR AROMATIC-RICH BIO-OIL PRODUCTION* Silpakorn University].

Trivedi, Y., Sharma, M., Mishra, R. K., Sharma, A., Joshi, J., Gupta, A. B., Achintya, B., Shah, K., & Vuppaladadiyamd, A. K. (2025). Biochar potential for pollutant removal during wastewater treatment: A comprehensive review of separation mechanisms, technological integration, and process analysis. *Desalination, 600*, 118509. DOI: 10.1016/j.desal.2024.118509

Viell, J., Theis, S., Götz, R., Weber-Bernard, A., Echtermeyer, A. W. W., & Lindmeyer, M. (2025). Benefits and Challenges for Decentralized Use of Biomass as Feedstock for Chemicals. *Chemieingenieurtechnik (Weinheim), 97*(6), 00043. DOI: 10.1002/cite.202500043

Wang, Y., Wang, Z., Lin, Y., Qin, Y., He, R., Wang, M., Sun, Q., & Peng, Y. (2024). Nanocellulose from agro-industrial wastes: A review on sources, production, applications, and current challenges. *Food Research International, 192*, 114741. DOI: 10.1016/j.foodres.2024.114741 PMID: 39147548

Yaashikaa, P., Kumar, P. S., & Karishma, S. (2022). Review on biopolymers and composites–evolving material as adsorbents in removal of environmental pollutants. *Environmental Research, 212*, 113114. DOI: 10.1016/j.envres.2022.113114 PMID: 35331699

Yang, J., Wei, Q., Tian, C., Li, D., Li, H., Qin, G., Hu, K., & Zhang, Q. (2023). Preparation of biomass carbon composites MgO@ ZnO@ BC and its adsorption and removal of Cu (II) and Pb (II) in wastewater. *Molecules (Basel, Switzerland), 28*(19), 6982. DOI: 10.3390/molecules28196982 PMID: 37836824

Yang, Z., Zhang, X., Zhang, J., Liu, H., & Yu, X. (2023). Application of Biomass-Based Nanomaterials in Energy. *Advanced Energy and Sustainability Research, 4*(12), 2300141. DOI: 10.1002/aesr.202300141

Zaman, C. Z., Pal, K., Yehye, W. A., Sagadevan, S., Shah, S. T., Adebisi, G. A., Marliana, E., & Rafique, R. F. (2017). Pyrolysis: A sustainable way to generate energy. *Pyrolysis*, 1.

Zhang, H., Wang, Z., Luo, X., Lu, J., Peng, S., Wang, Y., & Han, L. (2020). Constructing hierarchical porous carbons with interconnected micro-mesopores for enhanced CO2 adsorption. *Frontiers in Chemistry, 7*, 919. DOI: 10.3389/fchem.2019.00919 PMID: 32010669

Zhang, T., Wang, Q., & Rui, Y. (2025). The impact of nanomaterials on plant health: A review of exposure, toxicity, and control. *Environmental Science. Nano, 12*(6), 2965–2982. DOI: 10.1039/D5EN00037H

Zhang, Y., Qu, M., Zhang, K., Liu, W., Wang, J., Ren, L., & Wang, F. (2025). Effect mechanisms of microwave on CO2 adsorption with cellulosic and non-cellulosic biochar. *Separation and Purification Technology, 373*, 133610. DOI: 10.1016/j.seppur.2025.133610

Zhu, J., Guo, Y., Chen, N., & Chen, B. (2024). A Review of the Efficient and Thermal Utilization of Biomass Waste. *Sustainability (2071-1050), 16*(21).

Chapter 6
Green and Sustainable Nanomaterials:
Focus on Bio-Derived Carbon Nanomaterials and Low-Temperature Synthesis Routes to Reduce Environmental Impact

Apoorav Sharma
https://orcid.org/0009-0002-9966-7691
Lovely Professional University, India

Lovleen Marwaha
Lovely Professional University, India

Zahid Ahmed Thokker
https://orcid.org/0009-0002-6897-5835
Lovely Professional University, India

ABSTRACT

Green and sustainable nanomaterials are the new hopeful solutions on how to lighten the environmental impact of the sophisticated fuel purification process. The topic of this chapter is bio-derived carbon nanomaterials and low-temperature synthesis paths as alternatives to the traditional energy-intensive nanomaterial fabrication. Agricultural waste, algae, and industrial biowaste are biomass-based precursors that can provide renewable, cheap, and scalable sources of carbon. There is a focus on hydrothermal and microwave-assisted, and low-temperature chemical vapor deposition technologies, which consume less energy and contribute to fewer dangerous emissions. The chapter is a critical discussion of physicochemical properties,

DOI: 10.4018/979-8-3373-5966-3.ch006

functionalization strategies, and integration with the nanocomposites to desulfurize fuel efficiently. Green nanotechnology is also considered in terms of environmental, economic, and regulatory perspectives to emphasize the viability of the industry.

1. INTRODUCTION

The emissions of sulfur in fossil fuels have been one of the main causes of atmospheric pollution that poses great threats to the ecological systems as well as the health of humans. Sulfur oxides are also formed during fuel combustion and cause the formation of acid rain, fine particulate matter, and secondary aerosol that causes respiratory and cardiovascular diseases (Babich and Moulijn, 2003; Srivastava, 2012). Counter to this, strict global fuel quality standards have been implemented to restrict the levels of sulfur to ultra-low contents usually less than 10-15 ppm and mostly of transportation fuels (Song, 2003). Such regulatory requirements have aggravated the need to have highly efficient desulfurization processes that can eliminate refractory sulfur compounds like dibenzothiophene and alkylated derivatives of the same. The traditional hydrodesulfurization that is primarily used industrially works under harsh environments with high temperature and pressure, and needs huge amounts of hydrogen; therefore, it is energy consuming and economically challenging (Prins et al., 2006). More importantly, hydrodesulfurization does not have a high level of efficiency against the sterically hindered sulfur compounds, and thus, it requires further post-treatment processes. Other emerging technologies that include oxidative, extractive, and adsorptive desulfurization have been considered as alternative processes because of mild conditions of operation and high selectivity; nonetheless, the technologies rely heavily on the nanomaterials with designed surface characteristics (Rafiee et al., 2018; Li et al., 2016). The sustainability of such nanomaterials in the environment is, however, a critical issue since most of the synthesis processes have been associated with the use of harmful chemicals, large quantities of thermal energy and massive carbon release. The concept of the circular economy connecting biomass waste valorization, the production of eco-friendly nanomaterials, and the production of cleaner energy is also identified in Fig.1 below.

Figure 1. Schematic overview illustrating the transition from sulfur-polluting fossil fuels to green fuel desulfurization technologies enabled by biomass-derived carbon nanomaterials and low-temperature synthesis route

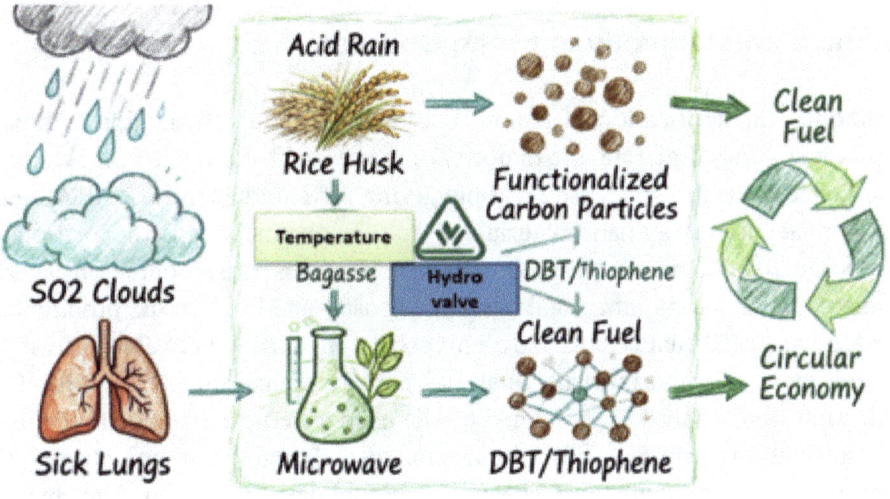

Emergence of Green Nanotechnology in Energy Applications

Green nanotechnology has become a strategic direction in order to overcome the environmental constraints of the traditional nanomaterials synthesis related to the promotion of renewable precursors, safer solvents, reduced reaction temperature, and waste-reduction solutions (Iravani, 2011; Dahlia et al., 2021). Its use in energy-related technologies has also grown fast in the past few years, with uses in catalysis, fuel purification, energy storage, and environmental remediation. In the fuel desulfurization context, green nanotechnology is a direction of the shift that is propelled by the necessity to harmonize high-performance sulfur removal with manufacturing paths without an impact on the climate. The most promising result of this transition is bio-derived carbon nanomaterials. These are produced using agricultural residues, algal biomass, and industrial biowaste as sustainable feedstocks to provide renewable, inexpensive, and scalable fossil-based carbons (Sevilla & Fuertes, 2011; Jain et al., 2016). Carbons obtained in biomass have a large surface area, controllable porosity, and a variety of surface functional groups, which qualify them as being especially favorable as adsorbents in an adsorptive desulfurization run. Simultaneously, synthesis methods with low temperatures, like hydrothermal carbonization, microwave-assisted, and plasma-assisted deposition, have greatly lowered the energy consumption and greenhouse gas emissions linked to the synthesis of nanomaterials (Zhang et al., 2018). In combination with low-energy synthesis and sustainable selection of feedstock, green nanotechnology has been placed as one of the enabling tools of green fuel purification technologies.

Rationale and Objectives of the Chapter

Although the application of nanotechnology-based desulfurization techniques has grown at a very high rate, the majority of the research conducted focuses on the performance of the materials used without giving sufficient attention to the environmental impact of making nanomaterials. Pyrolysis, chemical vapor deposition, and strong acid activation remain the predominant approaches in the production of carbon nanomaterials on a laboratory scale, which is costly and leads to the production of hazardous waste (Kumar et al., 2020). This escalating tension between performance and ecology is an important sustainability paradox concerning the study of fuel purification. This chapter is explained by the need to redress this disproportion by drawing attention to bio-derived carbon nanomaterials and low-temperature synthesis pathways as combined approaches to green fuel desulfurization. The chapter is organized based on the necessity to uncouple the concept of material efficiency and environmental degradation, and sustain industrial scalability and economic viability. These are the critical examination of renewable biomass-based carbon nanomaterials as a sustainable adsorbent and catalyst carrier, analysis of low-energy synthesis methods that are environmentally friendly, assessment of green functionalization used to enhance a desulfurization process, and determination of the techno-economic, environmental, and regulatory feasibility of using these green nanocomposites in practical desulfurization systems. This chapter will offer an overarching approach to next-generation green desulfurization technologies by combining performance indicators with the concept of life-cycle sustainability.

2. FUNDAMENTALS OF GREEN AND SUSTAINABLE NANOMATERIALS

Concept and Principles of Green Nanomaterials

Green nanomaterials can be described as nanostructured materials prepared, fabricated, and put to use by means of processes that, to a significant degree, lower or remove the usage and generation of hazardous substances as well as reduce energy usage and environmental impact. Green nanotechnology is conceptually based on the principles of green chemistry and green engineering, including the focus on the prevention of waste, the economy of atoms, the use of safer solvents, renewable feedstock, and energy efficiency (Anastas and Warner, 1998; Iravani et al., 2014). Regarding nanomaterials, the principles are applicable throughout the entire life cycle, including the choice of precursors and synthesis pathways to the application, reuse, and disposal of products at the end of their life. At the nano scale, surface reactivity, quantum confinement, high surface to volume ratios, and other material characteristics allow the material to have superior catalytic, adsorptive, and electronic output relative to bulk materials. Nevertheless, the very qualities may also increase the intensity of interactions between the environment and biology, which predetermines the necessity of sustainable design (Klaine et al., 2012). Green nanomaterials consequently incorporate performance and environmental accountability. It is done by using non-toxic solvents, biogenic reducing agents, synthesis at room temperature, and biodegradable or recyclable nanostructures. Green carbon nanomaterials are low examples of this philosophy in fuel desulfurization, providing high sulfur adsorption efficiency without using fossil-derived precursors or undergoing extreme conditions.

Sustainability Metrics for Nanomaterial Synthesis

Nanomaterials synthesis should be assessed with consideration of environmental, economic, and social performance elements to be considered among sustainability indicators. The assessment of the environmental footprint of nanomaterials based on life cycle assessment (LCA) has become the most popular quantitative tool that takes into consideration energy input of a product, greenhouse gas emissions, water consumption, toxicity potential, and waste (Finkbeiner et al., 2010; Upadhyayula et al., 2012). In the traditional methods of nanomaterials fabrication, like chemical vapor deposition, arc discharge, and high-temperature pyrolysis, LCA analyses have all indicated unreasonable energy usage and carbon emission. Along with environmental-related metrics, the techno-economic assessment parameters (that is, production yield, scalability, feedstock cost, and process integration potential) are also an important part of the industrial feasibility (Piccinno et al., 2012). The sustainability framework is further expanded with social sustainability issues, such as occupational safety and risks to the community exposure. Multi-criteria decision analysis has recently been used to compare green synthesis routes to conventional processes on the basis of combined environmental and economic indicators (Arvidsson et al., 2014). In the case of green nanomaterials, biomass-based feedstocks and low-temperature synthesis pathways are both shown to be much better in sustainability performance. Research indicates that hydrothermal and microwave-assisted technology consumes 40-70 per cent less energy than traditional high-temperature carbonization, and also lessens solvent toxicity and the formation of solid waste (Jain et al., 2016; Zhang et al., 2018). Such quantitative sustainability benefits render the concept of green nanomaterials especially desirable to be implemented on a large scale in fuel purification systems.

Environmental and Health Implications of Conventional Nanomaterials

Although the traditional nanomaterials have technological advantages, their environmental and health impacts during their life cycle are dangerous. Strong acids, organic solvents, high-temperature furnaces, and other substances during synthesis yield dangerous emissions and chemically polluted effluents (Kumar et al., 2020). Nanoparticles can be emitted to air, soil, and water during use and disposal, where they are highly reactive, causing ecotoxicological concerns (Klaine et al., 2012; Nel et al., 2006). Nanoparticles may enter cellular membranes, cause oxidative stress, impair the work of mitochondria, and cause inflammatory reactions at the biological level (Oberdorster et al., 2005). Nanomaterials, especially metals, have the potential of cytotoxicity, genotoxicity, and bioaccumulation hazards. It has also been reported that carbon nanomaterials that have been synthesized through harsh chemical activation do harbor remnants of metal contaminants and surface defects, all of which enhance toxicity (Aschberger et al., 2010). Regulatively, the environmental emission of engineered nanomaterials is challenging to govern as the data on long-term toxicity is inadequate, standard test protocols are not in place, and post-deployment surveillance mechanisms are poor. These issues raise more obstacles to regulatory acceptance and general acceptance of nanotechnological fuel purification systems. In turn, increased nanoparticle toxicity reduction by green nanotechnology and bio-compatible surface engineering has emerged as a key goal of sustainable nanomaterials.

Advantages of Bio-Derived and Low-Temperature Routes

The solution to environmental and health constraints of the conventional methods of fabrication of nanomaterials lies in bio-derived and low-temperature synthesis routes. Carbon sources derived through agricultural residues, forestry wastes, algae, and food processing by-products can be biomass-sourced carbons that have a lower carbon footprint (Sevilla & Fuertes, 2011; Jain et al., 2016). These feedstocks are abundant, cheap, and they are widely available, hence they are especially useful in a scalable fuel desulfurization. Hydrothermal carbonization, microwave-assisted synthesis, and low-energy sol-gel processing are low-temperature synthesis methods that have been shown to considerably lower thermal energy requirements and do not demand extreme reaction conditions (Zhang et al., 2018). These pathways allow the growth of nanostructures in controlled amounts with temperatures typically below

250 °C- 300 °C as opposed to over 800 °C in standard carbonization. Besides saving energy, the design of reactors is also made easier, safer in terms of protecting the process, and it is also scalable to increase production. Functionally speaking, bio-derived carbon nanomaterials have dense surface oxygen and nitrogen functionalities that contribute to sulfur adsorption by π-π, hydrogen bonding, and Lewis acid-base interactions. Their porosity and density of defects also enhance active site accessibility and hence their special reliance on adsorptive fuel desulfurization (Li et al., 2016; Rafiee et al., 2018). In addition, their lower toxicity, biodegradability, and recyclability have great benefits in regulatory compliance and environmental protection in the long term.

3. BIO-DERIVED CARBON NANOMATERIALS: RENEWABLE PRECURSORS AND SOURCES

Agricultural Waste as Carbon Feedstock

Agricultural residues have been classified as one of the renewable precursors that have been most exploited to produce bio-derived carbon nanomaterials because of their availability on the earth, low cost, and richness in carbon. Rice husk, wheat straw, sugarcane bagasse, coconut shell, corn cobs, and fruit peels are some of the wastes generated in tremendous amounts across the globe and normally present a challenge of disposal and pollution when not managed. Their conversion to carbon nanomaterials is both carbon-neutral and offers a waste management solution, and at the same time produces functional materials of high value (Jain et al., 2016; Ioannidou and Zabaniotou, 2007). These lignocellulosic biomasses contain a lot of cellulose, hemicellulose, and lignin, which break down under controlled thermal or hydrothermal processes to generate porous carbon structures. Carbon produced in agricultural wastes is often well-developed as microporous and mesoporous, and is thus very useful in adsorption-based processes, including fuel desulfurization. Further improvement of pore volume and surface functionality is done by chemical activation with environmentally friendly activating agents or hydrothermal treatments. Studies have shown that rice husk- and bagasse-derived carbons can achieve surface areas exceeding 1000 $m^2\ g^{-1}$, with abundant oxygen-containing functional groups that promote sulfur adsorption through π–π interactions and polar binding mechanisms (Li et al., 2016; Rafiee et al., 2018).

Algal Biomass-Derived Carbon Nanomaterials

The algal biomass has become quite a promising third-generation bio-feedstock in carbon nanomaterials creation owing to its high growth, high carbon-fixation efficiency, and low land and freshwater demand. Microalgae and macroalgae species contain lipids, proteins, and carbohydrates, which allow their transformation into a variety of carbon nanostructures, including carbon nanosheets, porous carbons, carbon dots, and graphene-like structures (Deng et al., 2015; Sevilla et al., 2017). In contrast to land biomass, algal cultivation will not interfere with food production but can be combined with the wastewater treatment and CO 2 capture, and this approach will improve the environmental sustainability. Carbons produced by algae have distinctive structural features, such as hierarchical porosity, nitrogen self-doping, and high defect density, which have a significant positive impact on adsorption behavior and surface reactivity. Amino functional groups that are added by algal proteins enhance the Lewis basicity and facilitate selective reactions with sulfur-based products (Sevilla et al., 2017). Also, algal precursors permit the synthesis of highly graphitized structures at relatively mild conditions and have therefore been of interest in the energy-efficient production of nanomaterials. All these benefits make algal biomass an important renewable source of the next-generation green carbon nanomaterials.

Industrial and Municipal Biowaste Resources

Industrial and urban biowastes are a new and extremely influential category of renewable biomasses for bio-derived carbon nanomaterials. The waste streams referred to food processing wastes, sewage sludge, paper mill sludge, textile waste, and plastic-contaminated organic waste contain a significant amount of organic carbon that can be readily converted into functional carbons by using thermochemical and hydrothermal processes (Zhang et al., 2017; Wei et al., 2020). The solution of these toxic wastes into high-value nanomaterials has a twofold environmental advantage; it reduces the waste disposal difficulties, and at the same time, it produces high-quality functional materials that are sustainable. Carbons obtained by sewage sludge, e.g., have intrinsic heteroatom doping (N, S, P) due to biological proteins and inorganic salts, which greatly increases adsorption and catalytic activity. Likewise, the carbons derived by the hydrolysis of food waste have been shown to possess high surface area and adjustable pore structures that can be used to tackle pollution control and subsequently utilized as energy (Wei et al., 2020). Nevertheless, the non-homogeneous character of industrial and municipal biowastes presents the issues of variability in compositions, heavy metals contamination, and standardization of the process. However, given the right pretreatment and purification methods, these waste materials have untapped potential for nanomaterial production based on the circular economy.

Physicochemical Characteristics of Bio-Derived Carbon Nanomaterials

The nature of the precursor, synthesis pathway, and activation are the key factors that strongly dictate the physicochemical properties of bio-derived carbon nanomaterials. They are usually characterized by high specific surface area, tunable micro- and mesoporous framework, large amounts of surface functional groups, and tunable crystallinity (Sevilla & Fuertes, 2011; Jain et al., 2016). Functional groups that include oxygen and nitrogen (hydroxyl, carboxyl, carbonyl, pyridinic-N, graphitic-N, etc.) impact on the adsorption affinity and surface polarity. Structurally, bio-derived carbons tend to have a partially graphitized structure with high defect density, thereby giving bio-derived carbons a high density of active sites to perform adsorption and catalytic reactions. These characteristics are especially beneficial in the work of fuel desulfurization, during which the interaction of sulfur and carbon is determined by surface heterogeneity and π-electron density (Li et al., 2016). Also, the distribution of pore sizes is highly important in the availability of large molecules of sulfur like dibenzothiophene and alkylated variants. The hierarchical porosity comprised of micro-, mes-, and macropores guarantees the high adsorption capacity and fast kinetics of mass transfer, thus enhancing the total desulfurization.

Economic and Scalability Aspects

Economically and industrially, bio-derived carbon nanomaterials have great cost benefits compared to fossil-derived and synthetic carbon nanostructures. Waste biomass is a cost-effective method of decreasing raw materials, in addition to value-adding low- or negative-value waste materials (Ioannidou and Zabaniotou, 2007; Jain et al., 2016). Moreover, low temperature synthesis methods like hydrothermal carbonization and microwave-assisted processing require minimum consumption of energy and minimal operating costs than the traditional high temperature carbonization and chemical vapor deposition pathways. The simplified nature of biomass-derived carbons synthesis protocols and the ability to use continuous-flow reactor systems also makes them scalable. Hydrothermal and activation-based carbon production already demonstrated with pilot-scale demonstrations has shown a high level of feasibility to be deployed commercially. Nevertheless, issues concerning feedstock variability, seasonal changes, impurity control, and reproducibility of the processes continue to hamper its complete industrialization. Standardized pre-processing, quality control and optimization of reactor design will be very important in overcoming these limitations to commercially translate bio-derived carbon nanomaterials in large-scale fuel desulfurization systems. It is also indicated in Table 1 below that there is a wide variety of renewable biomass feedstocks in producing bio-derived carbon nanomaterials as well as the carbon structure, surface functionality, and applicability to fuel desulfurization.

Table 1. Renewable Biomass Feedstocks for Bio-Derived Carbon Nanomaterials and Their Key Properties

Biomass Source	Major Origin	Dominant Carbon Structure	Surface Area Potential ($m^2 g^{-1}$)	Key Functional Groups	Primary Advantages	Desulfurization Relevance
Rice husk	Agricultural waste	Amorphous + silica-rich carbon	600–1200	–OH, –COOH, Si–O	High porosity, low cost	Excellent adsorption of thiophenes due to high surface area and surface polarity
Sugarcane bagasse	Agro-industrial waste	Microporous carbon	800–1500	–COOH, –C=O	Abundant, renewable, easy activation	High sulfur uptake capacity in liquid fuels after activation/ impregnation
Coconut shell	Plantation waste	Highly graphitized carbon	900–1600	–OH, aromatic C=C	High mechanical strength, low ash, good regenerability	Reusable adsorbent for thiophenic sulfur (thiophene, DBT) in fuels
Algal biomass	Aquatic bioresource	N-doped porous carbon	700–1400	$-NH_2$, pyridinic N, graphitic N	Self-doping, good CO_2 capture, tunable micro/ mesoporosity	High sulfur selectivity via heteroatom sites and π-complexation on doped carbons
Food waste	Municipal biowaste	Heterogeneous porous carbon	500–1300	–OH, –NH, –CO	Waste valorization, flexible feedstock, low cost	Dual waste management and sulfur removal in fuels and wastewater
Sewage sludge	Urban biowaste	Metal-embedded carbon matrix	400–900	N, S, P dopants; metal species (Fe, Ca, etc.)	Inherent heteroatom and metal doping, catalytic activity	Catalytic sulfur oxidation and adsorptive desulfurization through metal and heteroatom sites

4. LOW-TEMPERATURE AND ENERGY-EFFICIENT SYNTHESIS ROUTES

Limitations of High-Temperature Conventional Methods

Conventional synthesis routes for carbon nanomaterials, including high-temperature pyrolysis, arc discharge, laser ablation, and traditional chemical vapor deposition (CVD), typically operate at temperatures ranging from 700 to over 1200

°C. These methods demand substantial energy input, making them inherently carbon-intensive and economically burdensome (Kumar et al., 2020; Zhang et al., 2018). In addition to excessive energy consumption, such processes frequently require inert atmospheres, flammable reducing gases, and expensive reactor-grade materials, increasing both operational complexity and safety risks. From an environmental perspective, high-temperature processes generate large quantities of greenhouse gas emissions and often involve corrosive activating agents such as KOH, $ZnCl_2$, and strong acids, leading to toxic effluents and secondary waste streams (Ioannidou & Zabaniotou, 2007). Furthermore, these aggressive conditions promote structural over-carbonization and pore collapse, which can reduce adsorption efficiency for bulky sulfur molecules. The ecological burden of these conventional methods significantly undermines the sustainability objectives of nanotechnology-enabled fuel desulfurization. These limitations have driven the shift toward low-temperature and energy-efficient synthesis strategies capable of producing high-performance carbon nanomaterials with a lower environmental footprint.

Hydrothermal and Solvothermal Synthesis Techniques

Hydrothermal and solvothermal synthesis represent cornerstone low-temperature routes for producing bio-derived carbon nanomaterials. These methods involve the transformation of biomass precursors in subcritical water or organic solvents at temperatures typically between 150 and 300 °C under autogenous pressure (Jain et al., 2016; Titirici & Antonietti, 2010). Unlike conventional carbonization, hydrothermal processes enable direct conversion of wet biomass without requiring energy-intensive drying steps, making them particularly energy efficient. Hydrothermal carbonization promotes dehydration, decarboxylation, aromatization, and polymerization reactions that yield carbon-rich hydrochar with tunable porosity and oxygen-containing surface functionalities. These surface oxygen groups enhance the polarity and adsorption selectivity toward sulfur-containing molecules during fuel desulfurization. Solvothermal routes further expand structural control by adjusting solvent polarity, precursor solubility, and reaction kinetics. These processes also allow one-step heteroatom doping using nitrogen- or sulfur-containing solvents. From a sustainability perspective, hydrothermal and solvothermal synthesis significantly reduce energy demand, eliminate harsh chemical activators, and simplify waste management. Life cycle studies have demonstrated that hydrothermal processing can reduce net energy consumption by more than 50% compared with conventional high-temperature carbonization (Jain et al., 2016). These attributes make hydrothermal and solvothermal routes among the most industrially viable green synthesis techniques for bio-derived carbon nanomaterials.

Microwave-Assisted Green Synthesis

Microwave-assisted synthesis has emerged as a highly efficient green processing technique for carbon nanomaterial fabrication. In contrast to conventional heating, which relies on thermal conduction and convection, microwave irradiation enables rapid volumetric heating through dipolar polarization and ionic conduction mechanisms (Kappe, 2004). This results in dramatically reduced reaction times, often from several hours to a few minutes, while maintaining precise temperature control. For biomass-derived carbons, microwave heating accelerates dehydration, bond cleavage, and carbonization reactions at significantly lower apparent temperatures than conventional furnaces. Studies have shown that microwave-assisted carbonization produces highly porous carbon structures with well-distributed micro- and mesopores, high surface areas, and uniform heteroatom doping (Zhang et al., 2018). Additionally, microwave synthesis minimizes thermal gradients, improves product homogeneity, and significantly lowers electricity consumption. From an environmental standpoint, microwave-assisted methods reduce both energy input and thermal emissions, while eliminating prolonged high-temperature exposure. These advantages make microwave processing particularly suitable for rapid-scale production of adsorptive carbons for fuel purification. The main constraints remain reactor scalability and uniform field distribution at industrial scales, although recent continuous-flow microwave reactor developments are progressively addressing these limitations.

Reduced-Temperature Chemical Vapor Deposition

Chemical vapor deposition is traditionally associated with ultra-high temperatures required for growing carbon nanotubes, graphene, and related nanostructures. However, recent advances in catalyst engineering, plasma enhancement, and precursor optimization have enabled reduced-temperature CVD routes operating below 500–600 °C (Zhang et al., 2017). Plasma-enhanced CVD (PECVD), in particular, supplies additional energy through ionized gas species, thereby lowering the thermal budget required for carbon nucleation and growth. Reduced-temperature CVD offers several sustainability advantages, including lower power consumption, reduced thermal stress on substrates, enhanced material uniformity, and improved defect control. When combined with bio-derived gaseous precursors such as ethanol, methane derived from biogas, or biomass pyrolysis vapors, reduced-temperature CVD enables partially renewable pathways for producing high-quality carbon nanostructures. Despite these advantages, CVD processes still require vacuum systems, precise gas-flow control, and catalyst preparation, introducing higher capital costs relative to hydrothermal and microwave routes. Nevertheless, for applications requiring graphitic precision, such as sulfur-selective catalytic nanocomposites, reduced-temperature CVD represents a promising compromise between material quality and sustainability.

Comparative Energy and Emission Analysis

Comparative energy and emissions analyses consistently demonstrate that low-temperature synthesis routes outperform conventional high-temperature methods in sustainability metrics. Life cycle assessment studies indicate that hydrothermal and microwave-assisted carbonization can reduce total energy consumption by 40–70% and greenhouse gas emissions by more than 50% relative to conventional furnace-based carbonization and CVD processes (Arvidsson et al., 2014; Upadhyayula et al., 2012). Hydrothermal routes exhibit particularly favorable energy balances due to direct wet biomass processing, elimination of drying steps, and lower operational temperatures. Microwave-assisted routes further reduce processing times and electrical energy demand. In contrast, traditional high-temperature activation and CVD routes remain dominated by fossil-derived electricity and large heat losses. From a fuel desulfurization perspective, these sustainability advantages translate into a lower embedded carbon footprint for adsorbents and catalysts, significantly improving the environmental performance of entire fuel purification chains. As regulatory pressure increases on both fuel composition and industrial emissions, the integration of low-temperature and energy-efficient nanomaterial synthesis routes will be indispensable for the long-term commercialization of green desulfurization technologies.

Table 2. Comparison of Conventional and Green Low-Temperature Synthesis Routes for Carbon Nanomaterials

Synthesis Method	Typical Temperature (°C)	Energy Demand	Chemical Usage	Carbon Footprint	Scalability	Environmental Impact	Suitability for Green Desulfurization
High-temperatureare pyrolysis	800–1200	Very high	KOH, $ZnCl_2$, and H_2SO_4 commonly used as activating agents	Very high due to high heat input and flue gas emissions	High (mature technology, large reactors)	Toxic effluents and CO_2-intensive operation	Not suitable for green desulfurization because of the energy and emission burdens
Conventional CVD	700–1000	Very high (continuous high-temperature operation)	CH_4, H_2, metal catalysts (Ni, Fe, Co, etc.)	Very high due to fossil carbon precursors and energy use	Medium (capital-intensive, complex control)	High greenhouse gas emissions and catalyst waste	Limited suitability; mainly for high-end nanocarbons, not green fuels
Hydrothermal carbonization	180–250	Low (sub-300 °C, autogenous pressure)	Water only; no strong acids/ alkalis required	Low relative carbon footprint; potential net GHG reduction when integrated with waste valorization	High (slurry processing, compatible with wet biomass)	Minimal direct air emissions; liquid effluent requires management but is treatable	Excellent for green desulfurization adsorbents from wet biowastes
Microwave-assisted synthesis	150–300	Very low owing to rapid, volumetric heating and short residence times	Mild activators (e.g., low-dose KOH, $ZnCl_2$ or bio-activators)	Low process carbon footprint due to reduced energy consumption	Medium; scalable with modular microwave reactors	Low process emissions and smaller equipment; careful shielding needed	Excellent; energy-efficient route to high-surface-area carbons for desulfurization
Reduced-temperature CVD (PECVD)	400–600	Moderate (lower set-point than conventional CVD)	Bio-methane or diluted CH_4 plus catalysts	Moderate; reduced energy use and potential for biogenic carbon feedstocks	Medium (specialized equipment, but industry-proven)	Reduced direct emissions compared with conventional CVD	Good option where thin carbon coatings or nanostructures are needed for sulfur-selective surfaces
Solvothermal synthesis	150–300	Low (closed reactors, moderate temperatures)	Green or low-toxicity solvents and soft templates	Low when using benign solvents and renewable precursors	Medium (batch processes, scalable with reactor volume)	Generally eco-compatible; solvent recovery is important	Good for tailored porous carbons for green desulfurization and carbon capture

5. FUNCTIONALIZATION AND SURFACE ENGINEERING OF GREEN CARBON NANOMATERIALS

Need for Functionalization in Desulfurization Applications

While bio-derived carbon nanomaterials inherently possess high surface area, hierarchical porosity, and defect-rich frameworks, their native surfaces alone are often insufficient for achieving high selectivity and ultra-deep sulfur removal from complex fuel matrices. Sulfur compounds such as thiophene, benzothiophene, and dibenzothiophene exhibit different steric and electronic characteristics, requiring tailored surface chemistries for effective adsorption (Li et al., 2016; Rafiee et al., 2018). Functionalization is therefore essential to introduce specific chemical moieties that enhance interaction strength, selectivity, and adsorption kinetics. Surface functional groups such as hydroxyl, carboxyl, amine, and heteroatom-doped sites (N, S, P, B) act as active centers for sulfur binding through hydrogen bonding, Lewis acid–base interactions, and π–π stacking. In adsorptive desulfurization, nitrogen-doped carbons display enhanced affinity toward electron-deficient sulfur compounds, while metal-functionalized carbons provide additional coordination and redox-active sites (Zhao et al., 2018). Without surface engineering, even highly porous carbons may exhibit limited adsorption efficiency under realistic desulfurization conditions. Thus, functionalization serves as the critical link between green material synthesis and high-performance sulfur removal. Fig.2 below also illustrates key surface modifications (–OH, –COOH, $-NH_2$, N-doping, metal nanoparticles) and their interaction mechanisms with sulfur compounds (thiophene, DBT) through hydrogen bonding, π–π stacking, and coordination, alongside eco-friendly functionalization routes and adsorption-regeneration cycles

Figure 2. Schematic of green functionalization enhancing biomass-derived carbon desulfurization selectivity and reusability.

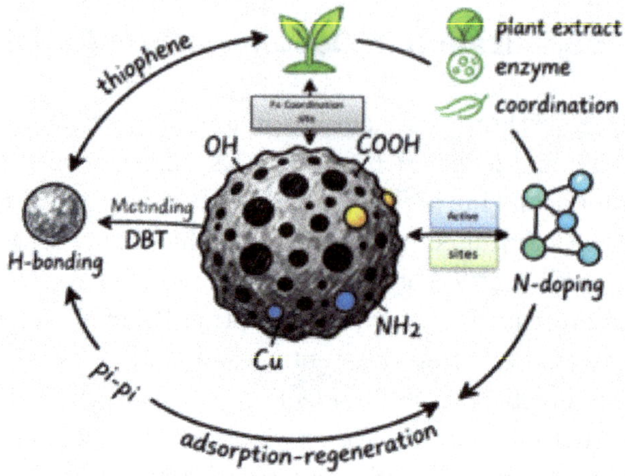

Eco-Friendly Chemical and Biological Functionalization Strategies

Conventional carbon functionalization methods often rely on harsh oxidants such as nitric acid, sulfuric acid, and potassium permanganate, which generate large volumes of acidic effluents and compromise environmental safety. In contrast, eco-friendly functionalization strategies emphasize mild oxidants, enzymatic processes, plant-derived reducing agents, and biological doping to achieve surface modification under benign conditions (Iravani, 2011; Anastas & Warner, 1998). Green chemical functionalization includes treatments using hydrogen peroxide, citric acid, ascorbic acid, and amino acids to introduce oxygen- and nitrogen-containing groups without producing toxic waste. Biomolecule-assisted doping using proteins, urea, chitosan, and plant extracts has emerged as a powerful strategy for nitrogen-rich functionalization of biomass-derived carbons (Sevilla et al., 2017). These biogenic dopants promote electron donation, improve surface charge density, and enhance sulfur affinity. Biological functionalization through microbial processing and enzymatic treatment further offers a low-energy route for introducing surface heteroatoms and defects. Certain microbial species have been shown to modify biomass surfaces prior to carbonization, enhancing pore development and active site density (Wei et al., 2020). These biological approaches significantly reduce chemical consumption and improve occupational safety while preserving high desulfurization efficiency

Surface Modification for Enhanced Adsorption Selectivity

Adsorption selectivity toward sulfur compounds depends not only on surface area but also on the electronic structure and surface polarity of carbon nanomaterials. Nitrogen-doped carbons exhibit superior selectivity due to enhanced basicity and strong electron-donating ability, which strengthens interactions with electron-deficient sulfur species such as dibenzothiophene (Zhao et al., 2018; Li et al., 2016). Sulfur- and phosphorus-doped carbons further introduce redox-active sites that facilitate charge-transfer interactions during adsorption. Metal and metal oxide decoration of green carbon supports is another highly effective surface engineering strategy. Eco-compatible deposition of iron, copper, zinc oxide, and cerium oxide nanoparticles creates synergistic adsorption–catalytic interfaces that dramatically improve sulfur uptake through coordination bonding and surface oxidation (Rafiee et al., 2018). Importantly, when bio-derived carbon supports are used, the environmental burden of metal loading is significantly reduced compared with conventional carbon nanostructures.

Stability, Reusability, and Regeneration Performance

For industrial desulfurization applications, long-term stability, reusability, and regeneration capacity are as critical as initial adsorption performance. Bio-derived carbon nanomaterials functionalized through green routes have demonstrated excellent structural stability under repeated adsorption–desorption cycles due to their robust carbon frameworks and chemically anchored surface groups (Jain et al., 2016; Rafiee et al., 2018). Regeneration is typically achieved through low-temperature thermal treatment, solvent washing, or mild oxidative desorption. Nitrogen- and metal-functionalized carbons generally maintain over 80–90% of their original sulfur adsorption capacity after five to ten regeneration cycles, indicating strong resistance to pore blockage and functional group degradation. The mild regeneration conditions required for green carbons further reduce energy input and material loss compared with conventional high-temperature reactivation of commercial activated carbons. From a sustainability perspective, stable reusability directly lowers material consumption, operational expenditure, and life-cycle environmental impact. Moreover, the biodegradability and low toxicity of bio-derived carbon platforms significantly reduce risks associated with material disposal after end-of-life.

6. GREEN CARBON-BASED NANOCOMPOSITES FOR FUEL DESULFURIZATION

Role of Nanocomposites in Advanced Desulfurization

Single-component carbon nanomaterials, despite their high surface area and tunable porosity, often exhibit limited sulfur selectivity and moderate adsorption capacity when exposed to complex real-fuel matrices containing competing aromatic hydrocarbons. To overcome these intrinsic limitations, green carbon-based nanocomposites have emerged as highly efficient platforms for advanced desulfurization. In these systems, bio-derived carbon acts as a sustainable support matrix, while secondary functional components such as metal nanoparticles, metal oxides, or polymers impart specific catalytic, redox, or selective adsorption properties (Rafiee et al., 2018; Li et al., 2016). Nanocomposite architectures enable synergistic interactions between the carbon framework and the functional phase, enhancing sulfur removal efficiency through cooperative adsorption–oxidation or adsorption–coordination mechanisms. Moreover, the dispersion of active phases on porous bio-carbons prevents particle agglomeration, increases the density of accessible active sites, and improves mass transfer kinetics. From a sustainability perspective, the use of renewable carbon substantially reduces the overall environmental footprint of nanocomposite synthesis while retaining high desulfurization performance.

Metal and Metal Oxide–Carbon Green Nanocomposites

Metal- and metal oxide–decorated green carbon nanocomposites represent the most widely investigated class of advanced desulfurization materials. Eco-compatible deposition of metals such as Fe, Cu, Zn, Ag, and Ce onto biomass-derived carbons generates highly active adsorption–catalytic interfaces for sulfur capture (Rafiee et al., 2018; Zhao et al., 2018). These metals function either as Lewis acid sites for sulfur coordination or as redox centers for oxidative transformation of sulfur compounds into sulfoxides and sulfones. Metal oxide–carbon nanocomposites, particularly ZnO, TiO_2, CeO_2, and Fe_3O_4 supported on bio-carbons, demonstrate superior selectivity toward thiophenic sulfur compounds due to strong metal–sulfur interactions and surface oxygen mobility. Magnetite (Fe_3O_4)–carbon nanocomposites additionally offer magnetic separation capability, enabling easy recovery and regeneration of spent adsorbents without filtration (Li et al., 2016). When synthesized through low-temperature impregnation and green reduction routes, these nanocomposites maintain a low carbon footprint while delivering adsorption capacities comparable to or exceeding those of conventional petroleum-derived carbons.

Polymer–Carbon Sustainable Nanocomposites

Polymer–carbon sustainable nanocomposites constitute another important category of green desulfurization materials. In these systems, bio-derived carbons are embedded within biodegradable or low-toxicity polymer matrices such as chitosan, polyaniline, polypyrrole, cellulose derivatives, and alginate-based networks (Wei et al., 2020; Zhao et al., 2018). The polymer component enhances mechanical strength, processability, and surface functional group density while preventing carbon particle dispersion losses. Conducting polymer–carbon nanocomposites, particularly polyaniline–carbon hybrids, exhibit enhanced sulfur adsorption due to combined π–π interactions, protonation-induced electrostatic attraction, and redox activity. Chitosan-based nanocomposites introduce abundant amine and hydroxyl groups that serve as strong coordination sites for sulfur species. These polymer–carbon platforms enable flexible shaping into beads, membranes, or monolithic adsorbents compatible with fixed-bed desulfurization reactors. From a sustainability standpoint, polymer–carbon green nanocomposites benefit from low-temperature fabrication, aqueous processing, and partial biodegradability. Their easy regeneration through mild solvent washing or low-temperature thermal treatment further reduces life-cycle energy demand. These properties make polymer–carbon nanocomposites particularly suitable for decentralized and modular fuel purification systems.

Mechanistic Pathways of Sulfur Removal

The sulfur removal performance of green carbon-based nanocomposites is governed by multiple synergistic mechanistic pathways. In pure adsorptive systems, sulfur compounds such as thiophene, benzothiophene, and dibenzothiophene interact with surface functional groups via π–π stacking, hydrogen bonding, and Lewis acid–base interactions (Li et al., 2016). Nitrogen-doped and oxygen-rich bio-carbons enhance electron density on the carbon framework, strengthening interactions with electron-deficient sulfur rings. In metal- and metal oxide–based nanocomposites, additional coordination and catalytic oxidation mechanisms dominate. Sulfur atoms donate lone-pair electrons to metal centers, forming strong σ-complexes that stabilize adsorption. Under oxidative conditions, these metals catalyze the conversion of sulfur compounds into more polar sulfoxides and sulfones, which are subsequently removed by adsorption or extraction (Rafiee et al., 2018). Polymer–carbon nanocomposites further introduce electrostatic and hydrogen-bonding mechanisms through protonated polymer chains and polar functional groups. In many hybrid systems, these mechanisms operate simultaneously, leading to multi-mode sulfur capture and significantly enhanced desulfurization efficiency even at ultra-low sulfur concentrations.

Performance Evaluation and Comparative Efficiency

Performance evaluation of green carbon-based nanocomposites is typically assessed through sulfur adsorption capacity, removal efficiency, selectivity, kinetics, and reusability. Numerous studies report sulfur uptake values exceeding 40–100 mg g^{-1} for metal-functionalized bio-carbons, rivaling or surpassing commercial activated carbons (Li et al., 2016; Zhao et al., 2018). Magnetic nanocomposites maintain high adsorption efficiency over multiple regeneration cycles with minimal performance loss. Polymer–carbon sustainable nanocomposites demonstrate slightly lower maximum sulfur capacities but offer superior mechanical durability, shape stability, and regeneration under milder conditions. Comparative studies indicate that green nanocomposites synthesized through hydrothermal, microwave-assisted, or biogenic routes exhibit comparable or superior performance to fossil-derived counterparts while offering major reductions in embedded energy and toxic emissions. From an industrial perspective, the balance between adsorption capacity, regeneration stability, synthesis cost, and environmental impact determines the overall competitiveness of green nanocomposites. Current evidence clearly demonstrates that green carbon-based nanocomposites provide an optimal convergence of performance, sustainability, and scalability for next-generation fuel desulfurization systems.

7. ENVIRONMENTAL, ECONOMIC, AND LIFE-CYCLE SUSTAINABILITY ASSESSMENT

Life-Cycle Assessment of Green Nanomaterials

Life-cycle assessment (LCA) is a comprehensive methodological framework used to quantify the environmental impacts associated with all stages of a material's life cycle, from raw material extraction through synthesis, application, and end-of-life disposal. In the context of green carbon-based nanomaterials for fuel desulfurization, LCA provides a critical tool for benchmarking sustainability performance against conventional fossil-derived nanomaterials (Arvidsson et al., 2014; Upadhyayula et al., 2012). Key impact indicators typically include cumulative energy demand, global warming potential, water footprint, human toxicity, and ecotoxicity. LCA studies consistently demonstrate that biomass-derived carbons synthesized via hydrothermal, microwave-assisted, and other low-temperature routes exhibit substantially lower environmental burdens than conventional high-temperature carbons and CVD-grown nanostructures. Reductions of 40–70% in cumulative energy demand and more than 50% in greenhouse gas emissions have been reported for hydrothermal carbonization pathways when wet biomass feedstocks are directly utilized without pre-drying (Jain et al., 2016; Arvidsson et al., 2014). Furthermore, the elimination of corrosive chemical activating agents significantly lowers acidification and aquatic toxicity potentials. From a use-phase perspective, the high reusability and regeneration stability of green nanocomposites further improve their life-cycle performance by reducing material replacement frequency and waste generation. End-of-life scenarios such as low-temperature reactivation, safe landfill disposal, or controlled thermal degradation introduce substantially lower environmental risk than metal-heavy conventional nanomaterials.

Carbon Footprint and Energy Payback Analysis

Carbon footprint analysis and energy payback time are essential indicators for evaluating the climate sustainability of nanomaterial production. Conventional high-temperature carbonization and CVD routes depend heavily on fossil-derived electricity and natural gas, resulting in substantial CO_2 emissions per unit mass of material produced (Upadhyayula et al., 2012; Kumar et al., 2020). In contrast, green synthesis routes rely on moderate thermal inputs, shorter processing times, and, in many cases, renewable biomass-derived internal energy sources. Hydrothermal and microwave-assisted carbonization typically operate below 300 °C and achieve carbonization within minutes to hours rather than days. As a result, their carbon payback period—the time required for environmental benefits during application to offset synthesis-related emissions—is significantly shorter than that of conventional nanomaterials (Jain et al., 2016). Studies further indicate that when biomass waste is used as feedstock, the avoided emissions from waste decomposition and open burning further offset production-related emissions, often resulting in near-carbon-neutral or even carbon-negative material cycles. In fuel desulfurization systems, the improved sulfur removal efficiency of green nanocomposites also contributes indirectly to emission reductions by enabling cleaner fuel combustion and reduced SO_2 release. Thus, both direct manufacturing-phase and indirect use-phase climate benefits combine to enhance the overall carbon sustainability of green nanomaterials.

Cost–Benefit and Techno-Economic Considerations

Economic feasibility is a determining factor for large-scale adoption of green nanomaterials in industrial desulfurization. The techno-economic performance of bio-derived carbon nanomaterials is strongly influenced by raw material cost, process energy consumption, chemical input requirements, equipment complexity, and regeneration efficiency (Piccinno et al., 2012; Kumar et al., 2020). Biomass-based feedstocks such as agricultural residues, algal biomass, and municipal biowaste typically have low or negative procurement costs, providing a strong economic advantage over petroleum-derived carbon precursors. Low-temperature synthesis routes further enhance economic competitiveness by significantly reducing electricity and fuel consumption, reactor material costs, and safety compliance expenses. Microwave-assisted and hydrothermal reactors operate at lower pressures and temperatures, enabling flexible modular design and rapid scale-up with reduced capital investment. In addition, the high regeneration efficiency and long service life of green nanocomposites reduce replacement costs during prolonged operation. Cost–benefit analyses indicate that, despite slightly higher initial development costs associated with process optimization and quality control, the long-term operational savings of green nanocomposites offset these early investments. When waste-derived feedstocks are integrated into a circular economy framework, the value addition from waste valorization further improves economic viability. Collectively, these factors demonstrate that green carbon-based nanomaterials are not only environmentally desirable but also economically sustainable for commercial desulfurization systems.

Industrial Feasibility and Process Integration

Industrial feasibility depends on the ability to integrate green nanomaterials into existing refinery infrastructure with minimal process disruption and operational risk. Bio-derived carbon nanocomposites exhibit excellent compatibility with fixed-bed, fluidized-bed, and slurry-based desulfurization reactors due to their tunable morphology, mechanical stability, and chemical robustness (Rafiee et al., 2018; Li et al., 2016). Their fabrication into beads, pellets, and monolithic structures further facilitates direct industrial deployment. Low-temperature synthesis routes provide additional advantages for scale-up because they require simpler thermal management, reduced corrosion-resistant construction materials, and lower explosion risk compared with high-temperature systems. Pilot-scale demonstrations of hydrothermal and microwave processes have already shown strong potential for continuous production under industrially relevant conditions (Jain et al., 2016). However, challenges related to feedstock variability, impurity content, and large-scale reactor optimization remain important technological bottlenecks. From a regulatory and safety standpoint, the reduced toxicity, low dust generation, and improved biodegradability of green nanomaterials ease occupational health compliance and environmental permitting procedures. As refinery emissions regulations continue to tighten worldwide, the industrial feasibility of green nanocomposites is expected to further improve due to their ability to simultaneously reduce fuel sulfur content and material-related environmental burdens.

8. REGULATORY, SAFETY, AND POLICY PERSPECTIVES

Environmental Regulations for Nanomaterial Manufacturing

The rapid expansion of nanotechnology in energy and environmental applications has intensified regulatory scrutiny over nanomaterial manufacturing, usage, and disposal. Environmental regulations governing nanomaterials primarily focus on limiting emissions, controlling effluent discharge, managing solid waste, and preventing unintended nanoparticle release into ecosystems (Klaine et al., 2012; Aschberger et al., 2010). Conventional nanomaterial production, particularly high-temperature carbonization and chemical vapor deposition, is associated with significant greenhouse gas emissions, solvent discharge, and chemically hazardous by-products. These processes often fall under strict industrial emission control standards and hazardous waste management laws. Green synthesis routes based on biomass feedstocks and low-temperature processing exhibit substantially lower regulatory burden due to reduced chemical input and lower emission profiles. Hydrothermal and microwave-assisted synthesis methods generate minimal airborne pollutants and avoid strong acids and corrosive salts, easing compliance with air and water pollution regulations (Jain et al., 2016; Kumar et al., 2020). Furthermore, the use of waste biomass as a precursor supports national waste-to-wealth and circular economy policies, increasingly promoted by environmental agencies worldwide. However, despite these sustainability advantages, regulatory frameworks specific to nanomaterials remain fragmented across regions. Many jurisdictions classify nanomaterials under existing chemical regulations rather than establishing nano-specific standards, creating inconsistencies in approval processes and environmental risk evaluation. Harmonization of regulatory standards for green nanomaterial manufacturing remains an urgent global policy requirement.

Occupational and Public Health Safety

Occupational exposure to engineered nanomaterials represents a critical safety concern due to the ability of nanoparticles to penetrate respiratory, dermal, and gastrointestinal barriers. Inhalation of ultrafine particles has been linked to pulmonary inflammation, oxidative stress, and systemic toxicity, particularly for metal and metal oxide nanoparticles (Nel et al., 2006; Oberdörster et al., 2005). During conventional nanomaterial manufacturing, worker exposure risks are high due to particle aerosolization, solvent evaporation, and high-temperature thermal emissions. Green carbon nanomaterials offer substantial improvements in occupational safety due to their lower metal content, reduced use of toxic chemicals, and inherently lower volatility. Biomass-derived carbons synthesized through wet-chemical or hydrothermal routes minimize dust generation and drastically reduce airborne nanoparticle exposure (Klaine et al., 2012). Polymer–carbon green nanocomposites further enhance handling safety by immobilizing nanoparticles within stable matrices, thereby reducing accidental release during transport, storage, and reactor loading. From a public health perspective, concerns also extend to downstream exposure through accidental material leakage, improper disposal, or combustion residues. Bio-derived carbons demonstrate lower cytotoxicity and improved biocompatibility compared with fossil-derived and metal-heavy nanomaterials (Aschberger et al., 2010). These attributes improve risk profiles during accidental environmental release and significantly enhance public acceptance of green nanotechnology in fuel purification applications.

Global Standards for Sustainable Nanotechnology

The absence of universally harmonized standards for sustainable nanotechnology presents a major barrier to industrial-scale commercialization. International organizations such as ISO and OECD have initiated frameworks for nanomaterial characterization, environmental risk testing, and occupational exposure limits; however, implementation remains inconsistent across national jurisdictions (Arvidsson et al., 2014). Current testing protocols often focus on acute toxicity and short-term exposure, whereas chronic ecological effects at ultra-low concentrations remain insufficiently understood. For green nanomaterials, the challenge lies in establishing a standardized sustainability certification that accounts not only for material performance but also for feedstock renewability, synthesis energy demand, life-cycle emissions, and end-of-life impact. Life-cycle assessment, carbon footprint labeling, and environmental product declarations are increasingly being recognized as essential certification tools for sustainable nanotechnologies (Upadhyayula et al., 2012). Several countries have begun integrating nanomaterial sustainability into clean energy policies, low-carbon technology roadmaps, and green manufacturing incentives. However, the lack of globally unified benchmarking protocols for green nanomaterials restricts their free trade and international technology transfer. The development of internationally accepted sustainability validation frameworks will therefore be essential for widespread global adoption of green nanocomposite-based desulfurization systems.

Policy Challenges and Future Governance

Policy governance for nanotechnology currently faces a dual challenge: encouraging technological innovation while simultaneously safeguarding environmental and public health. Conventional nanomaterial policies frequently struggle to keep pace with rapid scientific advancement, leading to regulatory lag and enforcement gaps (Piccinno et al., 2012). This policy inertia often results in uncertainty for industrial stakeholders and delays large-scale deployment of advanced nanomaterial systems. Green nanomaterials present an opportunity to reform nanotechnology governance by embedding sustainability at the core of policy design. Incentive-based instruments such as carbon credits, renewable materials subsidies, low-carbon manufacturing tax exemptions, and waste valorization credits can significantly accelerate industrial adoption. Policy alignment with broader climate action commitments, clean fuel standards, and circular economy missions will further enhance the market penetration of green carbon nanocomposites. Future governance frameworks must also incorporate transparent risk communication, community engagement, and open-access toxicity databases to strengthen public trust. Integrating precautionary principles with innovation-friendly regulatory pathways will be essential to ensure that green nanomaterial technologies for fuel desulfurization develop under responsible, ethical, and socially acceptable conditions. As fuel quality regulations become increasingly stringent, green nanotechnology is expected to receive stronger policy endorsement as both an environmental protection and industrial decarbonization strategy.

9. CURRENT CHALLENGES AND TECHNOLOGICAL BOTTLENECKS

Feedstock Variability and Quality Control

One of the most significant challenges in the large-scale deployment of bio-derived carbon nanomaterials is the inherent variability of biomass feedstocks. Agricultural residues, algal biomass, and municipal biowastes differ widely in their chemical composition, moisture content, ash fraction, and inorganic impurity levels depending on geographic origin, seasonal availability, and processing history (Ioannidou & Zabaniotou, 2007; Jain et al., 2016). This variability directly affects carbon yield, pore development, surface chemistry, and adsorption performance, leading to batch-to-batch inconsistency in nanomaterial quality. For fuel desulfurization applications, even subtle variations in pore size distribution and surface functional group density can significantly influence sulfur uptake capacity and selectivity. The presence of residual metals, chlorine, or phosphates in waste-derived feedstocks may also interfere with surface functionalization or introduce secondary contamination risks. Although pretreatment strategies such as washing, drying, and chemical conditioning can improve feedstock uniformity, they add processing steps, cost, and additional environmental burden. Developing standardized biomass preprocessing protocols and robust quality control frameworks thus remains a critical bottleneck for consistent production of high-performance green carbon nanomaterials.

Scale-Up Limitations of Green Synthesis Routes

While hydrothermal, microwave-assisted, and low-temperature sol–gel techniques show excellent sustainability at the laboratory scale, their translation to industrial-scale continuous production remains technically challenging. Hydrothermal processes require high-pressure autoclaves that are difficult to scale beyond pilot capacity due to engineering complexity, pressure safety regulations, and heat-transfer limitations (Titirici & Antonietti, 2010; Jain et al., 2016). Similarly, uniform microwave heating becomes increasingly difficult as reactor volume increases, leading to non-uniform temperature fields and product inconsistency. Reduced-temperature chemical vapor deposition routes, although lower in thermal demand than traditional CVD, still require precise gas-phase control, catalyst preparation, and vacuum systems that increase capital and maintenance costs (Zhang et al., 2017). At present, most industrial nanomaterial production infrastructure is optimized for fossil-derived carbons and high-temperature activation, creating a technological lock-in effect that hampers rapid transition to greener alternatives. Moreover, continuous-flow processing of bio-derived nanomaterials introduces additional challenges related to residence time control, raw material feeding, clogging in viscous biomass slurries, and real-time product monitoring. Overcoming these scale-up barriers will require advances in reactor design, digital process control, and modular production systems specifically tailored for low-temperature green synthesis.

Material Stability and Long-Term Performance

Long-term operational stability remains a critical technical bottleneck for green carbon-based nanocomposites in fuel desulfurization systems. Although many bio-derived carbons demonstrate high initial sulfur adsorption capacity, prolonged exposure to complex fuel matrices containing aromatics, nitrogen compounds, and trace metals can lead to pore blocking, surface fouling, and functional group deactivation (Li et al., 2016; Rafiee et al., 2018). This gradual decline in adsorption efficiency limits the operational lifespan of the material. Metal- and metal oxide–carbon nanocomposites face additional challenges related to metal leaching, sintering, and redox instability during repeated adsorption–regeneration cycles. Even trace metal release can pose product contamination risks and increase downstream purification requirements. Polymer–carbon sustainable nanocomposites, while mechanically robust, may suffer from polymer swelling, chemical degradation, or structural fatigue under prolonged exposure to hydrocarbon fuels and regeneration solvents. Thermal and solvent-based regeneration, although effective, can progressively alter pore architecture and surface chemistry, leading to gradual performance decay. At present, most reported stability studies remain limited to fewer than 10–15 regeneration cycles, which is insufficient to establish long-term industrial durability. Developing regeneration-resistant surface architectures and self-healing functional interfaces remains a crucial research gap.

Commercialization Barriers

Despite strong laboratory-scale performance and favorable sustainability metrics, commercialization of green nanomaterials for fuel desulfurization faces multiple non-technical barriers. One of the most significant obstacles is the lack of standardized benchmarking protocols for evaluating the performance, safety, and life-cycle sustainability of bio-derived nanocomposites (Arvidsson et al., 2014; Upadhyayula et al., 2012). Without harmonized test standards, industrial users face uncertainty in comparing green materials with established commercial activated carbons. Economic risks related to supply chain reliability of biomass feedstocks, seasonal availability, and logistics further complicate commercial adoption. Additionally, industrial stakeholders often exhibit conservative risk behavior toward new material technologies due to concerns about process disruption, catalyst poisoning, and product compatibility with existing refinery infrastructure. The absence of long-term field data under continuous industrial operating conditions further limits investor and regulatory confidence. Regulatory uncertainty also presents a significant commercialization bottleneck. Although green nanomaterials exhibit improved environmental and health safety profiles, nano-specific regulatory pathways remain underdeveloped in many regions. This results in prolonged approval timelines and compliance ambiguity. Furthermore, the lack of targeted financial incentives for green nanomaterial adoption in fuel processing limits market pull mechanisms. Collectively, these technological, regulatory, and economic bottlenecks underscore the need for integrated demonstration projects, long-duration pilot studies, and policy-supported commercialization programs to bridge the gap between laboratory innovation and full-scale industrial deployment.

10. FUTURE PROSPECTS AND RESEARCH DIRECTIONS

Next-Generation Bio-Derived Carbon Nanomaterials

The next phase of green carbon nanomaterial development will focus on designing next-generation bio-derived carbons with precisely tailored pore architecture, surface chemistry, and electronic structure for ultra-deep fuel desulfurization. Emerging strategies involve the use of engineered biomass precursors, genetically optimized algae, and hybrid bio-feedstocks to achieve controlled heteroatom doping and hierarchical porosity in a single-step synthesis process (Sevilla et al., 2017; Wei et al., 2020). Such designer biomass platforms offer unprecedented control over nitrogen, sulfur, and phosphorus content in the final carbon framework. Advanced templating approaches using biopolymers, natural minerals, and self-assembling biomolecules are also expected to play a major role in creating ordered mesoporous and macroporous bio-carbons. These structures will enhance mass transport of bulky sulfur molecules such as alkylated dibenzothiophenes that remain difficult to remove at ultra-low sulfur levels. Future research is also expected to focus on hybrid carbons combining graphitic and amorphous domains to balance electrical conductivity, surface reactivity, and mechanical stability. Additionally, integration of multifunctionality—such as simultaneous sulfur adsorption, nitrogen compound removal, and trace metal capture—into a single green carbon platform will be a major innovation frontier. These multifunctional systems would enable consolidated fuel purification processes, reducing reactor complexity, processing time, and overall environmental footprint.

Hybrid Green Nanocomposites for Ultra-Deep Desulfurization

Future desulfurization technologies will increasingly rely on hybrid green nanocomposites that integrate bio-derived carbons with multiple functional phases to achieve ultra-deep sulfur removal below 1 ppm. Research efforts are shifting toward multi-component systems where metal oxides, redox-active polymers, and heteroatom-doped carbons operate synergistically within a single nanocomposite architecture (Rafiee et al., 2018; Zhao et al., 2018). Such hybridization enables simultaneous adsorption, oxidation, and coordination-based sulfur capture. Emerging directions include bio-carbon-supported bimetallic catalysts for oxidative desulfurization, magnetically recoverable hybrid nanocomposites for continuous-flow reactors, and photo-assisted green oxidation systems that operate under ambient conditions. Visible-light-driven oxidation of sulfur compounds using bio-carbon-supported photocatalysts is particularly attractive for minimizing thermal energy input and enabling modular decentralized fuel purification. Another promising research direction involves coupling green nanocomposites with ionic liquids, deep eutectic solvents, and bio-solvent-based extractants to develop fully green integrated desulfurization systems. Such hybrid process designs can substantially reduce hydrogen demand, eliminate harsh oxidants, and enable low-temperature sulfur removal with near-zero secondary waste generation.

Digital Tools and AI in Green Nanomaterial Design

Digitalization and artificial intelligence (AI) are expected to play transformative roles in accelerating the discovery, optimization, and scale-up of green nanomaterials for fuel desulfurization. Data-driven materials design frameworks can analyze large experimental and computational datasets to identify optimal biomass precursors, synthesis parameters, and functionalization strategies with minimal experimental trial-and-error (Butler et al., 2018). Machine learning models are increasingly being used to predict adsorption capacity, selectivity, stability, and regeneration performance based on structural and surface descriptors of carbon nanomaterials. These tools can rapidly screen thousands of candidate bio-carbons and nanocomposites to identify the most promising formulations for targeted sulfur compounds. Digital twins of hydrothermal and microwave reactors are also emerging as powerful tools for process optimization, enabling real-time monitoring and control of green synthesis at scale. In the future, fully autonomous AI-driven synthesis platforms could integrate precursor selection, synthesis condition tuning, performance testing, and regeneration optimization into closed-loop systems. Such platforms would greatly reduce development time, cost, and material waste while maximizing sustainability performance.

Roadmap for Sustainable Industrial Deployment

The sustainable industrial deployment of green nanomaterials for fuel desulfurization will require coordinated progress across scientific, engineering, regulatory, and economic dimensions. At the process level, future research must prioritize continuous-flow low-temperature reactor design, standardized biomass preprocessing strategies, and robust quality-control systems to ensure reproducible large-scale production (Jain et al., 2016; Piccinno et al., 2012). From a systems perspective, integration of green nanocomposites into existing refinery infrastructure must focus on compatibility with fixed-bed and slurry reactor configurations, low pressure drop, and long-term mechanical stability. Field-scale pilot demonstrations under real refinery operating conditions will be critical for establishing industrial reliability, economic viability, and regulatory confidence. On the policy front, research must increasingly engage with life-cycle certification, green labeling, and carbon pricing mechanisms to strengthen market incentives for sustainable desulfurization technologies. Cross-sector collaboration among material scientists, chemical engineers, refinery operators, environmental agencies, and policymakers will be essential for overcoming remaining commercialization barriers. Ultimately, the future of fuel desulfurization lies in achieving a convergence of ultra-high sulfur removal efficiency, minimal environmental footprint, and economic competitiveness. Green carbon-based nanomaterials and nanocomposites, guided by digital design and supported by progressive.

11. CONCLUSION

The increasing need for ultra-clean fuels due to strict international environmental regulations and environmental sustainability has made fuel desulfurization the focus of innovation in the modern refinery. As has been revealed in this chapter, green and sustainable nanomaterials, and in particular bio-derived carbon nanomaterials produced by low temperature and energy-efficient methods, provide an opportunity to revolutionize the process of fuel purification in an environmentally responsible manner. Valorizing renewable biomass, agricultural residues, algae, and industrial biowaste also do not only reduces reliance on fossil-derived carbon but also facilitates the existence of circular economy models and waste-to-wealth policies. The combination of low-temperature technologies like hydrothermal, microwave-assisted, and reduced-temperature chemical vapor deposition has been demonstrated to significantly reduce energy consumption, greenhouse gas emissions, and toxic by-products. Moreover, surface functionalization and nanocomposite engineering can provide good control of the sulfur selectivity, the adsorption capacity, and regeneration stability, and green carbon-based nanocomposites can compete with or even outperform the functionality of traditional desulfurization materials. Life-cycle, techno-economic, and regulatory analyses are evident that the advantages of these green materials are long-term environmental and economic, and an enhanced occupational and public safety profile.

Although great strides have been achieved, there are still difficulties in the feedstock variability, scale-up, material stability over time, and harmonization of the regulations. Nevertheless, the overlapping of digital solutions, artificial intelligence-based material designing, hybrid nanocomposite structures, and enabling policies of sustainability are good bases to overcome them. In general, nanomaterials based on green carbon are an essential enabling technology of the future of low-carbon, high-efficiency fuel desulfurization, in assisting the world to move to cleaner energy systems and sustainable industrial processes.

REFERENCES

Anastas, P. T., & Warner, J. C. (1998). Green chemistry: Theory and practice. *Green Chemistry, 1*(1), 1–12. DOI: 10.1039/A902355P

Arvidsson, R., Molander, S., & Sandén, B. A. (2014). Review of life cycle assessment of nanomaterials. *Journal of Cleaner Production, 63*, 1–12. DOI: 10.1016/j.jclepro.2013.09.038

Aschberger, K., Johnston, H. J., Stone, V., Aitken, R. J., Hankin, S. M., Peters, S. A. K., & Tran, C. L. (2010). Review of carbon nanotubes' toxicity and exposure. *Toxicology and Applied Pharmacology, 243*(3), 316–325. DOI: 10.1016/j.taap.2009.11.020 PMID: 20860524

Babich, I. V., & Moulijn, J. A. (2003). Science and technology of novel processes for deep desulfurization of oil refinery streams: A review. *Fuel, 82*(6), 607–631. DOI: 10.1016/S0016-2361(02)00324-1

Butler, K. T., Davies, D. W., Cartwright, H., Isayev, O., & Walsh, A. (2018). Machine learning for molecular and materials science. *Nature, 559*(7715), 547–555. DOI: 10.1038/s41586-018-0337-2 PMID: 30046072

Dahlia, S., Singh, B., Singh, N., & Rai, A. K. (2021). Green nanotechnology: A sustainable approach in energy and environmental applications. *Environmental Nanotechnology, Monitoring & Management, 16*, 100504. DOI: 10.1016/j.enmm.2021.100504

Deng, J., Xiong, T., Xu, F., Li, M., Han, C., Gong, Y., Wang, H., & Wang, Y. (2015). Inspired by bread leavening: One-step synthesis of hierarchically porous carbons. *Green Chemistry, 17*(7), 4053–4060. DOI: 10.1039/C5GC00523J

Finkbeiner, M., Inaba, A., Tan, R., Christiansen, K., & Klüppel, H. J. (2010). The new international standards for life cycle assessment. *The International Journal of Life Cycle Assessment, 11*(2), 80–85. DOI: 10.1065/lca2006.02.002

Ioannidou, O., & Zabaniotou, A. (2007). Agricultural residues as precursors for activated carbon production. *Renewable & Sustainable Energy Reviews, 11*(9), 1966–2005. DOI: 10.1016/j.rser.2006.03.013

Iravani, S. (2011). Green synthesis of metal nanoparticles using plants. *Green Chemistry, 13*(10), 2638–2650. DOI: 10.1039/c1gc15386b

Jain, A., Balasubramanian, R., & Srinivasan, M. P. (2016). Hydrothermal conversion of biomass waste to activated carbon. *Chemical Engineering Journal, 283*, 789–805. DOI: 10.1016/j.cej.2015.08.014

Kappe, C. O. (2004). Controlled microwave heating in modern organic synthesis. *Angewandte Chemie International Edition, 43*(46), 6250–6284. DOI: 10.1002/anie.200400655 PMID: 15558676

Klaine, S. J., Alvarez, P. J. J., Batley, G. E., Fernandes, T. F., Handy, R. D., Lyon, D. Y., Mahendra, S., McLaughlin, M. J., & Lead, J. R. (2012). Nanomaterials in the environment. *Environmental Toxicology and Chemistry, 27*(9), 1825–1851. DOI: 10.1897/08-090.1 PMID: 19086204

Kumar, V., Goyal, A., & Gupta, R. (2020). Sustainable approaches in carbon nanomaterial synthesis. *Materials Science for Energy Technologies, 3*, 293–312. DOI: 10.1016/j.mset.2019.11.005

Li, W., Zhu, X., Chao, Y., & Wang, X. (2016). Adsorptive desulfurization of fuels using novel nanomaterials. *Energy & Fuels, 30*(8), 6604–6621. DOI: 10.1021/acs.energyfuels.6b00874

Nel, A., Xia, T., Mädler, L., & Li, N. (2006). Toxic potential of materials at the nanolevel. *Science, 311*(5761), 622–627. DOI: 10.1126/science.1114397 PMID: 16456071

Oberdörster, G., Oberdörster, E., & Oberdörster, J. (2005). Nanotoxicology. *Environmental Health Perspectives, 113*(7), 823–839. DOI: 10.1289/ehp.7339 PMID: 16002369

Piccinno, F., Hischier, R., Seeger, S., & Som, C. (2012). Scale-up framework for chemical processes. *Journal of Cleaner Production, 28*, 108–116. DOI: 10.1016/j.jclepro.2011.12.035

Prins, R., Bus, E., & van Bokhoven, J. A. (2006). Catalytic hydrodesulfurization. *Catalysis Today, 117*(1–3), 177–189. DOI: 10.1016/j.cattod.2006.05.012

Rafiee, E., Noori, E., & Eavani, S. (2018). Nanoadsorbents in fuel desulfurization. *Journal of Industrial and Engineering Chemistry, 61*, 1–16. DOI: 10.1016/j.jiec.2017.12.021

Sevilla, M., & Fuertes, A. B. (2011). Sustainable porous carbons. *Energy & Environmental Science, 4*(5), 1765–1771. DOI: 10.1039/c0ee00784f

Sevilla, M., Mokaya, R., & Fuertes, A. B. (2017). Nitrogen-doped porous carbons from algae. *Energy & Environmental Science*, *4*(8), 2930–2936. DOI: 10.1039/c1ee01608c

Song, C. (2003). New approaches to deep desulfurization. *Catalysis Today*, *86*(1–4), 211–263. DOI: 10.1016/S0920-5861(03)00412-7

Srivastava, V. C. (2012). Evaluation of desulfurization technologies. *RSC Advances*, *2*(3), 759–783. DOI: 10.1039/C1RA00309G

Titirici, M. M., & Antonietti, M. (2010). Sustainable carbon materials by hydrothermal carbonization. *Chemical Society Reviews*, *39*(1), 103–116. DOI: 10.1039/B819318P PMID: 20023841

Upadhyayula, V. K. K., Meyer, D. E., Curran, M. A., & Gonzalez, M. A. (2012). LCA of carbon nanotube products. *Journal of Cleaner Production*, *26*, 37–47. DOI: 10.1016/j.jclepro.2011.12.018

Wei, W., Yu, D., Jiang, Z., & Zhou, J. (2020). Food waste-derived carbon materials. *Journal of Cleaner Production*, *256*, 120678. DOI: 10.1016/j.jclepro.2020.120678

Zhang, Q., Uchaker, E., Candelaria, S. L., & Cao, G. (2018). Nanomaterials for energy conversion and storage. *Chemical Society Reviews*, *42*(7), 3127–3171. DOI: 10.1039/c3cs00009e PMID: 23455759

Zhang, Y., Lin, C., Li, J., & Wang, X. (2017). Low-temperature CVD of carbon nanomaterials. *Carbon*, *117*, 328–343. DOI: 10.1016/j.carbon.2017.02.063

Zhao, C., Li, X., Cheng, J., & Yang, Y. (2018). Nitrogen-doped porous carbons for sulfur adsorption. *Fuel*, *225*, 446–455. DOI: 10.1016/j.fuel.2018.01.010

Chapter 7
Green-Synthesized Nanocomposites for Sustainable Fuel Desulfurization:
Advances and Challenges

Abdullah Ozkan
https://orcid.org/0000-0002-6799-2396
Iskenderun Technical University, Turkey

Vildan Ozkan
https://orcid.org/0000-0001-8719-9099
Iskenderun Technical University, Turkey

Pinar Bozbeyoglu
Biomedical Device Technology Program, Electronics and Automation Department, Gümüşhane University, Turkey

Elif Esra Altuner
https://orcid.org/0000-0001-7663-6898
Kocaeli Health and Tecnology University, Turkey

ABSTRACT

The mounting global demand for energy signifies that we must depend on fossil fuels, which engenders environmental degradation and precipitates the emission of substantial quantities of sulfur oxides (SOx). Protecting themselves from environmental pollution is a top priority for many countries. Proper removal of sulfur-containing compounds from fossil fuels is becoming a major focus. Fuels such as diesel, fuel

DOI: 10.4018/979-8-3373-5966-3.ch007

oil, and gasoline are particularly important in this regard. Fuels such as diesel, fuel oil, and gasoline are particularly important in this regard. The numerous negative consequences of traditional hydrodesulfurization (HDS) methods are what researchers focus on, and these methods are limited. Some examples of these negative effects include higher costs, the use of too much pressure and temperature, and using too much hydrogen. Scientific studies show that biosourced nanocomposite materials are good for desulfurization methods because they can reduce the potential negative consequences. Using raw materials such as microorganisms, biopolymers, and plant extracts is one of the most important ways to increase energy efficiency and eliminate chemical emissions. The large surface areas, enhanced active site homogeneity, and diverse functional groups of nanocomposite materials obtained from such biosourced raw materials make them promising candidates for use in adsorptive and oxidative desulfurization processes. Metal-metal oxide (Fe, Ni, Cu, TiO_2) and carbon-based (graphene, carbon nanotube, and biochar-supported) nanocomposites are particularly preferred for achieving high sulfur removal with low energy consumption. The provision of information on the production methods, chemical properties, sulfur removal methods, and recent scientific research on biosourced nanocomposites is the purpose of this section. In addition, the environmental benefits, economic benefits, and potential roles of biosourced materials in alternative energy systems are discussed.

ABBREVIATIONS

Table 1 summarizes the abbreviations and acronyms used throughout this chapter for ease of reference.

Table 1. List of abbreviations.

Abbreviation	Full Form	Description / Equivalent Term
HDS	Hydrodesulfurization	Conventional catalytic desulfurization using hydrogen under high temperature and pressure
ODS	Oxidative Desulfurization	Oxidation-based sulfur removal using mild oxidants and catalysts
PDS	Photocatalytic Desulfurization	Light-driven oxidative desulfurization employing photocatalysts
ADS	Adsorptive Desulfurization	Sulfur removal through physical or chemisorptive interaction on porous materials

continued on following page

Table 1. Continued

Abbreviation	Full Form	Description / Equivalent Term
BDS	Biodesulfurization	Microbial or enzymatic desulfurization under ambient conditions
DES / NaDES	(Natural) Deep Eutectic Solvent	Green solvent systems composed of hydrogen bond donors and acceptors
MOF / COF	Metal–Organic Framework / Covalent Organic Framework	Porous crystalline hybrid materials used as catalysts or supports
POM	Polyoxometalate	Metal-oxide cluster catalysts with redox and acid properties
ROS	Reactive Oxygen Species	Active oxidative intermediates (•OH, $O_2•^-$, H_2O_2) in ODS/PDS
BT / DBT / 4,6-DMDBT	Benzothiophene / Dibenzothiophene / 4,6-Dimethyldibenzothiophene	Representative model sulfur compounds in fuel desulfurization studies
LCA / TEA	Life Cycle Assessment / Techno-Economic Analysis	Tools for evaluating sustainability and cost-performance of processes
QCL	Quantum Cascade Laser	Advanced light source used for in situ monitoring of sulfur compounds

1. INTRODUCTION

Although studies on alternative energy systems are increasing daily, fossil fuels are still the primary energy source worldwide. The organic sulfur compounds contained in fossil fuels have been a topic of interest for researchers for many years due to both their environmental impacts and the chemical resistance they develop during processing. Numerous scientific studies have demonstrated that aromatic sulfur compounds, such as benzothiophene (BT), dibenzothiophene (DBT), and 4,6-dimethyldibenzothiophene (4,6-DMDBT), which are commonly found in petroleum processes, cannot be purified using conventional methods due to their high thermochemical resistance (Shi and Wu, 2021). Sulfur content has knock-on effects on human and environmental problems, as well as causing reduced engine performance in vehicles, catalyst durability and deterioration in fuel quality.

Recent field studies conducted worldwide have revealed that, despite environmental protocols for sulfur removal, the sulfur content in diesel fuels is still not at the desired levels (Kwao-Boateng et al., 2024; Singh et al., 2023). In reality, it has been determined that the sulfur oxide compounds released into the environment are much higher than the values determined in laboratory experiments (Zhang et al., 2025). These studies indicate that sulfur oxide compound controls should be carried out at regular intervals, not only during the process but also during the storage and distribution processes of fuels after the process. Studies on total sulfur determinations should be in compliance with the regulations made at intervals to ensure the clarity of the numerical data (Reis Medeiros et al., 2024). In recent years, state-of-the-art laser-based spectroscopy systems have become important for the determination of small amounts of sulfur (Li et al., 2025).

Sulfur emissions are not limited to land-based vehicles. Furthermore, as per IMO 2020 regulations, ship fuels must contain a maximum of 0.5% sulfur oxide, demonstrating that the necessary attention is being paid to sulfur emissions in maritime transport (Xiao et al., 2025; Dickson and Huang, 2025). It has been noted that the small amount of sulfur in fuel in maritime transport also has a significant impact on engine performance (Sagin and Sagin, 2023; Melnyk et al., 2024). Studies on solid fuel-powered energy systems have shown that sulfur and ash reduction processes produce significant gains in SOx emissions. Furthermore, process optimizations require ongoing cost-performance balances (Lakhmir et al., 2025). It is stated that hybrid wet limestone-oxidant systems have gained an important place as an alternative in terms of SOx/NOx ratio control (Rizwan et al., 2025).

These results suggest that in order to minimize the sulfur compounds formed as a result of fossil fuel processes, it is necessary to start the necessary processes at the source. Although hydrodesulfurization (HDS) has been the main method used in enterprises in recent years, it does not offer hope in terms of continuity due to its large energy requirements, hydrogen consumption, complex catalyzed reactions, and low effectiveness for sterically limited compounds (Zhang et al., 2024; Wu et al., 2025). Thermodynamic limits during HDS make the complete conversion processes of compounds such as 4,6-DMDBT difficult (Rios et al., 2023; Leal et al., 2024). In robotic-assisted catalyst analyses, even when the phase arrangements of transition metal sulfides in the active role are standardized, deep desulfurization processes still require challenging processes (Ding et al., 2024). For these reasons, the trend towards new methods is increasing. Examples of these methods are: Adsorptive desulfurization, biological desulfurization, or photocatalytic oxidative desulfurization methods can be given as examples (Beshtar et al., 2025; Ma et al., 2025; Qiu et al., 2024; Alviar et al., 2024). Among these methods, the most interesting ones are the sulfur removal methods with bio-sourced nanocomposite materials. Nanocomposites of plant origin, obtained from biopolymers, or derived from microorganisms, attract attention not only due to their positive aspects such as low energy requirements, lack of toxic substance content, and support with biocompatible by-products, but also due to their ability to contribute to the local economy and their pro-recycling nature (Mishra and Devi, 2023; Radulescu et al., 2023; Rabiee, 2025).

2. GREEN SYNTHESIS APPROACHES

The use of bio-derived nanomaterials is cited in the literature as a new approach that will ensure sustainability due to their low chemical content and their perceived environmental benefits rather than their harmful effects. This approach is increasingly focused on new methods derived from phytochemical compounds, microorganisms, and biopolymers, replacing traditional methods using chemical solvents and harmful reducers at high temperatures and pressures. Recent studies indicate that the environmental impacts of biogenic reduction and optimization processes can be eliminated by producing and incorporating hybrid nanocomposites with more morphologically homogeneous, high surface area, and functional group content (Álvarez-Chimal & Arenas-Alatorre, 2023). In recent years, promising studies on bio-derived materials and fuel refining from bio-derived materials have been increasing to achieve the "cleaner production, low cost, high performance" goal.

2.1. Plant Extracts, Microorganism-Based Synthesis, and Biopolymer-Assisted Approaches

Plant Extracts: Plant-based processes are the most important part of biosynthesis. Phytochemical components such as flavonoids, polyphenols, alkaloids, terpenoids, and organic acids, as well as organic capping agents, are used to reduce metal ion content and optimize morphology (Rani et al., 2023). A review of studies in the literature shows that bio-derived materials positively enable the production of many nanomaterials (such as Ag, ZnO, Fe_3O_4, Se, and CeO_2) in one go and at low temperatures (Tan et al., 2023; Lazcano-Ramírez et al., 2023). Nanomaterials with different morphological structures have been produced from bio-derived, chemical-free materials. This diversity in morphology has been used in many areas such as environmental pollution, photocatalysis, biomedical applications, and sensor applications (Pirsaheb et al., 2024)

One of the most important aspects of biosource syntheses is their ability to facilitate the recycling of agricultural waste. In addition to being an important carbon source, such biosourced wastes can also serve as high stabilizers for polyphenols (Kar et al., 2023; Raza et al., 2023).

Biopolymer-assisted synthesis: We can obtain biopolymers using biological macromolecules such as cellulose and gelatin from chitosan. These polymers generally provide improved mechanical and chemical strength by bonding with metal ions (Bhagyaraj & Krupa, 2023). Bacterial cellulose is pretty special because it's able to trap metal nanoparticles. This is all thanks to its sponge-like structure and dense crystallinity. Composites comprising BC-TiO2, BC-Au, and BC-GO are extensively favoured for decreasing ecological contamination and in biomedical applications

(Candra et al., 2024; Khikani et al., 2024; Challa et al., 2025). Biopolymers are among the materials most frequently used by researchers for producing bio-sourced materials (Özkan & Bozbeyoğlu, 2025).

2.2. The Role of Green Reductants and Stabilizers

Polyphenols, flavonoids, terpenoids, carbohydrates, and proteins, which have bio-derived components, facilitate reduction reactions with metal ions by increasing the rate of electron uptake. As a result of this reaction, surface areas are covered, preventing agglomeration (Kulak & Gülmez Samsa, 2023). The structure of these compounds affects many important parameters, such as nucleation in the reaction, crystalline phase selectivity, and final morphological properties. The chemical structure of the plant material can vary depending on many parameters, such as temperature, pH, and the reaction method. The reactivity of the nanomaterial changes as a result of the chemical structure affected by changes in these parameters (Maheshwaran et al., 2024).The organic sugar and phenolic structure in honey, as well as its reducing and stabilizing properties, are important in enabling it to be synthesized in one go, such as Au, Ag, and ZnO (Bahari et al., 2023).

The surrounding organic shell that can form on metal oxide surfaces of flavonoid and phenolic molecules prevents aggregation and also provides the material with adsorption properties. To achieve these properties, the nanoparticle structure must be chemical-free (Hosseinzadeh et al., 2023). In addition, amino and hydroxyl groups in biopolymers such as chitosan and cellulose also have an effect on metal ions. They provide morphological homogeneity by coordinating metal ions (George et al., 2023). In summary, the properties of bio-derived materials are an important parameter in determining the properties of the nanocomposite materials to be produced.

2.3. Comparisons on Energy and Solvent Savings

Conventional nanoparticle production often requires long reaction times. It also requires high temperatures, organic solvents, and lengthy purification steps. However, bio-based reactive applications do not require high temperatures and, because the process proceeds with short-term reactions in aqueous media, are cost-effective and energy-efficient (Kandav and Sharma, 2024; Shrivastav et al., 2025).

Biogenic methods: (i) do not require high temperatures. (ii) require less organic solvent. (iii) generate less waste. (iv) offer the possibility of a solvent-free system.

When studies on solvent-free systems or those obtained with bio-derived solvents were examined, it was seen that a significant decrease in carbon emissions and an increase in energy consumption could be achieved (Pantoja-Romero et al., 2025; Wu et al., 2025). The use of bio-derived wastes reduces raw material and chemical consumption while contributing to the national economy (; Qasim et al., 2023). Life cycle analysis (LCA) supports the advantages of using nanomaterials obtained from bio-derived materials. Bio-derived low-temperature reaction processes provide lower environmental damage and higher scalability compared to older methods (Ramadan et al., 2024; Riaz et al., 2024).

In summary, bio-derived nanocomposites provide environmental protection and an increase in active surface diversity, while also reducing costs, chemical use, energy requirements, and solvent addition. Bio-derived synthesis types and reaction conditions are listed in Table 2.

Table 2. Green synthesis routes for nanocomposites and their typical reaction conditions Introductory sentence.

Method / Source	Reducing & Stabilizing Agent	Typical Conditions (T/ Time)	Resulting Nanostructure	Reference Example
Plant extract-mediated synthesis	Polyphenols, flavonoids	60–90 °C, 1–4 h	Spherical AuNPs or Fe_3O_4 nanocomposites	Zhang et al., 2025
Microbial synthesis	Enzymes, proteins	Ambient, 24–72 h	Metal oxide–biopolymer hybrid	Rios et al., 2023
Biopolymer templating	Chitosan, cellulose, starch	≤100 °C, 2–8 h	Porous carbon or metal oxide networks	Zhao et al., 2025
Ionic liquid / DES-assisted	Choline chloride–glycerol	80–120 °C, solvent-free	Uniform nanosheets	Pan et al., 2025

3. NANOCOMPOSITE TYPES

The performance of nanocomposite materials necessary for the extraction of sulfur from fossil fuels is influenced by the chemical composition of the active phase in addition to factors such as morphology, the content of functional groups, and the structure of the pore system. There are three main categories of nanocomposites, according to scientific studies. (i) metal/metal oxide nanocomposites; (ii) carbon-based nanocomposites; and (iii) hybrid or bio-based nanocomposites. So, in this section, we're going to take a look at nanocomposites, and we'll be focusing on three main things: what they do, how they work, and their impact on desulfurization. Table 3 provides examples of these composite materials, along with their desulfurization methods and advantages.

Table 3. Classification of green-synthesized nanocomposites and their functional roles in fuel desulfurization.

Nanocomposite Type	Composition / Example	Primary Function in Desulfurization	Advantages
Metal oxide-based	Fe_3O_4–TiO_2, ZnO–CeO_2	Catalytic oxidation (ODS/ PDS)	High redox activity, recyclable
Carbon-based	Graphene–Fe_3O_4, CNT–Au	Adsorptive removal (ADS)	High surface area, conductivity
MOF-derived	ZIF-8, MIL-101 composites	Oxidative catalysis + adsorption	Tunable porosity, multifunctional sites
Biopolymer-based	Chitosan–Fe, BC–TiO_2	Selective adsorption, biocompatibility	Renewable, low toxicity

This section details biosourced nanocomposites that can be used for sustainable desulfurization, addressing their functionality in desulfurization.

3.1. Metal/Metal Oxide Nanocomposites

Researchers are really into bio-sourced nanocomposite materials containing metal-metal oxides because of their properties, like their high surface energy, Lewis bonds in their structures, and their high potential for sulfur removal thanks to their oxygen-vacancy structures. At the molecular level, sulfur species are generally activated by systems based on Fe, Ni, Cu, and Ti.

Among these, Fe_3O_4 and its derivatives stand out for their electron transport capacity and magnetic recovery (Karpuz et al., 2025). In a 2023 study conducted by Rashid et al. on Ni/Fe_3O_4/rGO nanocomposites, it was reported that oxidative activity was increased by Fe_3O_4, but this was primarily due to rGO acting as an electron transfer channel in the oxidative conversion of DBT (Rashid et al., 2023). Another study (Kaeed et al., 2024) emphasized the potential for significant transformation at low temperatures of MoO_3-Fe_3O_4 nanocomposites, highlighting their surface acidity and oxygen vacancies (Guosheng et al., 2024). Researchers conducted a discrete investigation into Fe_2O_3–CeO_2 nanocomposites. The removal of sulfur is said to be significantly impacted by the Ce^{4+}/Ce^{3+} redox couples present in these nanocomposites (Radwan et al., 2023).

What makes TiO_2-based nanocomposites special is their photocatalytic performance. This is especially efficient under visible light when they are used with conductive materials like g-C_3N_4 or polyaniline (PANI). Impressive photoactivity has been demonstrated by the Anderson–polyoxometalate/TiO_2/g-C_3N_4 heterostructure, with electron–hole recombination during DBT oxidation being effectively suppressed (Zhang et al., 2023). Reviews point out that TiO_2 systems perform better because of changes to the surface and the addition of two types of metal oxide (Hamza et al., 2024).

At the same time, nanocomposites made from CuO and NiO work via two separate processes. These are called oxidative and adsorptive pathways. So, the $MgMoO_4$ co-catalyst in CuO–MgO/$MgMoO_4$ composites works as an electron-withdrawing agent, which makes the oxidative efficiency of CuO better (Zeng et al., 2024). A strong affinity for aromatic sulfur species is exhibited by CuO/NiO heterostructures thanks to their high surface polarity (Agbadaola et al., 2025). The mix of additives in these oxide systems is usually known to increase the rate of both oxidation and adsorption, which leads to efficient removal of sulfur in less extreme conditions.

3.2. Carbon-Based Nanocomposites

Carbon-containing nanomaterials possess high surface areas. In addition, their substantial mechanical strength and functionalizable surface chemistry offer considerable potential for sulfur removal. These materials enhance the dispersion of active metal oxide particles, thereby improving catalytic activity and stability during regeneration cycles.

Graphene-doped composites, similar to other carbon-doped composites, exhibit high electron mobility, which enhances the interaction between the catalyst and substrate during desulfurization. Fe_3O_4/GO and Fe_3O_4/rGO hybrid composites demonstrate significant magnetic separation and adsorption capabilities (Shadmehri et al., 2023; Lellala et al., 2024). Graphene-doped metal oxide composites also increase the rate constants for DBT oxidation reactions (Ahmed and Karim, 2025). Studies indicate that GO-doped Fe_2O_3 composites minimize environmental impacts during the oxidation of sulfur-containing compounds (Mugisha et al., 2025). The thick tubular structure of CNT-doped composites facilitates increased electron transfer rates and promotes homogeneous distribution of metal oxide particles. Hybrid CNT-doped metal oxide structures show considerable potential for adsorptive desulfurization due to their high surface area and strong adsorption capacity (Ali et al., 2025).

Composites derived from bio-additive solid fuels offer economic and environmental advantages. N-biochar composites containing $MgFe_2O_4$ additives demonstrate high desulfurization performance (Yuan et al., 2023). Biochar produced from eucalyptus waste has been shown to adsorb aromatic sulfur derivatives effectively (Liu et al., 2024). Additionally, carbon-based composites synthesized from tire waste exhibit high desulfurization efficiency (Tsai et al., 2025).

Scientific studies indicate that biochar-containing materials are widely utilized for sulfur removal. A key characteristic of these nanocomposite materials is their ability to inhibit metal agglomeration while facilitating electron transfer. This property enhances reaction rates during oxidative desulfurization processes (Altuner et al., 2023).

3.3. Hybrid and Bio-Based Composite Systems

The literature states that a key method for removing sulfur is the use of hybrid nanocomposites that contain both organic and inorganic additives. Significant ground is expected to be gained by them in the future, as is indicated by the increasing number of studies in this area in recent years. This group includes MOF-added composites, bacterial cellulose-added composites, polymer-metal composites, and bio-based composites (Altuner et al., 2022).

Due to their sponge-like structure, high pore capacity, and chemical functionalization capabilities, MOF (metal–organic framework) heterocomposites can exhibit selectivity in sulfur removal. Attention is being attracted by MOF-rGO and MOF-metal oxide heterocomposites with their high adsorption capacities and kinetic reaction rates during sulfur removal (Matloob et al., 2023; Salih, 2023). So, there's been a lot of research recently on how stable MOF-doped metal oxide composites are during regeneration, and it turns out they've got a pretty high acidic activation density too (Pan et al., 2025; Saeed et al., 2023).

Bacterial cellulose-doped nanocomposites have attracted the attention of researchers due to their homogeneous nanofibrous structures and high mechanical strength. ZnO–MOF-doped bacterial cellulose heterocomposites have demonstrated excellent performance in removing sulfur from diesel fuels (Matloob et al., 2025). The integration of metalloporphyrins into the hydrogel structure of bacterial cellulose and their decomposition of H_2S under light have been reported, thus enabling sulfur removal. Another notable feature is the high energy storage capacity of Fe catalysts located on carbon nanofiber structures derived from bacterial cellulose. This feature shows promise for fuel treatment (Lin et al., 2023).

The offer of composite matrices combining sustainability and high surface functionality is made by bio-based polymer–metal oxide hybrids. So, what we can see here is that the high performance of the modified carbon adsorbents made from chitosan–Fe (Boostani et al., 2024) in DBT removal and the high conductivity achieved by the lignosulfonate–bacterial cellulose composites after carbonization (Rao et al., 2023) demonstrate the broad application potential of bio-based composites.

So, to sum things up, it's pretty clear that both hetero- and bio-sourced composites are really promising. They've got some great benefits, like stopping pollution and cutting down on carbon emissions, and they're cheap and can get rid of a lot of sulfur.

4. MECHANISMS OF DESULFURIZATION

People use many different desulfurization techniques, which involve catalysts or biological processes that convert or get rid of the aromatic and heteroaromatic sulfur molecules that are found in fossil fuels. The main differences between these methods are the substances used (hydrogen/air/oxidant), the active substances (metal sulfides, radical species, Lewis acid centers, enzyme cofactors), electron transport patterns, and the nature of the chemical stress on the sulfur-carbon bond. The following processes are examined in this section: hydrodesulfurization (HDS), oxidative desulfurization (ODS), adsorptive desulfurization (ADS), photocatalytic desulfurization (PDS), and biodesulfurization (BDS).

4.1. Thermocatalytic Hydrodesulfurization (HDS) Mechanism

Today, HDS is the most commonly used desulfurization method at the refinery scale. The focus here is on the reduction of aromatic sulfur compounds over metal sulfidic catalysts (CoMoS, NiMoS, NiWS) in a hydrogen atmosphere. So, there are two main ways the mechanism works: direct desulfurization (DDS) and hydrogenation (HYD).

4.1.1. The Differences Between DDS and HYD Pathways

Research on the change of dibenzothiophene (DBT) at high resolution shows that the DDS pathway is marked by the direct breaking of the C–S bond. Conversely, the HYD pathway facilitates the cleavage of the C–S bond subsequent to the hydrogenation of aromatic rings in a sequential process (Linares and Brunet, 2023).

In catalysts based on MoS_2, it is very important to have active centers that show S-edge and M-edge topologies. Controlling the DDS/HYD selectivity is dependent on these. Basically, what happens is that you can change the adsorption energies by using dopants, like Ni and Co, and these dopants move d-electrons to the edges of MoS_2 to do that. This alters the route selectivity (Wang et al., 2024).

4.1.2. The Functioning of Sterically Hindered Compounds

In molecules that are sterically demanding, such as 4,6-DMDBT, the presence of electron density and cyclic hindrance obstructs the DDS pathway, thereby increasing the importance of the HYD pathway (Li et al., 2025). New catalyst designs incorporating indium modification, Ni–B amorphous phases and micro- and mesoporous carrier topologies offer substantial advantages in the conversion of DMDBT (see, for example, Lewandowski et al., 2025; Zhang et al., 2025; and Xiao et al., 2023).

4.1.3. The Acidity and the Arrangement of the Active Phase

Acidity is very important for Fe-based catalysts because it affects the energies of adsorption and the speeds of electron transfer. It also directly affects the choice of route (Guosheng et al., 2024). NiXCoY boride alloys may help HYD work better by making it easier for hydrogen to activate on surfaces, according to some people (Linares and Brunet, 2023).

HDS is not as long-lasting since it demands high temperatures and pressures because of thermodynamic and steric limits. Because of this, other steps that make the HDS mechanism perform better are becoming more critical (Guosheng et al., 2024).

Desulfurization methods use a number of catalytic and biocatalytic processes to change or get rid of sulfur compounds in fossil fuels that are aromatic or heteroaromatic. The reaction environment (hydrogen, air, or an oxidant), the active species (metal sulfides, radical species, (Guosheng et al., 2024). Lewis acid centers, or enzyme cofactors), the way electrons travel, and the sort of chemical stress that is exerted on the sulfur-carbon bond are the primary things that set these approaches apart. This part explains how hydrodesulfurization (HDS), oxidative desulfurization (ODS), adsorptive desulfurization (ADS), photocatalytic desulfurization (PDS), and biodesulfurization (BDS) all work together.

4.2. The ODS (Oxidative Desulfurization) Process

Many people think that ODS is a different or extra way to accomplish HDS, and it works best when it's cold. The main step is turning sulfur molecules into sulfoxide and sulfone derivatives through oxidation, and then pulling them into the polar phase. The oxidative–adsorptive desulfurization process can be efficiently carried out using multifunctional reactor systems that integrate oxidation and adsorption steps under controlled temperature and mixing conditions, as illustrated in Figure 1.

Figure 1. Schematic depiction of the Digital Basket Baffle Reactor (DBBBR) employed for oxidative-adsorptive desulfurization of diesel fuel (derived from Humadi et al., 2025, Processes, 13(10), 3084, CC BY 4.0).

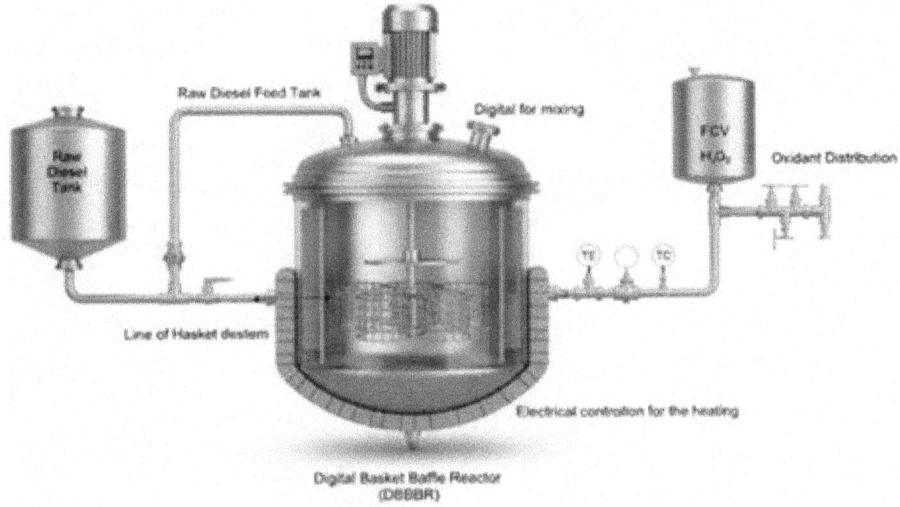

4.2.1. Creating Intermediate Species and Turning on Oxygen

The most critical part of ODS is turning on oxidants such H2O2, peroxymonosulfate (PMS), and polyoxometalates. According to several research (Hossain and Jhung, 2024; Li et al., 2025), metal centers (Mo, W, Cr, Nb) transform the oxidant into peroxy derivatives and oxidize the sulfur center by transferring electrons. Recent research indicates that metals supported by mesoporous carbon enhance the reactivity of oxidants and significantly increase the efficiency of electron transport (Gao et al., 2023). The efficacy of metal-based catalysts is significantly influenced by the synthesis method; nanocomposites produced using biological bottom-up approaches offer enhanced active-site accessibility and redox stability, as seen in Figure 2.

Figure 2. Conceptual diagram showing "Top-Down" and "Bottom-Up" nanoparticle synthesis approaches, highlighting biological routes involved in green synthesis (adapted from Álvarez-Chimal & Arenas-Alatorre, 2023, IntechOpen, CC BY 4.0).

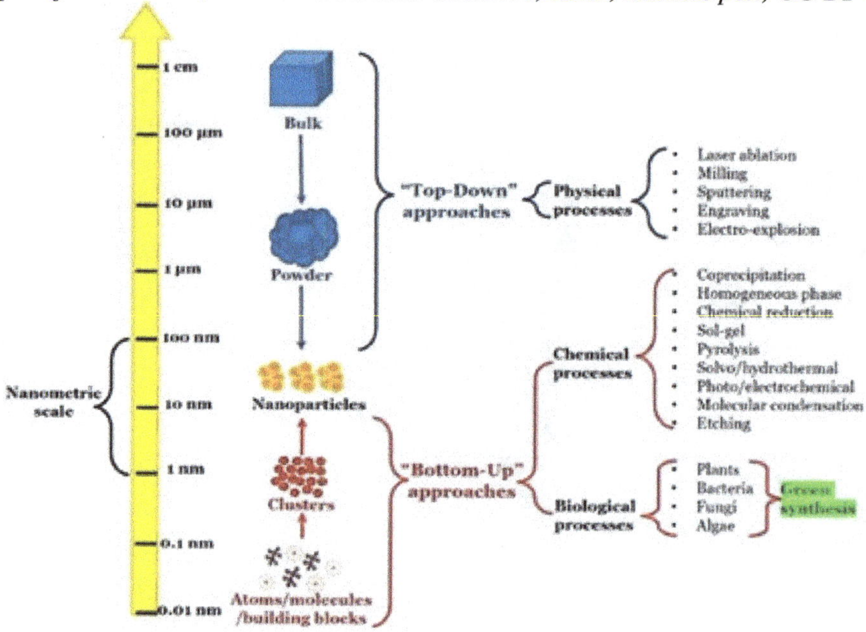

4.2.2. Regulation of Lewis Acids and Bases

Stabilizing intermediate species is vital, and this can be achieved by using Lewis acid centers. The π-electron density of thiocompounds is altered by this, which makes oxidation easier. The operation of ODS is dependent on the correlation between Lewis acidity and active surface polarity (Vermeersch et al., 2023; Wang et al., 2024).

4.2.3. Mechanisms of Emulsion and Interfacial Phase Interaction

Increasing the distance between the oxidant and fuel phases in Pickering emulsions and similar systems accelerates the oxidation process. So, there's been some research on these things. They're called amphiphilic polyoxometalate emulsion catalysts. The interface, which is the area where different things meet, is actually important for making these radical and peroxy species, which are pretty cool molecules. At least, that's what Liu et al. said in 2023.

4.3. Adsorptive Desulfurization (ADS) Mechanism

ADS is one of the prominent methods in sustainable fuel refining, especially because it can operate at low temperatures and pressures. The mechanism is fundamentally based on the selective adsorption of sulfur compounds onto metal, carbon, MOF, or heteroatom-modified surfaces.

Π-Complexation and Metal-centered Adsorption: It is known that Cu(I), Ag(I), or Ni(I) centers exhibit high affinity for DBT and its derivatives through a π-complexation mechanism. Recent studies have shown that conformational flexibility in novel Cu(I)-containing adsorbents enhances selectivity (Ma et al., 2025; Ju et al., 2025).

Surface Functionality and DFT Analyses: DFT modeling confirms that the electron density distribution in BN nanosheets, functionalized carbon nanotubes, and bio-based adsorbents is a determinant of DBT adsorption (Khaleghian et al., 2024; Sehhatigdiri and Pourhossein, 2024). It has been reported that microporosity and the adsorption capacity of oxygenated functional groups on heterogeneous surfaces like biochar significantly increase (Liu et al., 2024). The synergistic electron transfer between carbon matrices and embedded metal oxide domains promotes both oxidative and adsorptive interactions through radical formation, as illustrated in Figure 3.

Figure 3. Proposed mechanism of oxidative and adsorptive interactions in a carbon-supported green nanocomposite system, showing electron–hole transfer and radical formation under light irradiation (adapted from Bartoli et al., 2023, Catalysts, 13(10), 1336, CC BY 4.0).

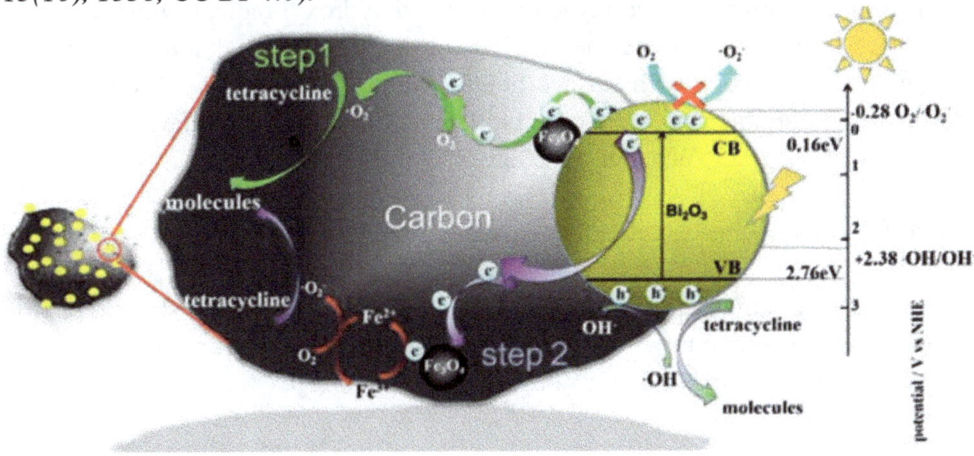

In addition to adsorption and oxidation mechanisms, the sustainability of these nanocomposite systems can be comprehensively evaluated through life-cycle assessment (LCA) and techno-economic analysis (TEA) methods.

Competitive Adsorption Mechanism: In real fuel mixtures, sulfur compounds compete with aromatic hydrocarbons for adsorption. Molecular dynamics studies show that surface competition between small sulfur compounds and DBT determines the adsorption efficiency (Zheng et al., 2024).

4.4. Mechanism of Photocatalytic Desulfurization (PDS)

PDS is the light-activated version of ODS and can attain high oxidation efficiency at low temperatures. The basic mechanism is based on the formation of electron-hole pairs on the catalyst surface upon light excitation and their conversion into reactive oxygen species (ROS).

4.4.1. Belt Structure Engineering and Carrier Separation

Modern studies on TiO_2, g-C_3N_4, and hybrid structures show that band gap engineering determines the oxidation rate by increasing photon utilization and electron-hole separation (Huang et al., 2024; Liu et al., 2025). The separation of reduction/oxidation zones accelerates the selective oxidation of sulfur species.

4.4.2. Production of Reactive Oxygen Species (ROS)

In sulfur oxidation, singlet oxygen (^$1O_2$) and hydroxyl radical (•OH) are particularly effective intermediate species. Recent studies confirm that the production of singlet oxygen, along with surface-bound radicals, plays a critical role in photooxidation during the activation of persulfate and peroxymonosulfate (Yue et al., 2023; Mao et al., 2025; Wang et al., 2025).

4.4.3. Advanced Heterojunction Design

Some types of junctions, such as Ag_2O/g-C_3N_4, In/TiO_2, and MOF-hybrid photocatalysts, help charge to move around more quickly, making sulfur oxidation much more efficient (Zhou et al., 2022; Zbuzant, 2022).

4.5. Biodesulfurization (BDS) Mechanism

Selective removal of sulfur is possible with BDS, without affecting the carbon backbone. Microorganisms carry out this process, which involves the most common 4S pathway.

4.5.1. 4S-Pathway Enzymatic Mechanisms

DszC, DszA, and DszB enzymes catalyze the sequential oxidation of sulfur and the eventual formation of sulfinate. Recent enzymatic modeling has shown that DszA facilitates the cleavage of the C–S bond via the N5-hydroperoxide intermediate (Ferreira et al., 2024). Studies on DszB have shown that kinetic efficiency is consistent with specific conformations in the active region (Sousa et al., 2020).

4.5.2. Microbial Metabolism and Genetic Engineering

The sulfur metabolism of Rhodococcus species forms the fundamental biochemical backbone of the BDS process. Recent genomic and metabolic engineering studies have shown that the efficiency of biodesulfurization can be increased through the overexpression of dsz genes, cofactor optimization, and enzyme stability (Bagchi and Srivastava, 2024; Hou et al., 2023).

4.5.3. Analytical Monitoring in Biodesulfurization

High-resolution SERS analyzes are used for real-time monitoring of DMDBT conversions and enable the verification of intermediate species (Ghaffar et al., 2024).

4.6. Comparative Assessment of Mechanisms

There exist notable distinctions among desulfurization mechanisms:

- HDS → elevated temperature/pressure, exceptional industrial capacity, although restricted efficacy on aromatic sulfur compounds
- ODS & PDS → low temperature, elevated selectivity, sustainability benefit
- ADS → little energy demand, although constrained capacity owing to competing adsorption
- BDS safeguards the carbon framework; yet, it exhibits sluggish kinetics and scalability challenges.

Table 4 summarizes the operating parameters, target sulfur species, and performance characteristics of these desulfurization methods for a clearer comparison.

Table 4 presents a succinct comparison of traditional and novel desulfurization methods, emphasizing their reaction conditions, target sulfur compounds, catalytic systems, and practical benefits or drawbacks.

Table 4. Comparative analysis of desulfurization techniques and operational attributes.

Mechanism	Operating Conditions	Target Sulfur Compounds	Typical Catalyst/ Material	Advantages / Limitations
HDS	300–400 °C, H_2, high pressure	DBT, 4,6-DMDBT	$CoMo/Al_2O_3$	High efficiency but energy-intensive; high H_2 consumption
ODS	≤120 °C, oxidant (H_2O_2 or organic peroxides)	BT, DBT	MoO_3, TiO_2 composites	Mild conditions, selective; limited oxidation depth
PDS	UV/visible irradiation	DBT, 4,6-DMDBT	TiO_2, ZnO–graphene, CdS	Solar-driven, eco-friendly; requires charge-separation improvement
ADS	Ambient, solvent-based	BT, DBT	Carbon, MOF, clay composites	Simple, regenerable; moderate adsorption capacity
BDS	30–40 °C, microbial	DBT, alkyl-DBT	*Rhodococcus* sp., *Gordonia* sp.	Environmentally benign, but slow kinetics and scale-up limits

In comparison to conventional hydrodesulfurization, oxidative and photocatalytic approaches offer more favorable reaction conditions characterized by enhanced softness and cleanliness. Table 4 clearly demonstrates this observation. However, biological routes are still showing promise for low-temperature integration, despite encountering certain kinetic challenges.

The following sections of this book chapter will explain how green-synthesized nanocomposites can improve these mechanisms.

5. RECENT DEVELOPMENTS IN GREEN-SYNTHESIZED NANOCOMPOSITES FOR DESULFURIZATION

Green-synthesized nanocomposites have garnered considerable attention in recent years in research aimed at eliminating aromatic sulfur compounds from fossil fuels (Altuner et al., 2021). This trend arises from the increasing interest in sustainable catalysis methods and the benefits of low energy consumption, biocompatible materials, less reliance on toxic chemicals, and facile regeneration. Current literature shows a significant performance leap in metal/metal oxides, carbon-based hybrids, and MOF/COF-based nanocomposites.

5.1. Advances in Metal and Metal Oxide-Based Green Nanocomposites

The green synthesis approach in metal oxide-based catalysts has shown significant advancements, particularly in ZnO, $CoFe_2O_4$, $NiFe_3O_4$, and MnO_2 systems. Matloob and colleagues (2025) provided a strong example of the industrial-scale applicability of green nanocomposites by developing a ZnO–Zn-MOF/bacterial nanocellulose hybrid for ultra-ODS in real diesel fuel. Similarly, Metawea et al. (2023) achieved high oxidation yield with very low energy input using ultrasound-assisted one-pot synthesis of the MIL-53(Al) structure.

The performance increase is also remarkable in magnetically recoverable catalysts. Wu et al. (2025) revealed that $CoFe_2O_4$ nanoparticles serve as a highly efficient and readily recoverable catalyst in aerobic oxidative desulfurization. Again, Wu and colleagues (2023) reported that high-entropy metal oxide-based magnetic catalysts are stable and highly efficient in aerobic ODS reactions.

It has also been reported that biocalcite-based composites with MnO_2 have shown successful results in both oxidative and adsorptive desulfurization in a digital basket reactor (Humadi et al., 2025). Additionally, Ni/Fe_3O_4/rGO hybrid nanocomposites have demonstrated enhanced desulfurization performance due to both their high surface area and superior electron transfer capability (Rashid et al., 2023).

5.2. Recent Developments in Carbon-Based Nanocomposites

Green carbon derivatives—particularly graphene, biochar, and CNT hybrids—are increasingly preferred in both adsorptive and catalytic desulfurization. It has been reported that polystyrene/graphene nanocomposite membranes offer selective permeability and high stability in the desulfurization of crude oil (Oleiwi et al., 2025). The sustainable recovery of graphene-based nanocomposites from waste sources is also a trend that strengthens the performance-cost relationship (Ikram et al., 2023).

The high efficiency of the biochar-supported $ZnFe_2O_4$ system in H_2S removal under anaerobic conditions (Yu et al., 2025) highlights the potential of carbon materials for broad-spectrum fuel purification. Studies on the removal of dyes using potassium-containing biochar/Fe_3O_4 composites (Nour et al., 2025) indicate that such materials are amenable to modifications for desulfurization.

In CNT-based hybrids, surface functionalization significantly improves performance. High-efficiency adsorbents suitable for desulfurization applications have been developed thanks to the high surface area, multiple active sites, and electron conductivity of CNT-metal hybrids (Abdullah et al., 2024; Satpathy and Moharana, 2023). The improvement in the electrochemical performance of NiS-CNT hybrids

(Batool et al., 2025) suggests that similar structures could increase ROS production in ODS mechanisms.

5.3. New Approaches in MOF/COF and Hybrid Systems

Metal-organic framework (MOF)-based nanocomposites have become one of the fastest-growing areas in green desulfurization research in recent years. Encapsulation of $H_4SiW_{12}O_{40}$ into an amide-functionalized MOF resulted in high oxidative desulfurization efficiency in diesel fuel (Sun et al., 2023). MOF-808@SBA-15 hibritleri, yüksek gözeneklilik ve çoklu aktif merkez avantajlarıyla oksidatif katalitik etkinliği daha da yükseltmiştir (Fernandes et al., 2024).

Optimizing one-pot oxidative-extractive processes with ZIF-67/activated carbon hybrids (Moghaddam and Shahrak, 2024) demonstrates that MOF-based composites can be used in multifunctional desulfurization processes. Additionally, the development of reusable chitosan-based adsorbents (Zavareh and Doori, 2025) indicates that biopolymer-MOF hybrids can offer economical and sustainable solutions.

Studies on POM-supported MOF systems have demonstrated the effectiveness of next-generation ROS production mechanisms in ultra-deep desulfurization. Green-friendly fiber-supported P-Mo-W polyoxometalate structures (Xue et al., 2025), composites performing aerobic ODS through flexible encapsulation within MOFs (Chu et al., 2024), and biomimetic POM-supported MOF structures (Li et al., 2025) represent the most advanced synthesis strategies in this field.

Similarly, peroxotungstate structures embedded in 3DOM-ZIF-8 (Gao et al., 2025) and MOF-UiO-66 systems modified with rare earth elements (Liao et al., 2024) have shown dramatic increases in catalytic activity. It has also been reported that hierarchical load separation modulation in MOF-on-MOF structures enhances ROS production in desulfurization processes (Priyadarshini et al., 2025).

5.4. Industrial Integration, Regeneration, and Economic Processes

Research on the industrial use of green-synthesized nanocomposites is on the rise. Elevated turnover rates with magnetically recoverable catalysts (Jarullah et al., 2024), ultrasonically enhanced minimal solvent usage (Metawea et al., 2023), and reduced raw material expenses in biomass-derived catalysts (Humadi et al., 2025) facilitate the development of economically sustainable systems. The incorporation of deep eutectic solvents (DES) into desulfurization processes (Wang et al., 2025) indicates a substantial reduction in environmental impact.

6. ENVIRONMENTAL AND ECONOMIC IMPACTS OF GREEN-SYNTHESIZED NANOCOMPOSITES IN FUEL DESULFURIZATION

6.1. Environmental Impacts and Life Cycle Perspective

The greatest environmental advantage of green synthesized nanomaterials is the use of non-toxic reductants, biomass-based precursors, and low-temperature energy regimes in their production processes (Karpuz et al., 2025). For example, in a study conducted with bacterial nanocellulose-supported ZnO–Zn-MOF composites, high oxidative activity was achieved with a process that produced almost no waste, unlike traditional metal oxide syntheses (Matloob et al., 2025). Similarly, numerous studies have shown a dramatic reduction in the carbon footprint during the production of nanocomposites derived from agricultural waste, woody biomass, and food industry byproducts (Motaleb et al., 2025; Chadha and Berry, 2025).

Comparisons in terms of ecotoxicity have shown that green ZnO, Fe_3O_4, $CoFe_2O_4$, and biochar-supported nanocomposites offer a significantly lower biological risk compared to conventional metal oxides in terms of surface charge, metal ion release, and ROS production (Thakur and Kumar, 2023; Liu et al., 2024). The heterogeneous structure of nanocomposites, especially those derived from bio-waste, minimizes heavy metal leaching, creating a safer profile in environmental receiving environments (Patil et al., 2023; Shetti et al., 2025).

Life cycle assessments (LCAs) are the analytical tool that most clearly demonstrates the environmental superiority of green-synthesized nanocatalysts (Borj et al., 2025; Wang et al., 2025). In the LCA analysis of oxidative desulfurization (ODS) systems developed for gas condensate, it was reported that low-temperature green catalysts reduced energy consumption by 30–45% (Borj et al., 2025). Similarly, large-scale LCA analyzes conducted on microalgae-based biofuel production lines show that recyclable nanocatalysts significantly reduce total greenhouse gas emissions (Wiatrowski et al., 2025). To provide a comparative perspective, key life-cycle and techno-economic indicators for conventional and green nanocomposite-assisted desulfurization processes are summarized in Table 5.

Table 5 outlines representative life-cycle assessment (LCA) and techno-economic analysis (TEA) metrics comparing conventional hydrodesulfurization (HDS) with emerging green nanocomposite-assisted processes.

Table 5. Life-cycle and techno-economic indicators that compare green and traditional methods of desulfurization.

Process Type	Energy Demand (MJ kg^{-1} S removed)	CO_2 Emission (kg eq CO_2 kg^{-1} S)	Estimated Cost (USD t^{-1} fuel)	Remarks
Conventional HDS	120–150	15–20	25–40	High hydrogen consumption and energy intensity
ODS (green catalyst)	40–60	5–8	12–18	Moderate oxidant use; recyclable catalyst
PDS	25–45	3–6	10–15	Solar-driven; low OPEX
ADS	15–30	2–5	8–12	Simple setup; regeneration cost sensitive
BDS	20–40	2–4	10–20	Bioprocess integration feasible; slower kinetics

Table 5 illustrates that procedures employing green-synthesized nanocomposites significantly reduce energy consumption and carbon emissions relative to conventional hydrodesulfurization, while maintaining comparable operational costs.

The fact that nanocatalysts derived from biomass-based materials can exhibit a carbon-neutral or carbon-negative profile is a critical advantage for sustainable fuel production (Motaleb et al., 2025). Additionally, magnetically recoverable oxide composites (e.g., Fe_3O_4/rGO and $CoFe_2O_4$ can be used repeatedly through regeneration, which aligns well with the principles of a circular economy by reducing waste load (Rashid et al., 2023; Wu et al., 2025).

6.2. Ecological Risk Management

Although green-synthesized nanocomposites generally exhibit low toxicity profiles, current understanding of their interactions with ecosystems remains limited. Therefore, research highlights the importance of monitoring the following factors:

- The potential for reactive oxygen species generation.
- The durability of nanocomposite surface coatings.
- Biodegradability of the nanocomposites.
- The risk of nanocomposite accumulation in aquatic organisms (Thakur & Kumar, 2023; Liu et al., 2024).

7. CHALLENGES AND FUTURE OUTLOOK FOR GREEN-SYNTHESIZED NANOCOMPOSITES IN FUEL DESULFURIZATION

There's been some pretty big improvements in sustainable chemistry and refining technology, and it's all down to these nanocomposites that are made using green synthesis. They're great for getting rid of sulfur in gasoline. A great deal of interest has been attracted by plant extracts, biomass-derived carbons, metal-organic frameworks (MOFs), natural deep eutectic solvents (NaDES), and biopolymer-supported nanostructures due to their non-toxic production methods and low energy requirements (Khan et al., 2025; Rabiee, 2025). Before these technologies can be used in commercial refineries instead of only being tested in laboratories, several scientific, environmental, and financial challenges must be overcome. Research gaps and potential avenues for progress must be identified if green-synthesized nanocomposites are to be implemented.

7.1. Industrial Scale-Up and Material Reproducibility Challenges

It is hard to scale up green synthesis methods from the lab, which makes it harder to regulate the process and get reliable raw materials. The reproducibility of nanocomposites is profoundly influenced by the compositional variability of biomass-derived precursors, particle size, and the quantity of functional groups, leading to considerable uncertainty in scalable manufacturing (Gacem and Abd-Elsalam, 2022; Selvaraj et al., 2024).

Many eco-friendly synthesis procedures, even when performed at moderate temperatures, may not reliably attain an optimal energy-cost equilibrium. As more items are made, the economy is put under more stress by efforts to dry, filter, and change them (Adassooriya et al., 2025; Alvarez, 2023). Standardization will happen for raw material management, continuous flow reactors, and process automation in the next scaling methodologies. The catalyst's stability, how it becomes inactive, and how to reactivate it.

7.2. Catalyst Stability, Deactivation Mechanisms, and Regeneration

One of the most fundamental technical obstacles limiting the industrial applicability of green-synthesized nanocomposites is catalyst stability. Metal oxide-carbon hybrids, MOF-based structures, and graphitic-supported composites are susceptible to deactivation mechanisms such as surface passivation, pore blockage, metal ion leaching, and morphological degradation resulting from interaction with reactive intermediates in the fuel environment (Anekwe and Isa, 2025; Daligaux et al., 2021).

Most regeneration processes require high temperatures or can contain aggressive chemical agents, which contradicts the environmental sustainability claims of green technologies. Innovative approaches such as oxidative-photocatalytic regeneration, low-energy thermal activation, and self-healing surface functions will be at the forefront of future catalyst design (Saeed et al., 2025; Lim and Ong, 2021).

7.3. Environmental and Regulatory Considerations

The environmental impact of the nanomaterials produced by green synthesis is not yet fully understood, despite the fact that it reduces the use of toxic solvents. However, there are significant data gaps concerning the prolonged accumulation of nanoparticles in terrestrial and aquatic environments, biological stress reactions, and the toxicological effects of nanocomposite degradation by-products (Thakur & Kumar, 2023; Luo et al., 2021).

The thing is, we don't have any set ways to check the safety of bionanocomposites, metal-organic frameworks (MOFs) and deep eutectic solvents (DESs), which makes it hard to see how safe they are for the environment (Andreani et al., 2020; Riaz et al., 2024). Huang et al. (2024) argue that the limited scope of life cycle assessment studies results in an inadequate evaluation of environmental impacts relating to carbon footprints, process by-products, and regeneration cycles.

It is anticipated that, at some point in the future, worldwide regulatory frameworks will broaden to encompass toxicity threshold values for nanomaterials, assessments of compatibility with refineries, and nano-OSH (occupational safety and health) protocols.

7.4. Techno-Economic Limitations and Refinery Integration

So, despite how well green nanocomposites are performing, the most important thing is still how much it's going to cost to make them. As Afzal et al. (2023) and Kim et al. (2022) point out, the overall cost can be significantly impacted by process steps such as biomass processing, solvent consumption, energy requirements, and catalyst regeneration.

Refineries are known for being big, always running, and designed to use the best energy. As a result, factors such as pressure-temperature compatibility, solvent cycles, corrosion risk, and energy integration must be given appropriate consideration to allow the smooth incorporation of catalysts produced using eco-friendly methods with hydrodesulfurization (HDS), oxidative desulfurization (ODS), or adsorptive desulfurization modules (Alabdullah et al., 2020; Mehri et al., 2021; Fotiadis et al., 2024).

Table 6. Future research challenges and strategic directions for green-synthesized nanocomposite-assisted desulfurization.

Research Domain	Current Limitation	Future Challenge / Research Direction	Expected Impact
Catalyst design	Limited active-site exposure	Hierarchical nanoarchitectures and surface engineering for enhanced adsorption–oxidation synergy	Improved activity and selectivity
Green synthesis scalability	Batch-dependent reproducibility	Continuous-flow and automated green synthesis systems	Industrial scalability and process consistency
Mechanistic understanding	Incomplete kinetic and thermodynamic models	Integration of DFT and machine-learning-based mechanism mapping	Predictive catalyst design
Environmental compatibility	Limited long-term ecotoxicity data	Full LCA–TEA coupling and biodegradability profiling	Verified sustainability and regulatory compliance
Catalyst regeneration	Activity loss after multiple cycles	Green regeneration using bio-reductants or low-temperature annealing	Cost reduction and circular catalyst use

Table 6 demonstrates that advancements in catalyst stability, process automation, and computational modeling are critical to realizing the full potential of green-synthesized nanocomposites for sustainable fuel desulfurization.

8. FUTURE PERSPECTIVES FOR SUSTAINABLE FUEL DESULFURIZATION

Recent advancements in green-synthesized nanocomposites have demonstrated significant efficacy in removing fuel sulfur. The use of these materials highlights current environmental challenges and offers opportunities to transition toward energy sources with reduced environmental impact. These innovations reflect a movement toward lower energy consumption, reduced reliance on hazardous chemicals, the development of environmentally benign materials, and improved selectivity, collectively enhancing adsorption performance. The refining industry has recognized green-synthesized nanocomposites as essential due to their unique properties. It is necessary to advance beyond current laboratory-scale applications and consider future energy trends. A comprehensive evaluation is required to ensure their long-term safety and suitability for industrial implementation, as well as to address process design considerations.

This section examines the future prospects for applying eco-friendly nanocomposites in fuel desulfurization technologies. The discussion is organized around four principal considerations: advancements in scientific and technological domains, environmental and economic sustainability, the integration of digital technologies and artificial intelligence into process design, and the establishment of a comprehensive results evaluation framework.

Accordingly, this proposed roadmap outlines the anticipated trajectory of green-synthesized nanocomposites in sustainable desulfurization technologies, emphasizing four principal domains: (i) scientific and technological advancement, (ii) requirements for environmental and economic sustainability, (iii) digitization and artificial intelligence-assisted process design, and (iv) formulation of a comprehensive outcome assessment framework.

8.1. Scientific and Technological Divergences

The most critical factor determining the future success of green nanocomposites is the simultaneous consideration of material excellence and process excellence. Current studies show that a wide range of materials, such as metal oxides produced with plant extracts, biomass-derived carbon materials, DES-catalyst hybrids, and MOF-based structures, exhibit high efficiency in desulfurization processes (Khan et al., 2025; Gacem and Abd-Elsalam, 2022; Kirubakaran et al., 2026). However, most of these systems have not been scaled up beyond the laboratory level. The main trends that will be decisive in terms of materials science over the next decade are:

- Standardized green synthesis protocols: Controlled feedstock preparation and process parameters that minimize the natural variability of biomass components.
- Hybrid structures with enhanced thermal and chemical stability: Optimizing metal oxide–carbon, MOF–carbon, or MOF–DES combinations for long-term use (Saeed et al., 2025; Deshmukh et al., 2025).
- Design based on the actual fuel matrix: Active site engineering (Nawaf and Abdul Majeed, 2024; Fotiadis et al., 2024) exhibiting high surface stability and selectivity in multi-component diesel and gas condensate environments, rather than model fuel systems.

In this regard, the integration of experimental methods with artificial intelligence in material design and the real-time adaptation of catalyst performance to fuel components will be decisive in the maturation of the technology.

8.2. Environmental and Economic Sustainability Dimension

Green-synthesized nanocomposites have a number of environmental benefits in addition to their low toxicity. Reduced raw material costs, utilisation of waste biomass, and considerable energy savings are included (Motaleb et al., 2025; Sharma et al., 2024). To make sure we can keep going for the long term, it is vital to expand the range of life cycle assessments (LCA), combine carbon footprint calculations with refinery processes, and come up with ways to regenerate catalysts that do not harm the environment.

The sustainability of green nanocomposites must be considered in relation to material performance and chemical properties, as well as sustainable supply chain management, cost-benefit optimization, and principles of the circular economy. This comprehensive viewpoint is essential for minimizing environmental impacts and enhancing the economic feasibility of green nanotechnologies in alignment with green chemistry principles.

8.3. The Role of Digitalization and Autonomous Systems

The use of digital twin technologies, AI-driven catalyst design, and data-guided process management is expected to be very important for desulfurization in the future. AI-driven models have led to significant changes in how active sites are designed on catalyst surfaces and how reaction parameters are optimised in real time (Rojas et al., 2025; Orlova, 2022). We can expect the following technology breakthroughs to become more important in the near future:

- Autonomous catalyst design platforms: Generative AI models speed up the process of discovering new materials.
- Early prediction of catalyst deactivation: Monitoring and predictive fault detection based on digital twins (Mikołajewska et al., 2025).
- Real-time process control: Keeping an eye on the performance of DES-, MOF-, and biomass-based composite catalysts in the reactor all the time.

This technological change should improve energy efficiency and operational stability, as well as extend the catalyst's lifespan. This will be good for both the economy and the environment.

8.4. Conclusion: The Next Generation of Sustainable Desulfurization

Green-synthesised nanocomposites offer a new approach to fuel production that is environmentally friendly, cost-effective and uses AI to design the process in line with the principles of the circular economy. They are also a new type of material. The findings of this study suggest that the following factors will impact desulfurisation processes in the future:

- uses very little energy;
- improved catalytic selectivity;
- using feedstocks from renewable sources and waste;
- creating process methodologies that are environmentally friendly;
- putting in place digital and self-driving control systems.

Using green nanocomposites in refineries from 2030 to 2040 will benefit the environment and be a vital part of ensuring the longevity of the economy and technology. Research at the intersection of materials science, process engineering and digital technologies is likely to usher in a new era of efficient, flexible and environmentally friendly low-sulfur fuel production.

This chapter shows the scientific, environmental, and digital ideas that will allow us to start making fuel that is better for the environment and uses less sulfur and carbon. A smarter, more efficient, and more sustainable energy future is going to be made possible by this.

REFERENCES

Abdullah, T. A., Rasheed, R. T., Adnan, M., Qasim, B., Sajet, F., Mansoor, H. S., Talib, H., Kadhim, M. S., Kułacz, K., Meharban, F., & Abdul-Zahra, Z. (2024). A comprehensive review of the application of modified carbon nanotubes (CNTs) for the removal of metals from wastewater. *Materials Research Express*, *10*(12), 122005. DOI: 10.1088/2053-1591/ad1770

Adassooriya, N. M., Manipura, A., & Husen, A. (2025). Industrial Scale-Up Production of Nanoparticles—Challenges and Opportunities. In Adassooriya, N. M., Manipura, A., & Husen, A. (Eds.), *Industrial Scale Production of Nanoparticles. Smart Nanomaterials Technology* (pp. 1–7). Springer., DOI: 10.1007/978-981-96-9176-0_1

Afzal, S., Singh, A., Nicholson, S. R., Uekert, T., DesVeaux, J. S., Tan, E. C., Dutta, A., Carpenter, A. C., Baldwin, R. M., & Beckham, G. T. (2023). Techno-economic analysis and life cycle assessment of mixed plastic waste gasification for production of methanol and hydrogen. *Green Chemistry*, *25*(13), 5068–5085. DOI: 10.1039/D3GC00679D

Agbadaola, M. T., Adeojo, O. D., Michael-Agbadaola, O., & Babalola, J. O. (2025). Facile fabrication of a CuO/NiO-biomass heterostructure micro-/nano-composite for the adsorptive removal of methylene blue from wastewater. *Separation Science and Technology*, *60*(13), 1659–1672. DOI: 10.1080/01496395.2025.2514954

Ahmed, H. R., & Karim, A. R. (2025). Structural insights into metal oxide nanocomposite graphene frameworks as catalysts for oxidative desulfurization of light gasoil. *Journal of Molecular Structure*, *1325*, 141081. DOI: 10.1016/j.molstruc.2024.141081

Alabdullah, M. A., Gomez, A. R., Vittenet, J., Bendjeriou-Sedjerari, A., Xu, W., Abba, I. A., & Gascon, J. (2020). A viewpoint on the refinery of the future: Catalyst and process challenges. *ACS Catalysis*, *10*(15), 8131–8140. DOI: 10.1021/acscatal.0c02209

Ali, G., Mirza, M., & Mustafa, G. M. (2025). CNT–metal oxide composites as cathode materials for Li-ion batteries. In *Nanostructured Lithium-Ion Battery Materials* (pp. 319–339). Elsevier., DOI: 10.1016/B978-0-443-13338-1.00015-0

Altuner, E. E., Akin, M., Bayat, R., Bekmezci, M., Burhan, H., & Sen, F. (2022). Challenges in commercialization of carbon nanomaterial-based sensors. In *Carbon Nanomaterials-Based Sensors* (pp. 381–392). Emerging Research Trends in Devices and Applications., DOI: 10.1016/B978-0-323-91174-0.00020-2

Altuner, E. E., Arıkan, K., Burhan, H., Ozdemir, S., & Şen, F. (2021). Commercial aspects of direct alcohol fuel cells. In Nanomaterials for Direct Alcohol Fuel Cells (pp. 511–524). https://doi.org/ DOI: 10.1016/B978-0-12-821713-9.00012-3

Altuner, E. E., Erduran, V., & Sen, F. (2023). Green synthesized nanomaterials for bioimaging. In Synthesis of Bionanomaterials for Biomedical Applications: Micro and Nano Technologies (pp. 265–286). Elsevier. https://doi.org/ DOI: 10.1016/B978-0-323-91195-5.00023-4

Alvarez, M. (2023). Challenges in scaling up nanomaterial synthesis for industry: Technical, economic, and environmental perspectives. American Journal of Materials Engineering, 4(3), 12–18. E-ISSN: 2688-1780.

Álvarez-Chimal, R., & Arenas-Alatorre, J. A. (2023). Green synthesis and characterization of nanoparticles. In Kumar, P. (Ed.), *Green Nanotechnology—Overview and Further Prospects* (pp. 1–27). IntechOpen., DOI: 10.5772/intechopen.110751

Alviar, K. A. T., Arbilo, A. N. V., Correa, S. S., & Choi, A. E. S. (2024). Advanced desulfurization via ultrasonic-assisted oxidative methods in gasoline and crude oil towards sustainability. *Chemical Engineering Transactions, 114*, 55–60. DOI: 10.3303/CET24114010

Andreani, T., Nogueira, V., Gavina, A., Fernandes, S., Rodrigues, J. L., Pinto, V. V., Ferreira, M. J., Silva, A. M., Pereira, C. M., & Pereira, R. (2020). Ecotoxicity to freshwater organisms and cytotoxicity of nanomaterials: Are we generating sufficient data for their risk assessment? *Nanomaterials (Basel, Switzerland), 11*(1), 66. DOI: 10.3390/nano11010066 PMID: 33396620

Anekwe, I. M. S., & Isa, Y. M. (2025). Unlocking catalytic longevity: A critical review of catalyst deactivation pathways and regeneration technologies. *Energy Advances, 4*(9), 1075–1113. DOI: 10.1039/D5YA00015G

Bagchi, A., & Srivastava, P. (2024). Genetic and metabolic engineering approaches for enhanced biodesulfurization. *Frontiers in Bioengineering and Biotechnology, 12*, 1482270. DOI: 10.3389/fbioe.2024.1482270 PMID: 39530058

Bahari, N., Hashim, N., Abdan, K., Md Akim, A., Maringgal, B., & Al-Shdifat, L. (2023). Honey as reducing/capping agent in green nanoparticle synthesis. *Nanomaterials (Basel, Switzerland), 13*, 1244. DOI: 10.3390/nano13071244 PMID: 37049336

Bartoli, M., Giorcelli, M., & Tagliaferro, A. (2023). A comprehensive overview on biochar-based materials for catalytic applications. *Catalysts, 13*(10), 1336. DOI: 10.3390/catal13101336

Bartoli, M., Jagdale, P., Tagliaferro, A., & Rovere, M. (2023). Green-synthesized carbon-based nanocomposites for catalytic oxidative desulfurization: Mechanistic insights and performance evaluation. *Catalysts, 13*(10), 1336. DOI: 10.3390/catal13101336

Batool, Z., Rehman, A. U., Sindhu, K., Hassan, R. U., Almufarij, R. S., Arayro, J., Jumanazarov, D., Abduvokhidov, A., Ashraf, G. A., & Jabeen, R. (2025). Structure–performance correlation in NiS–CNT hybrid electrodes for advanced electrochemical energy storage. *Journal of Electroceramics*, •••, 1–17. DOI: 10.1007/s10832-025-00445-6

Beshtar, M., Ahmadi, S. M. M., Asgharinezhad, A. A., & Larimi, A. (2025). Catalysis and process parameters in photocatalytic oxidative desulfurization of liquid transportation fuels: An in-depth review. *Journal of Cleaner Production, 145462*. Advance online publication. DOI: 10.1016/j.jclepro.2025.145462

Bhagyaraj, S., & Krupa, I. (2023). Biopolymer-assisted synthesis of hybrid nanoflowers. *Inorganic Chemistry Communications, 158*, 111644. DOI: 10.1016/j.inoche.2023.111644

Boostani, F., Sharififard, H., Lashanizadegan, A., & Darvishi, P. (2024). Desulphurisation of liquid fuel using adsorption onto bio-based activated carbon modified with chitosan and Fe ions. *International Journal of Environmental Analytical Chemistry, 104*(13), 2956–2979. DOI: 10.1080/03067319.2022.2076222

Borj, B. P., Saeidi, S., As'adi, H., Naddaf, A., Fekri, H., & Sarkhoshkalat, M. (2025). Environmental life cycle assessment (LCA) of oxidative desulfurization of gas condensate in novel oxidation system. *Environmental Pollution, 371*, 125958. DOI: 10.1016/j.envpol.2025.125958 PMID: 40024511

Candra, A., Ahmed, Y. W., & Kitaw, S. L.. (2024). Gold-decorated bacterial cellulose nanocomposites. *Journal of Drug Delivery Science and Technology, 94*, 105477. DOI: 10.1016/j.jddst.2024.105477

Chadha, M., & Berry, S. (2025). A comparative investigation of bio-waste derived renewable catalysts for sustainable bis-enol derivative synthesis: Transforming waste biomass into value-added products. *Research on Chemical Intermediates*, 1–21. DOI: 10.1007/s11164-025-05813-z

Challa, A. A., Saha, N., Zhivkova, T., Alexandrova, R., & Saha, P. (2025). Bacterial Cellulose/Graphene Oxide/Hydroxyapatite Biocomposite: A Scaffold from Sustainable Sources for Bone Tissue Engineering. *ACS Applied Materials & Interfaces, 17*(1), 572–582. DOI: 10.1021/acsami.4c17306 PMID: 39698804

Chu, F., Wei, X., Lu, B., Zhao, G., Li, Y., Yang, K., & Lü, H. (2024). Flexible encapsulation of polyoxometalate into metal–organic frameworks for ultra-deep aerobic oxidation desulfurization of diesel via a biomimetic approach in deep eutectic solvents. *Chemical Engineering Journal*, *481*, 148549. DOI: 10.1016/j.cej.2024.148549

Daligaux, V., Richard, R., & Manero, M. H. (2021). Deactivation and regeneration of zeolite catalysts used in pyrolysis of plastic wastes—A process and analytical review. *Catalysts*, *11*(7), 770. DOI: 10.3390/catal11070770

Deshmukh, D. S., Singh, S., Badgujar, K. C., Humne, V. T., Korpe, G. V., & Bhanage, B. M. (2025). Green innovations in C–H bond functionalisation: Exploring homogeneous recyclable catalytic systems. *Green Chemistry*, *27*(20), 5667–5708. DOI: 10.1039/D5GC00278H

Dickson, U. J., & Huang, C. (2025). 2050 UK shipping emissions of NOx, SOx, PM10, and PM2.5: Is zero pollution by 2050 feasible? *Marine Pollution Bulletin*, *220*, 118371. DOI: 10.1016/j.marpolbul.2025.118371 PMID: 40644899

Ding, Y., Shang, H., Yang, C., Zhao, L., & Duan, A. (2024). Identifying efficient and inexpensive hydrodesulfurization catalysts through machine-learning-assisted analysis of metal–sulfur bonds in transition metal sulfides. *Chemical Engineering Science*, *298*, 120337. DOI: 10.1016/j.ces.2024.120337

Fernandes, S. C., Leo, P., Santos-Vieira, I., de Castro, B., Cunha-Silva, L., & Balula, S. S. (2024). Improving oxidative catalytic efficiency for fuels desulfurization using hybrid materials based on MOF-808@SBA-15. *ChemCatChem*, *16*(16), e202400355. DOI: 10.1002/cctc.202400355

Ferreira, P., Neves, R. P., Miranda, F. P., Cunha, A. V., Havenith, R. W., Ramos, M. J., & Fernandes, P. A. (2024). DszA catalyzes C–S bond cleavage through N5–hydroperoxyl formation. *Journal of Chemical Information and Modeling*, *64*(10), 4218–4230. DOI: 10.1021/acs.jcim.4c00301 PMID: 38684937

Fotiadis, K., Kostoglou, M., Baltzopoulou, P., Zaspalis, V., & Karagiannakis, G. (2024). Activated carbon modification for real diesel adsorptive deep desulfurization: Experiments and modeling. *Chemical Engineering Communications*, *211*(9), 1319–1335. DOI: 10.1080/00986445.2024.2341265

Gacem, M. A., & Abd-Elsalam, K. A. (2022). Strategies for scaling up of green-synthesized nanomaterials: Challenges and future trends. In *Green Synthesis of Silver Nanomaterials* (pp. 669–698). Elsevier., DOI: 10.1016/B978-0-12-824508-8.00008-3

Gallego-Ramírez, C., Chica, E., & Rubio-Clemente, A. (2023, September 4). Life Cycle Assessment of Raw and Fe-Modified Biochars: Contributing to Circular Economy. *Materials (Basel)*, *16*(17), 6059. DOI: 10.3390/ma16176059 PMID: 37687752

Gao, M., Wang, L., Yang, Y., Sun, Y., Zhao, X., & Wan, Y. (2023). Metal and metal oxide supported on ordered mesoporous carbon as heterogeneous catalysts. *ACS Catalysis*, *13*(7), 4060–4090. DOI: 10.1021/acscatal.2c05894

Gao, Y., Lia, X., Huang, S., Zhao, N., & Zhao, J. (2025). Dicationic ionic liquid modified peroxophosphotungstate encapsulated in ordered macro-microporous 3DOM-ZIF-8 for ultra-deep desulfurization of simulated fuel. *Small*, *2501667*(34), 2501667. Advance online publication. DOI: 10.1002/smll.202501667 PMID: 40590137

George, D., Palanisamy, U. M., & Begum, K. M. S. (2023). Chitosan nanocomposite hydrogel carriers. *Canadian Journal of Chemical Engineering*, *101*(7), 3937–3947. DOI: 10.1002/cjce.24782

Ghaffar, U., Habiba, U., Majeed, M. I., Nawaz, H., Rashid, N., Akhtar, N., & Hajab, H. (2024). Surface-enhanced Raman spectroscopy for real-time analysis of biodesulfurization of 4,6-DMDBT. *Analytical Letters*, *57*(14), 2182–2199. DOI: 10.1080/00032719.2023.2288695

Guosheng, L., Kunhong, L., Xiaohan, L., Xinrui, Y., Jiaxin, S., Rong, G., & Baojian, S. (2024). Acidity regulation of Fe-based catalysts and its effect on the selectivity of HDS reaction pathways. *Journal of Fuel Chemistry & Technology*, *52*(2), 234–247. DOI: 10.1016/S1872-5813(23)60389-5

Hamza, Z. A., Dawood, J. J., & Jabbar, M. A. (2024). Review of TiO_2 as desulfurization catalyst for petroleum. *Catalysts*, *14*(6), 381. DOI: 10.3390/catal14060381

Hossain, M. A., & Jhung, S. H. (2024). Supported chromium nitride: A new catalyst for oxidative desulfurization of liquid fuel. *Fuel*, *372*, 132178. DOI: 10.1016/j.fuel.2024.132178

Hosseinzadeh, E., Foroumadi, A., & Firoozpour, L. (2023). DFT insights into phytochemical capping agents. *Inorganic Chemistry Communications*, *147*, 110243. DOI: 10.1016/j.inoche.2022.110243

Hou, J., Deng, H. K., Liu, Z. X., Xu, P., & Wang, L. J. (2023). Sulfur metabolism in Rhodococcus species and applications in fuel desulfurization. *Journal of Applied Microbiology*, *134*(3), lxad048. Advance online publication. DOI: 10.1093/jambio/lxad048 PMID: 36893799

Huang, L., He, G., Yuan, Y., Zhang, T. C., Wang, Y., & Yuan, S. (2024). Trivalent metal ion (Al, Ga, In)-doped TiO_2 for enhanced photocatalytic desulfurization of H_2S: Band structure regulation and mechanism. *Industrial & Engineering Chemistry Research, 63*(16), 7154–7165. DOI: 10.1021/acs.iecr.4c00074

Humadi, J. I., Hamad, K. I., Abdulkareem, H. A., Ismael, M. N., Jarullah, A. T., Ahmed, M. A., & Mujtaba, I. M. (2025). Eco-friendly oxidative–adsorptive desulfurization for real diesel fuel using green MnO_2 biowaste-extracted calcite in a digital basket reactor. *Processes (Basel, Switzerland), 13*(10), 3084. DOI: 10.3390/pr13103084

Ikram, R., Mohamed Jan, B., Nagy, P. B., & Szabo, T. (2023). Recycling waste sources into nanocomposites of graphene materials: Overview from an energy-focused perspective. *Nanotechnology Reviews, 12*(1), 20220512. DOI: 10.1515/ntrev-2022-0512

Jarullah, A. T., Hussein, A. K., Al-Tabbakh, B. A., Hameed, S. A., Mujtaba, I. M., Saeed, L. I., & Humadi, J. I. (2024). Production of green fuel using a synthetic magnetite mesoporous nano-silica composite catalyst for oxidative desulfurization: Experiments and process modeling. *Catalysts, 14*(8), 529. DOI: 10.3390/catal14080529

Ju, H., Zhang, Y., Zhang, J., Yu, Z., Zhang, Y., Zhang, X., Guo, X., Liu, J., Mao, Q., Liu, Q., Zhao, Y., Qi, T., Jiang, X., Guo, Z., & Chen, S. (2025). Development of a novel Cu(I) π-complexation adsorbent for ultra-deep desulfurization from a carbon dioxide stream. *Smart Molecules, 3*(1), e20240027. DOI: 10.1002/smo.20240027 PMID: 40625567

Kaeed, Z., Shahhosseini, S., & Rezaei, M. (2024). Preparation of the MoO_3–Fe_3O_4 nanocomposites for deep oxidative desulfurization of model fuel: Optimization and experimental design. *Journal of Industrial and Engineering Chemistry, 139*, 267–280. DOI: 10.1016/j.jiec.2024.05.003

Kandav, G., & Sharma, T. (2024). Green synthesis overview. *Particulate Science and Technology, 42*(5), 874–894. DOI: 10.1080/02726351.2023.2281452

Kar, S., Dey, S., & Chowdhury, K. B.. (2023). CuO/biochar composites via phyto-assisted synthesis. *Environmental Research, 236*, 116824. DOI: 10.1016/j.envres.2023.116824 PMID: 37549783

Karpuz, Ö., Baltacı, C., & Gül, A. (2025). *ve diğerleri.* Karabuğday kabuğu atığının sulu özütünden yeşil demir ve gümüş nanopartiküllerinin sentezi: Antibakteriyel, sitotoksik ve boya giderici özellikleri. *Biomass Conversion and Biorefinery, 15*, 27679–27711. DOI: 10.1007/s13399-024-06287-6

Khaleghian, M., Sheikhi, M., Shahab, S., Yahyaei, H., & Ahmadianarog, M. (2024). Theoretical investigation of the adsorption behavior of dibenzothiophene molecule on pristine boron nitride nanosheet. *Main Group Chemistry*, *23*(4), 425–438. DOI: 10.3233/MGC-230002

Khan, M. K., Faheem, K., Khan, Z., & Qayyum, N. (2025). Green synthesis of polymeric materials and nanocomposites for effluent treatment. In Sustainable Effluent Treatment and Resource Recovery (Vol. 1, pp. 153–180). https://doi.org/ DOI: 10.1021/bk-2025-1517.ch005

Khikani, M., Isopencu, G. O., & Deleanu, I. M.. (2024). BC-based antibacterial nanocomposite membranes. *Journal of Composites Science*, *8*(11), 475. DOI: 10.3390/jcs8110475

Kim, J. R., Hu, Y., Zavala, V. M., & Karthikeyan, K. G. (2022). Techno-economic analysis of pretreatments to dairy manure biomass for enhanced biogas production. *Bioresource Technology Reports*, *20*, 101275. DOI: 10.1016/j.biteb.2022.101275

Kirubakaran, D., Wahid, J. B. A., Karmegam, N., Jeevika, R., Sellapillai, L., Rajkumar, M., & SenthilKumar, K. J. (2026). A comprehensive review on the green synthesis of nanoparticles: Advancements in biomedical and environmental applications. *Biomedical Materials & Devices*, *4*(1), 388–413. DOI: 10.1007/s44174-025-00295-4

Kulak, M., & Gulmez Samsa, C. (2023). Flavonoid-mediated nanomaterial synthesis. In *Secondary metabolites based green synthesis* (pp. 49–65). Springer., DOI: 10.1007/978-981-99-0927-8_3

Kwao-Boateng, E., Ankudey, E. G., Darkwah, L., & Danquah, K. O. (2024). Assessment of diesel fuel quality. *Heliyon*, *10*(2), e24733. Advance online publication. DOI: 10.1016/j.heliyon.2024.e24733 PMID: 38312546

Lakhmir, M. A., Akhter, F., Ahmed, J., Mushtaq, S., Arain, H. J., & Ahsan, M. J. (2025). Impact of sulfur-ash reduction treatment on combustion performance, SOx emissions, and flue gas temperatures of lignite coals. *Environmental Science and Pollution Research International*, *32*(9), 5574–5588. DOI: 10.1007/s11356-025-36064-5 PMID: 39934577

Lazcano-Ramírez, H. G., Garza-García, J. J., & Hernández-Díaz, J. A.. (2023). Plant-mediated selenium nanoparticles. *Antibiotics (Basel, Switzerland)*, *12*(1), 115. DOI: 10.3390/antibiotics12010115 PMID: 36671316

Leal, E., Torres-Mancera, P., Alonso, F., Luna, R., Muñoz, J. A., & Ancheyta, J. (2024). Optimization methodology of dual-bed catalyst stacking systems to produce ultralow-sulfur diesel. *Industrial & Engineering Chemistry Research, 63*(42), 17857–17867. DOI: 10.1021/acs.iecr.4c02846

Lellala, K., Behera, S. K., Srivastava, P., Saeed, W. S., Haidyrah, A. S., & Burile, A. N. (2024). Fe_3O_4 nanoparticles decorated on N-doped graphene oxide for heavy metal removal and desulfurization. *Diamond and Related Materials, 150*, 111746. DOI: 10.1016/j.diamond.2024.111746

Lewandowski, M., Janus, R., Wadrzyk, M., Jaroszewska, K., & Zaborowska, K. (2025). Activity of amorphous NiB/SiO_2 in hydrotreating model reactions: Hydrodesulfurization of 4,6-DMDBT. *ACS Omega, 10*(34), 38968–38984. DOI: 10.1021/acsomega.5c04966 PMID: 40918321

Li, H., Xiang, X., Zhang, Y., Cheng, H., Chen, Q., Li, X., Wu, F., & Liu, X. (2025). The green preparation of ZrO_2-modified WO_3–SiO_2 composite from rice husk and its excellent oxidative desulfurization performance. *Catalysts, 15*(10), 996. DOI: 10.3390/catal15100996

Liao, M., Xiao, Y., Pi, X., Zhang, L., Zheng, M., Liu, J., & Guo, J. (2024). Zr-based UiO-66 doped with rare-earth elements supported polyoxometalate liquid for oxidative desulfurization of dibenzothiophene. *Energy & Fuels, 38*(14), 12555–12568. DOI: 10.1021/acs.energyfuels.4c01125

Lim, X. B., & Ong, W. J. (2021). A current overview of the oxidative desulfurization of fuels utilizing heat and solar light: From materials design to catalysis for clean energy. *Nanoscale Horizons, 6*(8), 588–633. DOI: 10.1039/D1NH00127B PMID: 34018529

Lin, X., Li, W., Nguyen, V., Wang, S., Yang, S., Ma, L., & Fan, Z. (2023). Fe single-atom catalyst nanocages supported by bacterial cellulose-derived carbon aerogel. *Chemical Engineering Journal, 477*, 146977. DOI: 10.1016/j.cej.2023.146977

Linares, C. F., & Brunet, S. (2023). Improving the hydrogenation pathway from the dibenzothiophene (DBT) hydrodesulfurization (HDS) reaction using alloys of bulk NiXCoyborides. *Reaction Kinetics, Mechanisms and Catalysis, 136*(2), 901–917. DOI: 10.1007/s11144-023-02393-0

Liu, D., Zhang, G., Kang, X., Wu, W., Zhao, L., Gao, J., & Xu, C. (2025). Hydrodesulfurization and isomerization performance of model FCC naphtha over sulfided Co(Ni)–Mo/Y catalysts. *Chinese Journal of Chemical Engineering, 77*, 185–194. DOI: 10.1016/j.cjche.2024.10.001

Liu, M., Ye, Y., Ye, J., Gao, T., Wang, D., Chen, G., & Song, Z. (2023). Recent advances of Fe_3O_4-based magnetic materials in catalysis. *Magnetochemistry (Basel, Switzerland)*, *9*(4), 110. DOI: 10.3390/magnetochemistry9040110

Liu, Q., Chang, Q., Liao, X., Jiang, Y., Lyu, X., Zhao, Q., Diao, J., Wang, X., Huang, X., & Lyu, S. (2024). Adsorption desulphurization of biochar derived from eucalyptus waste. *Powder Technology*, *448*, 120322. DOI: 10.1016/j.powtec.2024.120322

Luo, S., Liu, Y., Zhu, Y., Niu, Q., Cheng, M., Ye, S., Yi, H., Shao, B., Shen, M., Wen, X., Zeng, G., & Liu, Z. (2021). Perspectives on palladium-based nanomaterials: Green synthesis, ecotoxicity, and risk assessment. *Environmental Science. Nano*, *8*(1), 20–36. DOI: 10.1039/D0EN01048K

Ma, L., Zhong, Y., Qu, L., Feng, X., Zhang, Y., Hou, H., & Yang, G. (2025). Conformational flexibility of a novel Cu(I)-based adsorbent enhancing the selectivity of adsorptive desulfurization. *Separation and Purification Technology*, *358*, 130324. DOI: 10.1016/j.seppur.2024.130324

Maheshwaran, L., Nadarajah, L., & Senadeera, S. P. N. N.. (2024). Phytochemical screening methods. *Asian Plant Research Journal*, *12*(5), 11–38. DOI: 10.9734/aprj/2024/v12i5267

Mao, X., Wang, Y., Ma, Z., Hao, R., Fu, D., Yuan, B., & Zhao, Y. (2025). Deep oxidation removal of NO and SO_2 via singlet oxygen and hydroxyl radicals in $LaCoO_3$–H_2O_2 system. *ACS Omega*, *10*(39), 45297–45308. DOI: 10.1021/acsomega.5c04716 PMID: 41078817

Matloob, A. M., Abd El-Hafiz, D. R., Saad, L., & Mikhail, S. (2023). Cr/Fe–MOF/graphene nanocomposite for sulfur removal from diesel. *Journal of Inorganic and Organometallic Polymers and Materials*, *33*(1), 254–265. DOI: 10.1007/s10904-022-02472-0

Matloob, A. M., Al-Hagar, O. E., Abol-Fotouh, D., & Salem, H. M. (2025). Green synthesis of ZnO–Zn–MOF/bacterial nanocellulose for ultra-oxidative desulfurization of actual diesel fuel. *Scientific Reports*, *15*(1), 29246. DOI: 10.1038/s41598-025-14377-2 PMID: 40784888

Mehri, M., Khosravi-Nikou, M. R., Aghajani, M., & Shariati, A. (2021). Process intensification of liquid hydrocarbon fuel desulfurization using catalytic–oxidative–adsorptive method. *Chemical Papers*, *75*(7), 3667–3678. DOI: 10.1007/s11696-021-01618-7

Melnyk, O. M., Onishchenko, O. A., Shyshkin, O. V., Volkov, O. M., Volyanskyy, S. M., Maulevych, V. O., & Kreitser, K. O. (2024). Enhancing shipboard technical facility performance through the utilization of low-sulfur marine fuel grades. *Journal of Chemistry and Technologies*, *32*(1), 233–245. DOI: 10.15421/jchemtech.v32i1.297916

Metawea, R., Farag, H. A., El-Ashtoukhy, E. S., El-Latif, M. M. A., & El-Sayed, E. M. (2023). Ultrasound-assisted one-pot oxidative desulfurization using green-synthesized MIL-53(Al). *Scientific Reports*, *13*(1), 13728. DOI: 10.1038/s41598-023-40955-3 PMID: 37608052

Mikołajewska, E., Mikołajewski, D., Mikołajczyk, T., & Paczkowski, T. (2025). Generative AI in AI-based digital twins for fault diagnosis for predictive maintenance in Industry 4.0/5.0. *Applied Sciences (Basel, Switzerland)*, *15*(6), 3166. DOI: 10.3390/app15063166

Mishra, P. M., & Devi, A. P. (2023). Biogenic synthesis of metal oxide nanocomposites using plant specimens and their application towards treatment of wastewater. *Environmental Science and Pollution Research International*, *30*(50), 108512–108524. DOI: 10.1007/s11356-023-29989-2 PMID: 37775638

Moghaddam, M. R. H., & Shahrak, M. N. (2024). Coupled one-pot oxidative extractive desulfurization of a model fuel catalyzed by ZIF-67/biomass-derived activated carbon composite. *Inorganic Chemistry Communications*, *165*, 112489. DOI: 10.1016/j.inoche.2024.112489

Motaleb, K. A., Pranta, A. D., Islam, M. R., Islam, S., & Janutėnienė, J. (2025). Bio-waste as a resource for sustainable nanocomposites: Strategies and multifunctional applications. *Functional Composites and Structures*, *7*(2), 022002. DOI: 10.1088/2631-6331/add9e0

Mugisha, A. I., Fauteux-Lefebvre, C., Thangarasu, P., & Huerta-Aguilar, C. A. (2025). 2D graphene oxide–iron oxide catalysts for oxidation of sulfides. *Journal of Molecular Structure*, *1339*, 142382. DOI: 10.1016/j.molstruc.2025.142382

Nawaf, A., & Abdul Majeed, B. (2024). Kinetics study of oxidative desulfurization of real diesel fuel over uncoated and coated nano-catalysts in an oscillatory helical baffled reactor. *Journal of Chemical and Petroleum Engineering*, *58*(2), 359–374. DOI: 10.22059/jchpe.2024.377539.1522

Nour, M. M., Tony, M. A., Fouad, M. K., & Nabwey, H. A. (2025). Biowaste-to-catalyst: Magnetite-functionalized potato shell as green magnetic biochar catalyst (PtS200–Fe3O4). *Catalysts*, *15*(10), 997. DOI: 10.3390/catal15100997

Oleiwi, A. H., Jabur, A. R., Alsalhy, Q. F., & Zeinedini, A. (2025). Crude oil desulfurization by polystyrene/graphene nanocomposite membranes. *Periodica Polytechnica. Chemical Engineering*, *69*(1), 1–10. DOI: 10.3311/PPch.38227

Orlova, E. V. (2022). Design technology and AI-based decision-making model for digital twin engineering. *Future Internet*, *14*(9), 248. DOI: 10.3390/fi14090248

Özkan, V., & Bozbeyoğlu, P. (2025). Production of Pure PVC, Graphene and Carbon Nanotube-Doped PVC Electrospun Nanofiber Membranes: A Highly Effective Method for Removing Acid Blue 74 Dye from Wastewater. *Black Sea Journal of Engineering and Science*, *8*(2), 517–523. DOI: 10.34248/bsengineering.1634974

Pan, Y., Huang, W., Liu, Z. H., Yang, Z. Z., Tan, J. J., & Xu, B. M. (2025). Recent progress of MOFs for liquid-fuel desulfurization. *Separation and Purification Reviews*, *54*(2), 128–148. DOI: 10.1080/15422119.2024.2367646

Pantoja-Romero, W., Lavín-Flores, A., & Morell, G.. (2025). Solvent-free MOF–quantum dot composites. *Chemistry (Weinheim an der Bergstrasse, Germany)*, *31*(34), e202500655. DOI: 10.1002/chem.202500655 PMID: 40326208

Patil, A., Chaudhari, V., Patil, S. R., Borse, G. P., & Patil, V. (2023). A bio-waste derived sustainable heterogeneous catalyst for Biginelli reaction. *Journal of the Indian Chemical Society*, *100*(9), 101080. DOI: 10.1016/j.jics.2023.101080

Pirsaheb, M., Gholami, T., & Seifi, H.. (2024). Plant extracts in green nanomaterial synthesis. *Environmental Science and Pollution Research International*, *31*(17), 24768–24787. DOI: 10.1007/s11356-024-32983-x PMID: 38523214

Priyadarshini, P., Sahoo, S. K., & Parida, K. (2025). Hierarchically structured MOF-on-MOF photocatalysts with engineered charge dynamics for sustainable green fuel generation. *RSC Sustainability*. Advance online publication. DOI: 10.1039/D5SU00595G

Qasim, M., Clarkson, A. N., & Hinkley, S. F. (2023). Biomass-derived carbon nanoparticles. *International Journal of Molecular Sciences*, *24*(2), 1023. DOI: 10.3390/ijms24021023 PMID: 36674532

Qiu, X., Wang, B., Wang, R., & Kozhevnikov, I. V. (2024). New adsorption materials for deep desulfurization of fuel oil. *Materials (Basel)*, *17*(8), 1803. DOI: 10.3390/ma17081803 PMID: 38673161

Rabiee, N. (2025). Green Nanomaterials: Sustainable Innovations and Diverse Applications. ACS Symposium SeriesVol. 1520, Chapter 1pp 1-27, eISBN. DOI: 10.1021/bk-2025-1520

Radulescu, D. M., Surdu, V. A., Ficai, A., Ficai, D., Grumezescu, A. M., & Andronescu, E. (2023). Green synthesis of metal and metal oxide nanoparticles: A review of the principles and biomedical applications. *International Journal of Molecular Sciences, 24*(20), 15397. DOI: 10.3390/ijms242015397 PMID: 37895077

Radwan, M. S., Aboutaleb, W. A., El Naggar, A. M., Medany, S. S., & Shehata, M. R. (2023). Synergetic effect of Fe_2O_3–CeO_2 nanocomposites prepared via different techniques on photocatalytic desulfurization of heavy gas oil. *Arabian Journal for Science and Engineering, 48*(12), 15837–15850. DOI: 10.1007/s13369-023-07633-y

Ramadan, B. S., Wibowo, Y. G., Anwar, D., & Maryani, A. T. (2024). A review of life cycle assessment of nanomaterials-based adsorbent for environmental remediation. *Global NEST Journal, 26*(7), 06216. DOI: 10.30955/gnj.06216

Rani, N., Singh, P., Kumar, S., Kumar, P., Bhankar, V., & Kumar, K. (2023). Plant-mediated synthesis of nanoparticles and their applications: A review. *Materials Research Bulletin, 163*, 112233. DOI: 10.1016/j.materresbull.2023.112233

Rao, J. J., Leal, E., Trejo, F., & Ancheyta, J. (2023). Kinetic models of deep hydrotreating reactions to produce ultralow sulfur diesel. *Energy & Fuels, 37*(15), 11216–11247. DOI: 10.1021/acs.energyfuels.3c01466

Rashid, T., Raza, A., Saleh, H. A., Khan, S., Rahaman, S., Pandey, K., AlDamen, M. A., Sama, F., Ahmad, A., Shahid, M., & Ahmad, S. A. (2023). Green synthesis of Ni/Fe3O4/rGO nanocomposites for desulfurization of fuel. *ACS Applied Nano Materials, 6*(20), 18905–18917. DOI: 10.1021/acsanm.3c03270

Reis Medeiros, K. A., da Costa, L. G., Bifano Manea, G. K., de Moraes Maciel, R., Caliman, E., da Silva, M. T., de Sena, R. C., & de Oliveira, E. C. (2024). Determination of total sulfur content in fuels: A comprehensive and metrological review. *Critical Reviews in Analytical Chemistry, 54*(8), 3398–3408. Advance online publication. DOI: 10.1080/10408347.2023.2249564 PMID: 37610291

Riaz, F., Rasul, I., Azeem, F., ... (2024). Environmental impact and LCA of bionanocomposites. In Advances in Bionanocomposites (pp. 387–406). DOI: 10.1016/B978-0-323-91764-3.00009-7

Rios, J. J., Leal, E., Trejo, F., & Ancheyta, J. (2023). Kinetic models of deep hydrotreating reactions to produce ultralow sulfur diesel. *Energy & Fuels, 37*(15), 11216–11247. DOI: 10.1021/acs.energyfuels.3c01466

Rizwan, M., Ali, M. F., Nawaz, M. A., He, M., Song, Y., Jiang, P., Ullah, S., Hassan, M. M. A., & Zhou, X. (2025). Synergistic effect of NaOH/NaClO absorbent in a novel wet FGD scrubber to control SOx/NOx emissions. *Environmental Monitoring and Assessment*, *197*(2), 170. DOI: 10.1007/s10661-024-13455-8 PMID: 39808324

Rojas, L., Peña, Á., & Garcia, J. (2025). AI-driven predictive maintenance in mining: A systematic literature review on fault detection, digital twins, and intelligent asset management. *Applied Sciences (Basel, Switzerland)*, *15*(6), 3337. DOI: 10.3390/app15063337

Saeed, M., Firdous, A., Zaman, M. S., Izhar, F., Riaz, M., Haider, S., & Tariq, S. (2023). MOFs for desulfurization: Recent advances. *Journal of the Chinese Chemical Society (Taipei)*, *70*(4), 789–824. DOI: 10.1002/jccs.202200546

Saeed, M., Waseem, A., & Intisar, A. (2025). Catalytic activity of MnWO4@g-C3N4 and CuWO4@g-C3N4 nanocomposites for green fuel production. *Catalysis Surveys from Asia*, *29*(1), 11–25. DOI: 10.1007/s10563-024-09437-y

Sagin, S., and Sagin, A. (2023). Managing risk factors for emergency situations when using low-sulfur content fuel in marine diesel engines. Technology Audit and Production Reserves, 5(1/73), 37–43. .DOI: 10.15587/2706-5448.2023.290198

Salih, Y. M. (2023). Adsorptive desulfurization of DBT using hybrid MOF–activated carbon. *Journal of Sulfur Chemistry*, *44*(6), 694–711. DOI: 10.1080/17415993.2023.2241594

Sehhatigdiri, A., & Pourhossein, M. (2024). Application of aminated chitosan/functionalized CNT nanocomposite in adsorption desulfurization of fuels containing dibenzothiophene. *Fullerenes, Nanotubes, and Carbon Nanostructures*, *32*(5), 483–492. DOI: 10.1080/1536383X.2023.2296124

Selvaraj, V., Sagadevan, S., & Karuppasamy, G. (2024). Scale-Up Process of Green Nanomaterials. In Sustainable Green Synthesised Nano-Dimensional Materials for Energy and Environmental Applications (pp. 107–122). CRC Press. DOI: 10.1201/9781003362241-6

Shadmehri, M. A., Housaindokht, M. R., & Nakhaei Pour, A. (2023). Metal oxides supported on GO for oxidative desulfurization. *Applied Organometallic Chemistry*, *37*(11), e7272. DOI: 10.1002/aoc.7272

Sharma, S., Kumar, R., and Kumar, K. (2024). Synthesis of biowaste-derived carbon nanomaterials: Efficiency, quality, and environmental impact. Indian Chemical Engineer, 1–20. .DOI: 10.1080/00194506.2024.2442719

Shetti, R. S., Shetti, S. M., Sreenivasulu, M., Alibrahim, K. A., Alodhayb, A. N., & Shetti, N. P. (2025). Microwave-assisted synthesis of Ni-doped carbon from bio-waste as a cost-effective electrocatalyst for effective and sustainable water electrolysis. *Materials Today Sustainability, 101236*, 101236. Advance online publication. DOI: 10.1016/j.mtsust.2025.101236

Shi, Q., & Wu, J. (2021). Review on sulfur compounds in petroleum and its products: State-of-the-art and perspectives. *Energy & Fuels, 35*(18), 14445–14461. DOI: 10.1021/acs.energyfuels.1c02229

Shrivastav, G., Prava Jyoti, T., Chandel, S., & Singh, R. (2025). Eco-friendly extraction innovations. *Separation and Purification Reviews, 54*(3), 241–257. DOI: 10.1080/15422119.2024.2381605

Singh, S., Kulshrestha, M. J., Rani, N., Kumar, K., Sharma, C., & Aswal, D. K. (2023). An overview of vehicular emission standards. *MPAN. Journal of Metrology Society of India, 38*(1), 241–263. DOI: 10.1007/s12647-022-00555-4

Sousa, J. P., Neves, R. P., Sousa, S. F., Ramos, M. J., & Fernandes, P. A. (2020). Reaction mechanism and determinants for efficient catalysis by DszB. *ACS Catalysis, 10*(16), 9545–9554. DOI: 10.1021/acscatal.0c03122

Sun, M., Abazari, R., Chen, J., Hussain, C. M., Zhou, Y., & Kirillov, A. M. (2023). Encapsulation of H4SiW12O40 into amide-functionalized MOF for oxidative desulfurization of diesel. *ACS Applied Materials & Interfaces, 15*(45), 52581–52592. DOI: 10.1021/acsami.3c12374 PMID: 37931039

Thakur, A., & Kumar, A. (2023). Ecotoxicity analysis and risk assessment of nanomaterials for environmental remediation. *Macromolecular Symposia, 410*(1), 2100438. DOI: 10.1002/masy.202100438

Tsai, M. L., Lo, A. Y., Liu, J. H., and Dai, Y. M. (2025). Adsorptive desulfurization using waste-tire-derived carbon. C, 11(3), 47. .DOI: 10.3390/c11030047

Vermeersch, L., De Proft, F., Faulkner, V., & De Vleeschouwer, F. (2023). Unravelling the mechanism and governing factors in Lewis acid and noncovalent Diels–Alder catalysis. *International Journal of Molecular Sciences, 24*(5), 4938. DOI: 10.3390/ijms24054938 PMID: 36902369

Wang, Q., Huang, W., Li, X., Lin, S., Li, Z., & Ma, X. (2024). Electronic regulation of MoS_2 edge sites by d electron transfer of Ni or Co to improve the activity of CO sulfur-resistant methanation. *Chemical Engineering Journal, 480*, 148242. DOI: 10.1016/j.cej.2023.148242

Wang, Q., Wu, Y., Shan, Y., & Dong, W. (2025). Synergizing surface-bound radicals with singlet oxygen for degradation of emerging contaminants. *Environmental Science & Technology*, *59*(42), 22901–22913. DOI: 10.1021/acs.est.5c04418 PMID: 41082611

Wiatrowski, M., Klein, B., Do, T. N., Ou, L., Cai, H., Carlson, N., & Davis, R. (2025). Technology Case Study: Techno-Economic and Life Cycle Analysis for Microalgae Conversion Pathways to Fuels and Products (No. NREL/TP-5100-91854). National Renewable Energy Laboratory (NREL), Golden, CO (United States).

Wu, F., Yuan, Q., Wang, J., Wang, X., Luo, J., Guo, Y., Xu, H., & Wei, X. (2025). Oxidative desulfurization catalyzed by magnetically recoverable CoFe2O4 nanoparticles. *Arabian Journal of Chemistry*, *18*(1), 106076. DOI: 10.1016/j.arabjc.2024.106076

Wu, P., Deng, C., Liu, F., Zhu, H., Chen, L., Liu, R., Zhu, W., & Xu, C. (2023). Magnetically recyclable high-entropy metal oxide catalyst for aerobic catalytic oxidative desulfurization. *Chinese Journal of Catalysis*, *54*, 238–249. DOI: 10.1016/S1872-2067(23)64541-3

Xiao, C. K., Zou, Y. T., Li, D. Z., Wang, E. H., Wang, A. C., Gao, D. W., Duan, A.-J., Zheng, P., & Wang, X. L. (2023). Optimization of dendritic TS-1/silica micro–mesoporous composites for efficient hydrodesulfurization of dibenzothiophene and 4,6-dimethyldibenzothiophene. *Petroleum Science*, *20*(4), 2521–2530. DOI: 10.1016/j.petsci.2023.07.004

Xiao, G., Amamoo-Otoo, C., Wang, T., Li, Q., & Biancardo, S. A. (2025). Evaluating the impact of ECA policy on sulfur emissions from the five busiest ports in America. *Frontiers in Marine Science*, *12*, 1609261. DOI: 10.3389/fmars.2025.1609261

Xue, L. L., He, Q., Li, Z. Z., Wu, J. Q., Li, S. W., & Zhao, J. S. (2025). Remarkable catalytic performance of supported P–Mo–W polyoxometalate on green-friendly fiber for ultra-deep oxidative desulfurization. *Applied Organometallic Chemistry*, *39*(11), e70410. DOI: 10.1002/aoc.70410

Yu, B., Mi, J., Cui, C., Xiang, S., & Qiu, K. (2025). Biochar-supported zinc ferrite for H2S removal at room temperature under anaerobic conditions. *Chemical Engineering Research & Design*, *222*, 286–297. DOI: 10.1016/j.cherd.2025.09.015

Yuan, Y., Huang, L., Yılmaz, M., Zhang, T. C., Wang, Y., & Yuan, S. (2023). $MgFe_2O_4$-loaded N-doped biochar for H_2S desulfurization. *Fuel*, *339*, 127385. DOI: 10.1016/j.fuel.2022.127385

Yue, X., Shan, Y., Jiao, W., Shen, K., & Zhang, Y. (2023). Direct current effects on peroxydisulfate activation: Singlet oxygen generation and accelerated degradation. *Journal of Hazardous Materials*, *452*, 131187. DOI: 10.1016/j.jhazmat.2023.131187 PMID: 36996542

Zbuzant, M. (2022). An overview of photocatalysts for desulfurization. Eurasian Journal of Chemical. *Medicinal and Petroleum Research*, *1*(1), 10–19. DOI: 10.5281/zenodo.7353454

Zeng, X., Lin, C., Ren, Q., Shan, S., & Wang, H. (2024). CuO–MgO enhanced with $MgMoO_4$ for oxidative desulfurization. *Fuel*, *360*, 130630. DOI: 10.1016/j.fuel.2023.130630

Zhang, H., Su, T., Yu, S., Liao, W., Ren, W., Zhu, Z., Yang, K., Len, C., Dong, G., Zhao, D., & Lü, H. (2023). Fabrication of Anderson-Polyoxometalates/TiO_2/g-C_3N_4 heterojunction composite for efficient visible-light-driven photooxidative desulfurization. *Molecular Catalysis*, *536*, 112916. DOI: 10.1016/j.mcat.2023.112916

Zhang, X., Cheng, X., Kang, J., Fan, F., Sun, H., Zhou, A., & Zhou, W. (2025). The influence of In-modified NiMoS active phase on the direct desulfurization reaction process of 4,6-dimethyldibenzothiophene. *Molecular Catalysis*, *581*, 115143. DOI: 10.1016/j.mcat.2025.115143

Zhang, X., & Wang, R. (2025). Phase-transfer catalysis for fuel desulfurization. *Catalysts*, *15*(8), 724. DOI: 10.3390/catal15080724

Zhang, Y., Liu, F., Chen, W., Han, W., Zhai, W., Lu, Y., & Li, M. (2024). Effective reduction of hydrogen consumption in ultra-deep hydrodesulfurization of diesel: Deep insights into thermodynamic limitations during hydrotreating. *Fuel*, *356*, 129640. DOI: 10.1016/j.fuel.2023.129640

Zheng, H., Lian, J., Liu, Y., Tang, Z., & Ji, J. (2024). Molecular mechanism of competitive adsorption desulfurization involving sulfur-containing molecules. *Colloids and Surfaces. A, Physicochemical and Engineering Aspects*, *690*, 133610. DOI: 10.1016/j.colsurfa.2024.133610

Zheng, S., Yang, S., Chen, J., Wu, D., Qi, B., Zhang, C., Deng, W., Li, J., Mei, T., Wang, S., & Wan, L. (2024). New progresses in efficient, selective, and environmentally friendly recovery of valuable metal from e-waste and industrial catalysts. *Advanced Sustainable Systems*, *8*(6), 2300512. DOI: 10.1002/adsu.202300512

Zhou, X., Wang, T., Zhang, L., Che, S., Liu, H., Liu, S., Wang, C., Su, D., & Teng, Z. (2022). Highly efficient Ag_2O/Na-g-C_3N_4 heterojunction for photocatalytic desulfurization of thiophene in fuel under ambient air conditions. *Applied Catalysis B: Environmental*, *316*, 121614. DOI: 10.1016/j.apcatb.2022.121614

Chapter 8
Green Nanotechnology in Action and Magnetic and Hybrid Bio-Nanocomposites for Recyclable Fuel Desulfurization

Nayab Ahmad
The University of Agriculture, Peshawar, Pakistan

Sidra Ahmad
https://orcid.org/0009-0004-4831-5906
The University of Agriculture, Peshawar, Pakistan

Micheal Ayodeji Sokan-Adeaga
https://orcid.org/0000-0001-8550-1960
Department of Community Health and Primary Care, Faculty of Clinical Sciences, College of Medicine, University of Lagos, Lagos, Nigeria

Irshad Ali Khan
Department of Agriculture, The University of Swabi, Khyber Pakhtunkhwa, Pakistan

Adewale Allen Sokan-Adeaga
https://orcid.org/0000-0002-0885-1526
Public Health Program, College of Health Sciences, Bowen University, Iwo, Nigeria

Sezai Ercisli
Department of Horticulture, Faculty of Agriculture, Ataturk University, Erzurum, Turkey & HGF Agro, Ata Teknokent, Erzurum, Turkey

ABSTRACT

This chapter explores magnetic and hybrid bionanocomposites as a novel class of recyclable, eco-friendly materials for green fuel desulfurization. It covers the

DOI: 10.4018/979-8-3373-5966-3.ch008

structural and functional engineering of bio-based core–shell frameworks and the incorporation of magnetic cores (for example, Fe_3O_3) into biogenic supports such as biochar, cellulose, or graphene oxide. These materials demonstrate superior efficiencies in both adsorptive and oxidative desulfurization under mild conditions, making them suitable for the removal of refractory sulfur compounds. The chapter also discusses green synthesis methods, magnetic separation, and reusability of materials over multiple cycles. Environmental and safety considerations are addressed from a life cycle sustainability perspective, including biodegradability, toxicity, and waste management. In addition, this chapter analyses the challenges and scalability of economic and process integration.

INTRODUCTION

However, desulfurization of fossil fuels remains a major issue in the energy sector, as it brings environmental concerns and the control of sulfur emissions. Thiophenes (thiophene, benzothiophene (BT), and dibenzothiophene (DBT)) are common sulfur oxide (SO^{-2}) precursors during the combustion of fuels derived from oil. These emissions are pure statements of the ills of the environment and acid rain, along with atmospheric air pollution and other human health risks such as respiratory disease and even cancerous exposures. Consequently, global regulations, such as the United States Environmental Protection Agency (U.S. EPA) and the European Union (EU), have imposed a very tight limitation on the sulfur content of transportation fuels, which compels refiners to employ more efficient desulfurization systems. In the past, hydrodesulfurization (HDS) was known as an industrial workhorse for fuel desulfurization (D'Adamo et al., 2023). Although it is excellent for most sulfur compounds, HDS is costly (both capital and the cost of utilities) because it operates at high temperatures and hydrogen pressures and employs transition-metal-sourced catalysts (Arevalo-Basañez, Jiménez-García, Villalón-López, & Maya-Yescas, 2025). Additionally, HDS has very poor efficiency for refractory sulfur-containing alkyl-substituted BT rings. Coking and sintering are also restrictions on the lifetime of such a process as well as its economic feasibility (Ippolito, Romano, Innocenzi, Ferella, & Giansante, 2023).

With the abovementioned limitations, recyclable desulfurization systems based on green chemistry and sustainable engineering have been increasingly developed. It also provides an added benefit in that recyclable materials are economical because they can be utilized in more than one cycle, and there is correspondingly less waste and weight on the environment. This is also why the desulfurization of nanotechnology has achieved great success in the lighter century (Abd Zaid, AbdulRazak, & Abid, 2022). This is largely because nanomaterials have a high area/surface-volume ratio, tenable functionalizability on surfaces, and structural flexibility that can be exploited for catalytic and adsorptive removal. Such compositions may likewise be driven by magnetic fields in the fuel matrix as well as in the recovery mechanisms in such a way that the effortless separation offered by the numerous processes, other than those mentioned previously, also contributes to processing efficiency (Zbuzant, 2022). The design goal of recyclability is explained by the following number of interdependent factors: mechanical strength, ability to be regenerated and cut during work, and capacity to be used after several cycles. In addition to being energy efficient when recycled using a green process, a high-performance recyclable desulfurization material must possess the ability to degrade slowly, both physically and chemically, during circulation. These guidelines play a key role in the design of future-generation technologies for sulfur removal based on green nanotechnology (Jadhav & Sangwai, 2020).

Magnetic nanomaterials, especially those based on spinel ferrites such as magnetite (Fe_2O_3), maghemite (γ-Fe_2O_3), cobalt ferrite ($CoFe_2O_3$), and manganese ferrite ($MnFe_2O_3$), have gained much attention as suitable materials for environmental remediation. These types of materials present some interesting characteristics (superparamagnetic, high-reactivity surface, chemical inertness, and easy manipulation in an external magnetic field) (Fahim, Hasani, & Mahmoud Ragab, 2025). In particular, Fe_3O_3 nanoparticles are widely used because of their good synthetic properties, as well as their low cost, and biocompatibility. However, magnetic nanoparticles without any modification are difficult to disperse owing to the high surface energy and dipole–dipole interactions among the nanoparticles themselves. As discussed above, this results in agglomeration, which further decreases the dispersion and active surface area of particles in the liquid (Fadel et al., 2025). In addition, their surface chemistries are generally narrow; thus, modification or combination is needed to achieve the selective adsorption of sulfur compounds. Special chemical groups grafted onto the surface, such as amines, carboxyl, or thiols, can further increase their specific attraction to polar sulfoxides and aromatic sulfur compounds. In addition, the generation of core–shell structures, that is, a magnetic core encapsulated within a porous or catalytic shell, has been proven to enhance stability and facilitate direct interaction with sulfur compounds (Zou et al., 2024).

Therefore, it is desirable to develop hybrid nanomaterials that incorporate numerous functions into a single platform. Such magnetosensors are associated with a range of structural, catalytic, and adsorptive elements, such as mesoporous silica, metal–organic frameworks (MOFs), graphene oxide, activated carbon, and bio-based support structures (Sehgal & Kunde, 2022). Chemical selectivity, catalytic activity, and adsorption capacity are thought to be the synergistic effects of the composite systems. As an illustration, the high surface area and excellent porosity of biochar and the magnetic separability of Fe_3O_4 could be exploited in Fe_3O_4- biochar composites to achieve effective and green desulfurization. Complexity and scalability: A compromise in the synthesis of hybrid materials. Some significant factors to consider include interfacial compatibility, retention of magnetic properties, retention of catalytic properties, and thermal and chemical resistances. These parameters can be tuned to prepare bifunctional materials that meet several requirements for oxidative desulfurization (ODS), adsorptive desulfurization (ADS), and other recent technologies for fuel purification (Senthil, 2024).

Green nanotechnology centers on the contribution of renewable, biodegradable, and nontoxic materials to synthesis and utilization. Biomass-derived products (Cellulose, Chitosan, Alginate and Lignin/Biochar) are biodegradable alternatives to nonbiodegradable polymers that serve as general precursors in nanocomposite systems for desulfurization. Their chemical properties, including hydroxyl, carboxyl, and amino functional groups, enable high affinities for metal ions and sulfur-containing compounds, making them useful for diverse applications as adsorbents, scaffolds, and active moieties (Madani et al., 2022). Cellulose, the most natural polymer, is a versatile material used for the preparation of magnetic nanocomposites. The hydrophilic properties, mechanical properties, gelling, and film formation of the oily properties make it a better-dispersed nanoparticle and enhance the longevity of the composite. Chitosan provides additional functionality to the system through its primary amine functional groups, which can coordinate metal centers or sulfur molecules (Bellanthudawa et al., 2023).

Therefore, chitosan-based composites are promising materials for adsorption and catalysis under mild conditions. Biochar, obtained from pyrolyzing lignocellulosic biomaterials, is a well-known carbonaceous material that exhibits a high surface area, tunable porosity, and good thermal stability. It is an excellent scaffolding matrix for metal nanoparticles and magnetic oxides (Yu, Wang, Li, Zhang, & Zhou, 2022). The surface chemistry of biochar can also be modified by chemical activation or doping to promote adsorption of sulfur-containing molecules. Furthermore, the use of naphthalene grating with biochar to produce nanocomposites decreased the total environmental impact of the material and increased its reusability. The integration of bio-based materials in magnetic and hybrid nanocomposites is not only environmentally friendly, but also requires green synthesis routes (Mehdi, Khoja, Naqvi, Gao, & Amin, 2022). Plant extracts, microbial metabolites, and biopolymers have been reported to be reducing and stabilizing agents for the production of metal and metal oxide nanoparticles. These methods are free of the use of toxic solvents, surfactants, or high-energy processes and thus conform to green chemistry and engineering principles. Bio-based components also provide environmental advantages from the end-of-life perspective. Their biodegradability ensures minimal environmental impact after use, and harmless degradation products do not represent any significant threats to the environment or human health. In addition, most bio-based nanocomposites can be subjected to thermal and chemical regeneration to fit recycling desulfurization systems (Zhang, Jiang, & Balasubramanian, 2021).

Figure 1 . Schematic illustration of green nanotechnology-enabled fuel desulfurisation using magnetic and hybrid bio-nanocomposites. This process integrates Fe_3O_4*-based magnetic cores with biogenic supports (e.g. biochar, cellulose, and graphene oxide) to form functional hybrid nanocomposites. These materials facilitate both oxidative and adsorptive desulfurisation, followed by efficient magnetic separation and regeneration. This approach exemplifies eco-friendly and recyclable strategies aligned with green chemistry principles for the removal of sulfur compounds from liquid fuels.*

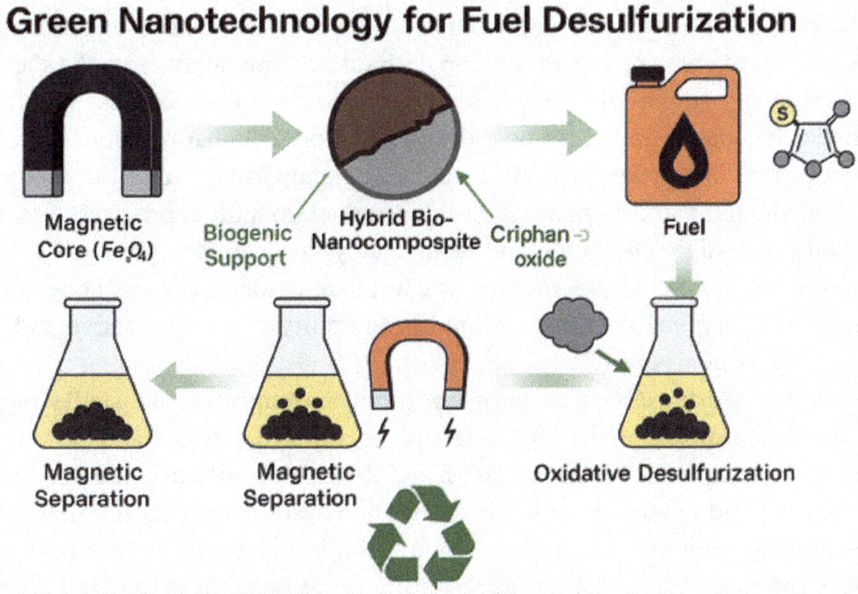

2. DESIGN PRINCIPLES OF MAGNETIC NANOCOMPOSITES

The development of magnetic nanocomposites for reclaimable fuel desulfurization relies on the proper incorporation of functionality, stability, and process adaptability in the design. An efficient nanocomposite system should not only exhibit highly efficient sulfur removal, but also sustain the structure and performance after numerous cycles. Structural assembly, magnetic core selection, surface functionalization, and synthetic protocols play crucial roles in achieving these expectations. This section is dedicated to the base design principles that include core–shell magnetic nanocomposites, ferrite materials, and green synthesis strategies (e.g., bio-templating) (Hamza, Dawood, & Jabbar, 2024).

2.1 Core-Shell Structures and Surface modification

Core–shell nanostructures are architectural designs that have been widely used in magnetic nanocomposites, particularly those based on applications requiring multifunctionality, stability, and surface control. A core–shell system is generally described as a magnetic nanoparticle in the core and an added layer (the shell) that encapsulates or superimposes physical or chemical properties (Rimmy Singh & Bhateria, 2021). Such a design enables the decoupled control of the magnetic properties of the core and the interaction with sulfur compounds from the shell. The core, which is made up of Fe_3O_4 (magnetite) or other spinel ferrites, supplies the magnetic character required for post-desulfurization reclamation using an external magnetic field. Nevertheless, naked magnetic nanoparticles tend to agglomerate via magnetic and van der Waals forces and do not contain appropriate functional groups for both high adsorption and facile catalysis. Therefore, the shell served as a protective and functional structure. It can block oxidation of the core, improve chemical stability, enable good solvent dispersibility, and add reactive and/or selective functional groups (Zou et al., 2024).

A number of materials have been used as shells in magnetic core–shell structures, such as silicon dioxide (SiO 2), polymers (e.g., polydopamine and polyethylene glycol) [3], metal oxides (such as TiO 2 and ZnO), and carbon-based counterparts (e.g., biochar and graphene oxide). In addition to desulfurization, the shell can also accommodate catalytic species, such as transition metal ions or heteropoly acids (HPAs) and oxidative agents, leading to oxidative desulfurization (ODS) routes (Tanimu et al., 2022). Surface modification approaches also broaden the applications of core–shell structures because they allow for the functionalization of different chemical moieties on the outer shell. 'For instance, sulfuric acid groups could raise the acidity causes DBT-like sulfurs oxidation.' Thiol or amine groups enhance the affinity for sulfur species through soft acid–base interactions. Furthermore, the introduction of porosity into the shell structure (e.g., using templating methods or selective etching) enhances the exposure of active sites and fast mass transport during desulfurization (Wu et al., 2022).

2.2 Fe_3O_4 and Other Ferrite Materials as MR Contrast Agents

Among them, magnetite (Fe_3O_3) is still the most widely applied type of magnetic core owing to its superior physicochemical properties, biocompatibility, and simple synthesis method. It is superparamagnetic at the nanoscale level, which means that the particles are capable of holding strong magnetism even with an applied field; however, they also have no bound magnetic polarization when the applied field is relieved (Nguyen, Tran, Xu, & Lee, 2021). Easy disassembly during the desulfurization process is of great importance for the adsorbent because it must be performed quickly and under reversible conditions to avoid aggregation during storage, making it easier to reuse again. Fe_3O_4 nanoparticles also exhibit a high saturation magnetization (~92 emu/g in the bulk), meaning they can be efficiently separated even under low magnetic fields. Furthermore, the external surface of Fe_3O_3 can be easily modified by hydroxyl groups for further functionalization or combination with shell materials (Io a et al., 2023).

Nevertheless, Fe_3O_4 is not the sole choice for the use of magnetic oxides in desulfurization. Other ferrites, including $CoFe_2O_3$, $MnFe_2O_3$, and $NiFe_2O_3$, possess potential for use in areas such as chemical stability, coercivity, and catalysis. For example, $CoFe_2O_3$ is chemically more stable in acidic media and is characterized by a higher coercivity, which can be favorable for high-temperature or harsh chemical environments (Shakirzyanov et al., 2023). However, $MnFe_2O_3$ has also been studied for its redox activity and catalytic oxidation. Typically, ferrite is selected depending on the pH, temperature, oxidative potential, and desired regeneration mode of operation. The type of synthetic method is also a key factor in the particle morphology, crystallinity, and magnetic behavior, which corresponds to the desulfurization performance (El-Khayatt, El Desouky, El-Khawas, & Azab, 2025).

2.3 Green Synthesis and Bio-Templating

Green chemistry concepts challenge the reduction of cytotoxic waste and energy consumption used in processing materials. In the case of magnetic nanocomposites, green synthesis methods that minimize the use of poisonous solvents, high temperatures, and reagents that are ecologically toxic are required. These methods would be especially effective in a final article that is anticipated to be utilized in high volumes within the environmental domain, such as fuel purification (Castiello et al., 2023). Biotemplating is a common green synthesis method. Under this method, natural biogenic materials (such as plant extracts, polysaccharides, proteins, or microorganisms) are employed as templates to control the structure or as reducing/stabilizing agents. For example, metal salts, such as polyphenols and flavonoids, can be used to form nanoparticles, which can reduce metal salts and/or prevent their aggregation, as well as some plant secondary metabolites. In this regard, polysaccharides, such as starch, alginate, and cellulose, offer three-dimensional networks that control the NP mouthing and support structure (Hessel et al., 2022).

Biochar derived by pyrolysis, where biomass such as wood chips, agricultural waste, or algae are heated in the absence of oxygen, serves as a source of carbon and template. GO is an ideal support matrix for magnetic nanoparticles owing to its inherent porosity, large surface area, and numerous surface functional groups (Chi et al., 2021). It coats the biochar with Fe_3O_4 to form an effective nanocomposite with a high adsorption desulfurization potential. In addition, the process does not contain any synthetic binders or surfactants, which restricts its environmental effects. Another new approach to green synthesis has emerged as microbial-assisted synthesis, in which the assistance of bacteria or fungi reduces and deposits magnetic particles. Certain microorganisms are known to autotrophically accumulate intrinsic biogenic magnetite by leaching ferric further to the surface, where Fe_3O_4 is precipitated in an anaerobic environment. All these approaches remain primarily in the developmental phase, yet they have the potential to realize affordable, easily scalable, and environmentally benign production of magnetic nanomaterials (El-Fawal et al., 2025).

2.4 Desulfurization Uses to Engineering Issues

When using a magnetic nanocomposite design to create an efficient desulfurization system, several parameters must be fine-tuned. These properties include size, working conditions, dispersibility, stability, and surface charge. The diameter of the nanoparticles must preferably be maintained at a value significantly less than hundred and fifty nm in order to possess a large surface area and reaction kinetics, yet not so large that it would result in irreversible aggregation or magnetic recovery problems (C. Wang et al., 2023). The zeta potential values may provide some insight into colloid stability, with higher (absolute) values between ±30 mV, possibly representing good dispersion in a polar environment. Thermal and chemical stability is also important, especially in (ODStive desulfurization ODS (), when the nanocomposite is in contact with oxidants, including H_2O2 or organic peroxide. Thus, the shell material must be resistant to deterioration by oxidation and leaching from the magnetic core (Kaeed, Shahhosseini, & Rezaei, 2024). The mechanical stability of this nanocomposite is crucial when applied in continuous-flow systems, where attrition or fragmentation can limit the recyclability of the materials. In real applications, the regeneration of nanocomposites is usually performed by washing, solvent extraction, and thermal treatment. In other words, the desorption of sulfur compounds should not damage the active sites or core structures in the design. Recent studies revealed that Fe_3O

$_3$–biochar or Fe_3O_3–MOF hybrids behaved for 5 to 10 cycles in use with no obvious activity decay, indicating their potential for long-term use (Rashid et al., 2023).

3.HYBRID NANOCOMPOSITE SYSTEMS

Hybrid nanocomposites are emerging functional materials that display the presence of two or more dissimilar phases (at least one in the nanometer range), where the organic phase is as important as the inorganic phase (C. Wang et al., 2023). Magnetic nanocomposites are an effective strategy for the desulfurization of recyclable fuel because they can provide a high surface area, chemical functions, and mechanical stability in the same system. It is believed that this combined effect provides higher desulfurization performance, better recyclability of the material, and compatibility with the environment. Recent studies have reported the synthesis of hybrids of magnetic nanostructures with porous, bio-based, or carbonaceous matrices (e.g., metal-organic frameworks (MOFs), biochar, graphene oxide, cellulose, and zeolites) (Kaeed et al., 2024). These materials are designed to make full use of the advantages of the contribution parts: magnetic cores applied for separation, porous matrices for adsorption, and functional shells or coatings for selectively interacting with sulfur compounds. In this section, we discuss the design and performance of three key classes of hybrid nanocomposites that are significant for recyclable desulfurization systems: MOF–biochar hybrids, zeolite–biocarbon composites, and GO/cellulose-based hybrids (Rashid et al., 2023).

3.1 MOF–Biochar Hybrids

Metal-organic frameworks (MOFs) are a class of ordered mesoporous materials formed by the linkage of metal-ion clusters via organic ligands. They are characterized by a high surface area (7000 and up to m^2/g), tunable pore size, and numerous chemical functionalities, which make them an attractive material for adsorption-catalysis studies, including desulfurization (Cai, Yan, Zhang, Zhou, & Jiang, 2021). However, some pure MOFs are fragile, expensive to prepare, and are difficult to recycle. To solve these problems, a combination of MOFs and biochar has recently emerged as a competitive alternative. Biochar, a carbonaceous solid material obtained from biomass pyrolysis, is porous, thermally stable, and abundant. When biochar is employed as a support for MOFs, not only are the mechanical strength and dispersion of MOF particles increased, but extra-adsorption sites and hydrophobicity are also introduced, which could optimize the interaction between adsorbents and target aromatic sulfur compounds such as dibenzo thiophene (DBT). Furthermore, the addition of magnetic nanoparticles such as Fe_3O_4 to MOF–biochar composites enhances magnetic recoverability and allows it to be easily isolated from liquid fuels without filtration (X. Zhou et al., 2021).

For instance, the Fe_3O_4 biochar composites ZIF-8 coated composites have shown noteworthy adsorption capacities for DBT through π–π interaction between the aromatic rings of DBT with organic linkers in MOFs and by physical confinement inside the pores. The introduction of biochar improves the thermal and chemical stability of the composites, and because of the magnetic core in which the active constituents are carried, regeneration and recovery are convenient (Tang, Wang, He, Wang, & Wei, 2024). Moreover, MOF–biochar hybrids can be functionalized with oxidative reagents (e.g. gaseous H_2O 2 and peracids) for oxidative desulfurization (ODS), where sulfur-contaminated products are oxidatively transformed into much more polar and extractable sulfoxides or sulfones. The dual-mode property (adsorption and catalysis) of the heterojunction leads to a significant increase in the desulfurization capacity of this hybrid system, making it possible for more extensive applications (Mourdikoudis et al., 2025).

3.2 Zeolite–Biocarbon Composites

Zeolites are aluminosilicate materials with well-defined microporous structures, high cation-exchange capacities, and good thermal and chemical stabilities. These latter attributes make them well suited for adsorption-desulfurization applications, especially when size-discriminating interactions and strong electrostatic binders are desired. However, conventional zeolites have the disadvantages of poor dispersion in nonpolar fuels and difficulty recycling. Hybridizing zeolites with biocarbon-activated carbon or biochar is one such approach to address these limitations (Pérez-Botella, Valencia, & Rey, 2022). Biocarbon materials exhibit high surface areas, porosities, surface functional groups (such as –OH and –COOH), and hydrophobicities, which are favorable for the adsorption of aromatic non-polar sulfur compounds. Such composites with zeolites exhibit a faster adsorption rate owing to the increased accessibility of the internal porous network and better mechanical characteristics. In addition, biocarbon in composites is also available for modulating the hydrophilic/hydrophobic properties of materials, and this can be valuable for the desulfurization of complicated fuels with diversified polarity. For example, a composite of HZSM-5 zeolite and biochar can absorb more DBT and its intermediate products than a single adsorbent, owing to synergistic adsorption mechanisms (Murrieta-Rico et al., 2022). In such a revised form system, zeolite provides microporous confinement and acidity support for the polar sulfur-functionalized biochar species, whereas biochar is able to physisorb hydrophobic species because of its specific surface area and increased dispersion. Moreover, to enable magnetic recovery, magnetic nanoparticles (e.g., Fe_3O_4 [1Ð6], $CoFe_2$O4) can be incorporated in zeolite–biocarbon hybrids. These multirole materials have been successfully applied in both flow and batch systems, with simple separation and reuse being one of their main operational benefits. The recyclability of magnetic zeolite–carbon composites is high, and they can be used for recycled desulfurization processes (Jacob, Alisi, & Surajo, 2025).

3.3 Graphene Oxide-Cellulose Hybrid Devices

Graphene oxide GO is a representative 2D O-containing-material with abundant oxygen groups (O-, –OH, –COOH) distributed on its basal planes and has a very large specific surface area and mechanical strength and it can tightly bind aromatic molecules by π–π stacking interaction [16]. These characteristics make GO a promising filler for fuel desulfurisation using hybrid nanocomposites. The dispersion, surface functionality, and recyclability of materials were enhanced by the incorporation of magnetic nanoparticles and/or biopolymers such as cellulose (Strassburg, Mayer, & Scheibel, 2022). Cellulose, the earth's most abundant biopolymer, has been used as a renewable and biodegradable matrix in composite networks. It has a backbone rich in hydroxyl groups and is easy to chemically modify, which can also be signalised by hydrophilic (as well as lipophilic) products. Both industrially available GO and cellulose can be fabricated into flexible high-capacity strong supports using a more environmentally friendly approach (Bianchini Silva et al., 2024).

2O 4 -GO-cellulose composites, for example, have been prepared via a co-precipitation or hydrothermal process, and these nano-composite materials exhibit remarkable hierarchical porosity and magnetic properties (Eskandarinezhad, Khosravi, Amarzadeh, Mondal, & Magalhães Filho, 2021). In these systems, π–π interactions with GO and other sulfur species (e.g., DBT) are supplemented by structural supports and sites for increased hydrogen bonding through cellulose. The retrieval and recycling of the magneto core were quick and easy. These hybrids may also be catalytically modified with catalyst components (MOx, M = Ti, Mo, and/or heteropoly acids) on the walls, which are optimized for mild oxidative desulfurization (SIDDIQUI, 2024). GO offers electron transfer and catalytic activity, whereas cellulose provides dispersibility to prevent agglomeration. Multifunctional composites possess the properties of adsorption and catalysis that unveil their capabilities to be used as a platform for desulfurization in laboratories and on a pilot scale (Dhanyashree et al., 2024).

4.SYNERGISTIC PROCESSES IN HYBRID SYSTEMS

Synergistic effects between two or more phases are regarded as one of the greatest advantages of hybrid nanocomposite materials. These effects occur mostly in performances that are greater than those of individual materials. For example, porous carbon will be closely attached to a magnetic core, which will ensure that more adsorption sites will be available compared with grain-like substances, and the separation process will be enhanced (Afolabi & Ndou, 2024). In addition, the increase in catalytically active sites of the hybrid type would realize the adsorption and oxidation of sulfur compounds simultaneously to improve the adsorptive desulfurization ability with high selectivity. Collaboration also affects structural integrity and recyclability. Hybrids are generally more resistant to fouling, thermal degradation, and chemical leaching than single-phase nanoparticles. Their hierarchical porosity allows fast mass transfer and prevents diffusion limitations, which are essential for treating real fuels with complex compositions. Furthermore, the modular nature of hybrid nanocomposites renders tunability to optimize their adsorption capacity towards selecting sulfur functionalities, fuel matrix effects, and process conditions (Tajdari, Babaei, GOudarzi, Partovi, & Rostami, 2021).

4.1 Synthesis Techniques

Overall, the chemical procedures involved in the synthesis of magnetic and hybrid bionanocomposites for fuel desulfurization have an essential effect on their recyclability and performance. To a large extent, the physical characteristics, surface chemistry, hydrodispersibility/stability, and functional behavior of nanocomposites are controlled by the conditions/techniques with/and which they are processed. Consequently, a controlled choice and optimisation of synthesis protocols are required to acquire desired physicochemical properties that promote high desulfurisation efficiency and recycling applications (Hosseini, Khoshsima, Sabzi, & Rostam, 2022). In this section, we discuss some of the important synthetic strategies employed in the fabrication of magnetic and hybrid bionanocomposites for desulfurization. Special attention is given to (i) one-pot versus multi-step syntheses, (ii) common chemical synthesis of materials such as sol–gel, solvothermal, and co-precipitation methods, and considerations for the design of materials compatible with magnetic recovery (Dembaremba, Majodina, Walmsley, Ogunlaja, & Tshentu, 2022).

4.2 One-Pot vs. Multi-Step Methods

For the synthesis of nanocomposites, the general approach can be roughly divided into one-pot and multi-step methods with their respective advantages, constraints, and scope. One-pot synthesis is an operation process that is more time-saving and economically efficient compared to multi-step methods, where all reagents are separately introduced into one reactor for composite structure formation (Yang et al., 2021). This process has the advantages of a simple and quick reaction time, low energy consumption, and low consumption of solvents and reagents, in good agreement with green chemistry concepts. One-pot methods are ideal for scale-up and are becoming the most popular methods, as are magnetic nanocomposites using bio-based or carbonaceous supports (Herman, Zhu, Ayepa, You, & Wang, 2023).

For instance, Fe_3O_4 nanoparticles could be co-precipitated in the presence of biochar or cellulose derivatives and homogeneous composites are prepared directly without post-synthesis blending. This one-pot co-precipitation allows the nanoparticles to be well-dispersed on the support surface, which helps avoid nanoparticle agglomeration or poor interface contact. Nonetheless, multistep synthesis remains important when specific site control in each phase or sequential functionalization is necessary (Tripathi, Singh, & Pant, 2021). In this method, each component of the nanocomposite is synthesized independently and further assembled via post-synthesis modification strategies (surface treatment or coating/loading with catalytic entities). Multi-step strategies are beneficial for building core–shell structures, starting with catalytic-site-selective doping or refining the porosity and surface charge (Abboud, Alnefaie, & Alhanash, 2022).

For example, a typical multiple-step process could include the preparation of Fe_3O_4 nanoparticles first and then coating with a silica shell, immobilization of a metal–organic framework (MOF), or oxidation catalyst (W. Li et al., 2022). This overcoat-based strategy provides greater control of the layer thickness, particle size distribution, and surface chemistry; however, it is generally time-, equipment-, and reagent-intensive. The selection of one-pot and multistep synthesis is based on application demand, target material complexity, and scalability. Although the one-pot approach is excellent for low-cost and green production, the multistep process is expected to achieve excellent performance (Akter et al., 2025).

4.3 Sol–Gel, Solvothermal and Co□Precipitation Schemes

Various chemical methods have been used to prepare magnetic nanocomposites, presenting different advantages in terms of material morphology, crystallinity, surface area, and scale-up. The most frequently used methods are sol–gel, solvothermal (and its modification hydrothermal), and co-precipitation methods (Mehmood et al., 2021).

4.3.1 Sol–Gel Method

Sol-gel synthesis consists of the hydrolysis and polycondensation of metal alkoxides or salts, leading to gel formation, after which thermal treatment is applied to transform into oxide networks. This approach is particularly useful for synthesizing uniform metal oxides and silica-like coatings (Rezaei et al., 2024). Magnetic sol–gel nanocomposites are frequently prepared by coating magnetic cores such as Fe_3O_4

particles with silica or other oxide shells to obtain defined core–shell structures. The silica shell not only protects the magnetic core from oxidation and aggregation but also offers a basis for the further immobilization of catalytic or adsorptive species. Furthermore, the sol–gel process offers the possibility of accurately controlling the porosity, surface area, and coating layer thickness. Nevertheless, this strategy usually requires a long reaction time and an aging process with alcohol solvents, leading to its limited application for scalable formation under green synthetic environments (Jia et al., 2021).

4.3.2 Solvothermal and Hydrothermal Methods

Solvothermal and hydrothermal syntheses were performed in sealed autoclaves under high-temperature, high-pressure conditions to obtain highly crystalline materials with the desired morphologies. Ideally, these strategies are used to fabricate nano-structured metal oxides, MOFs, and composites. Carbonaceous materials including GO and biochar have also been used as support for encapsulating Fe_3O_4 nanoparticles by solvothermal methods, resulting in the formation of well-dispersed and stable composites (Sabukuttan, 2025). Installation under high-pressure conditions improved the penetration of the precursors into the porous support, allowing homogeneous loading and strong interfacial attachment. Furthermore, the solvothermal parameters can be controlled to promote the growth of certain crystal phases, an aspect that is important if one wishes to integrate catalytic and photocatalytic components such as TiO_2 or MoO_3 into the nanocomposite. High-pressure equipment and long reaction times may limit fast or bulk manufacturing (Pavithra et al., 2024).

4.3.3 Co-Precipitation Method

According to the kaolin containing rare earth ore, undesired clay is removed through crushing and alternating sedimentation or Co-precipitation This [11] Co-precipitation This is the most popular and simple method for synthesis of magnetic nanoparticles especially Fe_3O_4. This entails the co-precipitation of Fe^{2+} and Fe^{3+} in an alkaline aqueous environment under inert or morphemic conditions. This approach is extremely effective, inexpensive, and expandable, and as a result, it can be applicable to large-scale industries (Nguyen et al., 2021). Implemented on or in the presence of a supporting matrix, such as cellulose, chitosan, or biochar, co-precipitation can directly obtain nanocomposites with good dispersion and interfacial contact. Determining factors in co-precipitation include the Fe^{2+}/Fe^{3+} ratio, pH, temperature, and stirring rate, which influence the size of the particles, crystalline nature, and magnetic properties. Moreover, it is more sustainable because biocompatible reducing agents or capping agents are used (Halvaeikhanekahdani, Zandi, Ahmad, & Payravand, 2025).

Magnetic separation of a swarm can be achieved using magnetic fabrication. Magnetic nanocomposites have the outstanding properties of easy separation and recycling, and external magnetic fields can be applied to achieve this objective. In this regard, synthetic methods should consider factors that support magnetic separation and simultaneously sustain functionality (Mehmood et al., 2021). To respond to external values, the magnetic core (typically Fe_3O_4 or its derivative ferrite) must first exhibit a large saturation magnetization (ms). Conditions that result in high crystallinity and controlled particle size (10–30 nm) are said to be the best for achieving superparamagnetic properties without aggregates. Second, surface exposure of the magnetic core is also important for a prompt response to the magnetic field. In the case of core–shell architectures, excessively thick shells or packed coatings can adversely affect the separation efficiency. Therefore, synthetic approaches need to find a compromise between protecting the center and minimizing the detrimental effects on the magnetic response (Ji, 2021).

Third, it is important that this function is maintained after the collection of magnetic particles. Materials should be able to tolerate the magnetic collection process and not lose activity with multiple collections and deployments. Approaches, such as covalent linkage of the magnetic and anion-functional cation components or full embedding of the core in chemically inert matrices, assist in avoiding leaching, mechanical degradation, or passivation by charging onto the surface. Finally, synthetic approaches must seek colloidal stability in both aqueous and organic solvents. Nanocomposites that tend to sediment or irreversibly aggregate cannot be collected and reused with a high efficiency. Surfactant-mediated synthesis, zeta potential manipulation, and surface modification are useful strategies to achieve long-term stability (Y. Wang et al., 2022).

4.4 Magnetic Recovery and Reusability

It is one of the main points to consider, and it develops in pursuing economic, environmentally friendly fuel purifying technology at high-level recycling for desulfurization materials. One class of such materials, magnetic nanocomposites, is well suited because of their ability for rapid and efficient external magnetic field-induced recovery. This characteristic allows for simpler operation at a reduced cost, eliminating the need for filtration or centrifugation and avoiding loss of material with concomitant environmental concerns (Duan et al., 2021). Therefore, magnetic recovery and reusability are the key functions of green nanotechnology for desulfurization. Magnetic nanocomposites commonly consist of a magnetic core based on magnetite (Fe_3O_3), maghemite (γ-Fe_2O_3), or spinel ferrites, including cobalt ferrite ($CoFe_2O_3$) and manganese ferrite ($MnFe_2O_3$). They are superparamagnetic or exhibit ferrimagnetic behavior and can be manipulated under a magnetic field. Upon exposure to an external magnetic field, these magnetic nanocomposites become aligned and move toward the direction of the field, thus promoting convenient recovery from the desulfurization medium (most probably in hydrocarbon-based liquid fuel). This separation process is fast, on the order of seconds to a few minutes, depending on the particle concentration, solvent viscosity, and intensity of the magnetic field used (GOyal, Meena, Shelare, & Kumar, 2025).

The performance of a magnetic separation process is controlled by various factors, among which material properties play an interrelated role. The saturation magnetization (Ms), which is the maximum magnetic response of any material, determines the efficiency at which the nanocomposite can be recovered. Fe_3O_4 particles have high Ms values in the bulk (up to 92 emu/g) but nanoparticles may show slightly lower ones because of surface effects (Rocha et al., 2021). The size and shape of the magnetic core are also important; particles 10-30 nm in diameter often behave as superparamagnetic materials, quickly responding to applied fields, but with no permanent magnetism when the field is removed. This prevents it from sticking together during storage and dispersibility in the fuel matrix. Dispersibility in the fuel phase is mandatory for homogeneous active nanocomposite - sulfur compound interactions. Surface functionalization, including surfactantlyanocomposites in which polar or charged groups are attached to their surfaces, leads to stabilization of the colloidal dispersion and avoids sedimentation/agglomeration. In addition, the intensity and mode of the external magnetic field (e.g., whether it is generated by permanent magnets or electromagnets) also affect the degree of completion and rate of separation. High-gradient magnetic separation (HGMS) systems have been investigated for continuous and scalable applications (Mehmood et al., 2021).

Apart from the fact that these nanocomposites can be recovered magnetically, another important issue addressed in previous studies is related to their recyclability over several cycles. For efficient material reuse with regard to ammo desulfurising processes and regeneration operations, nanomaterials should be able to retain their original forms and functions after repeated use (Ikram, Mohamed Jan, Nagy, & Szabo, 2023). This may prevent oxidation, leaching, fouling, or structural breakdown. Protective shells or matrices are commonly incorporated into many magnetic nanocomposites to improve their stability. For example, an Fe_3O_4 core can be encapsulated in a silica or carbonaceous shell to protect the core from chemical exposure and to maintain magnetic performance. Biochar/graphene oxide/polymeric-support composites serve to anchor nanomaterials and enhance their mechanical strength. The regeneration of magnetic nanocomposites involves the refreshment of their adsorption or catalytic ability after the removal of sulfur has been removed (Shaik & Franchetti, 2022). Currently, various regeneration techniques are available. Solvent washing, especially with polar solvents such as methanol or ethanol, is frequently used to desorb sulfur species from the adsorbent materials. This method is mild and unambiguously allows the retention of functional groups that chelate sulfur. Thermal regeneration, in which the spent adsorbent is heated under an inert or reducing atmosphere to decompose residual sulfur compounds, has been used to restore the active surface sites. However, thermal damage to organic or bio-based materials must be avoided. In the case of ODS, surface-bound sulfones can be eliminated, and catalytic activity can be restored by chemical regeneration with an oxidizing agent such as hydrogen peroxide. Some methods, including solvent and thermal treatments, have been used to achieve high regeneration efficiency (Sithole, 2024).

The regeneration performance is usually evaluated by the recovery of the desulfurization activity over a number of cycles. A material that remains stable in sulfur removal ability for at least 5-10 cycles has the potential for practical applications. Here, the stability of the magnetic core, durability of the functional groups, maintenance of a moderate surface area, and porosity are all important factors. Such covalently bound functional groups and cross-linked structures tend to be more durable than repeated regeneration (Chen et al., 2021). Realistic studies and laboratory tests demonstrate the possibility of using magnetic nanocomposites in practical recycling desulfurization systems. For instance, Fe_3O_3–biochar composites synthesized using agricultural wastes have shown a high removal ability of dibenzothiophene (DBT) in a model fuel and magnetic separation within less than one min (P. Li et al., 2021). These catalysts retained more than 85% of their initial activity even after seven cycles of eco-friendly ethanol washing, demonstrating the excellent compatibility of the bio-based support with green regeneration processes. In a different study, ZIF-8 (zinc-based metal–organic framework)/graphene oxide/ Fe_3O_4 nanoparticles magnetic hybrid was employed for oxidative desulfurization of DBT with hydrogen peroxide as an oxidant. It retained >90% activity in the tenth cycle, indicating that not only its water contact angle and pore volume but also its high sulfur capacity was still 95.4% compared with the first cycle. In the same context, chitosan adsorbent in the form of magnetic composites containing TiO 2 NPs has been applied for photocatalytic desulfurization under UV light, and a photoreaction was used to regenerate the material via oxidation of sulfur species adsorbed during the desulfurization process (Sang, Chen, Wang, Guo, & Fu, 2023).

However, a number of problems still exist that hamper the long-term performance of magnetic nanocomposites in industrial desulfurization applications. The loss of catalytic or functionalized ligand materials, such as metal catalysts and leaching, can limit performance and may be environmentally undesirable. Clogging of the surface with large hydrocarbons or N-containing moieties may block the active centers and decrease mass transfer. In addition, organic support structures or magnetic cores gradually decompose through multiple heating cycles or under strong oxidant conditions experienced during regeneration (Mehmood et al., 2021). By overcoming these limitations, material development strategies have shifted towards stronger chemical bonding between the magnetic and functional parts, such as covalent grafting or core-shell encapsulation with inert layers. Such hierarchical structures in a mesoporous framework have been designed to mitigate fouling and favor mass transfer. Both these technologies are being investigated to become greener and more sustainable through the application of mild, bio-regenerate regenerants such as sulfonic groups. LCA studies have also been applied to address green chemistry aspects (Hamza et al., 2024).

5. CATALYTIC AND ADSORPTIVE PERFORMANCE

This includes the catalytic and adsorptive properties of magnetic and hybrid bionanocomposites, which are essential for their activity toward fuel desulfurization. These characteristics affect the efficiency, selectivity, and practicality of the d-sulfurization of liquid hydrocarbon fuels. Desulfurization materials need to have a high loading capability of sulfur species in fuel, as well as strong catalytic activity with respect to the latter under a variety of physicochemical scenarios, because the design space of fuels is diverse and organosulfur compounds are complex. Magnetic hybrids and nanocomposites have been developed to take advantage of the adsorption and catalytic processes concurrently (Pervaiz et al., 2025). Adsorptive desulfurization (ADS) is based on physical or chemical interactions, such as S... nanocomposite interface adsorption/desorption processes to selectively remove sulfur compounds from hydrocarbons without changing their molecular structures. Catalytic desulfurization, in particular oxidative desulfurization (ODS), is characterized by the reaction of higher polarity and more easily separable species of sulfur compounds such as sulfoxides or sulfones. This section focuses on some of the applications of these nanocomposites in model fuels, their performance under various operative conditions, and the advantages gained from hybrid systems (Mitra, Racha, Shown, Mandal, & Das, 2023).

5.1 Model Fuel Desulfurization (e.g., DBT and BT)

Simulated fuels are widely used in laboratory studies to evaluate the activity of desulfurization materials. These model systems usually involve non-polar solvents, such as n-hexane, iso-octane, or dodecane, in which known amounts of sulfur-containing compounds (mainly dibenzothiophene (DBT), benzothiophene (BT), and their alkylated derivatives) are dissolved. In particular, DBT is commonly chosen as a model compound mainly because of its abundance in fossil fuels and because it represents other large DBTs that are relatively insensitive to conventional hydrodesulfurization (HDS) (Yeole & Parthasarthy, 2022). Magnetic and hybrid nanocomposites have shown considerably efficient performances in the desulfurization of model fuels as adsorbents and catalysts. In absorption systems, the acidic, basic, or π-electron-rich groups carried by adsorbents have interaction sites with the corresponding aromatic rings and sulfur atoms of DBT. This force can be π–π stacking, van der Waals forces, hydrogen bonding, and Lewis acid–base coordination. For example, the adsorption capacity of DBT over Fe_3O_3 nanoparticles supported on biochar or activated carbon is typically higher than 25–40 mg/g at room temperature. This is because oxygen, such as hydroxyl and carboxyl, with C =O in UT-APSs increases the polarity of the surface, thus favoring adsorption with DBT (Karimi & GOlmohammadi, 2024).

In the ODG process, the nanocomposite can not only take up sulfur, but also works as a catalyst to change the sulfur compounds. In general, oxidants (i.e., H_2O_2, TBHP, and organic peracids) have been employed as catalysts to convert DBT to DBT sulfoxide, and further to DBT sulfone. These oxidation products are more polar than their precursors and can be easily extracted using polar solvents. The conversion yields and selectivity for sulfone formation are high for hybrid nanocomposites in which catalytic species (such as molybdenum oxides, tungstates, transition metal cations, or heteropoly acids) are incorporated (El Kfoury, 2022). For instance, Fe_3O_3–ZIF-8–graphene oxide nanocomposites have been demonstrated to remove more than 95% of sulfur from model diesel in response to DBT exposure under mild conditions (room temperature and atmospheric pressure). These systems incorporate the magnetism of Fe_3O_4, catalysis-active nature of MOF structure and the large surface area and π–π interaction capacity of graphene oxide. Such synergistic systems allow for efficient desulfurisation, material regeneration, and reuse (Kargar, Ghahramaninezhad, Shahrak, & Balula, 2021).

5.2 Performance Under Varying Conditions

The catalytic and adsorptive behaviors of nanocomposites in desulfurization systems are highly dependent on their operational conditions. The efficiency and kinetics of sulfur removal depend on factors such as the temperature, oxidant concentration, contact time, pH, and fuel type. In desulfurization processes, the temperature has a two-fold effect. In the case of adsorption systems, increasing the temperature may weaken the physical bonds (e.g., van der Waals forces or hydrogen bonding), which in turn may lead to lower adsorption performance (Rahman et al., 2024). However, controlled heating can increase diffusion and mass transfer, enhancing the adsorption rates. However, for catalytic systems, a higher reaction temperature facilitates oxidation rates, resulting in a higher percentage of sulfur compound conversion.

For ODS reactions while working with nanocomposites, the temperature utilized is usually between 40°C and 80°C to achieve a good compromise between reactivity and material stability (Irshad et al., 2024).

Another factor is oxidant level. The ratio of oxidant to sulfur determines the amount of oxidative stress generated without complete conversion of DBT, but it requires an optimal molar ratio (between 2:1 and 4:1) for DBT and its analogs in ODS systems to avoid catalyst or support degradation. Over-oxidation can give rise to side reactions or leaching of active species, and under-oxidation also means decreased conversion efficiency (Akopyan et al., 2021). The equilibrium capacity and conversion efficiency also depend on contact time. The general equilibrium times ranged from 30 min to several hours, depending on the surface area and pore size distribution of the material, as well as the accessibility of the active sites. Magnetic stirring or ultrasonication can be used to decrease the mass transfer resistance and enhance the reaction rate (Fu & Turn, 2025).

The selectivity and efficiency can be influenced by the fuel composition (e.g., the type of aromatic hydrocarbons, nitrogen-containing compounds, and other impurities). Aromatics compete with sulfur species on adsorption sites and with nitrogen heterocyclic poisonous catalysts via strong coordination. Therefore, it is highly desirable to develop materials that are selective for sulfur over other heteroatoms. Therefore, in actual fuel systems, the interferences by these factors emphasize the significance of hybrid nanocomposites endowed with designed functionalities and multimodal interactions (Busca, 2021).

5.3 Synergy in Hybrid Systems

The enviable route towards enhancing the efficiency of desulfurization has been through the implementation of hybrid nanocomposite systems with a multiplicity of functional components within a single material system. All these cooperating mechanisms result in a significant improvement in adsorption compared to pure materials and a significantly better photocatalytic activity, not to mention stability (Salman et al., 2024). For example, the biochar structure of Fe_3O_4- biochar can provide a high specific area and skeleton porosity to evenly distribute Fe_3O_4 nanoparticles and other catalytically active constituents. The π-rich carbon framework of biochar has a higher adsorption capacity for aromatic sulfuric compounds in π-π interactions, and oxygen-containing functional groups are involved in hydrogen bonding and acid-base reactions. The catalytic capability of the composite to perform single-oil oxidative desulfurization can be both adsorptive and catalytic by coating the surface with a catalytic component such as molybdenum or tungsten oxides (Sadrara & Khorrami, 2023).

Similarly, high porosity, which is a property of the material, can be utilized by constructing composites with metalorganic frameworks (MOFs) or mesoporous silica shells encasing magnetic cores to achieve high porosity owing to their functional versatility and stability. MOFs possess certain channels that can be altered to enhance their specificity to sulfur compounds, and the MOFs may serve as hosts for active metal centers (Sadrara & Khorrami, 2023). These functions can be easily desulphurized and recycled by inserting them into a magnetically separable entity. Another potential form of synergistic improvement would be a strategy that involves using CCGO to make its hybrids with cellulose. The absorption of sulfur compounds to the large specific surface of GO and the electron-rich surfaces of graphene oxide are conducive to the catalytic reaction, which involves electron transfer. Owing to its mixture with cellulose or other biopolymers, the composite possesses excellent dispersing factors, mechanical strength, and environmental adaptability. Hybrid systems can also be used for the return of materials into cycles. Compounds that possess both mechanical and functional capacities were active after a high number of regeneration cycles. This ruggedness is necessary in practice and makes it simple to scale up this technique, where few experiments are available in the laboratory, to on-site experiments (Salman et al., 2024).

6. ASPECTS OF THE ENVIRONMENT AND SAFETY

The discovery of effective and ecologically sound materials for use in fuel purification has played a significant role in global conversion to clean and sustainable energy production. Green nanotechnology (especially magnetic and hybrid bio-nanocomposites in desulfurization technologies) has the potential to mitigate environmental harm while addressing the limitations of traditional technologies (Oloyede et al., 2024). However, the application of these technologies on a large scale should not be led only by their technical achievements but also by the benignity of their use and lifecycle. Nanomaterials, on the other hand, are characterized by concerns about their long-term relationship with the ecosystem and their toxicity to living organisms, not mentioning the matter of end-of-life disposal (Periasamy & Mohanta, 2024). Therefore, it is necessary to consider these factors in the study and design of magnetic and hybrid bionanocomposites that may be utilized to clean the environment or treat human health. Here, the toxicity and biodegradation of the green desulfurization materials are also discussed, with the disposal of the used adsorbents and catalysts inclusive of the currently approved requirements on the usage and disposal of the adsorbents and catalysts according to the regulations (Habib et al., 2023).

6.1 Material Toxicity and Biodegradability

Nanomaterials, based on their dimensionality and large surface-to-bulk ratio, frequently exhibit unique properties compared with bulk materials. Although they offer superior desulfurization performance, their toxicological properties must be assessed. The risk of environmental toxicity is related to multiple factors including composition, particle size, surface charge, solubility, and interactions with biological systems (Ameta, 2022). For magnetic nanocomposites, the core is usually made of iron oxide, which can be either magnetite (Fe_3O_3) or gamma-Fe_2O_3 (maghemite). Most of these materials are considered biocompatible and nontoxic at low concentrations, owing to the involvement of iron in biological systems. Nevertheless, under certain circumstances (e.g., acidic pH or long-term exposure), the degradation of iron-based nanoparticles leads to an enhancement in the dissolution rate and subsequent formation of Fe^{2+}or Fe^{3+}ions, acting as catalysts in ROS generation, which is responsible for the induction of oxidative stress in biological systems (SURESH, 2022).

The addition of other ferrites (e.g., $CoFe_2O_3$ and $NiFe_2O_3$) may pose more risks. Released cobalt and nickel ions have shown cytotoxic/genotoxic effects in some biological systems; thus, the leaching of nanoparticles requires control, as well as long-term exposure to the environment. Therefore, nanocomposites of such materials must be encapsulated in inert or biodegradable shells to avoid unwanted release (Mo, Zhang, & Zhang, 2024). Another very important aspect when environmental investigations of these materials are performed is biodegradability. Hybrid nanocomposites with natural polymers (e.g., cellulose, chitosan, alginate, and lignin) are highly environmentally friendly owing to their intrinsic biodegradation properties. These allotropes are completely biodegradable into innocuous end-products under both soil and aquatic conditions, thereby reducing long-term environmental risks. Similarly, biochar made from agricultural waste is considered environmentally safe and has been suggested as a soil conditioner in some situations (Kumar et al., 2025).

In addition, functionalization of nanocomposites with biodegradable ligands or surfactants enhances their dispersion in aqueous environments and reduces the incorporation of persistent synthetic compounds. However, care should be taken when choosing stabilizers and linkers, as some synthetic ones (e.g., polyethyleneimine or quaternary ammonium salts) may cause toxicity to particles and/or resistance to degradation (Sarkar, Sarkar, & Saha, 2025). The environmental sustainability of magnetic and hybrid nanocomposites depends on their use of biocompatible and biodegradable materials. However, complete toxicological studies, including in vitro and in vivo tests, are still required to measure their effects on diverse ecosystems and trophic levels (Rando et al., 2022).

6.2 Spent Adsorbents Waste Management

Similar to other functional materials, the useful lifetime of magnetic nanocomposite desulfurization agents is limited by saturation or breakdown, and they must be disposed of or regenerated. Appropriate end-of-life handling of these depleted materials is required to minimize environmental pollution and support the principles of the circular economy (Mehmood et al., 2021). When spent, adsorbents and catalysts can be loaded with sulfur (for example, DBT, BT, sulfoxides, and sulfones), oxidized organic moieties, residual oxidants, and potentially leached metals. If these materials are allowed to decompose naturally without proper treatment, toxic products may leach into soil, water, and air (Mugisha, 2024). Thus, the treatment approach should be tailored specifically to the nanocomposite constitution and nature of the sorbet pollutants. Typically, these materials can be regenerated by a thermal approach; for example, spent NMC can be heated under either an inert or reducing atmosphere to decompose the adsorbed species and renew its adsorption capability. Although such a method is effective, volatile organic compounds (VOCs) or sulfur oxides (SO_X), from which the VOCs and SO_X containing effluent gas may be released, may cause environmental air pollution, so that gas-liquid separation devices or toxic gas treatment apparatus must be further used (Rashid et al., 2023).

Chemical regeneration using solvents or mild oxidizers is another possibility that requires less energy. Polar solvents such as ethanol, methanol, and acetonitrile can remove adsorbed sulfur compounds from the surface of nanocomposites. Once pollutants are extracted, they can be recovered or processed according to standard waste management procedures (Di Carmine, Abbott, & D'agostino, 2021). Even if it is kinder to the support 'matrix,' this method may lead to partial leaching of the active species or even partial solubilization and loss of support. In other cases, these wastes can be stabilized in cementitious matrices or polymer composites for safe waste disposal or for use as building materials. Composites based on biochar, for instance, may be suitable for these methods because of their structural integrity and carbon content. This approach enables the sequestration of contaminants, with reduced leachability and increased material recovery (W. Zhou et al., 2021).

It is encouraging to recycle the metal parts (e.g., Fe, Co, and Mo) of spent nanocomposites whenever possible. These constitute environment-friendly and resource-efficient methods for metal recovery via acid leaching with precipitation or electrochemical methods. Finally, the LPAs methods can be used to compare the environmental implications of various disposal and recycling scenarios. These analyses can be used for process optimization and material design, ultimately leading to environmental footprint reduction throughout the product lifecycle (Anshassi & Townsend, 2021).

6.3 Compliance with Environmental Standards

The screening and implementation of magnetic and hybrid nanocomposites in desulfurization systems should meet national and international standards, regulations for chemical safety, air/water quality, and waste treatment/disposal. Adherence to these standards is critical to public health safety and leads to the faster market adoption of new technology. The production and use of chemical substances, including nanomaterials, are regulated in the EU by the Registration, Evaluation, Authorization, and Restriction of Chemicals (REACH) legislation (Vijayalakshmi, Florence, & Arunachalam, 2023). According to REACH, manufacturers must provide detailed safety data sheets (SDS) and risk analyses for all chemicals sold in the EU. Nanomaterials should be characterized in terms of size, surface area, solubility, and reactivity; ecotoxicological effects should also be assessed in terms of standard OECD test guidelines. In the U.S., nanomaterials are governed by the EPA under TSCA. A premanufacturing notification (PMN) of new nanomaterials is required by the EPA, which may include further testing or risk management based on their application, including environmental persistence. In addition, air quality standards restrict the discharge of sulfur oxides(SO_X),making efficient shifts necessary to avoid or eliminate secondary pollution during operation and regeneration (Tsoukleris et al., 2023).

At the international level, the ISO Technical Committee on Nanotechnologies (ISO/ TC 229) has published several standards for terminology, material characterization, and risk assessment, among others. These norms contribute to the harmonization of safety practices and regulatory clearance in different territories. Occupational safety standards should also be incorporated into green nanotechnology development for desulfurization. The exposure of workers engaged in the synthesis, processing, or regeneration of nanomaterials may occur through inhalation, dermal contact, and/ or ingestion. Suitable engineering controls, such as fume hoods, personal protective equipment, and waste collection facilities, are required to reduce the risk. In addition, it is obligatory for laboratory and industrial workers to undergo nanomaterial handling training and emergency response exercises (GOmez-Villalba, Salcines, & Fort, 2023).

7. SCALE-UP CHALLENGES AND OPPORTUNITIES

Scaling up from the small-scale point of discovery to the industrial scale for magnetic and hybrid bionanocomposites is crucial for sustainable fuel desulfurization. Although many studies have proven that these adsorbents are effective for adsorptive and catalytic desulfurization under controlled conditions, they are not widely utilized on a large scale due to a number of technological, economic, and prevailing policy barriers (Iroegbu & Ray, 2022). However, the growing need for cleaner and greener fuels in combination with more stringent environmental regulations, as well as trends towards green chemistry, has significantly broadened the outlook of industrial-scale applications of recyclable nanocomposite systems. This section deals with some critical issues for upgrading magnetic and hybrid nanocomposites for desulfurization, proposes possible methods to address these challenges, and presents an outlook on industrial and regulatory issues related to the implementation of such options in refinery processes (Ikram et al., 2023).

7.1 Process Integration with Refineries

For the nanocomposite-based desulfurization process to be successful, it must be integrated well with standard refinery units. Most, if not all, conventional desulfurization processes are hot and/or high-pressure systems incorporated as an integral part of a refinery process flow sequence (Hamza et al., 2024). Accordingly, any other system also needs to be adapted to these working conditions or provide a suitable alternative for throughput, energy requirements, and economy. A major hurdle is the physical and chemical compatibility of the nanocomposite systems with actual fuel streams. The model fuels employed in laboratory-based experiments generally do not possess some of the complexity and variation characteristics of industrial diesel or gasoline, which typically comprise various sulfur species, aromatic hydrocarbons, nitrogen-containing compounds, and metal impurities. These compounds may thus adsorb and/or poison the nanocomposites during catalysis, leading to the depletion of their activity or fouling of the active sites (Irshad et al., 2024).

nanocomposite materials with high selectivity for sulfur species, strong tolerance against competitive adsorption, and thermal chemical stability are needed. Moreover, reactor setups ranging from flows to fluidized and packed bed reactors via membranes are necessary for the continuous handling of fuels through systems where nanocomposites (dressing up) come into contact with the fuel, and these must be adjusted on a pilot/industrial scale (Yosefi, Khoshbin, & Karimzadeh, 2022). The separation and reuse of nanocomposites are also a concern. Although magnetic separation is successful in batch operation, continuous magnetic separation is feasible in such a high-flux refinery owing to sophisticated magnetic field applications and separation systems. High-gradient magnetic separation (HGMS), revolving/rotating magnetic drums, and filter-type magnets can be suitably modified to satisfy these requirements. The advancement of separation device technology that works at scale and in continuous flow presents a significant engineering challenge (Fard, Raeisi, Yousefipour, & Mosavin, 2024).

7.2 Techno-Economic Barriers

However, magnetic and hybrid bio-nanocomposites still have several significant techno-economic barriers that preclude their deployment in the marketplace. The obstacles to this progress are the cost of raw materials, complexity of methodologies, and restricted scalability of the existing production processes. (Batchu et al., 2022). The fabrication of a nanocomposite is generally a multi-step process that includes synthesis of the nanoparticles, surface functionalization, assembly into composites, and drying or shaping. Most laboratory techniques, such as solvothermal preparation, layer-by-layer deposition, and template synthesis, are not easily transferred to large-scale production at a low cost. In addition, the cost of production is significantly increased by expensive precursors (e.g., MOF ligands and metal catalysts), reagents, and specialized apparatus (Gu et al., 2025).

The low-cost (i.e., competitive with current HDS catalysts) production of nanocomposites is necessary to make them of industrial interest because of their good efficiency and recyclability. This requires green, economical synthetic approaches, such as the co-precipitation method, one-pot hydrothermal routes, and bio-templating in the form of agricultural waste or microbial systems. Employing abundant and renewable starting materials, especially bio-based supports, such as cellulose or biochar, provides value in terms of cost-effectiveness and sustainable development (Olaremu, Adedoyin, Ore, & Adeola, 2021). In addition to the cost of ingredients, economic assessments should also consider the following operational costs: regeneration, separation, and waste treatment. The energy used for the thermal regeneration or recovery of the solvent must be evaluated and compared with the energy required for conventional technologies. Both LCCA and TEA can be perceived as holistic approaches that can contribute to the assessment of the economic viability of large-scale nanocomposite systems in commercial applications (Damian, Devarajan, & T, 2025). There are figures in which data such as these analyses are structured. Such pilot systems can provide a better understanding of their real-world performance, regeneration rates, material stability, and system economics. However, there are a limited number of pilot-scale studies that have been published to date; therefore, it is necessary to invest and collaborate between academia, industry, and governmental organizations (Al-Attas et al., 2019).

7.3 Regulatory and Industry Outlook

The regulatory environment has shaped the future of desulfurization technologies. There is an increasing interest in the development of high-performance and friendly desulfurization technologies with advanced standards for fuel sulfur content. In this case, magnetic and hybrid bio-nanocomposites appear to be quite promising for fulfilling these two requirements on properties and environment (Ramesh, Rajeshkumar, Balaji, & Bhuvaneswari, 2023). The diesel sulfur concentration in most regions, such as the European Union and United States, should be very low (in the upper 10 ppm). Compliance with these requirements will necessitate new technologies, particularly those based on the removal of refractory sulfur species (unremovable) by HDS. The most promising alternatives which can potentially reduce energy usage and limit the emission of greenhouse gases are nanocomposite systems, especially those that can facilitate room temperature or mild desulfurisation (Nimbagal, Banapurmath, Sajjan, Patil, & Ganachari, 2021).

Nanomaterials are increasingly under regulatory radar (with regard to their environmental and human health toxicity). The U.S. The Environmental Protection Agency (EPA), European Chemicals Agency (ECHA), and various international agencies, including the Organization for Economic Co-operation and Development (OECD), have formulated regulations on the assessment, registration, and labelling of nanomaterials (Ragini Singh & Kumar, 2023). Hence, the commercial use of a nanocomposite should be exposed to a thorough safety assessment that involves ecotoxicity, bioaccumulation, and lifecycle effects. Furthermore, industry adoption depends on the perceived reliability and scalability of such solutions. Operators of oil refineries and petrochemical products are typically conservative and desire to test technologies in the field. Therefore, academic collaborators will be required to jointly develop pilot plants, longitudinal performance information, and operation concerns such as pressure drop, catalyst poisoning, and process integration into process control (Oksel Karakus, Bilgi, & Winkler, 2021). In addition, there is an increased interest in whether nanocomposites may be used in small- and distributed-scale fuel processors, such as would be applicable in remote or third-world use where large HDS systems are unlikely to be present. Market-specific markets or off-grid fuelling with recyclable modular nanocomposite units would encourage low-cost desulfurization using small quantities of energy. They may trigger premature commercialization and facilitate further market penetration in the future (Tovar-Lopez, 2023).

8. CASE STUDIES AND APPLICATIONS

Verification of 'green' nanotechnology applications into real-world problem-solving complements studies on the benefits and limitations of technology to issue-specific, targeted programs. Although the effectiveness of magnetic and hybrid bio-nanocomposites for fuel desulfurization has been proven at the laboratory scale, their movement to pilot and industrial levels requires efforts to investigate their performance reliability, perform impact analysis within operations, and conduct further cost-effective studies. In this section, some well-known case studies and comparison studies on oxidative/adsorptive desulfurization of olefins comparing to oxide-loading nanocomposites, as well as cooperation among universities and industries for emerging technology of desulfurization are introduced (Bibri, 2022).

8.1 Lab-to-Pilot Scale Implementations

Over the last decade, the evaluation of magnetic and hybrid nanocomposites has progressed beyond the laboratory scale. Pilot plants are essential for confirming process efficiency, material stability, regeneration ability, and compatibility with genuine fuel matrices. One successful example is Fe_3O_3-biochar nanocomposites made from agricultural waste (including coconut shells, rice husks, or sawdust) (Thomas & Nochehdehi, 2022). These composites prepared by low-temperature pyrolysis and subsequent co-technical precipitation of Fe^{2+} and Fe^{3+} ions exhibit exceptional adsorption capacities for sulfur-containing compounds such as dibenzothiophene (DBT) in model diesel systems. The nanocomposite was packed into a fixed-bed reactor and tested in a continuous-flow pilot-scale system with diesel as the reactant. After several cycles, the adsorption capacity was more than 85%, and most of its original efficiency was recovered after treatment with ethanol and mild thermal conditions. The magnetism of the composite facilitates the easy recollection of fins and separation of the liquid fuel stream (Pervaiz et al., 2025).

Another excellent example is a heteropoly acid (HPAs)-embedded hetero-nanocomposite of Fe_3O_4 cores encased by silica. This catalyst was tested in bench-scale oxidative desulfurization (ODS) of low-sulfur diesel. Extraction using hydrogen peroxide as the oxidant and a polar solvent (acetonitrile) enabled the system to achieve over 90 percent desulfurization of DBT, benzothiophene (BT) and 4,6-dimethyldibenzothiophene (4,6-DMDBT) under mild conditions (room temperature and atmospheric pressure). The material was also demonstrated to have good mechanical and catalytic stabilities within ten regeneration cycles and, therefore, could be utilized in the industry (Ren et al., 2021). This initial research suggests that material strength, recovery/isolation plans, and system design are vital requirements that must be considered in the practical application of green nanocomposites in desulfurization. Moreover, they showed that it is feasible to integrate such systems into portable treatment trains that do not need or can be connected to refinery facilities (Anekwe & Isa, 2025).

8.2 ODS vs. SADS Comparative Performance

The peculiarities of the mechanisms and work of oxidative desulfurization (ODS) and adsorptive desulfurization (ADS) should be known because it is possible to select the most appropriate nanocomposite system based on specific uses. The two have been successfully utilized with magnetic and hybrid bio-nanocomposites, yet they differ in working parameters, principle of energy, and rejuvenation behavior (Sahraei, 2023). The adsorptive desulfurisation (ADS) process relies on the preferential adsorption of sulfur compounds over other species on the surface or within nanocomposites through various interactions such as van der Waals forces, π–π stacking, hydrogen bonding, and acid–base coordination. Biochar-based composites, graphene oxide hybrids, and magnetic materials functionalized with amine or thiol groups have been shown to have high adsorption capacities for aromatic sulfur compounds, such as DBT and BT (Irshad et al., 2024). ADS was found to have potential because it is simple, requires low energy, and involves limited chemical additions. However, the limitations of this method are that the adsorbent is saturated with active sites, a decrease in efficiency with real fuel matrices having many competing species, and the requirement for regular regeneration (Cherubala, 2021).

Another approach is oxidative desulfurisation (ODS), which involves the deep oxidation of Sulphur compounds to obtain more polar compounds, sulfoxides, or sulfones that are extracted by a polar solvent or adsorbed on the support (Boshagh, Rahmani, Rostami, & Yousefifar, 2021). In addition, hybrid systems containing metal oxides (for example, MoO_3 and WO_3), HPAs, or transition metal ions immobilized in MOFs or silica templates have also demonstrated excellent catalytic performance. In addition, ODS is highly efficient in the removal of challenging sulfur species, such as 4,6-DMDBT ("heavy" dibenzothiophenes), which are relatively refractory to HDS and frequently reside in HDS-treated diesel (Rajendran et al., 2020). Comparative studies show that despite better recyclability and reduced process complexity, the sulfur removal efficiency of an ADS system is lower than that of an ODS system. Hybrid nanomaterials combining the adsorptive and catalytic roles, such as Fe_3O_4 biochar composite, are emerging to be a promising means of integrating the best of both worlds (Houda, Lancelot, Blanchard, Poinel, & Lamonier, 2018).

In addition, the selection of the ODS or ADS may depend on the processing objectives. For example, decentralized or mobile (trailer-mounted) fuel purification units may find ADS systems easier to operate, while central processing facilities may favor ODS systems for higher desulfurization capability with minimum operating downtime (Schnelle & Favaro, 2023).

8.3 Industrial and Academic Collaborations

The successful development of nanocomposites for desulfurization relies on the effective cooperation between academic research institutions, industrial companies, and government funds. "Collaborations such as these allow us to turn basic research into beneficial technologies, evaluate real fuel samples and data, and scale new technologies that we know will work in a commercial setting. "One exemplar work is the cooperation of a university research center focused on nanomaterials sciences and a state-owned oil company, focusing on magnetic MOF-derived nanoparticles composites for ultra-low sulfur diesel production (Hamza et al., 2024). The university team concentrated on material development and laboratory test work, with the commercial partner contributing process engineering skills and the pilot plant infrastructure. Through joint efforts, the consortium established a scalable approach for the synthesis of an Fe_3O_3 ZIF-8 composite followed by its application to real diesel samples with full techno-economic and life-cycle analysis. The purpose of the project was to demonstrate not only the technical performance of the material, but also any process modifications needed through integration in a refinery setting (Irshad et al., 2024).

In a different example, a startup originating from a materials science laboratory partnered with a chemical engineering company to design a cartridge-based modular mobile desulfurization unit based on magnetic nanocomposites. Replaceable or renewable magnetically recoverable nanocomposite cartridges are used in local or off-grid fuel-treatment systems. Focusing on commercialization, product design, and user aesthetics appealed to us, as this showed the possibility of nanocomposite systems in decentralized applications (military camps, countryside fuelling stations, or natural disaster platforms) (Mane, 2021). Research consortia and funding initiatives have also created international collaboration. Collaborative projects on green nanotechnology, sustainable fuels, and advanced separations are currently receiving funding from the European Union's horizon framework, the U.S. Department of Energy (DOE), and the Asian Research Council, among others. These activities lead to the pooling of knowledge, harmonization of testing platforms, and access to diverse expertise (Adams, 2024).

Cooperation between academics and the industry is another important point when discussing regulations. Collaboration allows safety data sheets to be developed, toxicological profiles to be created, and environmental impact assessments (required for compliance with organizations such as the EPA, REACH, and ISO) to be performed. Moreover, co-authored publications, patents, and conference presentations generated from such collaborations aid the understanding of nanocomposite-based desulfurization by the general scientific or industrial community (Rossoni, de Vasconcellos, & de Castilho Rossoni, 2024).

CONCLUSION

Magnetic and hybrid bio-nanocomposites offer a novel approach as an alternative green desulfurization technique owing to their high efficiency, magnetic recoverability, and compatibility with bio-based and biodegradable reactants. Their combined adsorptive and catalytic roles make them active during the removal of more refractory sulfur species under relatively mild process conditions, which is in good agreement with environmental and operational drivers. Although laboratory tests have demonstrated promising results, there are concerns about large-scale synthesis, long-term stability, and refinery integration. However, increasing pilot-scale applications and academic–industrial alliances indicate an increasing direction toward commercialization. Scalable green synthesis of these nanoparticles, the simultaneous removal of multi-contaminants, and full environmental appraisal are challenges for future work. Advanced nanocomposites that could be key solutions for cleaner fuels and contribute to global agendas, such as sustainability, will continue to develop in the coming years, particularly through interdisciplinary collaboration.

AI DISCLOSURE

During the preparation of this work, the author(s) used *Paperpal* solely for grammar and language refinement. After using this tool, the author(s) carefully reviewed and edited the content as needed and took full responsibility for the integrity and accuracy of the scientific content presented in this chapter.

REFERENCES

Abboud, M., Alnefaie, R., & Alhanash, A. (2022). Synthesis and application of unsupported and silica-supported Ni nanoparticles in catalysis. *Journal of Nanoparticle Research, 24*(2), 21. DOI: 10.1007/s11051-021-05350-x

Abd Zaid, S. M., AbdulRazak, A., & Abid, M. (2022). A review of nano-catalyst applications in kerosene desulfurization techniques. *Journal of Applied Sciences and Nanotechnology*, 2(2), 86–102. DOI: 10.53293/jasn.2022.4302.1094

Adams, Z. K. (2024). *Recycling of Rare Earth Magnets With Sulfur Based Chemistries and High Temperature Processing*. Massachusetts Institute of Technology.

Afolabi, O. A., & Ndou, N. (2024). Synergy of hybrid fillers for emerging composite and nanocomposite materials—A review. *Polymers, 16*(13), 1907. DOI: 10.3390/polym16131907 PMID: 39000762

Akopyan, A., Polikarpova, P., Anisimov, A., Lysenko, S., Maslova, O., Sen'ko, O., & Efremenko, E. (2021). Oxidation of dibenzothiophene with the subsequent bioconversion of sulfone. *Theoretical Foundations of Chemical Engineering, 55*(4), 778–785. DOI: 10.1134/S0040579521040035

Akter, T., Islam, A., Al Miad, A., Alam, M. K., Ahmed, S., & Sahadat Hossain, M. (2025). Morphological Modification of Metal Oxide Nanomaterials Using Different Types of Organic Modifiers. *ChemistryOpen, 14*(12), 202500342. DOI: 10.1002/open.202500342 PMID: 40944616

Al-Attas, T. A., Ali, S. A., Zahir, M. H., Xiong, Q., Al-Bogami, S. A., Malaibari, Z. O., Razzak, S. A., & Hossain, M. M. (2019). Recent advances in heavy oil upgrading using dispersed catalysts. *Energy & Fuels, 33*(9), 7917–7949. DOI: 10.1021/acs.energyfuels.9b01532

Ameta, S. C. (2022). Nanomaterials: An Introduction. In The Science of Nanomaterials (pp. 1-18): Apple Academic Press.

Anekwe, I. M. S., & Isa, Y. M. (2025). Unlocking catalytic longevity: A critical review of catalyst deactivation pathways and regeneration technologies. *Energy Advances, 4*(9), 1075–1113. DOI: 10.1039/D5YA00015G

Anshassi, M., & Townsend, T. G. (2021). Reviewing the underlying assumptions in waste LCA models to identify impacts on waste management decision making. *Journal of Cleaner Production, 313*, 127913. DOI: 10.1016/j.jclepro.2021.127913

Arevalo-Basañez, E., Jiménez-García, G., Villalón-López, U. A., & Maya-Yescas, R. (2025). The HDS Process: Origin, Process Evolution, Reaction Mechanisms, Process Units, Catalysts, and Health Risks. *Processes (Basel, Switzerland), 13*(9), 2817. DOI: 10.3390/pr13092817

Batchu, S. P., Hernandez, B., Malhotra, A., Fang, H., Ierapetritou, M., & Vlachos, D. G. (2022). Accelerating manufacturing for biomass conversion via integrated process and bench digitalization: A perspective. *Reaction Chemistry & Engineering, 7*(4), 813–832. DOI: 10.1039/D1RE00560J

Bellanthudawa, B., Nawalage, N., Handapangoda, H., Suvendran, S., Wijayasenarathne, K., Rathnasuriya, M., Wickramasinghe, P. G. M. U., Aberathna, A. A. A. U., Tennakoon, A., & Perera, I. (2023). A perspective on biodegradable and non-biodegradable nanoparticles in industrial sectors: Applications, challenges, and future prospects. *Nanotechnology for Environmental Engineering, 8*(4), 975–1013. DOI: 10.1007/s41204-023-00344-7

Bianchini Silva, M., Costa, U. O., Mattoso, L. H. C., Monteiro, S. N., de Souza, M. L., & Vitorazi, L. (2024). Magnetic Nanoparticles in Biopolymer Fibers: Fabrication Techniques and Characterization Methods. *Polymers, 16*(19), 2805. DOI: 10.3390/polym16192805 PMID: 39408514

Bibri, S. E. (2022). The social shaping of the metaverse as an alternative to the imaginaries of data-driven smart Cities: A study in science, technology, and society. *Smart Cities, 5*(3), 832–874. DOI: 10.3390/smartcities5030043

Boshagh, F., Rahmani, M., Rostami, K., & Yousefifar, M. (2021). Key factors affecting the development of oxidative desulfurization of liquid fuels: A critical review. *Energy & Fuels, 36*(1), 98–132. DOI: 10.1021/acs.energyfuels.1c03396

Busca, G. (2021). Production of gasolines and monocyclic aromatic hydrocarbons: From fossil raw materials to green processes. *Energies, 14*(13), 4061. DOI: 10.3390/en14134061

Cai, G., Yan, P., Zhang, L., Zhou, H.-C., & Jiang, H.-L. (2021). Metal–organic framework-based hierarchically porous materials: Synthesis and applications. *Chemical Reviews*, *121*(20), 12278–12326. DOI: 10.1021/acs.chemrev.1c00243 PMID: 34280313

Castiello, C., Junghanns, P., Mergel, A., Jacob, C., Ducho, C., Valente, S., Rotili, D., Fioravanti, R., Zwergel, C., & Mai, A. (2023). GreenMedChem: The challenge in the next decade toward eco-friendly compounds and processes in drug design. *Green Chemistry*, *25*(6), 2109–2169. DOI: 10.1039/D2GC03772F

Chen, Y., Wang, T., Tian, H., Su, D., Zhang, Q., & Wang, G. (2021). Advances in lithium–sulfur batteries: From academic research to commercial viability. *Advanced Materials*, *33*(29), 2003666. DOI: 10.1002/adma.202003666 PMID: 34096100

Cherubala, R. B. (2021). Desulphurization of diesel fuel using carbon-based metal oxide nanocomposites. Vaal University of Technology (South Africa), Chi, N. T. L., Anto, S., Ahamed, T. S., Kumar, S. S., Shanmugam, S., Samuel, M. S., . . . Pugazhendhi, A. (2021). A review on biochar production techniques and biochar based catalyst for biofuel production from algae. *Fuel*, *287*, 119411.

D'Adamo, I., Ippolito, N. M., Shalchian, H., Innocenzi, V., Ferella, F., & Veglio, F. (2023). A feasibility study for a circular approach in oil refining: Metals recovery from hydrodesulphurization catalysts. *Sustainable Materials and Technologies*, *36*, e00634. DOI: 10.1016/j.susmat.2023.e00634

Damian, C. S., Devarajan, Y., J, R., & T, R. (2025). Nanocatalysts in Biodiesel Production: Advancements, Challenges, and Sustainable Solutions. *ChemBioEng Reviews*, *12*(2), e202400055. DOI: 10.1002/cben.202400055

Dembaremba, T. O., Majodina, S., Walmsley, R. S., Ogunlaja, A. S., & Tshentu, Z. R. (2022). Perspectives on strategies for improving ultra-deep desulfurization of liquid fuels through hydrotreatment: Catalyst improvement and feedstock pre-treatment. *Frontiers in Chemistry*, *10*, 807225. DOI: 10.3389/fchem.2022.807225 PMID: 35936099

Dhanyashree, S., Jahnavi, H., Harini, H., Prasad, S. R., Nagaswarupa, H., & Naik, R. (2024). Design and development of novel nanoferrite and metal oxide composites for energy applications: a review. Multifunctional Inorganic Nanomaterials for Energy Applications, 96-119.

Di Carmine, G., Abbott, A. P., & D'agostino, C. (2021). Deep eutectic solvents: Alternative reaction media for organic oxidation reactions. *Reaction Chemistry & Engineering*, *6*(4), 582–598. DOI: 10.1039/D0RE00458H

Duan, Y., Mehariya, S., Kumar, A., Singh, E., Yang, J., Kumar, S., Li, H., & Kumar Awasthi, M. (2021). Apple orchard waste recycling and valorization of valuable product-A review. *Bioengineered, 12*(1), 476–495. DOI: 10.1080/21655979.2021.1872905 PMID: 33472503

El-Fawal, E. M., El Naggar, A. M., El-Zahhar, A. A., Alghandi, M. M., Morshedy, A. S., El Sayed, H. A., & Mohammed, A. M. E. (2025). Biofuel production from waste residuals: Comprehensive insights into biomass conversion technologies and engineered biochar applications. *RSC Advances, 15*(15), 11942–11974. DOI: 10.1039/D5RA00857C PMID: 40264955

El Kfoury, M.-T. (2022). Novel Process for Organic Peroxides Synthesis. Polytechnique Montréal, Eskandarinezhad, S., Khosravi, R., Amarzadeh, M., Mondal, P., & Magalhães Filho, F. J. C. (2021). Journal of Composites and Compounds. *Journal of Composites and Compounds, 3*, 43–56.

El-Khayatt, A. M., El Desouky, F. G., El-Khawas, E., & Azab, A. (2025). Nanoferrite-modified CuO/SnO_2 nanocomposites with tunable magnetic and electrical properties for prospective applications. *Inorganic Chemistry Communications, 180*, 115019. DOI: 10.1016/j.inoche.2025.115019

Fadel, M., Salvador Fernández, M., Pilati, V., Martínez-García, J. C., Alvarez-Alonso, P., & Rivas, M. (2025). A holistic view of magnetic nanoparticles for advanced water treatment and cleanup. *Journal of Physics. D, Applied Physics, 58*(40), 403001. DOI: 10.1088/1361-6463/ae01b2

Fahim, Y. A., Hasani, I. W., & Mahmoud Ragab, W. (2025). Promising biomedical applications using superparamagnetic nanoparticles. *European Journal of Medical Research, 30*(1), 441. DOI: 10.1186/s40001-025-02696-z PMID: 40452035

Fard, N. E., Raeisi, I., Yousefipour, Z., & Mosavin, R. (2024). Review of metal sulfide-based MXene nanocomposites for environmental applications, gas sensing, energy storage, and photothermal therapy. *ACS Applied Nano Materials, 7*(6), 5698–5728. DOI: 10.1021/acsanm.3c05455

Fu, J., & Turn, S. Q. (2025). Oxidation mechanism of sulfur-containing compounds and antioxidant depletion dynamics: Insights into interactions. *Fuel, 381*, 133341. DOI: 10.1016/j.fuel.2024.133341

Gomez-Villalba, L. S., Salcines, C., & Fort, R.GOmez-Villalba. (2023). Application of inorganic nanomaterials in cultural heritage conservation, risk of toxicity, and preventive measures. *Nanomaterials (Basel, Switzerland), 13*(9), 1454. DOI: 10.3390/nano13091454 PMID: 37176999

Goyal, A., Meena, P. K., Shelare, S., & Kumar, S.GOyal. (2025). Advancing biofuel production through nanomaterials: A review of efficiency improvements, challenges, and future directions. *Nanocomposites, 11*(1), 151–172. DOI: 10.1080/20550324.2025.2518755

Gu, H., Kong, Y., Huang, D., Wang, Y., Raghavan, V., & Wang, J. (2025). Scaling cultured meat: Challenges and solutions for affordable mass production. *Comprehensive Reviews in Food Science and Food Safety, 24*(4), e70221. DOI: 10.1111/1541-4337.70221 PMID: 40635127

Habib, U., Ahmad, F., Awais, M., Naz, N., Aslam, M., Urooj, M., & Sajid, M. (2023). Sustainable catalysis: Navigating challenges and embracing opportunities for a greener future. *J. Chem. Environ, 2*(2), 14–53. DOI: 10.56946/jce.v2i2.205

Halvaeikhanekahdani, P., Zandi, S., Ahmad, Q., & Payravand, H. (2025). Smart magnetic nanoparticles under electromagnetic stimulation: A synergistic approach to targeted chemotherapy and hyperthermia. *Nanomedicine (London), 20*(16), 2143–2166. DOI: 10.1080/17435889.2025.2534324 PMID: 40685946

Hamza, Z. A., Dawood, J. J., & Jabbar, M. A. (2024). Review of TiO2 as desulfurization catalyst for petroleum. *Catalysts, 14*(6), 381. DOI: 10.3390/catal14060381

Herman, R. A., Zhu, X., Ayepa, E., You, S., & Wang, J. (2023). Advances in the One-Step Approach of Polymeric Materials Using Enzymatic Techniques. *Polymers, 15*(3), 703. DOI: 10.3390/polym15030703 PMID: 36772002

Hessel, V., Tran, N. N., Asrami, M. R., Tran, Q. D., Long, N. V. D., Escribà-Gelonch, M., & Sundmacher, K. (2022). Sustainability of green solvents–review and perspective. *Green Chemistry, 24*(2), 410–437. DOI: 10.1039/D1GC03662A

Hosseini, A., Khoshsima, A., Sabzi, M., & Rostam, A. (2022). Toward application of ionic liquids to desulfurization of fuels: A review. *Energy & Fuels, 36*(8), 4119–4152. DOI: 10.1021/acs.energyfuels.1c03974

Houda, S., Lancelot, C., Blanchard, P., Poinel, L., & Lamonier, C. (2018). Oxidative desulfurization of heavy oils with high sulfur content: A review. *Catalysts, 8*(9), 344. DOI: 10.3390/catal8090344

Ikram, R., Mohamed Jan, B., Nagy, P. B., & Szabo, T. (2023). Recycling waste sources into nanocomposites of graphene materials: Overview from an energy-focused perspective. *Nanotechnology Reviews, 12*(1), 20220512. DOI: 10.1515/ntrev-2022-0512

Io a, M.-A., Cursaru, L.-M., chiopu, A.-G., Tudor, I. A., Motoc, A.-M., & Piticescu, R. M. (2023). Fe_3O_4 Core–shell nanostructures with anticancer and antibacterial properties: A Mini-review. *Processes (Basel, Switzerland), 11*(7), 1882. DOI: 10.3390/pr11071882

Ippolito, N. M., Romano, P., Innocenzi, V., Ferella, F., & Giansante, G. (2023). The Circular Economy Approach in Oil Refining: Recycling of Spent FCC and HDS Catalysts. Paper presented at the Offshore Mediterranean Conference and Exhibition.

Iroegbu, A. O. C., & Ray, S. S. (2022). Recent developments and future perspectives of biorenewable nanocomposites for advanced applications. *Nanotechnology Reviews, 11*(1), 1696–1721. DOI: 10.1515/ntrev-2022-0105

Irshad, T., Saeed, M., Munir, M., Kousar, R., Intisar, A., Din, M. I., Ahsan, H., Abid, M. A., Irfan, M., & Osman, S. M. (2024). Excellent catalytic performance over reduced graphene-boosted novel nanoparticles for oxidative desulfurization of fuel oil. *Nanotechnology Reviews, 13*(1), 20240116. DOI: 10.1515/ntrev-2024-0116

Jacob, A. G., Alisi, I. O., & Surajo, J. M. (2025). Kaolinite clay as green and sustainable raw material for zeolites production: A review. *FUDMA Journal of Sciences, 9*, 18–35. DOI: 10.33003/fjs-2025-09(AHBSI)-3496

Jadhav, R. M., & Sangwai, J. S. (2020). Interaction of heavy crude oil and nanoparticles for heavy oil upgrading. In *Nanotechnology for Energy and Environmental Engineering* (pp. 231–255). Springer. DOI: 10.1007/978-3-030-33774-2_10

Ji, F. (2021). *Collective Behaviors of Small-Scale Magnetic Building Blocks in Fluid*. The Chinese University of Hong Kong.

Jia, Y., Yang, C., Chen, X., Xue, W., Hutchins-Crawford, H. J., Yu, Q., Topham, P. D., & Wang, L. (2021). A review on electrospun magnetic nanomaterials: Methods, properties and applications. *Journal of Materials Chemistry. C, Materials for Optical and Electronic Devices, 9*(29), 9042–9082. DOI: 10.1039/D1TC01477C

Kaeed, Z., Shahhosseini, S., & Rezaei, M. (2024). Preparation of the MoO3-Fe_3O_4 nanocomposites for deep oxidative desulfurization of model fuel: Optimization and experimental design. *Journal of Industrial and Engineering Chemistry, 139*, 267–280. DOI: 10.1016/j.jiec.2024.05.003

Kargar, H., Ghahramaninezhad, M., Shahrak, M. N., & Balula, S. S. (2021). An effective magnetic catalyst for oxidative desulfurization of model and real fuels: Fe_3O_4/ZIF-8/TiO_2. *Microporous and Mesoporous Materials, 317*, 110992. DOI: 10.1016/j.micromeso.2021.110992

Karimi, R., & Golmohammadi, B. (2024). Simulation and optimization of heavy fuel oil (HFO) refining platform to low sulfur marine fuel (LS-FO) by oxidative desulfurization. *Scientific Reports*, *14*(1), 29830. DOI: 10.1038/s41598-024-81242-z PMID: 39616214

Kumar, S., Singh, H., Singh, H., Soni, H., Bechelany, M., & Singh, J. (2025). Extraction of Metallic Nanoparticles from Electronic Waste for Biomedical Applications: Pioneering Sustainable Technological Progress. *Sustainability (Basel)*, *17*(5), 2100. DOI: 10.3390/su17052100

Li, P., Jiang, H., Barr, A., Ren, Z., Gao, R., Wang, H., Fan, W., Zhu, M., Xu, G., & Li, J. (2021). Reusable polyacrylonitrile-sulfur extractor of heavy metal ions from wastewater. *Advanced Functional Materials*, *31*(51), 2105845. DOI: 10.1002/adfm.202105845

Li, W., Guo, Z., Yang, J., Li, Y., Sun, X., He, H., Li, S., & Zhang, J. (2022). Advanced strategies for stabilizing single-atom catalysts for energy storage and conversion. *Electrochemical Energy Reviews*, *5*(3), 9. DOI: 10.1007/s41918-022-00169-z

Madani, M., Hosny, S., Alshangiti, D. M., Nady, N., Alkhursani, S. A., Alkhaldi, H., Al-Gahtany, S. A., Ghobashy, M. M., & Gaber, G. A. (2022). Green synthesis of nanoparticles for varied applications: Green renewable resources and energy-efficient synthetic routes. *Nanotechnology Reviews*, *11*(1), 731–759. DOI: 10.1515/ntrev-2022-0034

Mane, M. B. (2021). Studies in newer methodologies and materials/material modifications for environmental, pollution control applications specifically for water treatment and desulfurization.

Mehdi, R., Khoja, A. H., Naqvi, S. R., Gao, N., & Amin, N. A. S. (2022). A review on production and surface modifications of biochar materials via biomass pyrolysis process for supercapacitor applications. *Catalysts*, *12*(7), 798. DOI: 10.3390/catal12070798

Mehmood, A., Khan, F. S. A., Mubarak, N. M., Tan, Y. H., Karri, R. R., Khalid, M., Walvekar, R., Abdullah, E. C., Nizamuddin, S., & Mazari, S. A. (2021). Magnetic nanocomposites for sustainable water purification—A comprehensive review. *Environmental Science and Pollution Research International*, *28*(16), 19563–19588. DOI: 10.1007/s11356-021-12589-3 PMID: 33651297

Mitra, S., Racha, S. M., Shown, B., Mandal, S., & Das, A. K. (2023). New insights on oxidative desulfurization for low sulfur residual oil production. *Sustainable Energy & Fuels*, *7*(1), 270–279. DOI: 10.1039/D2SE01363K

Mo, Y., Zhang, Y., & Zhang, Q. (2024). The pulmonary effects of nickel-containing nanoparticles: Cytotoxicity, genotoxicity, carcinogenicity, and their underlying mechanisms. *Environmental Science. Nano, 11*(5), 1817–1846. DOI: 10.1039/D3EN00929G PMID: 38984270

Mourdikoudis, S., Dutta, S., Kamal, S., Gómez-Graña, S., Pastoriza-Santos, I., Wuttke, S., & Polavarapu, L. (2025). State-of-the-Art, Insights, and Perspectives for MOFs-Nanocomposites and MOF-Derived (Nano) Materials. *Advanced Materials, 37*(52), 2415399. DOI: 10.1002/adma.202415399 PMID: 40255059

Mugisha, A. I. (2024). Synthesis and application of multifunctional iron oxide-graphene oxide nanocomposite materials for sulfur pollutant removal. Université d'Ottawal University of Ottawa, Murrieta-Rico, F. N., Antúnez-García, J., Yocupicio-Gaxiola, R. I., Galván, D. H., GOnzález, J. C., & Petranovskii, V. (2022). Zeolites as initial structures for the preparation of functional materials. *Journal of Applied Research and Technology, 20*(1), 92–116.

Nguyen, M. D., Tran, H.-V., Xu, S., & Lee, T. R. (2021). Fe_3O_4 nanoparticles: Structures, synthesis, magnetic properties, surface functionalization, and emerging applications. *Applied Sciences (Basel, Switzerland), 11*(23), 11301. DOI: 10.3390/app112311301 PMID: 35844268

Nimbagal, V., Banapurmath, N., Sajjan, A. M., Patil, A. Y., & Ganachari, S. V. (2021). Studies on hybrid bio-nanocomposites for structural applications. *Journal of Materials Engineering and Performance, 30*(9), 6461–6480. DOI: 10.1007/s11665-021-05843-9

Oksel Karakus, C., Bilgi, E., & Winkler, D. A. (2021). Biomedical nanomaterials: Applications, toxicological concerns, and regulatory needs. *Nanotoxicology, 15*(3), 331–351. DOI: 10.1080/17435390.2020.1860265 PMID: 33337941

Olaremu, A. G., Adedoyin, W. R., Ore, O. T., & Adeola, A. O. (2021). Sustainable development and enhancement of cracking processes using metallic composites. *Applied Petrochemical Research, 11*(1), 1–18. DOI: 10.1007/s13203-021-00263-1

Oloyede, C. T., Itabiyi, O. E., Popoola, O. A., Jekayinfa, S. O., Olaniyan, M. A., Adebayo, A. O., Ogunkunle, O., Zamri, M. F. M. A., & Fattah, I. M. R. (2024). Navigating prospects and challenges for green fuels for achieving economical, environmental and ecological resilience: A scientific review. *Biofuels, 15*(7), 929–941. DOI: 10.1080/17597269.2023.2299090

Pavithra, S., Sakunthala, A., Vinofia, S., Pramoda, K., Kathirvel, V., & Keri, R. S. (2024). Superior performance of hierarchical 3D microflower CuS grown on carbon cloth as binder free electrode for supercapacitor applications. *Materials Research Bulletin, 180*, 113010. DOI: 10.1016/j.materresbull.2024.113010

Pérez-Botella, E., Valencia, S., & Rey, F. (2022). Zeolites in adsorption processes: State of the art and future prospects. *Chemical Reviews, 122*(24), 17647–17695. DOI: 10.1021/acs.chemrev.2c00140 PMID: 36260918

Periasamy, P., & Mohanta, Y. K. (2024). Global energy crisis: need for energy conversion and storage. In *Green Nanomaterials in Energy Conversion and Storage Applications* (pp. 45–73). Apple Academic Press.

Pervaiz, S., Javed, M., Shah, A., Latif, A., Nasir, S., & Shah, I. (2025). Environmental applications of magnetic nanohybrid materials. *RSC Advances, 15*(25), 19899–19936. DOI: 10.1039/D5RA03470A PMID: 40503319

Rahman, K. M., Amiri, O., Ahmed, S. S., Ismael, S. J., Rasul, N. S., Babakrb, K. A., Dadkhah, M., & Jamal, M. A. (2024). Mechanism and kinetic of piezo-catalytic desulfurization of model and actual fuel samples over CexOy/SrO nanocomposite at room temperature. *Heliyon, 10*(2), e24707. DOI: 10.1016/j.heliyon.2024.e24707 PMID: 38304796

Rajendran, A., Cui, T.-y., Fan, H., Yang, Z., Feng, J., & Li, W.-y. (2020). A comprehensive review on oxidative desulfurization catalysts targeting clean energy and environment. *Journal of Materials Chemistry. A, Materials for Energy and Sustainability, 8*(5), 2246–2285. DOI: 10.1039/C9TA12555H

Ramesh, M., Rajeshkumar, L., Balaji, D., & Bhuvaneswari, V. (2023). Sustainable and renewable nano-biocomposites for sensors and actuators: A review on preparation and performance. *Current Analytical Chemistry, 19*(1), 38–69. DOI: 10.2174/1573411018666220421112916

Rando, G., Sfameni, S., Galletta, M., Drommi, D., Cappello, S., & Plutino, M. R. (2022). Functional Nanohybrids and Nanocomposites development for the removal of environmental pollutants and bioremediation. *Molecules (Basel, Switzerland), 27*(15), 4856. DOI: 10.3390/molecules27154856 PMID: 35956804

Rashid, T., Raza, A., Saleh, H. A., Khan, S., Rahaman, S., Pandey, K., & Shahid, M. (2023). Green synthesis of Ni/Fe_3O_4/rGO nanocomposites for desulfurization of fuel. *ACS Applied Nano Materials, 6*(20), 18905–18917. DOI: 10.1021/acsanm.3c03270

Ren, J., Cao, J.-P., Yang, F.-L., Liu, Y.-L., Tang, W., & Zhao, X.-Y. (2021). Understandings of catalyst deactivation and regeneration during biomass tar reforming: A crucial review. *ACS Sustainable Chemistry & Engineering*, *9*(51), 17186–17206. DOI: 10.1021/acssuschemeng.1c07483

Rezaei, B., Yari, P., Sanders, S. M., Wang, H., Chugh, V. K., Liang, S., & Gómez-Pastora, J. (2024). Magnetic nanoparticles: A review on synthesis, characterization, functionalization, and biomedical applications. *Small*, *20*(5), 2304848. DOI: 10.1002/smll.202304848 PMID: 37732364

Rocha, L. S., Sousa, É. M., Pereira, D., Gil, M. V., Otero-Irurueta, G., Gallo, M. J. H., & Calisto, V. (2021). Sustainable and recoverable waste-based magnetic nanocomposites used for the removal of pharmaceuticals from wastewater. *Chemical Engineering Journal*, *426*, 129974. DOI: 10.1016/j.cej.2021.129974

Rossoni, A. L., de Vasconcellos, E. P. G., & de Castilho Rossoni, R. L. (2024). Barriers and facilitators of university-industry collaboration for research, development and innovation: A systematic review. *Management Review Quarterly*, *74*(3), 1841–1877. DOI: 10.1007/s11301-023-00349-1

Sabukuttan, D. (2025). Hydrothermal Synthesis and Physicochemical Characterization of Carbon Quantum Dots from Marine Cyanobacteria Leptolyngbya sp. nov. KMBMA-1. *BioNanoScience*, *15*(3), 1–13.

Sadrara, M., & Khorrami, M. K. (2023). Designing an efficient organic–inorganic hybrid nanocomposite for simultaneous oxidative/adsorptive desulfurization of model and real fuel oils. *Scientific Reports*, *13*(1), 15134. DOI: 10.1038/s41598-023-42392-8 PMID: 37704702

Sahraei, S. (2023). Assessment of reaction parameters in the oxidative desulfurization reaction. *Energy & Fuels*, *37*(20), 15373–15393. DOI: 10.1021/acs.energyfuels.3c02169

Salman, M., Ur Rahman, A., Usman Farooq, M., Shakir, M., Subhan, F., Akram, F., Khan, K., Ullah, M., & Yaseen, M. (2024). Advances in the applications of mixed matrix membranes for desulfurization of transportation fuels. *Separation and Purification Reviews*, *53*(4), 448–474. DOI: 10.1080/15422119.2024.2322425

Sang, P., Chen, Q., Wang, D.-Y., Guo, W., & Fu, Y. (2023). Organosulfur materials for rechargeable batteries: Structure, mechanism, and application. *Chemical Reviews*, *123*(4), 1262–1326. DOI: 10.1021/acs.chemrev.2c00739 PMID: 36757873

Sarkar, M., Sarkar, C., & Saha, S. (2025). Catalytic activity of biodegradable polymeric colloidal particles in ground water remediation: recent progress and challenges. Journal of Macromolecular Science, Part A, 1-36.

Schnelle, S., & Favaro, F. M. (2023). ADS standardization landscape: Making sense of its status and of the associated research questions. arXiv preprint arXiv:2306.17682.

Sehgal, B., & Kunde, G. (2022). Recent Advances in the Catalytic Applications of Magnetic Nanomaterials. Emerging Applications of Low Dimensional Magnets, 9-31.

Senthil, K. P. (2024). Employing carbon nanocomposites for the removal of hazardous contaminants from aqueous solution. *Malaysian NANO-An International Journal, 4*(1), 24–44. DOI: 10.22452/mnij.vol4no1.2

Shaik, S., & Franchetti, M. J. (2022). Recycling and sustainability of nanocomposites. In *Applications of Nanocomposites* (pp. 236–259). CRC Press. DOI: 10.1201/9781003247074-11

Shakirzyanov, R. I., Kozlovskiy, A., Zdorovets, M. V., Zheludkevich, A., Shlimas, D., Abmiotka, N., Kazantsev, P. A., Zubar, T. I., Trukhanov, S. V., & Trukhanov, A. (2023). Impact of thermobaric conditions on phase content, magnetic and electrical properties of the $CoFe_2O4$ ceramics. *Journal of Alloys and Compounds, 954*, 170083. DOI: 10.1016/j.jallcom.2023.170083

SIDDIQUI, M. T. H. (2024). Fabrication of magnetic carbon nanocomposites using rice husk via one pot solvothermal carbonization coprecipitation synthesis. RMIT University, Singh, R., & Bhateria, R. (2021). Core–shell nanostructures: A simplest two-component system with enhanced properties and multiple applications. *Environmental Geochemistry and Health, 43*(7), 2459–2482.

Singh, R., & Kumar, S. (2023). Regulatory and safety concerns regarding the use of active nanomaterials in food industry. In *Nanotechnology advancement in agro-food industry* (pp. 269–306). Springer. DOI: 10.1007/978-981-99-5045-4_8

Sithole, T. (2024). A review on regeneration of adsorbent and recovery of metals: Adsorbent disposal and regeneration mechanism. *South African Journal of Chemical Engineering, 50*(1), 39–50.

Strassburg, S., Mayer, K., & Scheibel, T. (2022). Functionalization of biopolymer fibers with magnetic nanoparticles. *Physical Sciences Reviews, 7*(10), 1091–1117. DOI: 10.1515/psr-2019-0118

SURESH, C. A. (2022). Nanomaterials: An Introduction. The Science of Nanomaterials: Basics and Applications, 1.

Tajdari, A., Babaei, A., Goudarzi, A., Partovi, R., & Rostami, A. (2021). Hybridization as an efficient strategy for enhancing the performance of polymer nanocomposites. *Polymer Composites*, *42*(12), 6801–6815. DOI: 10.1002/pc.26341

Tang, S., Wang, Y., He, P., Wang, Y., & Wei, G. (2024). Recent Advances in Metal–Organic Framework (MOF)-Based Composites for Organic Effluent Remediation. *Materials (Basel)*, *17*(11), 2660. DOI: 10.3390/ma17112660 PMID: 38893925

Tanimu, A., Tanimu, G., Ganiyu, S. A., Gambo, Y., Alasiri, H., & Alhooshani, K. (2022). Metal-free catalytic oxidative desulfurization of fuels— A review. *Energy & Fuels*, *36*(7), 3394–3419. DOI: 10.1021/acs.energyfuels.1c04411

Thomas, S., & Nochehdehi, A. R. (2022). Handbook of magnetic hybrid nanoalloys and their nanocomposites: Springer Nature.

Tovar-Lopez, F. J. (2023). Recent progress in micro-and nanotechnology-enabled sensors for biomedical and environmental challenges. *Sensors (Basel)*, *23*(12), 5406. DOI: 10.3390/s23125406 PMID: 37420577

Tripathi, K., Singh, R., & Pant, K. K. (2021). Tailoring the physicochemical properties of Mg promoted catalysts via one pot non-ionic surfactant assisted co-precipitation route for CO_2 Co-feeding syngas to methanol. *Topics in Catalysis*, *64*(5), 395–413. DOI: 10.1007/s11244-020-01410-z

Tsoukleris, D. S., Gatou, M.-A., Lagopati, N., Sygellou, L., Christodouleas, D. C., Falaras, P., & Pavlatou, E. A. (2023). Chemically modified TiO2 photocatalysts as an alternative disinfection approach for municipal wastewater treatment plant effluents. *Water (Basel)*, *15*(11), 2052. DOI: 10.3390/w15112052

Vijayalakshmi, K., Florence, J. A. K., & Arunachalam, P. (2023). Industrial wastewater sources and treatment strategies by using bionanomaterials. In *Nanobiomaterials* (pp. 255–266). CRC Press. DOI: 10.1201/9780429057038-21

Wang, C., Lu, W., Wu, W., Zhang, L., Guo, W., Huang, G., & Xu, S. (2023). Mechanical grinding of FeNC nanomaterial with Fe_3O_4 to construct magnetic adsorbents for desulfurization. *Separation and Purification Technology*, *306*, 122574. DOI: 10.1016/j.seppur.2022.122574

Wang, Y., Fens, M. H., Van Kronenburg, N. C., Shi, Y., Lammers, T., Heger, M., van Nostrum, C. F., & Hennink, W. E. (2022). Magnetic beads for the evaluation of drug release from biotinylated polymeric micelles in biological media. *Journal of Controlled Release*, *349*, 954–962. DOI: 10.1016/j.jconrel.2022.07.044 PMID: 35931210

Wu, Z., Li, Y., Zhang, C., Huang, X., Peng, B., & Wang, G. (2022). Recent advances in metal-organic-framework-based catalysts for thermocatalytic selective oxidation of organic substances. *Chem Catalysis*, *2*(5), 1009–1045. DOI: 10.1016/j.checat.2022.02.010

Yang, G., Liang, J., Hu, X., Liu, M., Zhang, X., & Wei, Y. (2021). Recent Advances on Fabrication of Polymeric Composites Based on Multicomponent Reactions for Bioimaging and Environmental Pollutant Removal. *Macromolecular Rapid Communications*, *42*(6), 2000563. DOI: 10.1002/marc.202000563 PMID: 33543565

Yeole, N. R., & Parthasarthy, V. (2022). Design of experiments (DOE) for adsorptive desulfurization (ADS) of liquid fuels–A review. *Materials Today: Proceedings*, *57*, 1613–1618. DOI: 10.1016/j.matpr.2021.12.230

Yosefi, L., Khoshbin, R., & Karimzadeh, R. (2022). Beneficial incorporation of metal-sulfur interaction in adsorption capacity of boron nitride based adsorbents used in highly selective sulfur removal. *Fuel*, *310*, 122277. DOI: 10.1016/j.fuel.2021.122277

Yu, S., Wang, L., Li, Q., Zhang, Y., & Zhou, H. (2022). Sustainable carbon materials from the pyrolysis of lignocellulosic biomass. *Materials Today Sustainability*, *19*, 100209. DOI: 10.1016/j.mtsust.2022.100209

Zbuzant, M. (2022). An overview of the use of photo catalysts for desulfurization. Eurasian Journal of Chemical. *Medicinal and Petroleum Research*, *1*(1), 10–19.

Zhang, B., Jiang, Y., & Balasubramanian, R. (2021). Synthesis, formation mechanisms and applications of biomass-derived carbonaceous materials: A critical review. *Journal of Materials Chemistry. A, Materials for Energy and Sustainability*, *9*(44), 24759–24802. DOI: 10.1039/D1TA06874A

Zhou, W., Meng, X., Gao, J., Zhao, H., Zhao, G., & Ma, J. (2021). Electrochemical regeneration of carbon-based adsorbents: A review of regeneration mechanisms, reactors, and future prospects. *Chemical Engineering Journal Advances*, *5*, 100083. DOI: 10.1016/j.ceja.2020.100083

Zhou, X., Jin, H., Xia, B. Y., Davey, K., Zheng, Y., & Qiao, S. Z. (2021). Molecular cleavage of metal-organic frameworks and application to energy storage and conversion. *Advanced Materials*, *33*(51), 2104341. DOI: 10.1002/adma.202104341 PMID: 34605072

Zou, Y., Sun, Z., Wang, Q., Ju, Y., Sun, N., Yue, Q., & Wang, Z. (2024). Core–Shell Magnetic Particles: Tailored Synthesis and Applications. *Chemical Reviews*, *125*(2), 972–1048. DOI: 10.1021/acs.chemrev.4c00710 PMID: 39729245

Chapter 9
Case Studies in Fuel Purification and Emission Control

Rupam Hazra
https://orcid.org/0009-0003-0332-2128
Global Institute of Management and Technology, Krishnanagar, India

ABSTRACT

This chapter examines the evolving landscape of fuel purification and emission control, emphasizing innovative desulfurization approaches using nanocomposites and advanced materials. As environmental regulations tighten and global efforts to combat pollution and climate change intensify, cleaner fuel technologies have become critical. Conventional methods like hydrodesulfurization, though widely used, often fail to remove refractory sulfur compounds and involve high energy and operational costs. In response, researchers are turning to nanostructured materials—such as metal–organic frameworks (MOFs), graphene derivatives, mesoporous silica, and functionalized zeolites—for their superior selectivity and reactivity. The chapter features case studies, pilot projects, and material science insights, covering green desulfurization, material regeneration, emissions control, and the challenges of implementing sustainable purification strategies. It serves as a key resource for researchers, students, and professionals aiming to advance cleaner, more resilient energy systems.

1. INTRODUCTION

The global energy landscape is undergoing a rapid transformation, driven by growing concerns over environmental degradation, public health, and climate change.

DOI: 10.4018/979-8-3373-5966-3.ch009

Among the most pressing challenges is the need to reduce Sulfur emissions from fossil fuels, which are major contributors to air pollution, acid rain, and equipment corrosion. Traditional desulfurization methods, such as hydrodesulfurization (HDS), although effective, are energy-intensive, costly, and often inadequate for removing refractory Sulfur compounds. This has spurred an urgent search for cleaner, more efficient fuel purification strategies. At the forefront of this innovation are nano-composites and other advanced materials, which offer remarkable advantages in selectivity, surface activity, and operational flexibility. By enabling low-temperature, low-pressure Sulfur removal and enhancing emission control mechanisms, these materials are redefining what is technically and economically feasible in fuel treatment. This chapter introduces a series of case studies and technological in-sights that highlight the evolving role of material science in achieving sustainable desulfurization. It aims to bridge the gap between laboratory breakthroughs and industrial-scale applications, offering researchers, practitioners, and policymakers a comprehensive view of how cutting-edge fuel purification strategies can drive cleaner energy futures.

2. FUNDAMENTALS OF FUEL DESULFURIZATION AND EMISSION CONTROL

Fuel desulfurization and emission control form the cornerstone of clean energy initiatives aimed at minimizing the harmful environmental and health impacts associated with Sulfur-based pollutants. Sulfur compounds, commonly present in fossil fuels such as diesel, gasoline, and crude oil, contribute significantly to the formation of Sulfur oxides (Sox) during combustion, which in turn cause acid rain, respiratory issues, and equipment corrosion. Desulfurization eliminates elemental sulfur and its compounds from solids and liquids. and gases (Wang et al., 2005). Traditional methods like hydrodesulfurization (HDS) have been widely used but are energy-intensive and less effective at removing refractory Sulfur compounds, especially from complex fuel matrices. Emerging techniques such as adsorptive, oxidative, and biological desulfurization offer more selective, energy-efficient, and environmentally friendly alternatives. In parallel, increasingly stringent global emission standards are driving the need for advanced purification strategies that go beyond basic compliance. Understanding the chemical nature of Sulfur compounds, the mechanisms of their removal, and the evolving regulatory landscape is critical for developing next-generation desulfurization technologies, particularly those leveraging nanocomposites and other advanced materials.

A. Types of Sulfur Compounds in Fuels

Sulfur compounds in fuels occur in various chemical forms and are a major concern due to their negative environmental and operational impacts. These compounds are present in both fossil fuels like crude oil and natural gas and must be treated or removed during refining. One of the main responsibilities in petroleum refining is removing sulfur from petroleum products since sulfur compounds provide significant geological information(Shi & wu, 2021). The combustion of sulfur-containing fuels leads to the release of sulfur oxides (SOx), which contribute to air pollution, acid rain, and respiratory issues. The sulfur content also affects fuel quality and can corrode pipelines and refining equipment. Depending on their chemical nature, sulfur compounds are broadly categorized into inorganic and organic forms.

a) Inorganic Sulfur Compounds (e.g., Hydrogen Sulfide - H_2S)

Inorganic sulfur compounds are generally simpler in structure and are often found as gases in natural gas and crude oil. Hydrogen sulfide (H_2S) is the most common inorganic sulfur compound and is known for its rotten egg smell and high toxicity. It is corrosive to metals and poses serious safety hazards during extraction and processing. Removing H_2S is typically the first step in fuel treatment, often done through gas sweetening or amine treatment processes. If not removed, H_2S can cause equipment failure and environmental hazards.

b) Organic Sulfur Compounds (e.g., Thiols, Thiophenes, Sulfides)

Organic sulfur compounds are chemically bonded to hydrocarbon molecules and are much harder to remove than inorganic ones. Thiophenes, benzothiophenes, and dibenzothiophenes are examples of stable organic sulfur compounds that survive conventional refining processes. These compounds require advanced treatment methods like hydrodesulfurization (HDS) to be effectively reduced. If left untreated, they contribute heavily to sulfur oxide emissions during combustion, affecting air quality and human health.

B. Emission Regulations and Global Standards

Emission regulations and global standards for sulfur in fuels are critical to minimizing environmental pollution and protecting public health. These rules are enforced by governments and international organizations to control the amount of sulfur oxides (SOx) released during fuel combustion. High sulfur content in fuels contributes to acid rain, smog formation, and respiratory problems, especially in densely populated or industrial areas. However, the current emission limits that have been established by several countries differ in a number of ways (Singh et al., 2023). To address this, many countries have implemented strict sulfur limits in transportation fuels, power generation, and marine fuels. These regulations also push refineries to invest in cleaner technologies and adopt low-sulfur fuel production methods.

a) International Maritime Organization (IMO) Regulations

The International Maritime Organization (IMO) sets global limits on sulfur content in marine fuels to reduce emissions from ships. As of January 1, 2020, the IMO 2020 regulation mandates that the sulfur content in marine fuel must not exceed 0.5% by mass, a significant reduction from the previous 3.5%. In Emission Control Areas (ECAs), the limit is even stricter at 0.1%. These regulations aim to reduce sulfur oxide emissions, which are a major contributor to acid rain and particulate pollution near coastlines and ports. Compliance requires the use of low-sulfur fuels, scrubber systems, or alternative fuels like LNG (liquefied natural gas).

b) Euro and EPA Standards for Road Fuels

The European Union (EU) and the U.S. Environmental Protection Agency (EPA) have established stringent sulfur limits for gasoline and diesel fuels. The EU's Euro 6 regulation restricts sulfur in road fuels to no more than 10 parts per million (ppm), and the U.S. Tier 3 regulations similarly impose low-sulfur requirements. These regulations are designed to reduce emissions from vehicles, including SOx, particulate matter, and other toxic compounds. By limiting sulfur, they also improve the performance of catalytic converters and other emission control technologies. Compliance with these standards has significantly reduced urban air pollution and contributed to improved public health outcomes.Bottom of Form

C. Overview of Adsorptive, Oxidative, and Biological Techniques

The removal of sulfur compounds from fuels—known as desulfurization—is a critical process for meeting environmental standards and improving fuel quality. Beyond conventional hydrodesulfurization, three promising alternative techniques are adsorptive, oxidative, and biological desulfurization. These methods are particularly useful for removing refractory (hard-to-remove) sulfur compounds such as thiophenes and dibenzothiophenes that are resistant to traditional processes. Different techniques, including chemical methods like Fenton oxidation and electrochemical oxidation, physical processes such as adsorption and membrane filtration, and numerous biological methods, have been identified for wastewater treatment(Rashid et al., 2021). Each technique operates on a different scientific principle, offering unique advantages in efficiency, cost, and environmental impact. The selection of the most suitable method depends on factors like fuel type, sulfur compound complexity, and processing conditions.

a) Adsorptive Desulfurization

Adsorptive desulfurization involves the physical or chemical binding of sulfur compounds onto the surface of solid materials known as adsorbents. The process is typically conducted at ambient or mild temperatures, making it energy-efficient and suitable for deep desulfurization of fuels. Adsorptive techniques are especially effective for removing specific sulfur species from liquid fuels without altering the hydrocarbon structure. However, adsorbent regeneration and saturation are challenges that affect long-term viability.

b) Oxidative Desulfurization (ODS)

Oxidative desulfurization alters sulfur compounds chemically into more polar sulfoxides or sulfones with the use of oxidizing agents like hydrogen peroxide, ozone, or peracids. These oxidized sulfur species are more soluble in polar solvents and can be easily separated from fuel through extraction. ODS is particularly effective for removing refractory sulfur compounds like dibenzothiophenes that are difficult to eliminate through hydrodesulfurization. It operates under milder conditions and does not require hydrogen gas, reducing operational complexity. However, it may require post-treatment steps and proper disposal of the oxidized sulfur waste.

c) Biological Desulfurization

Biological desulfurization employs microorganisms or their enzymes to selectively remove sulfur from organic molecules without breaking the hydrocarbon chains. This technique is environmentally friendly and energy-efficient, operating under mild conditions (ambient temperature and pressure). Certain bacteria, such as *Rhodococcus erythropolis*, can metabolize sulfur atoms in compounds like dibenzothiophene (DBT), converting them into harmless sulfate forms. Biological methods are selective and do not produce toxic byproducts, but they are generally slower and may face scale-up challenges in industrial applications. Nonetheless, this approach holds great promise for sustainable and green fuel processing.

3. NANOCOMPOSITES IN FUEL TREATMENT: A MATERIAL SCIENCE PERSPECTIVE

Nanocomposites have emerged as a transformative class of materials in the field of fuel purification, offering unique physicochemical properties that address the limitations of conventional desulfurization techniques. These materials, typically composed of nanoscale particles embedded within a matrix (metallic, ceramic, polymeric, or carbon-based), exhibit high surface area, tenable porosity, and enhanced reactivity—key attributes for capturing and degrading Sulfur-containing compounds in fuels. From a material science standpoint, the effectiveness of nanocomposites lies in their customizable structure and functionalization, which allow for selective adsorption and catalytic activity against targeted Sulfur species. Hybrid systems combining metal oxides, carbon nanotubes, graphene derivatives, or metal–organic frameworks (MOFs) with active sites can dramatically improve desulfurization efficiency under milder operational conditions. Significant efforts are underway, but the prospect of bulk structural super nanocomposites appears further away now than it was a few years ago (Dzenis,2008). Moreover, their renderability, stability under thermal stress, and compatibility with industrial systems make them ideal candidates for scalable applications. This section explores the fundamental material characteristics that govern nanocomposite behavior in fuel treatment, laying the groundwork for understanding their role in next-generation emission control technologies.

A. Properties and Classifications of Nanocomposites

Nanocomposites are multiphase materials that incorporate nanoscale components (typically less than 100 nanometers) into a matrix material to significantly enhance physical, chemical, mechanical, and thermal properties. The inclusion of nanoparticles leads to superior performance compared to conventional composites due to the large surface area and unique interactions at the nanoscale. The definition of nanocomposites has expanded considerably to include a wide range of systems like one-dimensional, two-dimensional, three-dimensional, and amorphous substances, composed of clearly different elements and combined at the nanometer level(Okpala,2013). These materials exhibit enhanced strength, conductivity, flame resistance, and barrier properties even at low filler concentrations. Depending on the combination of matrix and nanofiller materials, nanocomposites can be engineered for use in various industries such as automotive, aerospace, electronics, energy, and environmental remediation. The performance of a nanocomposite depends not only on the nature of the nanomaterials used but also on their dispersion and interfacial bonding within the matrix.

a) Classification Based on Matrix Material

Nanocomposites are mainly categorized according to the kind of matrix that contains the nanoparticles. The three main categories are:

- **Polymer Matrix Nanocomposites (PMNCs):**

These consist of nanoparticles dispersed in a polymer matrix (e.g., thermoplastics or thermosets). They are lightweight, flexible, and commonly used in packaging, automotive parts, and electronics. Nanoclays, carbon nanotubes, and graphene are popular fillers used to enhance strength, barrier properties, and electrical conductivity.

- **Metal Matrix Nanocomposites (MMNCs):**

In these, metal matrices (such as aluminum, magnesium, or titanium) are reinforced with ceramic or carbon-based nanofillers. MMNCs offer superior mechanical strength, thermal stability, and wear resistance. They are widely utilized in aerospace and defense sectors. Processing techniques consist of powder metallurgy and liquid metal infiltration.

- **Ceramic Matrix Nanocomposites (CMNCs):**

These consist of ceramic materials as the matrix (e.g., alumina, zirconia) with nanoscale reinforcements. CMNCs provide high temperature resistance, corrosion resistance, and toughness, making them ideal for structural and high-temperature applications. However, they are more brittle and challenging to process compared to other types.

b) Classification Based on Nanofiller Morphology

Nanocomposites can also be categorized by the shape and dimensionality of the nanofillers used:

- **0D Nanofillers (Nanoparticles):**

These are spherical or quasi-spherical particles (e.g., metal oxides, silica nanoparticles) with dimensions in all directions confined to the nanoscale. They are frequently employed to improve optical, thermal, and mechanical characteristics.

- **1D Nanofillers (Nanotubes, Nanorods):**

Examples include carbon nanotubes (CNTs) and nanowires, which have high aspect ratios and provide excellent mechanical reinforcement and electrical conductivity.

- **2D Nanofillers (Nanosheets):**

These include graphene, nanoclays, and MXenes. Due to their high surface area and layered structure, they significantly improve barrier properties, mechanical strength, and conductivity.

- **3D Nanostructures:**

These are relatively rare and include porous nanostructured materials or networks. They offer complex functionality such as tunable porosity and enhanced catalytic activity.

B. Mechanisms of Sulfur Adsorption and Oxidation

The removal of sulfur compounds from fuels is essential for reducing harmful emissions and meeting environmental standards. Among the alternative desulfurization methods, adsorption and oxidation are widely studied for their effectiveness, especially in removing refractory sulfur species such as thiophenes and dibenzothiophenes. From the kinetics of this autocatalytic reaction, we derived a redox mechanism (Steijns et al., 1976). These techniques can be applied under mild operating conditions and often serve as complementary or standalone approaches to conventional hydrodesulfurization. Understanding the mechanisms behind sulfur adsorption and oxidation helps optimize materials and processes for cleaner fuel production. The efficiency of these methods depends on factors such as sulfur compound type, material surface properties, reaction conditions, and catalyst presence.

a) Mechanism of Sulfur Adsorption

Sulfur adsorption involves capturing sulfur-containing molecules from fuels onto the surface of solid adsorbents. This process can occur through physical or chemical interactions depending on the material and sulfur compound involved. In physisorption, weak van der Waals forces hold the sulfur molecules onto porous surfaces like activated carbon or mesoporous silica. In chemisorption, stronger chemical bonds form between sulfur atoms and active sites, such as metals (e.g., Cu^+, Ag^+, Zn^{2+}), resulting in more selective and permanent adsorption. Another important mechanism is **π**-complexation, where aromatic sulfur compounds like thiophenes interact with transition metal Centers via electron donation, enhancing selectivity and efficiency.

b) Mechanism of Sulfur Oxidation

Sulfur oxidation converts sulfur compounds into more polar forms (e.g., sulfoxides or sulfones), which are easier to remove through extraction or adsorption. The process begins with an oxidizing agent—such as hydrogen peroxide, ozone, or peracids—that generates reactive oxygen species. These species selectively attack the sulfur atom in compounds like dibenzothiophene (DBT), modifying its structure without breaking the carbon-sulfur bonds. Catalysts, such as molybdenum, tungsten, or heteropoly acids, often enhance the reaction by facilitating oxidant activation and improving reaction rates. After oxidation, the polar sulfur products can be separated from the fuel using polar solvents or adsorbents, making the process efficient under mild, hydrogen-free conditions.

C. Benefits over Traditional Catalysts

Advanced desulfurization techniques—such as adsorptive, oxidative, and biological methods—often employ novel materials and catalysts that offer significant advantages over traditional hydrodesulfurization (HDS) catalysts like Co-Mo or Ni-Mo on alumina. While HDS is effective for removing simple sulfur compounds under high temperature and pressure, it struggles with refractory sulfur species such as dibenzothiophene and its derivatives. Modern catalysts, including nanostructured materials, metal-organic frameworks (MOFs), and biocatalysts, overcome these limitations by enabling efficient sulfur removal under milder, more energy-efficient conditions. Zeolites possess specific benefits and some drawbacks compared to traditional solid catalysts in various applications (Perot & Guisnet 1990). These innovative approaches not only improve sulfur removal efficiency but also contribute to lower operational costs and reduced environmental impact. Their selectivity, reusability, and versatility across different fuel types make them increasingly attractive for sustainable fuel refining.

a) Higher Efficiency and Selectivity

Modern catalysts used in adsorption and oxidation processes demonstrate greater selectivity toward refractory sulfur compounds, such as thiophenes and benzothiophenes. Materials like doped mesoporous silica, modified zeolites, or MOFs can be tailored to interact specifically with sulfur-containing molecules. Unlike traditional HDS catalysts, which may require high hydrogen pressure and elevated temperatures, these new systems often work under ambient or mild conditions. This not only enhances desulfurization performance but also preserves the structure of valuable fuel hydrocarbons, preventing undesired hydrogenation. As a result, better fuel quality is achieved with minimal energy input.

b) Environmental and Economic Advantages

New-generation catalysts typically allow for hydrogen-free desulfurization, reducing reliance on high-pressure hydrogen—a costly and energy-intensive input in traditional HDS. Additionally, many of these materials are regenerable, meaning they can be reused multiple times without significant loss of activity, lowering overall operational costs. The reduced temperature and pressure requirements also contribute to lower energy consumption and carbon emissions, aligning with global sustainability goals. Furthermore, biocatalysts used in biological desulfurization are biodegradable and operate under environmentally benign conditions, making them ideal for green refining technologies. Overall, these benefits contribute to safer, cleaner, and more cost-effective desulfurization processes compared to conventional catalytic systems.

4. METAL–ORGANIC FRAMEWORKS (MOFS) FOR FUEL DESULFURIZATION

Metal–Organic Frameworks (MOFs) represent a highly promising class of porous materials for advanced fuel desulfurization due to their exceptional surface area, tenable pore sizes, and chemical versatility. Composed of metal ions coordinated with organic ligands, MOFs offer highly ordered structures that can be engineered for selective adsorption of Sulfur-containing compounds such as thiophenes, pentathiophenes, and dibenzothiophenes—particularly the refractory species that are challenging to remove via conventional hydrodesulfurization. Metal–organic frameworks (MOFs), a new category of porous materials, have demonstrated significant promise in numerous applications (Hao et al., 2019). Functionalization of MOFs with polar groups or catalytic centres further enhances their selectivity and adsorption capacity, enabling efficient capture even at low Sulfur concentrations. MOFs such as ZIF-8, UiO-66, and MIL-series frameworks have demonstrated excellent performance in both liquid-phase and gas-phase desulfurization processes. Moreover, their modular architecture allows for post-synthetic modifications tailored to specific fuel types and operating conditions. While challenges remain regarding their thermal stability, cost, and large-scale implementation, MOFs are at the forefront of material innovations that could revolutionize clean fuel processing. This section discusses key MOF structures, their Sulfur adsorption mechanisms, and case studies showcasing their application in real-world fuel treatment scenarios.

A. MOFs as High-Performance Adsorbents

Metal-Organic Frameworks (MOFs) are a class of porous materials composed of metal ions coordinated to organic linkers, creating a highly ordered three-dimensional structure. Their extremely high surface area and adjustable pore sizes make MOFs excellent candidates for adsorbing various molecules, including sulfur compounds found in fuels. Metal-organic frameworks (MOFs), an innovative category of porous substances, have been extensively studied over the last twenty years (Peng et al., 2018). The chemical functionality of MOFs can be tailored to enhance their selectivity toward specific sulfur species such as thiophenes and sulfides. Additionally, MOFs are known for their stability and ability to be regenerated, which makes them suitable for repeated use in desulfurization processes. Due to these unique features, MOFs have gained significant attention as efficient and sustainable adsorbents in advanced fuel purification technologies.

a) High Selectivity and Capacity

MOFs can be precisely designed with specific pore sizes and chemical groups that target sulfur-containing molecules like thiophenes, benzothiophenes, and sulfides. This tailored structure allows MOFs to selectively adsorb these sulfur compounds even in complex fuel mixtures, minimizing the capture of non-target hydrocarbons. The high surface area of MOFs provides numerous active sites for adsorption, which increases their capacity to hold larger amounts of sulfur molecules. This selective adsorption helps achieve deep desulfurization, reducing sulfur content to very low levels and improving fuel quality. As a result, MOFs outperform many conventional adsorbents in both efficiency and specificity.

b) Regenerability

A key benefit of MOFs is their capability to be regenerated and reused with minimal loss of adsorption capacity. After adsorbing sulfur compounds, MOFs can be treated through processes such as thermal desorption, solvent washing, or pressure changes to release the captured sulfur species. This regeneration capability reduces the need for frequent replacement of adsorbent materials, lowering operational costs. Additionally, it makes the desulfurization process more sustainable by minimizing waste generation. The durability and recyclability of MOFs contribute significantly to their appeal in industrial applications where continuous operation is necessary.

B. Functionalization Techniques for Selectivity Enhancement

Selectivity is an extremely important characteristic of a chemical reaction(Shul'pin,2010). Functionalization techniques involve modifying the surface or structure of adsorbents to improve their selectivity toward specific sulfur compounds in fuels. By introducing various chemical groups, metals, or ligands onto the adsorbent material, the interaction between sulfur molecules and the adsorbent can be significantly enhanced. This targeted modification helps in distinguishing sulfur compounds from other hydrocarbons, ensuring more efficient removal during desulfurization. Functional groups such as amines, thiols, or transition metal ions are commonly used to increase affinity for sulfur species like thiophenes and sulfides. These tailored adsorbents not only boost the capacity but also reduce the energy requirements by enabling adsorption under milder conditions.

a) Chemical Group Attachment

Attaching specific functional groups such as amines, carboxyls, or thiols onto the adsorbent surface improves the chemical affinity toward sulfur compounds. These groups create stronger interactions like hydrogen bonding or coordination bonds with sulfur atoms, enhancing selectivity and adsorption efficiency. For example, amine-functionalized materials can selectively adsorb acidic sulfur compounds, improving the overall desulfurization performance. This approach also helps reduce interference from non-sulfur hydrocarbons, making the process more economical and effective.

b) Metal Ion Incorporation

Incorporating transition metal ions such as copper (Cu), silver (Ag), or zinc (Zn) into the adsorbent framework promotes π-complexation with aromatic sulfur compounds like thiophenes. These metals form reversible coordination bonds with sulfur atoms, facilitating selective adsorption even in complex fuel mixtures. Metal-functionalized adsorbents often exhibit enhanced stability and regeneration capabilities, making them practical for industrial applications. This technique is widely used in materials like Metal-Organic Frameworks (MOFs) and zeolites to significantly improve desulfurization efficiency.

C. Case Study: Diesel Desulfurization Using Zn-Based MOFs

Zinc-based Metal-Organic Frameworks (Zn-MOFs) have shown promising results in the desulfurization of diesel fuel due to their unique structural and chemical properties. These MOFs combine zinc ions with organic linkers to create porous frameworks that offer high surface area and tunable pore environments, ideal for trapping sulfur-containing molecules. Zn-based MOFs exhibit strong affinity toward sulfur compounds such as thiophenes and dibenzothiophenes, which are typically challenging to remove by conventional methods. The removal of sulfur-based compounds from fuel via the pervaporation method is a progress facilitated by membrane technology (Abdali et al., 2021). Their ability to selectively adsorb these refractory sulfur species allows for efficient reduction of sulfur content in diesel under mild operating conditions. Additionally, Zn-MOFs demonstrate excellent regenerability, enabling repeated use without significant loss in adsorption capacity, which is vital for sustainable fuel processing.

a) Adsorption Performance

Zn-based MOFs show high adsorption capacities for aromatic sulfur compounds found in diesel, effectively lowering sulfur levels to meet stringent environmental standards. Their tunable pore size allows selective adsorption of larger sulfur molecules while excluding non-sulfur hydrocarbons, enhancing overall efficiency. Studies reveal that these MOFs can achieve sulfur removal rates exceeding 90% under relatively mild temperatures and pressures, making them competitive alternatives to traditional hydrodesulfurization.

b) Regeneration and Reusability

The regeneration of Zn-MOFs is typically achieved through thermal treatment or solvent washing, which desorbs the captured sulfur compounds and restores the adsorbent's capacity. This recyclability reduces the operational costs and environmental impact associated with frequent adsorbent replacement. Moreover, Zn-MOFs maintain structural integrity and adsorption efficiency after multiple regeneration cycles, making them suitable for continuous industrial applications. Their durability and ease of regeneration support sustainable and cost-effective diesel desulfurization processes.

5. GRAPHENE-BASED NANOMATERIALS IN EMISSION CONTROL

Graphene-based nanomaterials have garnered significant attention in emission control due to their extraordinary electrical conductivity, mechanical strength, and high surface-to-volume ratio, which make them highly effective for capturing and transforming Sulfur-containing pollutants. Derivatives such as graphene oxide (GO) and reduced graphene oxide (go) offer abundant oxygen-containing functional groups and defect sites that can be engineered to adsorb Sulfur species selectively or serve as catalysts in oxidative desulfurization reactions. When incorporated with metal nanoparticles or functionalized with heteroatoms (e.g., nitrogen or Sulfur), these graphene hybrids exhibit enhanced reactivity and improved affinity for theophanic compounds commonly found in fuels. The distinctive physicochemical characteristics of graphene, particularly its remarkably high surface area, electron mobility, thermal conductivity, and mechanical strength, can result in innovative or enhanced technologies to tackle urgent global environmental issues (Perreault et al., 2015). In emission control systems, graphene-based materials are also being explored for catalytic converters, sensors, and membranes designed to minimize Sox release during fuel combustion.

A. Graphene Oxide for Sulfur Capture

Graphene oxide (GO) is a highly oxidized form of graphene that contains various oxygen-containing functional groups such as hydroxyl, epoxy, and carboxyl groups. These functional groups provide active sites that can interact strongly with sulfur-containing compounds in fuels, making GO an effective material for sulfur capture. The large surface area and layered structure of graphene oxide facilitate the adsorption of sulfur molecules like thiophenes and sulfides. Additionally, GO's surface chemistry can be easily modified to improve its selectivity and capacity for sulfur adsorption. Conversely, because of its significant size, sulfur doping causes bond polarization, lattice distortion, and modifications in the distribution of charge density (Wadi et al., 2024). Due to its excellent mechanical strength, chemical stability, and tunable properties, graphene oxide is gaining attention as a promising adsorbent in advanced fuel desulfurization technologies.

a) High Surface Area and Active Sites

Graphene oxide offers an exceptionally large surface area combined with abundant oxygen-containing groups, which serve as active sites for binding sulfur compounds. This combination enhances the adsorption capacity and allows for efficient capture of diverse sulfur species from complex fuel mixtures.

b) Tunable Surface Chemistry

The surface of graphene oxide can be chemically modified or functionalized to improve its affinity toward specific sulfur compounds. Such modifications enable targeted adsorption, increasing selectivity and efficiency in desulfurization processes. This adaptability makes GO a versatile adsorbent for various fuel types.

B. Hybrid Systems with Transition Metals

Hybrid adsorbent systems that combine materials like graphene oxide or metal-organic frameworks with transition metals have emerged as highly effective for sulfur removal from fuels. Transition metals such as copper (Cu), silver (Ag), and zinc (Zn) are incorporated into the adsorbent matrix to enhance interactions with sulfur compounds through mechanisms like π-complexation and coordination bonding. These metals act as active sites that selectively bind sulfur-containing molecules such as thiophenes and sulfides, significantly improving the selectivity and capacity of the hybrid adsorbents. The synergy between the porous support material and the metal ions allows for efficient sulfur capture under mild operating conditions. Additionally, these hybrid systems often demonstrate good stability and regenerability, making them attractive for practical industrial applications. Moreover, we contend that this information has not been regarded as significant up to now in system theory or computer science, nor in the domain of hybrid systems inquiry (Cuijpers et al., 2002).

a) Enhanced Selectivity via Metal Sites

Transition metals provide specific binding sites that preferentially interact with sulfur atoms, enabling selective adsorption of sulfur compounds even in the presence of other hydrocarbons. This targeted interaction reduces competitive adsorption and improves overall desulfurization efficiency.

b) Improved Regeneration and Stability

The robust nature of hybrid adsorbents allows them to undergo multiple regeneration cycles, typically via thermal or chemical treatment, without significant loss in performance. The presence of transition metals often enhances the structural stability of the adsorbent, ensuring long-term usability in continuous fuel processing operations.

C. Case Study: Sulfur Reduction in Biodiesel Blends

Biodiesel blends, though considered cleaner alternatives to fossil fuels, can still contain sulfur compounds that contribute to emissions and engine wear. Effective sulfur removal from biodiesel is essential to meet stringent environmental regulations and improve fuel quality. Recent studies have demonstrated that advanced adsorbents and catalytic methods can significantly reduce sulfur content in biodiesel blends without compromising their renewable nature. Techniques such as adsorption using modified materials and oxidative desulfurization have proven particularly effective in targeting sulfur species in biodiesel's complex matrix. The creation of insoluble substances in biodiesel mixtures has garnered significant interest since it leads to filter blockages (Zhang et al., 2011). These approaches help produce cleaner biodiesel blends that comply with global sulfur limits, supporting sustainable transportation fuels.

a) Adsorptive Techniques

Adsorption using materials like activated carbon, metal-organic frameworks (MOFs), and graphene oxide has shown promise in capturing sulfur compounds from biodiesel blends. Functionalization of adsorbents enhances selectivity and capacity, allowing effective removal of refractory sulfur species under mild conditions.

b) Oxidative Desulfurization

Oxidative methods convert sulfur compounds into sulfoxides and sulfones, which are more polar and easier to separate from biodiesel. Using mild oxidants and catalysts, these techniques achieve high sulfur reduction efficiency while maintaining the fuel's desirable properties.

6. ZEOLITES AND MESOPOROUS SILICA FOR CLEAN FUEL APPLICATIONS

Zeolites and mesoporous silica materials play a critical role in clean fuel applications, particularly in the adsorption and catalytic removal of Sulfur-containing compounds from petroleum-based fuels. Zeolites, with their crystalline aluminosilicate structures and well-defined microporous frameworks, offer high surface acidity, ion-exchange capabilities, and molecular sieving properties that are highly effective in targeting specific Sulfur species like thiophenes and pentathiophenes. Their tenable pore size and thermal stability make them ideal for integration into hydrodesulfurization (HDS) units and adsorptive desulfurization systems. Similarly, mesoporous silica materials—such as MCM-41 and SBA-15—possess larger pore sizes, high surface areas, and modifiable surface chemistry, enabling them to host active catalytic sites and facilitate the diffusion of bulky Sulfur molecules. In this situation, zeolites and mesoporous materials might assist chemists in creating new biofuels. creation procedures (Perego & Bosetti, 2011). Functionalization of these materials with transition metals or oxidative agents enhances their desulfurization performance, particularly in oxidative desulfurization (ODS) processes under mild conditions.

A. Structural Features and Surface Modification

The performance of adsorbents in sulfur removal largely depends on their structural characteristics and surface properties. Key structural features include high surface area, appropriate pore size distribution, and a stable framework, all of which facilitate efficient adsorption of sulfur compounds. Surface modification involves introducing functional groups or metal ions onto the adsorbent surface to enhance its chemical affinity toward sulfur molecules. The alteration of polymer materials using non-equilibrium gaseous plasma has garnered significant interest in recent decades because of industrial requirements(Mozetič,2019). These modifications improve selectivity and adsorption capacity by creating stronger interactions such as coordination bonds or hydrogen bonding with sulfur species. Tailoring both structure and surface chemistry enables the design of highly efficient adsorbents for advanced desulfurization processes.

a) High Surface Area and Porosity

A large surface area provides more active sites for sulfur molecules to bind, which directly increases the adsorption capacity of the material. Porous structures with well-defined pore sizes enable easier diffusion of sulfur compounds into the adsorbent, improving uptake rates. Materials such as activated carbons and metal-organic frameworks (MOFs) exemplify this feature with their interconnected pore networks. Proper pore size distribution ensures that sulfur compounds, especially larger molecules like dibenzothiophenes, can access internal adsorption sites. Therefore, optimizing surface area and porosity is critical for achieving efficient and deep desulfurization.

b) Functionalization for Enhanced Selectivity

Surface modification introduces specific chemical groups or metal ions that interact selectively with sulfur-containing molecules. For example, adding amine or thiol groups creates sites for strong chemical bonding with sulfur atoms, increasing selectivity over non-sulfur hydrocarbons. Transition metal ions like copper or zinc promote π-complexation with aromatic sulfur compounds, further enhancing adsorption efficiency. These functional groups not only improve adsorption capacity but also allow for targeted removal of refractory sulfur species that are difficult to eliminate by conventional methods. Functionalization thus plays a key role in developing high-performance, selective adsorbents for fuel desulfurization.

B. Regeneration and Reusability of Catalysts

Regeneration and reusability are critical factors in the practical application of catalysts used for sulfur removal, as they determine the economic viability and sustainability of the process. Over time, catalysts can become deactivated due to the accumulation of sulfur compounds or other impurities, which blocks active sites and reduces efficiency. Effective regeneration methods restore catalyst activity by removing these adsorbed species without damaging the catalyst structure. The regeneration or disposal of deactivated heterogeneous catalysts depends on chemical, economic and environmental factors(Trimm,2001). Additionally, catalysts that maintain their performance over multiple cycles reduce operational costs and environmental impact, making them more suitable for industrial applications.

a) Efficient Regeneration Techniques

Catalysts can be regenerated through several methods such as thermal treatment, solvent washing, or chemical oxidation to remove adsorbed sulfur species and restore active sites. Thermal regeneration involves heating the catalyst to decompose or desorb sulfur compounds, often under controlled atmospheres to prevent structural damage. Solvent washing uses suitable solvents to dissolve and remove sulfur residues without affecting the catalyst framework. Chemical oxidation converts sulfur species into more easily removable forms, facilitating catalyst cleaning. Choosing the right regeneration method ensures catalyst longevity and consistent desulfurization performance.

b) Importance of Catalyst Stability for Reusability

The structural and chemical stability of catalysts determines how well they can withstand repeated regeneration cycles without losing activity. Catalysts made from robust materials such as metal oxides, zeolites, or MOFs typically maintain their framework integrity during regeneration. Stability also prevents loss of active sites or sintering of metal particles, which can degrade catalytic efficiency. Reusable catalysts reduce the frequency of replacement, lowering material costs and waste generation. Therefore, designing catalysts with high stability is essential for sustainable and cost-effective sulfur removal processes.

C. Case Study: FCC Fuel Gas Purification Using Modified Zeolites

In this case study, the use of modified zeolites for fluid catalytic cracking (FCC) fuel gas purification is examined as an effective strategy for reducing Sulfur emissions in petroleum refining. FCC units are widely used in refineries to convert heavy hydrocarbons into lighter fractions like gasoline and olefins, but they also generate fuel gas streams rich in hydrogen sulphide (H_2S) and other Sulfur compounds. To mitigate this, zeolites modified with metal ions such as copper (Cu), zinc (Zn), or rare-earth elements have been deployed as sorbents or catalysts for selective Sulfur capture. In one industrial-scale application, Y-type zeolites functionalized with Cu and Ce demonstrated high efficiency in adsorbing H_2S from FCC off-gases, with removal efficiencies exceeding 90% under optimal conditions. These modified zeolites also showed excellent thermal and hydrothermal stability, allowing for multiple regeneration cycles without significant loss of performance.D. Bottom of Form

7. EMERGING GREEN TECHNIQUES IN DESULFURIZATION

As environmental regulations tighten and sustainability becomes a central goal in fuel processing, emerging green desulfurization techniques are gaining traction as alternatives to conventional hydrodesulfurization (HDS). These eco-friendly methods aim to reduce energy consumption, eliminate the need for high-pressure hydrogen, and minimize environmental impact while effectively removing refractory Sulfur compounds. Among the most promising are ionic liquids (ILs) and deep eutectic solvents (DESs), which offer tenable physicochemical properties and high selectivity for Sulfur compounds without generating hazardous byproducts. Photocatalytic desulfurization, using materials like Tio_2 or doped semiconductors, utilizes light energy to oxidize Sulfur species into more polar forms for easy extraction. Similarly, ultrasonic-assisted desulfurization enhances mass transfer and reaction kinetics, enabling faster and more efficient Sulfur removal under milder conditions. Creating strong adsorbents and nanomaterials with optimal regeneration methods is essential to enhance the effectiveness of desulfurization processes(Saleh,2022). These techniques are not only more environmentally benign but also adaptable to decentralized or small-scale operations. While still facing scalability and cost challenges, such green approaches represent the next generation of sustainable fuel purification technologies, with potential to complement or replace traditional methods in the near future.

A. Ionic Liquids and Deep Eutectic Solvents

Ionic liquids (ILs) and deep eutectic solvents (DESs) are innovative classes of green solvents gaining attention for their exceptional ability to remove sulfur compounds from fuels. These solvents can dissolve a wide range of sulfur-containing molecules, making them effective in desulfurization processes. Their design flexibility allows customization to target specific sulfur species, improving selectivity and efficiency.

a) Ionic Liquids for Selective Sulfur Extraction

Ionic liquids are salts in liquid form at relatively low temperatures, composed of bulky organic cations and various anions, which create highly customizable solvent environments. This versatility enables ILs to selectively interact with sulfur compounds like thiophenes and mercaptans through mechanisms such as hydrogen bonding and π-π interactions. Their negligible vapor pressure minimizes solvent loss and environmental pollution during the desulfurization process. Additionally, ILs can be designed to facilitate easy separation and recycling, enhancing economic feasibility.

b) Deep Eutectic Solvents as Green Alternatives

Deep eutectic solvents are formed by mixing two or more components, typically a hydrogen bond donor and acceptor, which interact to create a eutectic mixture with a melting point lower than either component alone. DESs share many advantages with ILs, such as low toxicity, biodegradability, and tunable properties, but are often easier and cheaper to prepare. Their unique solvation abilities allow efficient extraction of sulfur-containing compounds from fuel blends. Furthermore, DESs can be recycled and reused multiple times, contributing to sustainable desulfurization technologies. Due to these benefits, DESs are emerging as attractive solvents in fuel purification processes.

B. Photocatalytic and Ultrasonic-Assisted Desulfurization

Photocatalytic and ultrasonic-assisted desulfurization are innovative techniques that enhance sulfur removal from fuels by using light energy and ultrasonic waves, respectively. These methods provide environmentally friendly alternatives to conventional desulfurization, often working under mild conditions and reducing the need for harsh chemicals or extreme temperatures. Photocatalysis uses light-activated catalysts to oxidize sulfur compounds into more easily removable forms, while ultrasonic treatment generates cavitation bubbles that improve mass transfer and accelerate chemical reactions. Combining these approaches can lead to more efficient and sustainable sulfur reduction processes in fuel refining.

a) Photocatalytic Desulfurization

In photocatalytic desulfurization, semiconductor substances like TiO_2 or ZnO capture light energy to produce reactive species like hydroxyl radicals. These radicals oxidize sulfur-containing compounds like thiophenes into sulfoxides or sulfones, which can then be separated from the fuel more easily. This method operates under ambient temperature and pressure, reducing energy consumption and environmental impact. Additionally, photocatalysts can be reused multiple times, making the process cost-effective.

b) Ultrasonic-Assisted Desulfurization

Ultrasonic-assisted desulfurization utilizes high-frequency sound waves to create microscopic cavitation bubbles in the fuel mixture. When these bubbles collapse, they generate localized high temperatures and pressures that enhance the breakdown of sulfur compounds and improve their interaction with catalysts or oxidants. This effect increases mass transfer rates and accelerates desulfurization reactions without the need for high external energy input. Ultrasonic treatment is often combined with other desulfurization methods to boost overall performance and achieve deeper sulfur removal. Its simplicity and effectiveness make it a promising adjunct technology in fuel processing.

8. INDUSTRIAL-SCALE CASE STUDIES AND PILOT PROJECTS

Industrial-scale case studies and pilot projects serve as critical proof points for validating the real-world applicability of advanced desulfurization technologies, particularly those involving nanomaterials and green methods. In recent years, several refineries and energy companies have tested nanocomposite-based adsorbents and catalysts—such as metal-doped zeolites, MOFs, and graphene hybrids—within existing hydrodesulfurization units to enhance Sulfur removal efficiency without major infrastructure changes. One notable pilot project implemented a continuous flow oxidative desulfurization (ODS) system using a hybrid catalyst composed of mesoporous silica and transition metal oxides, achieving over 95% Sulfur removal in diesel under mild temperature and pressure conditions. Another example involves a petrochemical plant integrating ionic liquid-based desulfurization into its pre-treatment line, successfully reducing Sulfur content below regulatory thresholds with reduced hydrogen consumption and waste generation. These industrial implementations underscore both the scalability and challenges of deploying advanced materials, including issues related to material regeneration, cost-efficiency, and long-term durability. Nevertheless, such projects demonstrate a shift from laboratory innovation to commercial readiness, paving the way for cleaner fuel technologies to be widely adopted in mainstream refining operations.

A. Retrofitting Existing Plants with Nanocomposite Systems

Retrofitting existing fuel processing plants with nanocomposite systems offers a promising approach to enhance sulfur removal efficiency without the need for entirely new infrastructure. Nanocomposites combine nanoparticles with traditional materials to create adsorbents or catalysts that possess superior surface area, reactivity, and selectivity. Integrating these advanced materials into current setups can improve desulfurization performance, reduce energy consumption, and extend the operational lifespan of equipment. This approach provides a cost-effective pathway to meet increasingly stringent sulfur regulations while utilizing existing facilities. Additionally, nanocomposites often allow for easier regeneration and improved durability compared to conventional materials.

a) Enhanced Adsorption and Catalytic Efficiency

Nanocomposite materials exhibit enhanced adsorption and catalytic properties due to their high surface-to-volume ratio and synergistic effects between the nanoparticles and the host matrix. When introduced into existing reactors or adsorbers, these materials increase the number of active sites for sulfur compound interaction, leading to faster and deeper desulfurization. Their tailored surface chemistry allows for selective removal of stubborn sulfur species like dibenzothiophenes. This enhancement results in improved fuel quality with minimal modifications to the plant's operational conditions, making the retrofit seamless and effective.

b) Cost-Effective Implementation and Sustainability

Retrofitting plants with nanocomposite systems minimizes capital expenditure compared to building new desulfurization units from scratch. Moreover, nanocomposites typically offer better regenerability and longer service life, which lowers replacement frequency and waste generation. This contributes to a more sustainable desulfurization process, aligning with environmental goals and regulatory compliance. Hence, retrofitting with nanocomposites is a practical strategy to upgrade refinery capabilities efficiently.

B. Cost-Effectiveness and Scalability Challenges

While advanced desulfurization technologies like nanocomposites and novel catalysts offer significant performance benefits, their widespread adoption is often hindered by cost-effectiveness and scalability issues. The high initial investment for developing and producing these materials, combined with challenges in scaling laboratory successes to industrial volumes, poses barriers to commercialization. Addressing these challenges requires optimizing synthesis methods, improving material durability, and ensuring that the operational costs remain competitive with traditional technologies. Furthermore, practical concerns such as raw material availability and integration into existing infrastructure must be carefully managed to realize large-scale deployment.

a) High Production and Material Costs

Many advanced adsorbents and catalysts involve complex synthesis processes using expensive precursors, which increase their production costs. Nanomaterials and metal-organic frameworks (MOFs), for example, often require precise control over reaction conditions and costly raw materials, limiting affordable mass production. These high costs translate to increased capital expenditure for refineries, making decision-makers hesitant to adopt new technologies without clear long-term economic benefits. Developing simpler, scalable, and cost-efficient synthesis routes is critical to overcoming this hurdle.

b) Scalability and Integration Issues

Scaling up from laboratory or pilot-scale processes to full industrial implementation presents significant challenges, including maintaining consistent material quality and performance. Nanocomposites and specialized catalysts must be produced in large quantities without loss of activity or selectivity. Additionally, retrofitting or integrating these advanced systems into existing fuel processing plants requires careful engineering to avoid operational disruptions. Ensuring compatibility with current process conditions, safety standards, and regulatory requirements adds further complexity. Addressing these scalability and integration issues is essential for successful commercialization.

9. CHALLENGES, RISKS, AND FUTURE PROSPECTS

While advanced desulfurization technologies—particularly those based on nanocomposites and green chemistry—offer significant promise, their widespread adoption faces several challenges and risks. Material cost and scalability remain key barriers, as many nanomaterials and designer catalysts are expensive to produce and may require specialized equipment for integration into existing refinery infrastructure. Regeneration and long-term stability of these materials also pose concerns, especially under harsh industrial conditions where thermal and chemical degradation can reduce performance over time. Moreover, toxicological and environmental impacts of certain nanomaterials are still not fully understood, raising questions about their safe handling, disposal, and potential ecological effects. On the regulatory front, the lack of standardized testing protocols for novel desulfurization materials can slow down commercialization and deployment. Despite these obstacles, the future prospects are optimistic, driven by increasing environmental regulations, policy incentives, and demand for low-Sulfur fuels. Innovations in material design, such as low-cost bio-based nanocomposites and recyclable solvents, are expected to improve economic viability. Continued interdisciplinary collaboration among researchers, industry leaders, and policymakers will be essential to overcome current limitations and realize the full potential of sustainable, next-generation fuel purification technologies.

A. Environmental and Toxicological Risks of Nanomaterials

Nanomaterials, despite their promising applications in fuel desulfurization and other fields, raise concerns regarding their environmental and toxicological impacts. Their potential toxicity to humans, animals, and plants necessitates careful assessment and management. Additionally, the release of nanomaterials into the environment during production, use, or disposal could lead to accumulation and unforeseen ecological effects.

a) Potential Human Health Impacts

Nanoparticles can penetrate biological membranes and enter cells due to their small size, potentially causing oxidative stress, inflammation, or DNA damage. Exposure can occur through inhalation, skin contact, or ingestion, particularly for workers handling nanomaterials. Long-term effects remain insufficiently studied, raising concerns about chronic toxicity and bioaccumulation. Hence, proper safety protocols and exposure limits are essential to protect human health during manufacturing and application processes.

b) Environmental Fate and Ecotoxicity

Once released into air, water, or soil, nanomaterials may persist or transform into other forms, impacting microbial communities, aquatic life, and plants. Their high surface reactivity can disrupt natural biochemical processes and food chains, potentially leading to ecosystem imbalances. Additionally, nanoparticles may adsorb or transport other pollutants, complicating their environmental behavior. Monitoring and controlling nanomaterial emissions and waste are vital to minimize ecological risks and ensure sustainable use.

B. Lifecycle Assessment and Regulatory Considerations

Lifecycle assessment (LCA) and regulatory frameworks play vital roles in ensuring that the development and use of nanomaterials in fuel desulfurization are sustainable and safe. LCA evaluates the environmental impacts of nanomaterials from raw material extraction, synthesis, application, to disposal or recycling, providing a holistic understanding of their ecological footprint. Regulatory considerations focus on establishing guidelines for safe handling, usage, and disposal, protecting both human health and the environment. Together, these approaches help balance technological innovation with safety and environmental responsibility.

a) Comprehensive Lifecycle Assessment (LCA)

LCA involves analyzing each stage of a nanomaterial's lifecycle to identify potential environmental impacts such as energy consumption, greenhouse gas emissions, waste generation, and toxicity. This assessment helps in optimizing manufacturing processes, selecting greener materials, and designing for recyclability. By understanding the full environmental cost, developers can minimize adverse effects and improve the sustainability of nanomaterial-based desulfurization technologies. Incorporating LCA early in the design phase supports informed decision-making and responsible innovation.

b) Regulatory Frameworks and Safety Guidelines

Regulatory bodies worldwide are increasingly focusing on nanomaterial safety, creating standards and guidelines to control production, usage, and disposal. These regulations address aspects such as permissible exposure limits, labeling requirements, and environmental release controls. Compliance ensures that nanomaterial applications in fuel processing do not pose undue risks to workers, consumers, or ecosystems. Moreover, proactive regulatory oversight encourages the development of safer nanomaterials and fosters public trust in nanotechnology-based solutions.

C. The Future of Sustainable Fuel Purification

The future of sustainable fuel purification lies in developing advanced technologies that not only efficiently remove sulfur compounds but also minimize environmental impact and energy consumption. Innovations such as nanomaterials, green solvents, and hybrid catalytic systems are driving this transformation. Additionally, integrating renewable energy sources and circular economy principles into fuel processing will further enhance sustainability. The goal is to create cleaner fuels with reduced emissions while maintaining cost-effectiveness and scalability, ultimately supporting global efforts to combat air pollution and climate change.

a) Integration of Green and Nanotechnology-Based Solutions

Emerging purification methods increasingly utilize environmentally friendly materials like ionic liquids, deep eutectic solvents, and nanocomposites that offer high selectivity and regenerability. These technologies reduce the reliance on harsh chemicals and minimize waste production. Nanotechnology, in particular, enables precise control over adsorbent and catalyst properties, leading to more efficient sulfur removal at lower energy costs. The combination of these green technologies is expected to revolutionize desulfurization processes, making them cleaner, safer, and more sustainable.

b) Circular Economy and Energy-Efficient Processes

Sustainable fuel purification also involves adopting circular economy principles, such as recycling spent catalysts and solvents, and recovering sulfur byproducts for industrial use. Energy-efficient techniques like photocatalysis and ultrasonic-assisted desulfurization reduce overall energy consumption and carbon footprint. Incorporating renewable energy sources, such as solar or wind, into purification processes further enhances environmental benefits. These integrated approaches will help fuel producers meet stricter regulations while promoting resource conservation and reducing environmental impacts.

10. CONCLUSION

The advancement of fuel desulfurization and emission control technologies is not merely a technical challenge—it is a critical component of the global transition toward cleaner, more sustainable energy systems. As demonstrated throughout this volume, the integration of nanocomposites and other advanced materials offers transformative potential in addressing the environmental and regulatory demands of modern fuel processing. From metal–organic frameworks and graphene-based systems to green solvents and photocatalytic techniques, innovative material science is reshaping how we approach Sulfur removal. However, realizing these benefits at scale requires a clear understanding of the associated risks, lifecycle impacts, and implementation challenges. The case studies and pilot projects presented here underscore that success lies in the synergy between innovation, environmental responsibility, and industrial applicability. Moving forward, sustained collaboration among academia, industry, and government will be vital in accelerating adoption and ensuring these technologies contribute meaningfully to climate goals. By embracing material innovation within a framework of ethical practice and ecological stewardship, the fuel sector can lead the way in building a cleaner, safer, and more resilient energy future.

REFERENCES

Abdali, A., Mahmoudian, M., & Nozad, E. (2021). Desulfurization of a model fuel using pervaporation membranes containing Zn-MOFs. *Journal of Polymer Research*, *28*(7), 271. DOI: 10.1007/s10965-021-02472-7

. Cuijpers, P. J. L., Reniers, M. A., & Heemels, W. P. M. H. (2002). Hybrid transition systems.

Dzenis, Y. (2008). Structural nanocomposites. *Science*, *319*(5862), 419–420. DOI: 10.1126/science.1151434 PMID: 18218884

. Hao, L., Hurlock, M. J., Ding, G., & Zhang, Q. (2019). Metal–organic frameworks towards desulfurization of fuels. *Metal-Organic Framework: From Design to Applications*, 175-202.

Mozetič, M. (2019). Surface modification to improve properties of materials. *Materials (Basel)*, *12*(3), 441. DOI: 10.3390/ma12030441 PMID: 30709009

Okpala, C. C. (2013). Nanocomposites–an overview. *Int. J. Eng. Res. Dev*, *8*(11), 17.

Peng, Y., Zhang, Y., Huang, H., & Zhong, C. (2018). Flexibility induced high-performance MOF-based adsorbent for nitroimidazole antibiotics capture. *Chemical Engineering Journal*, *333*, 678–685. DOI: 10.1016/j.cej.2017.09.138

Perego, C., & Bosetti, A. (2011). Biomass to fuels: The role of zeolite and mesoporous materials. *Microporous and Mesoporous Materials*, *144*(1-3), 28–39. DOI: 10.1016/j.micromeso.2010.11.034

Perot, G., & Guisnet, M. (1990). Advantages and disadvantages of zeolites as catalysts in organic chemistry. *Journal of Molecular Catalysis*, *61*(2), 173–196. DOI: 10.1016/0304-5102(90)85154-A

Perreault, F., De Faria, A. F., & Elimelech, M. (2015). Environmental applications of graphene-based nanomaterials. *Chemical Society Reviews*, *44*(16), 5861–5896. DOI: 10.1039/C5CS00021A PMID: 25812036

Rashid, R., Shafiq, I., Akhter, P., Iqbal, M. J., & Hussain, M. (2021). A state-of-the-art review on wastewater treatment techniques: The effectiveness of adsorption method. *Environmental Science and Pollution Research International*, *28*(8), 9050–9066. DOI: 10.1007/s11356-021-12395-x PMID: 33483933

Saleh, T. A. (2022). Global trends in technologies and nanomaterials for removal of sulfur organic compounds: Clean energy and green environment. *Journal of Molecular Liquids*, *359*, 119340. DOI: 10.1016/j.molliq.2022.119340

Shi, Q., & Wu, J. (2021). Review on sulfur compounds in petroleum and its products: State-of-the-art and perspectives. *Energy & Fuels, 35*(18), 14445–14461. DOI: 10.1021/acs.energyfuels.1c02229

Shul'pin, G. B. (2010). Selectivity enhancement in functionalization of C–H bonds: A review. *Organic & Biomolecular Chemistry, 8*(19), 4217–4228. DOI: 10.1039/c004223d PMID: 20593075

Singh, S., Kulshrestha, M. J., Rani, N., Kumar, K., Sharma, C., & Aswal, D. K. (2023). An overview of vehicular emission standards. *MPAN. Journal of Metrology Society of India, 38*(1), 241–263. DOI: 10.1007/s12647-022-00555-4

Steijns, M., Derks, F., Verloop, A., & Mars, P. (1976). The mechanism of the catalytic oxidation of hydrogen sulfide: II. Kinetics and mechanism of hydrogen sulfide oxidation catalyzed by sulfur. *Journal of Catalysis, 42*(1), 87–95. DOI: 10.1016/0021-9517(76)90094-4

Trimm, D. L. (2001). The regeneration or disposal of deactivated heterogeneous catalysts. *Applied Catalysis A, General, 212*(1-2), 153–160. DOI: 10.1016/S0926-860X(00)00852-8

Wadi, V. S., Varghese, A. M., Kuppireddy, S., Shimoga, G., Karanikolos, G. N., & Alhassan, S. M. (2024). Three-dimensional graphene with the decoration of elemental sulfur-based sulfur-nitrogen dots for CO2 capture. *Journal of Environmental Chemical Engineering, 12*(4), 113130. DOI: 10.1016/j.jece.2024.113130

Wang, L. K., Williford, C., & Chen, W. Y. (2005). Desulfurization and emissions control. In *Advanced Air and Noise Pollution Control* (pp. 35–95). Humana Press. DOI: 10.1007/978-1-59259-779-6_2

Zhang, T., Chao, Y., Liu, N., Thompson, J., Garcia, M., He, B. B., Gerpen, J. V., & Chen, S. (2011). Case study of biodiesel-diesel blends as a fuel in marine environment. *Advances in Chemical Engineering and Science, 1*(2), 65–71. DOI: 10.4236/aces.2011.12011

Chapter 10
Artificial Intelligence Nanocomposite Synergy for Next-Generation Fuel Desulfurization

Rabia Yasa Kostas
https://orcid.org/0000-0003-4023-1850
Gumushane University, Turkey

Pinar Bozbeyoglu
Gumushane University, Turkey

Elif Esra Altuner
https://orcid.org/0000-0001-7663-6898
Kocaeli Health and Tecnology University, Turkey

Muzaffer Elmas
Kocaeli Health and Technology University, Turkey

ABSTRACT

Fuel desulfurization is crucial due to the harmful effects of acid rain, air pollution, and health issues associated with sulfur compounds from fossil fuel combustion. Stricter international regulations require reducing sulfur content in fuels to ultra-low levels (<15 ppm), which current hydrodesulfurization (HDS) technology struggles to achieve due to high energy costs and inefficiencies with refractory aromatic sulfur compounds, such as dibenzothiophene (DBT). This gap in technology is driving the development of new solutions. This study explores the integration of materials science, chemical engineering, and artificial intelligence (AI) to meet ultra-low sulfur

DOI: 10.4018/979-8-3373-5966-3.ch010

targets. It highlights alternative strategies, such as adsorptive (ADS) and oxidative (ODS) desulfurization, using advanced nanocomposite materials, including Metal-Organic Frameworks (MOFs), graphene hybrids, and zeolites. These innovative materials focus on "precision targeting" and "molecular recognition," offering greater efficiency under milder conditions compared to traditional HDS methods AI and computational science are crucial for identifying the most suitable materials among a vast array of candidates. Combining Density Functional Theory (DFT) with Machine Learning (ML) accelerates material discovery by transitioning from a trial-and-error approach to an "inverse design" method. Successful AI tools, including Artificial Neural Networks (ANNs) and Genetic Algorithms (GAs), have been proven effective in optimising desulfurization processes through real-world examples. In summary, achieving cleaner fuels requires innovative materials guided by AI. The study also addresses the challenges of scaling this interdisciplinary approach and envisions the potential for transforming desulfurization from a cost centre into a profit centre through precise molecular separation.

1. INTRODUCTION

Fuel desulfurisation is essential for protecting the environment and human health by removing harmful sulfur compounds from fuels. These compounds, released from burning fossil fuels, contribute to acid rain and form particulate matter (PM2.5), a major air pollutant that leads to respiratory and cardiovascular diseases.

International and national regulations have been implemented to reduce the sulfur content in fuels significantly. For example:

- IMO 2020: Reduced the sulfur limit in marine fuels from 3.5% to 0.50%.
- U.S. EPA Tier 3: Limited average sulfur content in gasoline to 10 ppm.
- EU Euro 5/6: Set maximum sulfur content for gasoline and diesel at 10 ppm.
- Canada: Adopted limits of 15 ppm for on-road diesel and 10 ppm for gasoline.

These stringent targets challenge traditional hydrodesulfurisation (HDS) technologies, particularly in removing refractory sulfur compounds like dibenzothiophene (DBT), due to their complex chemical structures. To address these challenges, an interdisciplinary approach is needed. This involves:

1. Developing new nanostructured materials that can selectively capture or convert sulfur.
2. Engineering efficient, scalable desulfurisation processes that utilize these materials with lower energy consumption.

3. Leveraging artificial intelligence to accelerate the discovery and optimisation of these materials, moving beyond traditional trial-and-error methods.

This study will explore the emerging technologies that combine nanocomposite materials and AI to tackle the challenges of fuel desulfurization. The discussion will begin by examining the limitations of HDS technology, followed by an introduction to promising adsorptive (ADS) and oxidative (ODS) strategies, along with the key materials involved. The discussion will encompass the role of AI and computational methods, with a focus on the integration of Density Functional Theory (DFT) and Machine Learning (ML). Concrete case studies will illustrate the practical success of these innovations, followed by an examination of the challenges and future directions for industry applications.

2. ENVIRONMENTAL AND HEALTH EFFECTS OF SULFUR

Fossil fuels, which include coal, oil, and natural gas, inherently contain organic sulfur compounds due to their formation from the remains of ancient organisms that lived millions of years ago. During the combustion of these fuels, which occurs at extremely high temperatures for purposes such as energy production, transportation, and industrial processes, the sulfur atoms present react with oxygen in the atmosphere. This reaction results in the release of significant quantities of sulfur oxides (SO_X), particularly sulfur dioxide (SO_2) (Jafarinejad, 2016), (Altuner et al., 2022)..

The environmental ramifications of these emissions are both extensive and profound, contributing to issues that stretch from local air quality deterioration to transcontinental ecological disruption. When sulfur oxides interact with water vapour and other atmospheric constituents, they produce acidic compounds, most notably sulfuric acid (H_2SO_4). This acid formation is the primary driver of acid rain, a phenomenon that poses severe threats to the environment.

Acid rain can cause widespread damage, leading to the deforestation of vast areas and severely impacting ecosystems. It adversely affects aquatic life in lakes and rivers by altering pH levels and introducing toxic metals leached from the soil. Additionally, acid rain degrades the chemical structure of soil, rendering it less productive and jeopardising agricultural outcomes. The corrosive nature of acid rain also leads to significant erosion of buildings and historical monuments, inflicting permanent harm on both natural ecosystems and cultural heritage.

One striking historical example of this devastation is found in the “Black Triangle” region, which encompasses parts of Germany, Poland, and the Czech Republic. During the 1970s and 1980s, this area experienced extensive forest loss, with tens of thousands of hectares of forested land destroyed as a direct consequence of acid rain originating from industrial sources. The environmental degradation witnessed in the Black Triangle serves as a sobering reminder of the lasting impacts of industrial emissions and the urgent need for sustainable practices to afeguard our planet and its natural assets (Grennfelt et al., 2020).

Figure 1. Schematic representation of sources (drivers) of air pollution and effects on human health

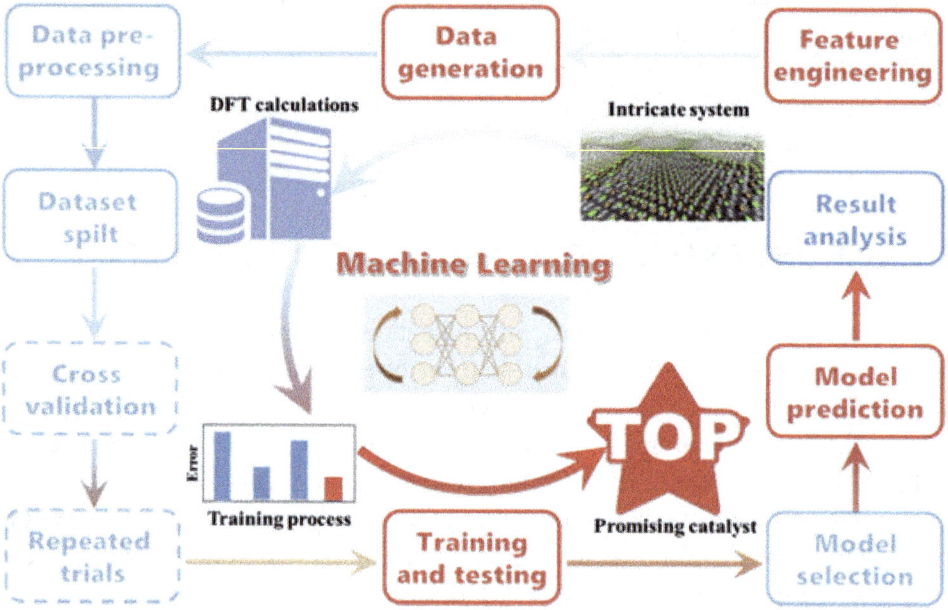

(Reprinted from Iram et al., 2025)

Sulfur emissions have serious effects on both the environment and human health. Sulfur oxides (SO_X) contribute to the formation of fine particulate matter ($PM_{2\cdot 5}$), which, when inhaled, can enter the lungs and worsen respiratory diseases like asthma and chronic bronchitis. Studies indicate that even short-term air pollution spikes lead to more hospital admissions for respiratory issues (Anderson et al., 2012).

Long-term exposure to $PM_{2\cdot 5}$ is linked to systemic inflammation, increased cardiovascular disease risk, and premature death (Brook et al., 2010). A study by the Finnish Meteorological Institute warns that if sulfur limits in marine fuels aren't reduced, over 570,000 premature deaths worldwide could result from this pollution between 2020 and 2025 (Sofiev et al., 2018).

Additionally, compounds like hydrogen sulfide (H_2S) found in crude oil are highly toxic and pose significant safety risks in oil and gas exploration, sewage, and pulp industries, potentially causing acute toxicity or death in case of leaks (Nag et al., 2020).

3. EFFECTS OF SULFUR ON INDUSTRIAL OPERATIONS

The negative impacts of sulfur extend beyond environmental and health concerns; they also pose significant operational risks for industrial processes. The corrosive properties of sulfur compounds lead to the deterioration and thinning of metals in critical refinery equipment, such as pipelines, pumps, valves, and reactors. This not only raises maintenance costs but also jeopardizes operational safety by increasing the likelihood of leaks and accidents (Skinner et al., 1960).

Figure 2. Environmental Health Risk Assessment and Acute Effects of Sulfur Dioxide

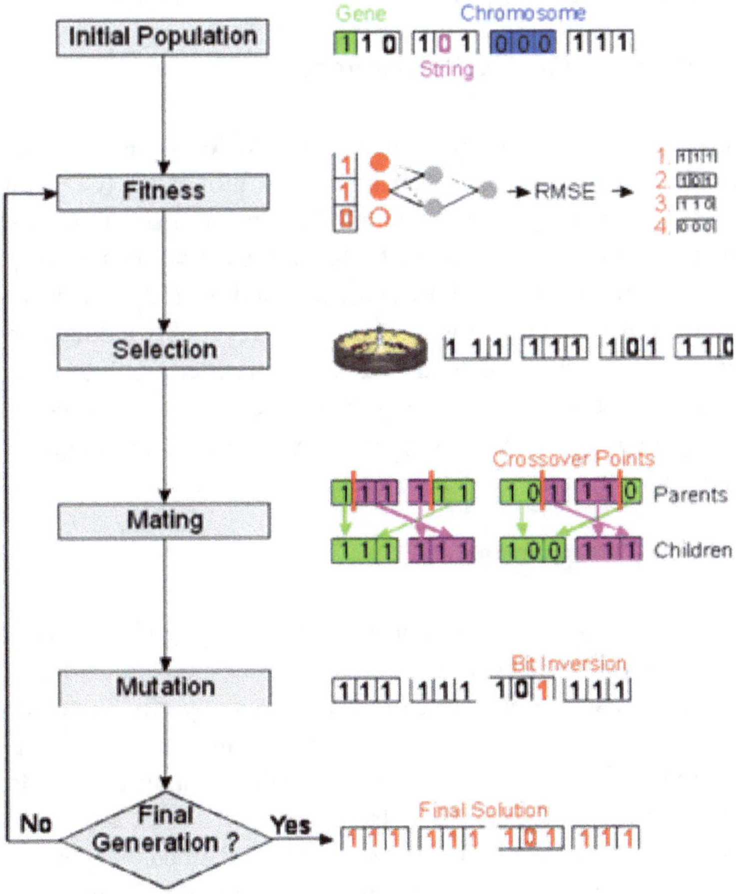

(Reprinted from Susetyo et al., 2024)

One of the most significant industrial impacts of sulfur is its detrimental effect on catalysts, which can be described as "poisoning." Catalysts are materials that substantially accelerate the speed and efficiency of chemical reactions without undergoing permanent changes themselves. This unique characteristic is vital in a wide range of applications, including petroleum refining, where it facilitates the conversion of crude oil into valuable fuels, and automotive exhaust systems, where it helps convert harmful emissions into less harmful substances.

Unfortunately, sulfur compounds can form strong chemical bonds with the active sites of catalysts, effectively blocking these sites and preventing essential chemical reactions from occurring. This interaction not only leads to a reduction in the catalyst's overall efficiency but can also necessitate more frequent catalyst regeneration or replacement, thereby increasing operational costs and downtime (Bartholomew, 2001). The implications of this issue are particularly pronounced in two key areas: the efficiency of hydrocarbon processing and the effectiveness of emissions control technologies.

3.1. Automotive Catalytic Converters

Catalytic converters are components in vehicle exhaust systems designed to transform harmful emissions, such as carbon monoxide (CO), nitrogen oxides (NOx), and unburned hydrocarbons, into less harmful substances. However, high levels of sulfur in fuel can permanently block the active sites of the precious metal catalysts—platinum (Pt), palladium (Pd), and rhodium (Rh)—used in these converters. This phenomenon, known as "poisoning," irreversibly reduces the effectiveness of the catalysts and shortens their lifespan (Epling et al., 2004). As a result, this situation often necessitates the premature replacement of these costly parts, which are typically expected to last for years, thereby imposing a significant financial burden on both consumers and manufacturers.

3.2. Refinery Processes

In refineries, processes like fluid catalytic cracking (FCC) and hydrocracking are used to convert heavy petroleum fractions into more valuable, lighter products such as gasoline and diesel. These processes typically rely on zeolite-based acidic catalysts (Wu, 2023). However, high sulfur content in the feedstock—specifically the crude oil being processed—can deactivate these catalysts. This deactivation not only reduces process efficiency but also increases the costs associated with catalyst replacement (Ancheyta et al., 2005). These costs encompass not only the purchase of new catalysts but also the extended downtime required for shutting down the entire unit, which can last several days. During this period, tons of deactivated catalyst must be removed from the reactor and replaced with new material. This downtime can lead to significant financial losses for the refinery due to decreased production, potentially amounting to millions of dollars.

4. GLOBAL HARMONY AND BORDERS

The documented harms of sulfur emissions have necessitated the implementation of strict legal regulations at global and national levels. These regulations include severe sanctions for non-compliance, such as detaining ships or imposing multi-million-dollar fines on industrial facilities. Countries adopt these rules primarily to protect the health of their citizens and reduce public health expenditures (Ausmeel et al., 2020). These regulations have also become the primary driving force for innovation in desulfurization technologies, compelling companies to invest in cleaner technologies. Indeed, without a legal requirement, there is insufficient economic incentive for companies to undertake these costly R&D processes.

4.1. International Maritime Organisation (IMO)

The maritime sector has historically been one of the most significant contributors to global SO_X emissions, primarily due to the use of high-sulfur heavy fuel oils (HFO). In response to this issue, the International Maritime Organisation (IMO) implemented a landmark regulation known as "IMO 2020," which took effect on January 1, 2020. This rule reduced the maximum allowable sulfur content in marine fuels worldwide from 3.5% to 0.50%.

One of the key outcomes of this regulation has been an improvement in air quality, particularly in coastal areas and heavily trafficked port cities by ships. In designated Emission Control Areas (ECAs), such as the Baltic Sea, the coasts of North America, and the English Channel, the sulfur limit is even stricter at just 0.10%. These regions receive additional protection due to their high population density and ecological sensitivity.

The IMO 2020 regulation has significantly transformed the global maritime and fuel supply chains. It has compelled refineries to produce a new type of Very Low Sulfur Fuel Oil (VLSFO) and has forced shipowners to either use this more expensive fuel or invest in costly technologies, such as exhaust gas cleaning systems, commonly referred to as "scrubbers."

Research utilising satellite data has confirmed a notable and measurable decrease in SO_2 concentrations along global shipping routes immediately following the implementation of the IMO 2020 regulation.

4.2. US Environmental Protection Agency (EPA)

In road transportation, the EPA's "Tier 2" and "Tier 3" programs have gradually reduced sulfur levels in gasoline and diesel fuels. This phased approach was strategically chosen to give both the refinery industry time to make billions of dollars in infrastructure investments and to allow automotive manufacturers to develop new engines and exhaust systems. The Tier 3 program, effective as of 2017, reduced the average sulfur content in gasoline to 10 ppm (parts per million) on an annual basis. For diesel fuels, the "Ultra Low Sulfur Diesel" (ULSD) standard has imposed an upper limit of 15 ppm for on-road vehicles since 2006. Before this standard, the sulfur limit in pre-2006 diesel fuels was 500 ppm, representing a 97% reduction (Helgeson et al., 2010). The foundation of the EPA's approach is the philosophy of treating the fuel and the vehicle as an integrated system. Low-sulfur fuels are a prerequisite for the effective operation of modern, high-efficiency emission control systems because the sulfur in the fuel, through the SO_X compounds formed during combustion, clogs diesel particulate filters (DPFs) and chemically poisons the surface of catalytic converters, rendering them permanently non-functional (Majewski et al., 2001)

This regulatory framework creates immense technological and financial pressure on refineries. Traditional desulfurization methods are becoming both technically inefficient, due to the extreme operational conditions (higher temperatures/pressures) required to meet these new, ultra-low sulfur targets, and economically unsustainable due to exorbitant energy and hydrogen costs (Kawatra, 2001). This situation is a concrete example of how scientific evidence leads to public pressure, which in turn leads to legal regulations, and these regulations become an industrial and economic necessity that triggers technological innovation. Environmental policies have now become one of the most important factors shaping research and development priorities in materials science and chemical engineering.

Figure 3. Schematic representation of zero-discharge desulfurization wastewater treatment processes, integrating three main approaches

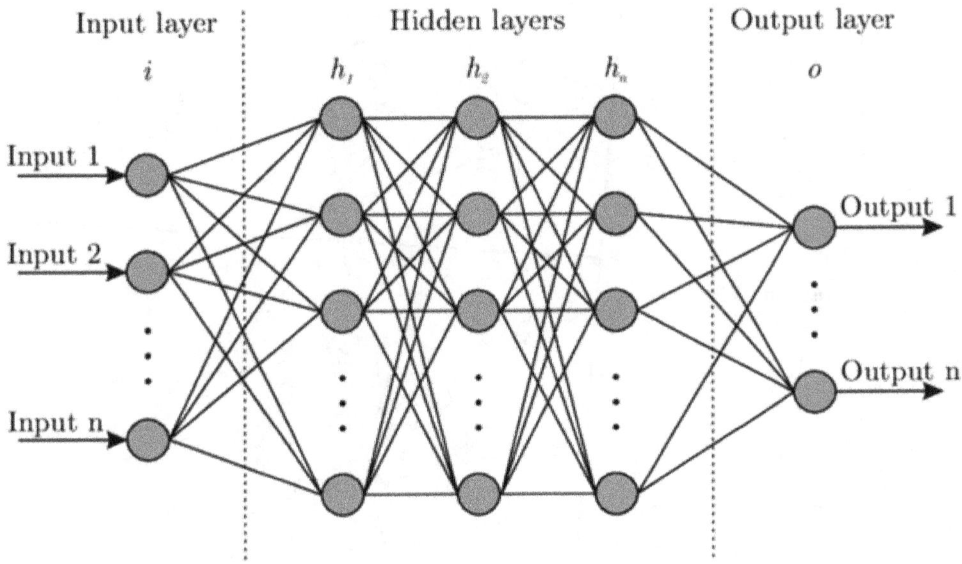

(Reprinted from Xu et al.,2024)

Zero-discharge desulfurization wastewater treatment process integrating flue gas evaporation, membrane distillation, and evaporative crystallisation. The system utilises flue gas heat and membrane separation to remove pollutants, while evaporation enables the recovery of by-products such as $NaHCO_3$ and Ca $(OH)_2$.

Figure 4. Desulphurization wastewater from power plants

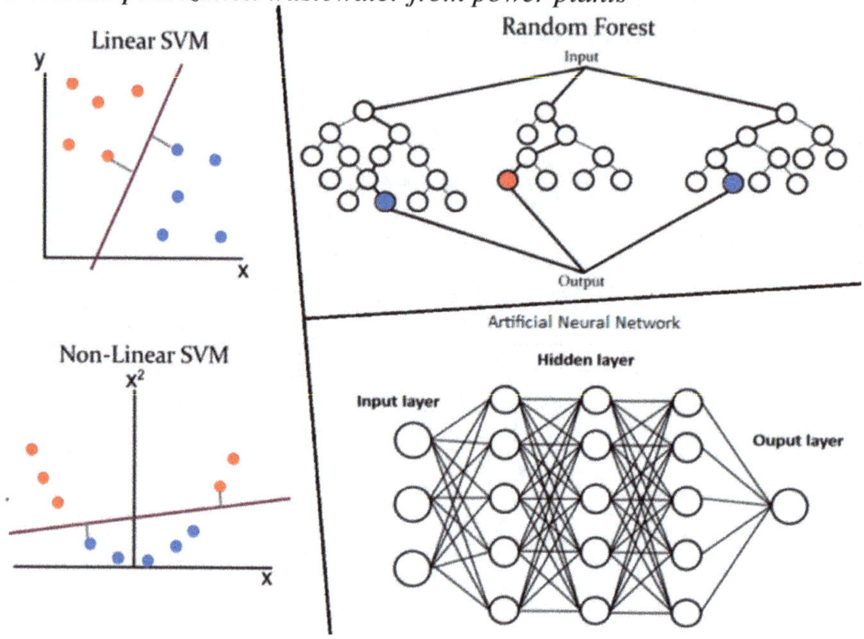

(Reprinted from Xu et al.,2024)

As illustrated in Figure 4, this technology lowers wastewater pollutant levels through advanced treatment processes, aiming for near-zero discharge. It enables resource recovery from wastewater while minimising environmental contamination. Nevertheless, the desulfurization process generates substantial amounts of complex wastewater, which contains water, sulfuric acid, sulfates, suspended solids, and various other compounds.

Furthermore, the desulfurization problem is no longer merely a quantitative issue of the total amount of sulfur to be removed. In other words, the problem has shifted from "how much" sulfur to remove to "which kind" of sulfur molecules must be removed. The nature of the problem is also changing with the global shift from easily processed "sweet" (low-sulfur) crude oil to more complex and high-sulfur "sour" crude. This trend is driven by the gradual depletion of easily accessible, high-quality "sweet" oil reserves worldwide, forcing the industry to turn to the more abundant but more challenging-to-process "sour" reserves in regions such as the Middle East and Venezuela (Speight, 2014). These sour crude oils not only contain more total sulfur but also higher proportions of large, complex, and sterically hindered "refractory" aromatic sulfur compounds, such as dibenzothiophene (DBT) and its alkylated derivatives (Song, 2003). This transforms the desulfurization problem from one of reducing total sulfur to a qualitative chemical challenge of targeting specific and difficult-to-remove molecules. It is this qualitative challenge that exposes a fundamental weakness of traditional technologies and explains why the surface chemistry and selectivity of next-generation nanomaterials have become so critical.

5. TECHNOLOGICAL LIMITS OF THE CONVENTIONAL HYDRODESULFURIZATION (HDS) PROCESS

For decades, hydrodesulfurization (HDS) has been the backbone of petroleum refineries and accepted as the industry standard for reducing sulfur content in fuels. While this technology has achieved great success in reducing sulfur levels from 1-2% (10,000-20,000 ppm) to 500 ppm, meaning it removes over 95% of the sulfur, it falls short of meeting today's ultra-low sulfur targets (<15 ppm). The issue is not that HDS cannot remove sulfur, but that its efficiency drops dramatically when attempting to remove the last few hundred parts per million of the most refractory sulfur compounds in the fuel (Topsøe et al., 1996). The near-zero sulfur levels demanded by governments have exposed the fundamental technological, economic, and chemical limitations of HDS technology. (Saleh, 2020).

HDS is fundamentally a process where chemical reactions are accelerated using a catalyst. The process is typically performed in a fixed-bed reactor in the presence of a catalyst composed of cobalt-molybdenum (CoMo) or nickel-molybdenum (NiMo) sulfides impregnated on a high-surface-area alumina (γ-Al_2O_3) support. The process operates by passing the fuel stream, fed with an abundance of hydrogen gas (H_2), through the reactor at high temperatures (300–400 °C) and high pressures (30–130 atmospheres). Under these harsh conditions, the carbon-sulfur (C-S) bonds in organic sulfur compounds are broken via a hydrogenolysis reaction on the catalyst surface. As a result, the sulfur atom is removed as hydrogen sulfide (H_2S) gas, while the hydrocarbon skeleton is desulfurised. The resulting H_2S gas is subsequently removed from the system in amine scrubbing units. Understanding the introductory chemistry and operational conditions of this process is critical for comprehending why it faces technological and economic bottlenecks in meeting today's ultra-low sulfur targets.

5.2. Operational and Economic Bottlenecks

The industrial-scale application of the HDS process entails a series of significant operational and economic restructuring efforts. This need for restructuring increases exponentially, especially as the goal of ultra-deep desulfurization (<15 ppm) is pursued.

5.2.1. High Energy and Hydrogen Consumption

The most prominent disadvantage of HDS is its high energy requirement. Heating the reactors to temperatures of 300-400 °C and pressures exceeding 100 atmospheres, and maintaining these conditions, makes HDS units one of the most energy-intensive operations in a refinery, often accounting for a significant portion of the facility's total energy budget (Chang et al., 2013). Additionally, the process stoichiometrically consumes large amounts of hydrogen. The production of hydrogen (typically via steam reforming of natural gas) is both a costly and carbon-intensive process (Nikolaidis & Poullikkas, 2017). These high operating costs are inevitably reflected in the pump price of the final products, gasoline and diesel. More importantly, in trying to solve one environmental problem (sulfur), this process creates an "environmental dilemma" by indirectly exacerbating another (greenhouse gas emissions) due to the carbon-intensive nature of hydrogen production.

5.2.2. Catalyst Life and Deactivation

HDS catalysts lose their initial reaction efficiency over time. The main reasons for deactivation are: (a) the clogging of the catalyst's porous structure by coke, a reaction byproduct, which prevents fuel from reaching the active sites; (b) the reduction of surface area through sintering, where active metal sulfide particles agglomerate at high temperatures, decreasing the number of sites where the reaction can occur; and (c) the permanent blocking (poisoning) of the active sites, the atomic points where the reaction takes place, by other impurities in the feedstock, such as nitrogen (Okunev et al., 2015). This deactivation typically necessitates catalyst replacement every two to three years. While periodic regeneration (cleaning) to remove coke buildup is possible, these procedures never restore the catalyst to its initial activity and only slightly extend its lifespan. Ultimately, catalyst replacement is a significant operational expense, not only due to the cost of the new catalyst but also the downtime, which means shutting down the entire HDS unit for days.

5.2.3. Octane Loss

In the desulfurization of high-octane fuels like gasoline, which are less prone to 'engine knock' (the premature and uncontrolled detonation of the air-fuel mixture in the engine cylinder that can damage the engine), the severe conditions of hydrodesulfurization (HDS) can lead to an undesirable side effect. Valuable components that enhance the octane number, such as olefins (alkenes) and aromatics, are saturated (hydrogenated) under high hydrogen pressure. This side reaction can reduce the gasoline's octane number by several points, resulting in a significant loss of product quality. Such a decrease may cause the fuel to fall below the legal octane standard (e.g., 95 octane) (Rana et al., 2007). Consequently, refiners must find a delicate and costly balance: they can either intensify the process to eliminate the last traces of sulfur, which results in the loss of octane, and then add expensive octane-boosting additives to restore it, or they can operate within the legal limits to minimise octane loss.

5.2.4. The Chemical Obstacle of HDS Technology for Stubborn Sulfur Compounds

Perhaps the most fundamental limitation of HDS is its weak chemical selectivity, which is its inability to distinguish between different types of molecules and act only on the intended targets. The process is not equally effective against all sulfur compounds; it easily breaks down simple molecules, but its efficiency drops for complex ones. For example, while over 99% of simple compounds like thiols can be easily removed under standard HDS conditions, only 20-30% of a refractory compound like 4,6-dimethyldibenzothiophene (4,6-DMDBT) might be removed under the same conditions (Mosca et al., 2023). Aliphatic and simple sulfur compounds like thiols, sulfides, and disulfides are easily decomposed under HDS conditions. However, the C-S bonds in heterocyclic compounds where the sulfur atom is part of an aromatic ring system, such as thiophene, benzothiophene (BT), and especially dibenzothiophene (DBT), are more stable and harder to break.

At the apex of this challenge are alkyl-substituted DBTs, such as 4,6-dimethyldibenzothiophene (4,6-DMDBT). In these molecules, the methyl groups (CH_3) on either side of the sulfur atom act as a "molecular shield," physically protecting the sulfur atom. This phenomenon, known as "steric hindrance" in chemistry, prevents the sulfur compound from approaching the active sites on the catalyst surface where the reaction occurs (Nakano et al., 2013). Just as a large piece of furniture cannot pass through a narrow doorway, this bulky molecule cannot fit into the active pocket of the catalyst to establish the interaction necessary to break the C-S bond. Consequently, removing these molecules requires making the HDS reaction conditions (temperature and pressure) even harsher, which demonstrates why HDS alone is insufficient for ultra-deep desulfurization. However, this approach leads to a vicious cycle that creates other problems while trying to solve one environmental issue: intensifying the reaction raises energy consumption and thus greenhouse gas emissions to exorbitant levels, increases octane loss, and makes the process economically unsustainable. In other words, the additional energy spent to remove sulfur creates another burden for the planet.

This situation shows that HDS technology is a victim of its own success. The technology has been highly effective in reducing sulfur levels from a standard 1-2%

(10,000-20,000 ppm) to "low-sulfur" levels (e.g., 500 ppm). However, as regulatory authorities mandate "ultra-low sulfur" (ULSD, <15 ppm) targets, almost all the sulfur remaining in the fuel after the simple molecules are easily removed consists of the aforementioned refractory molecules, such as dibenzothiophene (DBT) and its derivatives. The operational severity and cost required to remove these last few ppm of sulfur are disproportionately higher than what was needed to remove the initial 99%. This is a classic example of the "law of diminishing returns," indicating that HDS has reached its technological and economic limits for ultra-deep desulfurization.

Furthermore, the global shift of petroleum reserves from easily processed light and sweet crude to more challenging, heavy, and sour crude is systematically making this fundamental weakness of HDS more apparent. "Heavy and sour" oil is considered "bad" because it is more viscous, contains more metal impurities, and is harder to process, and the reason for this shift is the depletion of high-quality reserves (Babak, 2018). The weakness of HDS becomes more pronounced because these sour crude oils not only have a higher total sulfur content but also contain a higher proportion of large, complex, and sterically hindered DBT derivatives (Grey, 2015). This means the chemical nature of the feedstock—the crude oil to be processed in the refinery—is inevitably becoming more "challenging" over time. "Challenging" here means that the oil contains a higher proportion of the types of molecules that HDS struggles with the most. Therefore, the future sustainability of HDS is threatened not only by economic pressures but also by a fundamental chemical threat imposed by the nature of the industry's future primary raw material, i.e., the risk of the technology becoming chemically incompatible with its feedstock. This situation shows that minor improvements aimed at increasing HDS efficiency are no longer sufficient. Since the root of the problem is chemical, the solution must also be chemical. Therefore, developing next-generation technologies that serve as alternatives or complements to HDS, such as adsorptive desulfurization (ADS) and oxidative desulfurization (ODS), is not a choice but an industrial necessity (Özkan et al., 2022; Wu et al., 2025).

6. NANOCOMPOSITE-BASED ADVANCED DESULFURIZATION STRATEGIES

The technological and economic limitations of hydrodesulfurization (HDS) have led scientists and industrial research and development professionals to seek more efficient, selective, and sustainable methods for desulfurization (Whitesides, 2005). In this quest, the unique possibilities offered by nanotechnology have gained attention. Nanotechnology's distinctiveness lies in its ability to design materials at the atomic and molecular levels, resulting in properties that were previously unattainable. Unlike conventional materials, nanomaterials can be tailored to specifically recognise and capture target sulfur molecules (Saha et al., 2020).

Alternative strategies to HDS, particularly adsorptive desulfurization (ADS) and oxidative desulfurization (ODS), show revolutionary potential through the use of nanocomposite materials. This potential arises from their ability to function under mild conditions, such as room temperature, without the harsh requirements of high temperature, high pressure, and excessive hydrogen consumption associated with HDS (Shi et al., 2023). Such advancements could make the desulfurization process more cost-effective and environmentally friendly.

These innovative materials, characterised by high surface areas, tunable chemical properties, and unique reactivities, present promising solutions to the challenges that HDS faces. In this area of research, nanostructures such as Metal-Organic Frameworks (MOFs) (Hao et al., 2020), graphene-based hybrid nanocomposites (Roman et al., 2021), and ion-exchanged zeolites (Hernández-Maldonado & Yang, 2007) are being intensively studied due to their superior properties.

6.1. Adsorptive Desulfurization ADS Method

ADS operates on the principle of selectively adhering sulfur compounds in fuel to the surface of a solid material known as an adsorbent, similar to how a magnet works. This process separates the sulfur compounds from the rest of the fuel. Unlike Hydrodesulfurization (HDS), which requires high temperatures and hydrogen gas, and creates a significant carbon footprint due to hydrogen production, ADS typically functions at ambient temperatures and atmospheric pressure, eliminating the safety risks associated with handling flammable gases.

What makes ADS particularly appealing is not only its energy efficiency but also its high selectivity in targeting and capturing refractory aromatic sulfur compounds, such as dibenzothiophene (DBT) and its derivatives, that HDS often struggles to remove effectively (Naaz et al., 2022). The success of the ADS process heavily depends on selecting and designing the appropriate adsorbent material. Although this presents challenges, it also offers opportunities for nanotechnology to create the ideal adsorbent tailored for specific applications. In this context, nanostructured materials are at the forefront of innovation.

Figure 5. A comprehensive illustration of sulfur compounds in diesel fuel, their environmental impacts, and desulfurization technologies

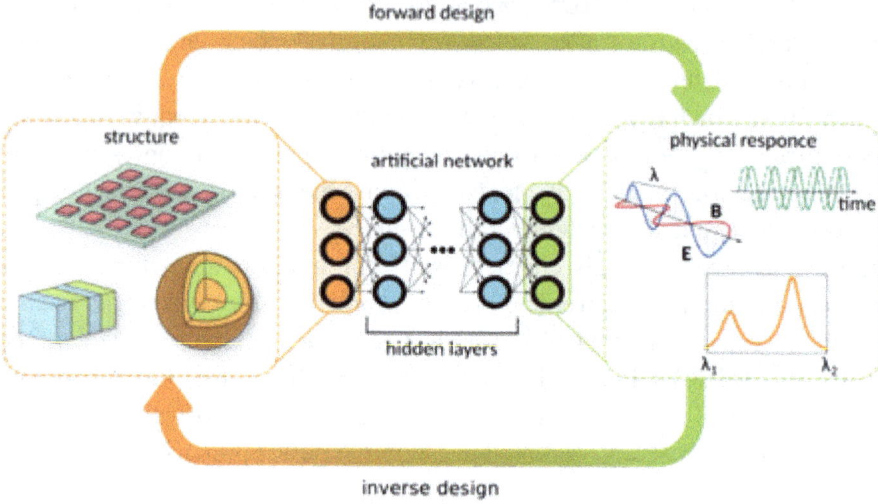

(Reprinted from Hu et al, 2025)

Sulfur, an inherent element of crude oil, primarily exists in diesel fuel as organic compounds, including mercaptans, thiophenes, benzothiophenes, and dibenzothiophenes (Figure 5). During combustion, these compounds generate harmful gaseous, liquid, and solid emissions (Table 1).

6.2. Metal-Organic Frameworks (MOFS)

MOFs are a class of materials in which metal ions or clusters (nodes) are linked by organic molecules (linkers) to form three-dimensional, crystalline, and highly porous structures. The most notable features of MOFs include their exceptionally high surface areas (in some cases, one gram can cover an area larger than a football field), precisely tunable pore sizes, and chemically modifiable inner surfaces (Furukawa et al., 2013). Scientists can precisely tune the pore sizes by selecting longer or shorter organic linkers during synthesis and can modify the inner surfaces by adding different chemical groups to these linkers. These properties make them ideal candidates for selectively capturing sulfur compounds. The adsorption efficiency of MOFs relies on the synergistic power of multiple interaction mechanisms that simultaneously capture the sulfur molecule from several points. These mechanisms are:

S-M Coordination: The metal centres in the MOF structure, which are open to forming new chemical bonds (open metal sites), form a strong chemical bond by creating a direct "docking" bond (coordination bond) with the electron pairs that the sulfur atom can share (Altuner, 2023).

π-Complexation: The negative electron clouds (π-electron clouds) above and below the ring structure of aromatic sulfur compounds establish a weak but selective bond through a type of electrical attraction (π-complexation) with positively charged regions in the MOF structure (Grande et al., 2004).

Acid-Base Interactions: In chemistry, substances that accept electron pairs are called Lewis acids, and those that donate them are called Lewis bases. The natural attraction between the electron-seeking metal centres (Lewis's acid) in the MOF and the electron-rich regions on the sulfur (Lewis's base) also contributes to adsorption. By choosing different metal types or organic linkers during synthesis, scientists can tune the strength of these interactions and thus "program" the MOF to capture a particular sulfur molecule, like DBT, specifically.

6.3. Other Porous Nanomaterials

Besides MOFs, other nanostructured materials also show promise for ADS. Zeolites, being crystalline aluminosilicate structures, can be customised for ADS through their ion-exchange capacity (Altuner et al.,2022). In this process, sodium ions within the zeolite's porous structure are removed and replaced with copper(I) (Cu^+) ions. Copper(I) ions are highly effective at attracting the electron clouds of aromatic sulfur compounds. This allows the zeolite to find its target with high selectivity (recognising only the sulfur-containing molecules) among millions of molecules in the fuel and capture them with high capacity (storing a large number of sulfur molecules) (Altuner et al.,2021; Hernández-Maldonado & Yang, 2007). Activated carbon and other carbon-based nanomaterials are notable for their high surface areas, chemical stability (their ability to maintain their structure even under harsh industrial conditions), and low cost (Bozbeyoglu et al., 2023). Coating the surfaces of these carbon materials with metal oxide nanoparticles or adding different atoms, such as nitrogen, into the carbon structure (doping), modifies the surface's chemical properties. These modifications can significantly increase the material's natural attraction (affinity) for sulfur compounds and thus its adsorption capacity (the amount of sulfur it can hold per unit area) (Gundogdu et al., 2022; Azeez et al., 2023; Kampouraki et al., 2019).

6.4. Chemical Conversion of Sulfur Oxidative Desulfurization (ODS) Strategy

ODS is based on a completely different philosophy from HDS. Instead of breaking the strong Carbon-Sulfur (C-S) bond, it aims to chemically transform the sulfur atom into a form that is much more easily separable from the fuel (Altuner et al.,2021). The process involves the oxidation of sulfur compounds to their corresponding sulfoxides (R_2SO) and sulfones (R_2SO_2) in the presence of an oxidant (typically hydrogen peroxide, H_2O_2, or molecular oxygen, O_2) and a catalyst.

In chemistry, polarity refers to a molecule having positive and negative charge centres, causing it to act like a tiny magnet. The fuel itself is primarily composed of nonpolar (i.e., 'non-magnetic') hydrocarbons. The oxidised sulfone products, however, are highly polar ('magnet-like'). Just as oil and water do not mix, this polarity difference allows polar sulfones to be easily separated from nonpolar fuels through energy-efficient and straightforward methods, such as extraction (removal with a solvent) or adsorption.

The most significant advantage of ODS is its high reactivity towards sterically hindered DBT derivatives, which HDS struggles with the most. The reason is as follows: the methyl groups that physically shield the sulfur atom in HDS also create an electron-rich region around it. The oxidants used in the ODS process have an 'electron-seeking' (electrophilic) nature. Therefore, the very structure that acts as a 'shield' for HDS becomes a bright target for ODS, as if inviting oxidation. This chemical compatibility makes the molecules that are most difficult for HDS among the easiest targets for ODS (Ishihara et al., 2005).

6.4.1. Graphene and Its Hybrids

Graphene, a two-dimensional carbon allotrope—that is, a different crystalline form of the same element (carbon)—and its derivatives (graphene oxide (GO), reduced graphene oxide (rGO)), offer a versatile and powerful potential for ODS. Thanks to their high surface areas, excellent electron transport properties, and oxygen-containing chemical arms (functional groups) on their surfaces, they can function both as catalysts on their own and as catalyst support materials (Ameen, et al.,2023). Graphene sheets provide an ideal support platform for metal oxide nanoparticles such as manganese dioxide (MnO_2), titanium dioxide (TiO_2), or cobalt oxide (CoO). In these graphene/metal oxide hybrids, the graphene support prevents the metal oxide nanoparticles from clumping together to form large, non-functional clusters (agglomeration), thereby ensuring more active surface area is accessible for reaction (Karpuz et al.,2024). Furthermore, the high conductivity of graphene facilitates electron transfer during the catalytic reaction, synergistically enhancing catalytic activity by creating an effect more potent than either material could achieve alone (Qu et al., 2014). Similarly, combining MOF structures with graphene can significantly improve performance by integrating the adsorptive and catalytic properties of both materials.

6.4.2. Other Nanocatalysts

Various nanocatalysts besides graphene are being investigated in the field of ODS:

Polyoxometalates (POMs): These materials are molecular anions composed of metal-oxygen clusters. They are highly effective for ODS due to their strong acidic properties and their ability to accept and donate electrons during a reaction (redox properties) (Kozhevnikov, 1998). Their flexibility in being used both dissolved in the fuel (homogeneous) and immobilised on a solid support (heterogeneous) is an advantage.

Metal Oxide Nanoparticles: Metal oxide nanoparticles such as tungsten oxide (WO_3), molybdenum oxide (MoO_3), and vanadium pentoxide (V_2O_5) are also widely used. These materials react with H_2O_2 to form highly reactive peroxo-metal complexes, which are species that react very quickly and easily with other molecules. These intermediates are the primary agents that efficiently oxidise the sulfur compounds.

6.5. Paradigm Shift from Brute Force to Precision Targeting

Next-generation methods represent a significant change in how we tackle desulfurization. The previous hydrodesulfurization (HDS) method used high temperatures and pressures to break strong carbon-sulfur (C-S) bonds. The new approach, focusing on "precision targeting," requires much less energy, resulting in improved energy efficiency and lower costs (Saha et al., 2020). Two key strategies stand out in this approach:

ADS Strategy (Molecular Recognition): ADS focuses on selectively capturing the sulfur compound. Here, the lock is the target sulfur molecule, characterised by a specific geometric shape and electronic structure (e.g., DBT). The key is a nanometer-sized MOF pore designed to match the shape and chemistry of this molecule perfectly. Just as the wrong key will not open a lock, other hydrocarbon molecules in the fuel cannot enter these "custom-designed" pores, which provides high selectivity.

ODS Strategy (Chemical Tagging): ODS works by chemically "tagging" (oxidising) the sulfur atom, changing its polarity, and making it easy to separate from the system.

This philosophy shift transforms the desulfurization problem from a thermodynamic bond-breaking problem to one of surface science and molecular recognition. Nanocomposites are the fundamental tools that enable the implementation of this new paradigm because only nanotechnology offers the ability to build "keys" and "tags" precise enough to recognise or specifically transform molecules (Kobayashi, 2020).

To present all this information more clearly, we have summarised the discussed technologies and materials in the following tables: Table 1 and Table 2.

Table 1. Comparison of Major Desulfurization Technologies

Technology	Operating conditions (temp./pressure)	Hydrogen requirement	Effect on refractory compounds (dbt, 4,6-dmdbt)	Advantages	Disadvantages
HDS	High (300-400 °C / 30-130 atm)	Required	Low / Very Low	Industrially mature, high efficiency	High energy cost, H_2 consumption, octane loss, catalyst deactivation
ADS	Ambient / Low	Not Required	High	Low energy consumption, simple operation	Limited adsorbent capacity, requires regeneration, and competition with aromatics.
ODS	Moderate (Ambient - 80 °C / 1 atm)	Not Required	Very High	High efficacy against refractory compounds, mild conditions	Oxidant cost, sulfone separation required, potential for fuel quality degradation
BDS	Low (Ambient)	Not Required	High	Environmentally friendly, high selectivity	Very slow reaction rate, industrial scaling challenges

Table 2. Properties of Prominent Nanocomposite Classes for Desulfurization

Nanocomposite Class	Typical Surface Area (m^2/g)	Main Mechanism	Targeted Sulfur Compounds	Reported Performance Example	Key Advantages	Key Challenges
MOFs	1000 - 7000+	π-complexation, s-m coordination, acid-base	aromatics (thiophene, bt, dbt)	umcm-150: 40 mg-s/g (bt)	tunable pore structure, very high surface area, functionalization capability	water stability, synthesis cost, scaling up
Graphene /Metal Oxide	50 - 1500	catalytic oxidation	all sulfur compounds, especially DBT derivatives	mno_2/rgo: 80% efficiency for dbt	high catalytic activity, synergistic effects, good electron transfer	synthesis complexity, long-term stability
Metal-Loaded Zeolites	500 - 900	π-complexation	aromatics (thiophene, BT)	cu(i)-y zeolite: 23.2 mg-s/g (jet fuel)	high selectivity, thermal stability, and established industrial use	metal ion leaching, competitive adsorption with aromatics
Modified Activated Carbon	800 - 2000+	π-π interactions, acid-base, adsorption	aromatics (dbt)	ac-n-mo: 19.94 mg-s/g (dbt)	low cost, high surface area, easy regeneration	low selectivity (if unmodified), mechanical strength

7. A NEW ERA IN MATERIALS DISCOVERY WITH ARTIFICIAL INTELLIGENCE AND COMPUTATIONAL SCIENCE

The traditional material development process is based mainly on chemical prediction, hypotheses, and subsequent long "trial-and-error" cycles. This classic approach requires a rigid procedure that can take months, even for a single new material candidate: a researcher formulates a hypothesis, synthesises a material in the lab, changes a single factor (e.g., temperature), synthesises again, tests, and records the results. This cycle is repeated sometimes more than a hundred times. This process is not only slow and costly but also highly open to human errors due to unorganised data records and manual repetitions (Agrawal & Choudhary, 2016). This situation often leads to the discovery of new materials inefficiently, and many discoveries are made by chance, such as penicillin or Teflon. However, an efficient industrial strategy cannot be based on chance; on the contrary, it needs to be more controlled.

In recent years, however, the rise of computational science and artificial intelligence (AI) has ushered in a new era for approaches in materials science, transforming classical processes. This does not mean that the "trial-and-error" cycle is eliminated; instead, it makes this cycle more effective, allowing for a focus on the 5-10 most promising candidates among thousands, or even millions, of experiments (Jain et al., 2013). Now, through atomic-level simulations and intelligent algorithms that learn from many datasets, it is becoming possible to "design" materials for a specific purpose instead of discovering them. The difference between these two concepts is critical: 'Discovering' is trying to understand the properties of an existing material ("I made this material, what is it good for?"). 'Designing,' on the other hand, is planning a material at the atomic level before it is even synthesised to have a targeted property (e.g., "what is the best structure to capture the DBT molecule?") (Sanchez-Lengeling & Aspuru-Guzik, 2018).

This revolution has the potential to accelerate the development of next-generation desulfurization nanocomposites at an unprecedented speed. The main reasons for this speed provided by artificial intelligence are:

Massive Parallel Screening: While it takes a researcher days to synthesise a single material in the lab, a high-performance computer can simulate the potential performance of thousands of different materials in a virtual environment in the same amount of time (Sabbe et al., 2012).

Discovering Hidden Relationships: AI algorithms can capture the complex and hidden statistical relationships between a material's atomic structure and its desulfurization performance, as revealed in datasets (Ramprasad et al., 2017).

Reducing Physical Waste: AI helps to minimise the time, chemicals, and energy spent in the lab environments by eliminating experiments that are almost certain to fail from the outset.

Figure 6. Detailed demonstration of the production cycle of new materials based on data-driven methods

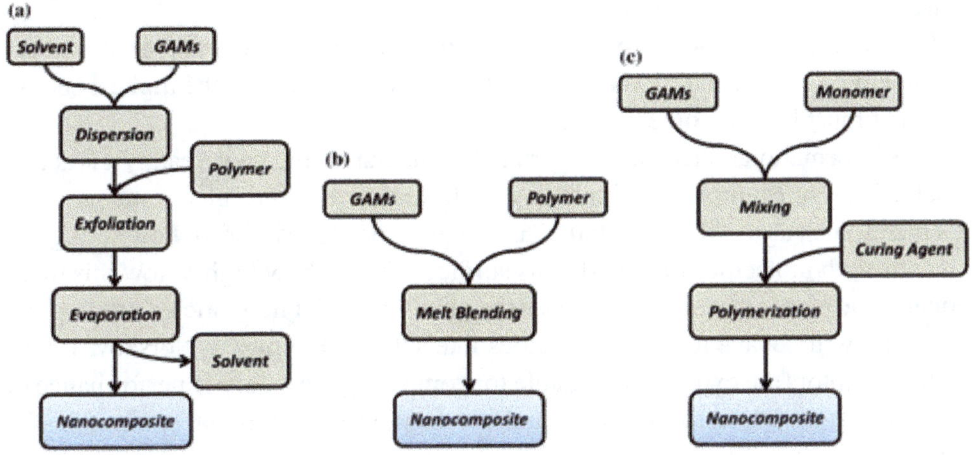

(Reprinted from Gomez et al, 2019)

7.1. Atomistic Modelling with Density Functional Theory (DFT)

The most fundamental level of understanding material behaviour is through the principles of quantum mechanics. Quantum mechanics states that in the subatomic world, particles (such as electrons) do not have fixed positions but rather behave according to wave functions that describe the probability of finding them in specific regions. Solving the wave function (Schrödinger equation) for every single electron in a material is practically impossible due to its extreme complexity. Density Functional Theory (DFT) is a brilliant approach developed to accomplish this impossible task. Instead of tracking each electron individually, DFT calculates the total electron density in the entire system (i.e., the probability of finding an electron at any given point). The theorem at the heart of the theory demonstrates that this total density contains sufficient information to determine all the fundamental properties of the material, including its energy and electronic structure (Jones, 2015). This approach has been a turning point for computational research in the field of catalysis. It has made it possible to make predictions at the atomic level that were previously impossible (Nørskov et al., 2009).

Using DFT, researchers can do the following in a virtual laboratory environment:

Calculate Adsorption Energy: This can be determined by calculating the energy difference between two states: when the molecule is bound to the surface and when it is far from the surface.

Map the Reaction Mechanism: They can draw the lowest energy path, the "molecular trail," by simulating all possible intermediate steps and transition states between the start and end points of a chemical reaction.

Predict Activation Energy Barriers: They can predict how fast or slow a reaction will be by calculating the energy of the "transition state," the highest energy peak that must be overcome for the reaction to start.

This information is critically important for understanding why a catalyst is active or selective. For example, DFT calculations have shown that an electronic property of a metal, such as its "d-band centre," is an indicator (descriptor) of how strongly it interacts with molecules that bind to its surface. That is, how high or low this theoretical value (d-band centre) is allows us to directly predict how 'sticky' or 'slippery' that metal will behave towards molecules like sulfur (Hammer & Nørskov, 1995). Such descriptors are extremely valuable for comparing the catalytic performance of different materials and for providing a theoretical basis for rational catalyst design (Jagadeesan et al., 2018).

However, the most significant weakness and limitation of DFT is its enormous computational cost. Simulating the adsorption of a single molecule on a complex nanocomposite surface can take hours or even days on high-performance computer clusters. This situation makes it practically impossible to systematically screen the vast chemical space of thousands or millions of potential material candidates with DFT (Curtarolo et al., 2013).

This raises a fundamental question in materials science: in a "chemical universe" with millions of potential candidates, is it smarter to test each candidate individually with such a costly method, or is a "map" needed to predict which candidates are worth testing more efficiently? It is at this point that Machine Learning (ML), which enables the transition from brute-force computation to thoughtful data analysis, comes into play.

Figure 7. Schematic representation of different methods for AI/ML-supported material discovery: a) Products produced in real-time simulation environments are supported by human experiments and subsequent algorithmic optimisation, providing feedback to the experiments. b) Large data sets from real-time experimental environments are scanned with algorithms, and hidden relationship predictions are derived. c) ML methods work with multiple data types to generate new sets of material representations.

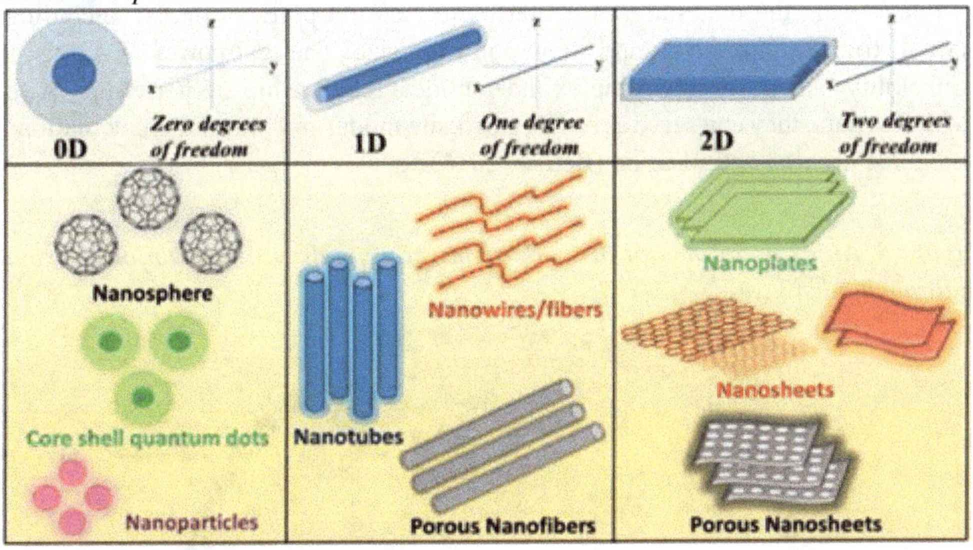

(Reprinted from Fowler et al, 2024)

7.2. Data-Driven Approaches and Artificial Intelligence

Machine Learning (ML), a subfield of Artificial Intelligence, is a set of algorithms that enable computers to learn from data without being explicitly programmed. In materials science, ML is used to uncover complex relationships between a material's structural and chemical properties (input) and its performance (output) that are difficult to express with traditional physical models. In this field, a wide range of models are used, differing according to the size and complexity of the dataset: Simple Models (e.g., Linear Regression), Ensemble Models that capture more complex relationships (e.g., Random Forest), and Deep Learning that learns the most complex data patterns (e.g., Neural Networks) (Butler et al., 2018). These models not only make predictions but can also optimise the data collection process by intelligently directing which DFT calculation will provide the most information, through strategies like "Active Learning" (Lookman et al., 2019). Two powerful approaches that stand out for complex tasks like material discovery are:

Artificial Neural Networks (ANN): ANNs are computational models consisting of interconnected mathematical processing units ("neurons") arranged in layers. Each connection has a "weight," and the model adjusts these weights best to represent the input-output relationship in the training data. In the field of desulfurization, an ANN model can be trained to predict an output, such as sulfur removal efficiency, by considering the material's structured numerical properties (inputs), including crystal structure, atomic composition, and pore size. The real power of ANNs is their ability to model highly complex and nonlinear relationships. When trained with sufficient data, they can serve as a "fast surrogate model" for slow DFT calculations and guide process optimisation (Behler, 2016).

Figure 8. An artificial neural network architecture with an input, an output and multiple hidden layers

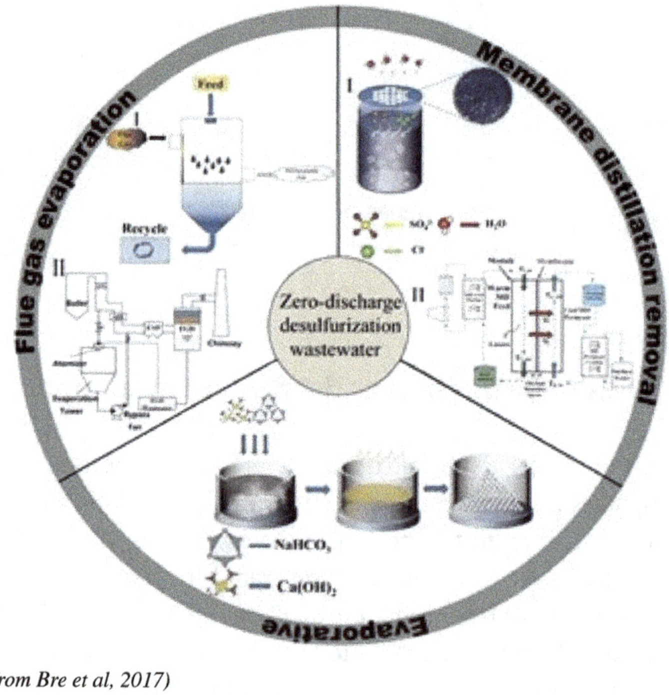

(Reprinted from Bre et al, 2017)

Genetic Algorithms (GA): GAs are a population-based optimisation technique that utilises many samples from target datasets. The process starts with a random sample from the population (dataset) of potential solutions, i.e., individuals (data). The performance of each individual is evaluated with a "fitness" score. Then, a new generation is created by combining the properties of the fittest individuals through "crossover" and adding small random changes through "mutation." This process allows the fittest solution to evolve and emerge over generations (Alshammari et al., 2016). The power of GAs lies not only in finding the best point in a known area, but also in their potential to discover entirely new and unexpected solutions by efficiently exploring a vast and complex search space, as seen in areas such as crystal structure prediction (Oganov & Glass, 2006). They are particularly of groundbreaking importance in hybrid (GA-ANN) methods, where they are combined with an ANN model. This synergy tremendously accelerates the discovery and optimisation process, improving success rates.

Figure 9. A flowchart of the genetic algorithm, illustrating how feature selection is performed in four main steps: fitness, evaluation, selection, crossover, and mutation.

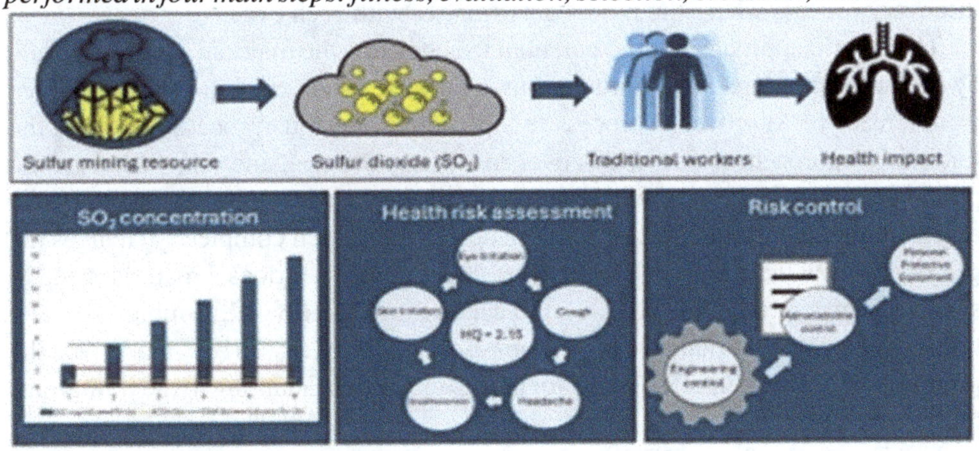

(Reprinted from Dieterle, 2003)

7.3. DFT and Machine Learning Integration

The stark contrast between the accuracy of DFT and the speed of ML has led to the birth of a hybrid strategy that combines the two approaches. This synergistic integration brings together the best aspects of both approaches (Schleder et al., 2019):

Data Generation: DFT is used to generate high-accuracy and physically meaningful "ground truth" data for a small number (dozens or hundreds) of material candidates. This data forms the high-quality initial **training set** required to train an ML model.

Surrogate Model Training: This "expensive" DFT data is used to train an ML model (e.g., a neural network or a random forest model). The trained ML model becomes a fast and computationally efficient surrogate model that can instantly predict the target property (e.g., adsorption energy) that would require DFT calculation, by taking the basic properties of the material as input (Montoya & Persson, 2017).

Accelerated Screening and Discovery: This trained ML surrogate model is then used to screen a massive database of tens of thousands, or even hundreds of thousands, of hypothetical materials. This screening process, which would take years with DFT, can be completed in a significantly shorter time with the ML model. The ML model identifies the most promising few candidates.

Validation and Feedback: Finally, this small number of candidates marked as best by ML is validated with high-accuracy DFT calculations or experimental synthesis. This validation step also serves as a test set to measure the model's generalisation ability. This new data can be added back to the training set to further improve the ML model. This creates a self-feeding and continuously improving discovery process where the model gets smarter with each cycle.

This hybrid approach has the potential to accelerate the material discovery cycle by one or more orders of magnitude. This synergy is transforming materials science from a reactive "discovery" process to a proactive "design" process. While in the traditional approach, scientists discover an existing material and ask, "What are the properties of this material we have, and how do we adapt to these properties?", the artificial intelligence-supported "inverse design" approach completely changes the paradigm. Now the question has become, "what properties does a material need to solve a specific problem I have (e.g., to capture a refractory sulfur molecule), and how can I create this material?" (Gómez-Bombarelli et al., 2018). This gives the power to design the perfect solution for needs, rather than adapting to the limitations of existing materials.

In this new paradigm, DFT and ML are two fundamental components in a truly symbiotic relationship, compensating for each other's weaknesses and combining their strengths. DFT calculations provide "high-value" data that is expensive to produce, but each represents a physical reality. ML models, on the other hand, produce "low-cost" predictions whose accuracy depends on the quality of the training data. The goal of the DFT-ML hybrid approach is to create a "computationally efficient" ML model that can screen the most expansive material space with the highest accuracy by performing the minimum number of "expensive" DFT calculations (Figure 10).

Advanced ML strategies such as Active Learning are at the centre of this optimisation (Lookman et al., 2019). In these strategies, the ML model actively guides the data collection process by intelligently predicting which new material's DFT calculation will add the most information to the model. This guidance gains tremendous power when combined with optimisation methods like Genetic Algorithms (GA). For example, while GA suggests the most promising new candidates, the Active Learning strategy determines which of these candidates, when calculated with DFT, will most accelerate the overall discovery process (Chen et al., 2022). This is an innovative collaboration that maximises the overall efficiency of the discovery process.

Figure 10. Schematic drawing of detailed procedures of catalytic applications by integrating ML and DFT methods

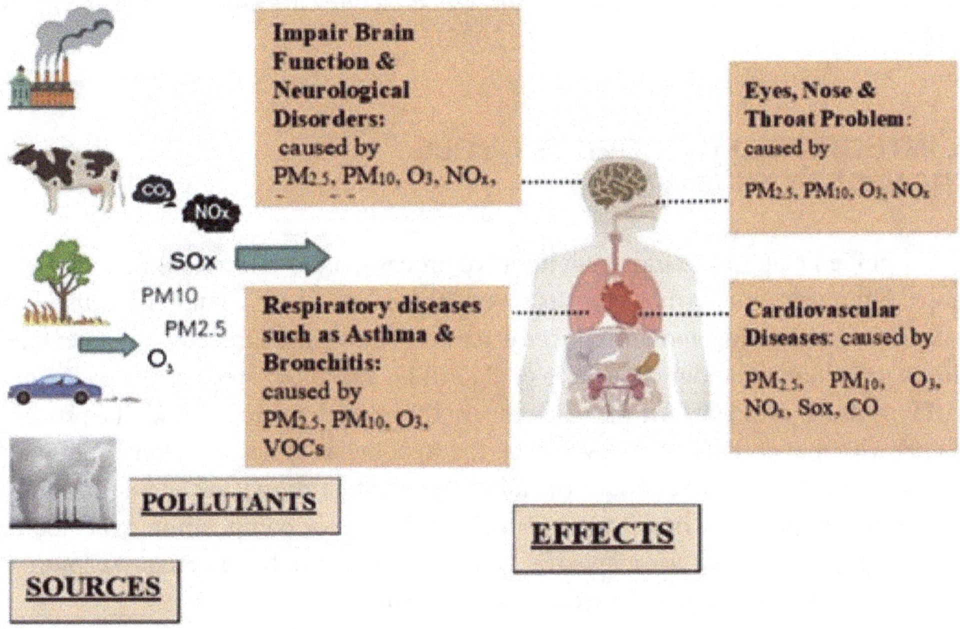

(Reprinted from Guan et al, 2021)

8. AI-GUIDED NANOCOMPOSITE DESIGN FOR DESULFURIZATION

When the concepts discussed in the previous sections—the necessity of desulfurization, the inadequacy of traditional methods, the potential of nanocomposites, and the revolutionary role of artificial intelligence (AI)—come together, a new and powerful approach emerges in materials science: AI-guided nanocomposite design. This approach goes beyond being a one-way tool, creating a win-win situation: on one hand, new materials are developed to target a specific problem; on the other, the artificial intelligence models themselves become more sophisticated by learning from this process. Each discovery cycle enables the AI to learn not only what works but also *why* it works, paving the way for the development of more powerful and context-specific design strategies for complex problems in the future (Kitano, 2021; Burger et al., 2020).

This approach surpasses isolated theoretical calculations or trial-and-error experiments, establishing an integrated discovery cycle that incorporates elements such as data, simulation, and autonomous robotic systems (Raccuglia et al., 2016; Zuo et al., 2021). Below, this cycle will be examined in detail, along with its application to the desulfurization problem, through concrete case studies and successful applications in this field (Seriket et al., 2025).

9. INTEGRATED DISCOVERY CYCLE FROM DATA TO MATERIAL

AI-guided material design is an iterative process, often referred to as a "closed-loop." This cycle aims to make the discovery process more efficient, repeatable, and scalable by minimising human-induced errors, slowness, and subjectivity, rather than eliminating human labour (Potyrailo et al., 2011). The basic steps of the cycle are:

Hypothesis and Design: Just as in the traditional scientific method, the process begins with a hypothesis. However, here the hypothesis arises not from a human's intuition, but from an AI model (usually a generative model or an optimisation algorithm) proposing a new nanocomposite structure or composition expected to show high performance for a specific task—for example, to selectively capture a refractory sulfur molecule—based on existing data.

Computational Screening and Validation: Thousands of proposed candidates are rapidly screened using faster computational methods (e.g., a machine learning surrogate model trained with density functional theory, DFT). The purpose of this stage is to sift through thousands of potential "haystacks" to find the candidates most likely to be the "needle." This stage narrows down the candidate pool to the most promising few. This narrowed list is then examined in more detail using costly but accurate methods, such as DFT.

Experimental Synthesis and Characterisation: This step represents the transition from the virtual world to the real world. The candidate or candidates confirmed to be most promising computationally are synthesised in the lab (often using robotic automation platforms). It is essential to remember that all computational predictions are merely assumptions until they are experimentally validated (Ishizuki et al., 2023). The desulfurization performance of the synthesised material is tested, and the results obtained are compared with both the predictions of the AI model and a reference point (baseline), which is the performance of the best-known material in the literature. This comparison is the final test that determines whether the hypothesis generated by the AI was successful.

Feedback and Learning: The new data obtained experimentally (both successful and unsuccessful results that performed worse than predicted) are used to retrain and improve the initial AI model. Failures, in particular, are invaluable data points for the model to learn which of its assumptions were wrong and to fill the gaps in its "knowledge map" (Kunal et al., 2025). This feedback enhances the model's predictions, enabling it to propose even better candidates in the next iteration of the cycle. Sometimes an unexpected failure can lead to the discovery of an entirely new chemical mechanism or direct researchers to explore a completely different class of materials (Aspuru-Guzik et al., 2018; Coley et al., 2019).

This cyclical approach intelligently converges towards the optimal material by increasing the system's "knowledge" with each iteration.

Table 3. Comparison of Case Studies of AI-Assisted Desulfurization

Case study no	Article (citation)	Desulfurization type/fuel	An artificial intelligence method used	Key result/success metric	Noteworthy feature
1	Gheisari, 2022	ODS / Fuel Oil	Artificial Neural Network (ANN) - (5:8:1 architecture)	R^2 = 0.938 (test set)	Demonstrating that an ANN alone can accurately model a complex chemical process.
2	Abdu et al., 2024	Adsorptive / Diesel (Zawia Refinery)	Artificial Neural Network (ANN) - (10-neuron structure)	R^2 = 0.986 (overall)	Showing how the optimal ANN architecture is selected through trials using real industrial plant data.
3	Pouladi Borj et al., 2024	ODS / Gas Condensate	Hybrid Models: Fuzzy, ANFIS, GA-Fuzzy, GA-ANFIS	$R^2 \approx 0.97$ (for hybrid models with GA)	Proving that the performance of other AI models (Fuzzy, ANFIS) can be dramatically increased by using a Genetic Algorithm (GA).
4	Alardhi et al., 2022	Desulfurization / Crude Oil (AL-Ahdab)	Artificial Neural Network (ANN) - (2 hidden layer structure)	R^2 = 0.999	Showing that even with a limited number of experiments (15), a multi-layer neural network can achieve almost perfect prediction accuracy.
5	Jorjani et al., 2007	Biological / Coal	Two Separate and Customised ANN Models	R^2=0.98 (for inorganic S) R^2=0.97 (for organic S)	Highlighting the power of developing two different ANN models, separately optimised for different sub-components of a problem (inorganic and organic sulfur).

Furthermore, this AI-guided design process enables multi-objective optimisation, a much more sophisticated approach than traditional single-objective optimisation. In traditional experimental methods, typically one condition (e.g., temperature) is changed while all other conditions are kept constant so that the effect of the variable can be clearly measured. However, this approach cannot reflect the complex nature of the real world. In real-world applications, a material needs to optimise multiple and often conflicting properties simultaneously (e.g., high capacity, easy regenerability, low cost, and high stability) (Lu et al., 2021). A material that offers high capacity may be more expensive or less stable. One of the most significant advantages brought by AI is its ability to manage this multidimensional and conflicting network of goals simultaneously.

Evolutionary methods, such as genetic algorithms, have a natural ability to handle such problems. The basis of this ability lies in their working not with a single solution, but with a whole set of solutions called a "population" at each step. This enables the algorithm to progress by evolving numerous good candidates that represent different trade-offs simultaneously (Deb et al., 2002; Tran & Ulissi, 2018). As a result, instead of a single "best" solution, they can produce a set of optimal solutions, known as a "Pareto front," that offer different trade-offs between various properties.

11. INDUSTRIAL APPLICABILITY, CHALLENGES AND FUTURE PERSPECTIVES

Transforming promising laboratory results and the theoretical potential of AI-guided design into practical and economically viable solutions at an industrial scale presents a series of significant challenges. Additionally, it is crucial to manage the production, use, and disposal of these next-generation nanomaterials within the existing framework of industrial safety and environmental regulations. Addressing the barriers to applying the synergy between nanocomposites and artificial intelligence in real-world scenarios is essential. We must explore strategies to overcome these obstacles and outline a future vision for the field.

11.1. Scaling Up From Lab to Industry and Cost

The production of complex nanocomposites, synthesised with perfect properties on a milligram or gram scale in the laboratory, in the quantities of tons needed by a refinery, is one of the biggest challenges of nanotechnology. This challenge is not just about increasing quantity; the real issue is maintaining precise control and quality in a high-volume laboratory production environment. A recipe that works perfectly in the laboratory often gives unexpected results on the massive scale of industrial reactors.

11.1.1. Scaling Up Nanocomposite Synthesis

The large-scale synthesis of complex metal-organic frameworks (MOFs) or precisely designed graphene-metal oxide hybrids presents significant challenges in terms of cost-effectiveness and reproducibility. One major issue is maintaining consistent control of temperature and pressure in industrial-scale reactors, which can result in variations in product quality. Additionally, the high cost of pure chemicals required for laboratory-scale production makes it difficult to scale up for larger applications (Teo et al., 2021).

Batch production processes tend to be inefficient on an industrial scale. This is akin to baking only one tray of cookies at a time; there are prolonged waiting times between each batch for filling, heating, cooling, and emptying. Therefore, developing continuous production methods, such as continuous flow reactors that allow for uninterrupted material production, is crucial for making these materials commercially viable (Kim et al., 2023).

The expense of specialised organic molecules or rare metals used in synthesis also impacts overall economic feasibility. This increased cost will ultimately be reflected in the price of the final product, such as fuel. Another significant challenge on an industrial scale is preventing agglomeration (Pietsch, 2004). Agglomeration occurs when nanoparticles stick together to form larger, non-functional clusters, similar to a bunch of grapes. The effectiveness of nanomaterials is mainly due to their high surface areas; however, agglomeration diminishes this surface area, rendering the material less effective.

11.1.2. Regeneration and Lifespan

For adsorbents used in processes like ADS to be industrially valuable, it is not enough for them to be just effective; they must also be durable, cheaply regenerable, and long-lasting. A refinery unit is targeted to operate continuously for years. Therefore, adsorbents must be able to withstand hundreds, or even thousands, of adsorption-desorption cycles without a significant drop in performance so that continuous and costly replacements do not stop production (Ma et al., 2018). The lifespan of the adsorbent directly affects the total operating cost of the process, as a short-lived material incurs not only the cost of purchasing new material but also the loss of millions of dollars due to production stoppages. The regeneration process itself (e.g., thermal swing, pressure swing, or solvent washing) must also be energy-efficient, fast, and low-cost (Velu et al., 2003). Otherwise, the cost of regenerating the adsorbent can negate the benefit it provides. These costs are not only financial; they also have an environmental dimension. For example, burning natural gas for thermal regeneration generates additional greenhouse gas emissions, whereas solvent washing produces a hazardous liquid waste stream that requires treatment. This situation also poses the risk of creating another waste problem while addressing one environmental issue (sulfur in fuel) (Srivastava, 2012).

11.2. Barriers and Reliability in Artificial Intelligence Models

Despite the great potential of artificial intelligence in material discovery, significant barriers exist to the reliable and effective use of these models in practical applications.

11.2.1. Data Scarcity and Quality

Machine learning models are, by their nature, data-hungry. Their performance is directly dependent on the amount and, more importantly, the quality of the data they are trained on. However, there is a critical point here: More data does not always mean better results. If the data is noisy, inconsistent, or erroneous, increasing the amount of data can worsen the model rather than improve it. Therefore, data problems are not limited to qualification or amount. In the field of materials science, high-quality, standardised, labelled, and publicly available datasets are still quite limited (Hill et al., 2016). The inconsistency of data obtained in different laboratories under different conditions weakens the generalisation ability of models. This situation can be summarised by the principle of "garbage in, garbage out"; that is, the more flawed and low-quality data you provide to a model, the more flawed and unreliable the results will be, regardless of how advanced the model is. The solution to this problem lies in increasing transparency across industry and academia, that is, creating standardised data infrastructures (such as FAIR data principles) where all used parameters and datasets are openly shared (Himanen et al., 2019).

11.2.2. Interpretability ("Black Box" Problem)

Complex ML models and intense neural networks often operate like a "black box." The term "black box" refers to a closed system where we can see what goes in and what comes out, but we cannot understand what or how it transforms this input into the output. The primary reason for this is that a decision in these models arises from a vast number (millions) of mathematical operations and interactions that the human mind cannot comprehend. It is therefore difficult to understand why a model makes a particular prediction or why it classifies a material as "good." This lack of interpretability is a significant obstacle to the scientific discovery process. The principle of reproducibility, which is fundamental to science, requires an experiment to produce the same result under certain conditions consistently. However, some randomness like AI models (mainly neural networks) can lead them to produce marginally different results each time, even if trained with the same data. This situation conflicts with the need for scientists to understand not only "what" works but also "why" it works to uncover fundamental scientific principles. Therefore, the development of Explainable AI (XAI) techniques that make the decision-making processes of models transparent is of vital importance (Oviedo et al., 2022). Furthermore, some traditional ML models like Random Forest are naturally more interpretable than neural networks because they can show how much each feature influenced the decision (Stanev et al., 2018).

11.2.3. Generalisation and Extrapolation

ML models are very successful at recognising patterns they have seen in their training data and making predictions within these patterns (interpolation). However, the actual value of a model is measured by its performance on data it has never seen before. The ability of models to make accurate predictions for truly new and unexpected material classes or chemical domains that are entirely outside the training set (extrapolation) is often limited (Westermayr et al., 2021). This situation can arise from two fundamental problems: a) Overfitting: This occurs when the model memorises the noise and random details in the data instead of learning the main pattern in the training data. Just like a student who only memorises the answer key for an exam, the model gives excellent results on the training set but cannot answer a new question (test data) that it has not seen before. b) Underfitting: This is the case where the model is too simple even to capture the basic pattern in the data (Westermayr et al., 2021). This model performs poorly on both training and testing. This can limit the potential of AI to make truly groundbreaking, "out-of-the-box" discoveries and can cause models to remain within the boundaries of the existing knowledge domain. Therefore, ensuring that the model is not just memorising, but is "learning" the fundamental physical and chemical principles, is one of the most important challenges of AI-guided research.

11.3. Future Studies

Despite these challenges, the future of the field is moving towards an even deeper integration of artificial intelligence, industrial automation, and data science. Given that the complex scientific problems of the 21st century can no longer be solved within the boundaries of a single discipline, the synergy of these fields forms the basis for future progress.

11.3.1. Autonomous Materials Discovery Platforms

The vision for the future includes fully autonomous "closed-loop laboratories" or "self-driving laboratories," where AI software directly manages robotic synthesis arms, automated characterization devices, and testing stations. The goal of these systems is not to eliminate the human element but to transform the role of humans in the lab. This approach aims to address limitations such as slowness, inconsistency, and fatigue, enabling robots to conduct thousands of experiments continuously, without interruption or error (Stach et al., 2021).

In these environments, scientists can delegate repetitive and tedious laboratory tasks to robots, enabling them to focus on higher-level activities, such as formulating creative hypotheses, interpreting results, and developing overall research strategies. Such platforms can independently design, produce, and test new materials while planning subsequent experiment cycles based on previous results.

However, there are significant challenges to realising this vision. Key obstacles include the substantial initial investment required to install these autonomous systems, the need for highly skilled personnel to operate and maintain them, and the lack of established hardware and software standards, all of which hinder widespread adoption in this field (Correa-Baena et al., 2018).

Looking ahead, it is anticipated that the most critical bottleneck will not be the algorithms' intelligence but rather the production rate of high-quality experimental data needed to train these algorithms. Additionally, challenges such as high energy consumption and ongoing maintenance demands of autonomous systems will also present obstacles. Therefore, autonomous laboratories are envisioned not merely as conveniences but as essential components of the AI-driven material discovery ecosystem, which comes with its own set of challenges to overcome.

11.3.2. Digital Twins

On an industrial scale, digital twin technology can revolutionise the optimisation of complex chemical processes, such as desulfurization. A digital twin is a virtual, dynamic copy of a real physical unit. For example, the digital twin of an ADS column instantly processes data from hundreds of sensors, such as instantaneous fuel flow rate, temperature, pressure, and sulfur concentration at the outlet. This virtual model is continuously updated with real-time data and simulates the current state and future behaviour of the physical unit. Its main benefit is the ability to conduct scenario analysis without risk: refinery engineers can test the answers to questions like “how will the catalyst life be affected if we switch to a sourer crude oil?” on the digital twin before risking production (Amthiou et al., 2023). AI algorithms running on this digital twin can adjust operational parameters in real-time to maximise process efficiency and predict potential failures before they occur (predictive maintenance). However, this technology also has its limitations. Creating a high-accuracy digital twin requires a vast amount of sensor data and in-depth process knowledge, making it an extremely costly and complex endeavour. Furthermore, as the physical plant ages over time, the digital model must be continuously calibrated to stay in sync with reality; otherwise, it can lead to incorrect decisions due to "model drift" (Attaran et al., 2023).

11.4. Precision Refining and Value Chain Transformation

This vision highlights the potential of combining AI with nanocomposites, not only to produce cleaner fuels but also to transform the entire petroleum refining value chain—from crude oil to final products—adding value at each step. Highly selective nanocomposites, designed with the aid of AI, can precisely separate sulfur and other valuable molecules in fuel streams, including aromatic compounds such as benzene and xylene, as well as nitrogen compounds (Ates et al., 2020). This means that a technology initially developed for sulfur removal can find unexpected and much more profitable applications.

The "desulfurization" process has the potential to evolve into a broader platform for "molecular separation" or "precision refining" (Sholl & Lively, 2016). Such advancements could enable refineries not only to produce cleaner fuels from low-value crude oil but also to generate high-value chemical raw materials for the petrochemical industry, including those used in the manufacture of plastics and polymers. This represents a transformative vision for the industry, allowing desulfurization to shift from being a cost centre to a potential profit centre.

CONCLUSION

This study examines the impact of artificial intelligence and nanocomposite materials on fuel desulfurization. Increasing environmental and health demands make traditional hydrodesulfurisation (HDS) inadequate, particularly against complex sulfur compounds. The need to lower sulfur levels below 15 ppm signals that new methods are essential.

Nanocomposite materials, including Metal-Organic Frameworks and graphene-based hybrids, are key to innovative desulfurisation strategies through adsorptive (ADS) and oxidative (ODS) processes that precisely target sulfur molecules. Traditional methods of selecting suitable materials are insufficient, and this is where artificial intelligence (AI) plays a critical role.

AI accelerates material discovery by combining atomistic simulations with machine learning, allowing for the proactive design of new materials. The goal is to develop solutions that perform well in the lab and can be scaled for industrial use. While challenges such as synthesis and AI reliability persist, future developments like autonomous discovery platforms offer promising solutions.

In summary, the shift to cleaner fuels depends on smart materials guided by AI. This requires collaboration across fields such as chemical engineering and data science, transforming researchers from lab practitioners to strategists who manage large datasets and automated systems for a sustainable energy future.

REFERENCES

Abdu, A., Eljugmani, A. S., Alghadamsi, A., Ejdedhom, H., & Alowayb, R. (2024). Artificial neural network modelling techniques employed to study desulfurization of Zawia refinery diesel. *Proceedings of the 10th Jordan International Chemical Engineering Conference (JIChEC10).*

Agrawal, A., & Choudhary, A. (2016). Perspective: Materials informatics and big data: Realisation of the "fourth paradigm" of science in materials science. *APL Materials*, *4*(5), 053208. DOI: 10.1063/1.4946894

Alardhi, S. M., Jabbar, N. M., Al-Jadir, T., Ibrahim, N. K., Dakhil, A. M., Al-Saedi, N. D., Al-Saedi, H. D., & Adnan, M. (2022). Artificial neural network model for predicting the desulfurization efficiency of AL-Ahdab crude oil. In. AIP Conference Proceedings: Vol. 2443. *Issue 1* (p. 030033). AIP Publishing., DOI: 10.1063/5.0091975

Alshammari, A., & Kalevaru, V. N. (2016). Supported gold nanoparticles as promising catalysts. Catalytic Application of Nano-Gold Catalysts, 57-81.

Altuner, E. E. (2023). Electrochemical sensitivity, selectivity, and sensory properties of a natural polymer-supported titanium nanocomposite electrode towards lysine amino acid. *Muş Alparslan Üniversitesi Fen Bilimleri Dergisi*, *11*(1), 28–32. DOI: 10.18586/msufbd.1268440

Altuner, E. E., Akin, M., Bayat, R., Bekmezci, M., Burhan, H., & Sen, F. (2022). Challenges in commercialization of carbon nanomaterial-based sensors. In *Carbon Nanomaterials-Based Sensors* (pp. 381–392). Emerging Research Trends in Devices and Applications., DOI: 10.1016/B978-0-323-91174-0.00020-2

Altuner, E. E., Arıkan, K., Burhan, H., Ozdemir, S., & Şen, F. (2021). Commercial aspects of direct alcohol fuel cells. In *Nanomaterials for Direct Alcohol Fuel Cells* (pp. 511–524). DOI: 10.1016/B978-0-12-821713-9.00012-3

Altuner, E. E., Bekmezci, M., Bayat, R., Akin, M., Isik, I., & Şen, F. (2021). Dendrimer-based nanocomposites for alcohol fuel cells. In *Nanomaterials for Direct Alcohol Fuel Cells* (pp. 337–352). DOI: 10.1016/B978-0-12-821713-9.00007-X

Altuner, E. E., Bekmezci, M., & Sen, F. (2022). Manufacturing techniques of magnetic polymer nanocomposites. In *Handbook of Magnetic Hybrid Nanoalloys and Their Nanocomposites* (pp. 303–318). DOI: 10.1007/978-3-030-90948-2_12

Altuner, E. E., Gulbagca, F., Tiri, R. N. E., Aygun, A., & Sen, F. (2023). Highly efficient palladium–zinc oxide nanoparticles synthesized by biogenic methods: Characterization, hydrogen production and photocatalytic activities. *Chemical Engineering Journal Advances, 14*, 100465. DOI: 10.1016/j.ceja.2023.100465

Altuner, E. E., Gur, T., & Şen, F. (2021). Ternary/quaternary nanomaterials for direct alcohol fuel cells. In *Nanomaterials for Direct Alcohol Fuel Cells* (pp. 157–172). DOI: 10.1016/B978-0-12-821713-9.00001-9

Altuner, E. E., Ozalp, V. C., Yilmaz, M. D., Sudagidan, M., Aygun, A., Acar, E. E., Tasbasi, B. B., & Sen, F. (2022). Development of electrochemical aptasensors detecting phosphate ions on TMB substrate with epoxy-based mesoporous silica nanoparticles. *Chemosphere, 297*, 134077. DOI: 10.1016/j.chemosphere.2022.134077 PMID: 35218784

Ameen, F., Altuner, E. E., Tiri, R. N. E., Gulbagca, F., Aygun, A., Sen, F., Majrashi, N., Orfali, R., & Dragoi, E. N. (2023). Highly active iron(II) oxide–zinc oxide nanocomposite synthesized using *Thymus vulgaris* plant as bioreduction catalyst: Characterization, hydrogen evolution and photocatalytic degradation. *International Journal of Hydrogen Energy, 48*(55), 21139–21151. DOI: 10.1016/j.ijhydene.2022.11.229

Amthiou, H., Arioua, M., & Benbarrad, T. (2023). Digital twins in Industry 4.0: A literature review. *ITM Web of Conferences, 52*, 01002. DOI: 10.1051/itmconf/20235201002

Ancheyta, J., Sánchez, S., & Rodríguez, M. A. (2005). Kinetic modelling of hydrocracking of heavy oil fractions: A review. *Catalysis Today, 109*(1–4), 76–92. DOI: 10.1016/j.cattod.2005.08.015

Aspuru-Guzik, A., Lindh, R., & Reiher, M. (2018). The matter of quantum chemistry. *ACS Central Science, 4*(2), 144–152. DOI: 10.1021/acscentsci.7b00550 PMID: 29532014

Ates, B., Koytepe, S., Ulu, A., Gurses, C., & Thakur, V. K. (2020). Chemistry, structures, and advanced applications of nanocomposites from biorenewable resources. *Chemical Reviews, 120*(17), 9304–9362. DOI: 10.1021/acs.chemrev.9b00553 PMID: 32786427

Attaran, M., & Celik, B. (2023). Digital twin: Benefits, use cases, challenges, and opportunities. *Decision Analytics Journal, 6*, 100165. DOI: 10.1016/j.dajour.2023.100165

Ausmeel, S., Eriksson, A., Ahlberg, E., Sporre, M. K., Spanne, M., & Kristensson, A. (2020). Ship plumes in the Baltic Sea Sulfur Emission Control Area: Chemical characterisation and contribution to coastal aerosol concentrations. *Atmospheric Chemistry and Physics*, *20*(15), 9135–9151. DOI: 10.5194/acp-20-9135-2020

Azeez, M. O., & Ganiyu, S. A. (2023). Review of biomass-derived activated carbon for production of clean fuels by adsorptive desulfurization: Insights into processes, modifications, properties, and performances. *Arabian Journal of Chemistry*, *16*(10), 105182. DOI: 10.1016/j.arabjc.2023.105182

Bartholomew, C. H. (2001). Mechanisms of catalyst deactivation. *Applied Catalysis A, General*, *212*(1–2), 17–60. DOI: 10.1016/S0926-860X(00)00843-7

Behler, J. (2016). Perspective: Machine learning potentials for atomistic simulations. *The Journal of Chemical Physics*, *145*(17), 170901. DOI: 10.1063/1.4966192 PMID: 27825224

Bre, F., Giménez, J., & Fachinotti, V. (2017). Prediction of wind pressure coefficients on building surfaces using artificial neural networks. *Energy and Building*, *158*, 1429–1441. Advance online publication. DOI: 10.1016/j.enbuild.2017.11.045

Brook, R. D., Rajagopalan, S., Pope, C. A.III, Brook, J. R., Bhatnagar, A., Diez-Roux, A. V., Holguin, F., Hong, Y., Luepker, R. V., Mittleman, M. A., Peters, A., Siscovick, D., Smith, S. C.Jr, Whitsel, L., & Kaufman, J. D. (2010). Particulate matter air pollution and cardiovascular disease: An update to the scientific statement from the American Heart Association. *Circulation*, *121*(21), 2331–2378. DOI: 10.1161/CIR.0b013e3181dbece1 PMID: 20458016

Burger, B., Maffettone, P. M., Gusev, V. V., Aitchison, C. M., Bai, Y., Wang, X., Li, X., Alston, B. M., Li, B., Clowes, R., Rankin, N., Harris, B., Sprick, R. S., & Cooper, A. I. (2020). A mobile robotic chemist. *Nature*, *583*(7815), 237–241. DOI: 10.1038/s41586-020-2442-2 PMID: 32641813

Butler, K. T., Davies, D. W., Cartwright, H., Isayev, O., & Walsh, A. (2018). Machine learning for molecular and materials science. *Nature*, *559*(7715), 547–555. DOI: 10.1038/s41586-018-0337-2 PMID: 30046072

Chang, A. F., Pashikanti, K., & Liu, Y. A. (2013). *Refinery engineering: Integrated process modelling and optimisation*. John Wiley & Sons., DOI: 10.1002/9783527666836

Chen, L., Tian, Y., Hu, X., Yao, S., Lu, Z., Chen, S., Zhang, X., & Zhou, Z. (2022). A universal machine learning framework for electrocatalyst innovation: A case study of discovering alloys for hydrogen evolution reaction. *Advanced Functional Materials*, *32*(47), 2208418. Advance online publication. DOI: 10.1002/adfm.202208418

Coley, C. W., Eyke, N. S., & Jensen, K. F. (2019). Autonomous discovery in the chemical sciences, Part I: Progress. *Angewandte Chemie International Edition, 59*(51), 22858–22893. DOI: 10.1002/anie.201909987 PMID: 31553511

Correa-Baena, J.-P., Hippalgaonkar, K., Duren, J., Jaffer, S., Chandrasekhar, V., Stevanovic, V., Wadia, C., Guha, S., & Buonassisi, T. (2018). Accelerating materials development via automation, machine learning, and high-performance computing. *Joule, 2*(8), 1410–1420. Advance online publication. DOI: 10.1016/j.joule.2018.05.009

Curtarolo, S., Hart, G. L., Nardelli, M. B., Mingo, N., Sanvito, S., & Levy, O. (2013). The high-throughput highway to computational materials design. *Nature Materials, 12*(3), 191–201. DOI: 10.1038/nmat3568 PMID: 23422720

Deb, K., Pratap, A., Agarwal, S., & Meyarivan, T. (2002). A fast and elitist multi-objective genetic algorithm: NSGA-II. *IEEE Transactions on Evolutionary Computation, 6*(2), 182–197. DOI: 10.1109/4235.996017

Dieterle, F. J. (2003). Multianalyte quantifications by means of integration of artificial neural networks, genetic algorithms and chemometrics for time-resolved analytical data.

Epling, W. S., Campbell, L. E., Yezerets, A., Currier, N. W., & Parks, J. E.II. (2004). Overview of the fundamental reactions and degradation mechanisms of NOx storage/reduction catalysts. *Catalysis Reviews. Science and Engineering, 46*(2), 163–245. DOI: 10.1081/CR-200031932

European Parliament and Council. (2009). Directive 2009/30/EC of the European Parliament and of the Council of 23 April 2009 amending Directive 98/70/EC as regards the specification of petrol, diesel and gas-oil and introducing a mechanism to monitor and reduce greenhouse gas emissions. *Official Journal of the European Union, L, 140*(88). https://data.europa.eu/eli/dir/2009/30/oj

Fowler, J., Kottwitz, M., Trask, N., & Dingreville, R. (2024). Beyond combinatorial materials science: The 100 prisoners problem. *Integrating Materials and Manufacturing Innovation, 13*(1), 83–91. Advance online publication. DOI: 10.1007/s40192-023-00330-6

Furukawa, H., Cordova, K. E., O'Keeffe, M., & Yaghi, O. M. (2013). The chemistry and applications of metal–organic frameworks. *Science, 341*(6149), 1230444. DOI: 10.1126/science.1230444 PMID: 23990564

Gheisari, M. (2022). Modelling of oxidative desulfurization process by artificial neural network. *EasyChair Preprint* No. 7943.

Gómez-Bombarelli, R., Wei, J. N., Duvenaud, D., Hernández-Lobato, J. M., Sánchez-Lengeling, B., Sheberla, D., Aguilera-Iparraguirre, J., Hirzel, T. D., Adams, R. P., & Aspuru-Guzik, A. (2018). Automatic chemical design using a data-driven continuous representation of molecules. *ACS Central Science*, *4*(2), 268–276. DOI: 10.1021/acscentsci.7b00572 PMID: 29532027

Government of Canada. (2015, July 29). Regulations amending the sulphur in gasoline regulations. *Canada Gazette*, *149*(15). https://gazette.gc.ca/rp-pr/p2/2015/2015-07-29/html/sor-dors187-eng.html

Grande, C., Araújo, J. D. P., Cavenati, S., Firpo, N., Basaldella, E., & Rodrigues, A. (2004). New π-complexation adsorbents for propane–propylene separation. *Langmuir*, *20*(13), 5291–5297. DOI: 10.1021/la036400s PMID: 15986665

Grennfelt, P., Engleryd, A., Forsius, M., Hov, Ø., Rodhe, H., & Cowling, E. (2020). Acid rain and air pollution: 50 years of progress in environmental science and policy. *Ambio*, *49*(4), 849–864. DOI: 10.1007/s13280-019-01244-4 PMID: 31542884

Guan, B., Jiang, H., Wei, Y., Liu, Z., Wu, X., Lin, H., & Huang, Z. (2021). Density functional theory research for atomic structure, properties prediction, and rational design of selective catalytic reduction catalysts: Current progresses and future perspectives. *Molecular Catalysis*, *510*, 111704. DOI: 10.1016/j.mcat.2021.111704

Gundogdu, A., Bozbeyoglu, P., Imamoglu, M., Baltaci, C., Duran, C., & Bulut, V. N. (2022). Characterization of the adsorption mechanism of cadmium(II) and methylene blue upon corncobs activated carbon. *Analytical Letters*, *56*(3), 433–448. DOI: 10.1080/00032719.2022.2098310

Hammer, B., & Nørskov, J. K. (1995). Electronic factors determining the reactivity of metal surfaces. *Surface Science*, *343*(3), 211–220. DOI: 10.1016/0039-6028(96)80007-0

Hao, L., Hurlock, M., Ding, G., & Zhang, Q. (2020). Metal–organic frameworks towards desulfurization of fuels. *Topics in Current Chemistry*, *378*(1), 17. DOI: 10.1007/s41061-020-0280-1 PMID: 31993845

Helgeson, N., Trueblood, D., Holden, B., Jack, J., & Griffin, G. (2010). *Diesel engine air emissions reduction technologies* (No. NFESCTR2328ENV).

Hernández-Maldonado, A., & Yang, R. (2007). Desulfurization of transportation fuels by adsorption. *Catalysis Reviews. Science and Engineering*, *46*(2), 111–150. DOI: 10.1081/CR-200032697

Hill, J., Mulholland, G., Persson, K., Seshadri, R., Wolverton, C., & Meredig, B. (2016). Materials science with large-scale data and informatics: Unlocking new opportunities. *MRS Bulletin*, *41*(5), 399–409. DOI: 10.1557/mrs.2016.93

Himanen, L., Geurts, A., Foster, A. S., & Rinke, P. (2019). Data-driven materials science: Status, challenges, and perspectives. *Advancement of Science*, *6*(21), 1900808. DOI: 10.1002/advs.201900808 PMID: 31728276

Hu, Q., Lu, J., Hui, J., Rao, Z., Ren, Y., & Wang, H. (2025). Artificial intelligence-driven development in rechargeable battery materials: Progress, challenges, and future perspectives. *Advanced Functional Materials*, *08438*(52), e08438. Advance online publication. DOI: 10.1002/adfm.202508438

Hu, Y., Li, N., Wang, M., Qiao, Z., Gu, D., Zhu, L., Yuan, D., & Wang, B. (2025). Overview of research status and development trends in diesel desulfurization technology. *Catalysts*, *15*(3), 251. DOI: 10.3390/catal15030251

International Maritime Organisation. (2020). IMO 2020 – Cutting sulphur oxide emissions. IMO Media Centre. https://www.imo.org/en/MediaCentre/HotTopics/Pages/Sulphur-2020.aspx

Iram, S., Qaisar, I., Shabbir, R., Pomee, M. S., Schmidt, M., & Hertig, E. (2025). Impact of air pollution and smog on human health in Pakistan: A systematic review. *Environments (Basel, Switzerland)*, *12*(2), 46. DOI: 10.3390/environments12020046

Ishihara, A., Wang, D., Dumeignil, F., Amano, H., Qian, E. W., & Kabe, T. (2005). Oxidative desulfurization and denitrogenation of a light gas oil using an oxidation/adsorption continuous flow process. *Applied Catalysis A, General*, *279*(1–2), 279–287. DOI: 10.1016/j.apcata.2004.10.037

Ishizuki, N., Shimizu, R., & Hitosugi, T. (2023). Autonomous experimental systems in materials science. *Science and Technology of Advanced Materials: Methods*, *3*(1), 2197519. DOI: 10.1080/27660400.2023.2197519

Jafarinejad, S. (2016). Control and treatment of sulfur compounds especially sulfur oxides (SOx) emissions from the petroleum industry: A review. *Chemistry International*, *2*(4), 242–253.

Jagadeesan, D., Vernekar, D., Gupta, S., & Jaiswal, G. (2018). New opportunities in heterogeneous catalysis. *Proceedings of the Indian National Science Academy*, 99. DOI: 10.16943/ptinsa/2018/49504

Jain, A., Ong, S. P., Hautier, G., Chen, W., Richards, W. D., Dacek, S., Cholia, S., Gunter, D., Skinner, D., Ceder, G., & Persson, K. A. (2013). Commentary: The Materials Project: A materials genome approach to accelerating materials innovation. *APL Materials*, *1*(1), 011002. DOI: 10.1063/1.4812323

Jones, R. O. (2015). Density functional theory: Its origins, rise to prominence, and future. *Reviews of Modern Physics*, *87*(3), 897–923. DOI: 10.1103/RevModPhys.87.897

Jorjani, E., Chehreh Chelgani, S., & Mesroghli, S. (2007). Prediction of microbial desulfurization of coal using artificial neural networks. *Minerals Engineering*, *20*(14), 1285–1292. DOI: 10.1016/j.mineng.2007.07.003

Kampouraki, Z.-C., Giannakoudakis, D. A., Nair, V., Hosseini-Bandegharaei, A., Colmenares, J. C., & Deliyanni, E. A. (2019). Metal–organic frameworks as desulfurization adsorbents of DBT and 4,6-DMDBT from fuels. *Molecules (Basel, Switzerland)*, *24*(24), 4525. DOI: 10.3390/molecules24244525 PMID: 31835616

Karpuz, Ö., Baltacı, C., Gül, A., Gülen, J., Bozbeyoğlu, P., & Aydoğan, N. (2024). Green synthesis of iron and silver nanoparticles from aqueous extract of buckwheat husk waste: Antibacterial, cytotoxic, and dye decolorization properties. *Biomass Conversion and Biorefinery*, •••, 1–33. DOI: 10.1007/s13399-024-06287-6

Kawatra, S. K., & Eisele, T. C. (2001). *Coal desulfurization: High-efficiency preparation methods*. Taylor & Francis.

Kim, E., Tomczak, K., Chandrasiri, H., Pálmai, M., Ghaznavi, A., Gritsenko, D., Xu, J., & Snee, P. (2023). Nucleation control of quantum dot synthesis in a microfluidic continuous flow reactor. *Frontiers in Nanotechnology*, *4*, 1096267. DOI: 10.3389/fnano.2022.1096267

Kitano, H. (2021). Nobel Turing Challenge: Creating the engine for scientific discovery. *npj Systems Biology and Applications*, *7*, 29. DOI: 10.1038/s41540-021-00189-3

Kobayashi, S. (2020). *Nanoparticles in catalysis*. DOI: 10.1007/978-3-030-56630-2

Kozhevnikov, I. V. (1998). Catalysis by heteropoly acids and multicomponent polyoxometalates in liquid-phase reactions. *Chemical Reviews*, *98*(1), 171–198. DOI: 10.1021/cr960400y PMID: 11851502

Lookman, T., Balachandran, P. V., Xue, D., & Yuan, R. (2019). Active learning in materials science with emphasis on adaptive sampling using uncertainties for targeted design. *npj Computational Materials*, *5*, 21. DOI: 10.1038/s41524-019-0153-8

Ma, X., Sakanishi, K., & Mochida, I. (1994). Hydrodesulfurization reactivities of various sulfur compounds in diesel fuel. *Industrial & Engineering Chemistry Research, 33*(2), 218–222. DOI: 10.1021/ie00026a007

Majewski, W. A. (2001). *Diesel particulate filters.* DieselNet, Ecopoint Inc., https://courses.washington.edu/cive494/DPF.pdf

Montoya, J., & Persson, K. (2017). A high-throughput framework for determining adsorption energies on solid surfaces. *npj Computational Materials, 3.* DOI: 10.1038/s41524-017-0017-z

Mosca, L. P., Gapan, A., Angeles, R., & Lopez, E. C. (2023). Stability of metal–organic frameworks: Recent advances and future trends. *Engineering Proceedings, 56*, 146. DOI: 10.3390/ASEC2023-16280

Naaz, A., Sheikh, A. R., & Arham, M. (2022). Oxidative and adsorptive desulfurization of fuel by using metal–organic frameworks: Recent progress. *IOSR Journal of Applied Chemistry, 15*, 34–53. DOI: 10.9790/5736-1510013453

Nakano, K., Ali, S., Kim, H.-J., Kim, T., Alhooshani, K., Park, J.-I., & Mochida, I. (2013). Deep desulfurization of gas oil over NiMoS catalysts supported on alumina-coated USY-zeolite. *Fuel Processing Technology, 116*, 44–51. DOI: 10.1016/j.fuproc.2013.04.012

Neira, M. P. (2019). Air pollution and human health: A comment from the World Health Organisation. *Annals of Global Health, 85*(1), 141. DOI: 10.5334/aogh.2712 PMID: 31871904

Nikolaidis, P., & Poullikkas, A. (2017). A comparative overview of hydrogen production processes. *Renewable & Sustainable Energy Reviews, 67*, 597–611. DOI: 10.1016/j.rser.2016.09.044

Nørskov, J. K., Bligaard, T., Rossmeisl, J., & Christensen, C. H. (2009). Towards the computational design of solid catalysts. *Nature Chemistry, 1*(1), 37–46. DOI: 10.1038/nchem.121 PMID: 21378799

Oganov, A. R., & Glass, C. W. (2006). Crystal structure prediction using *ab initio* evolutionary techniques: Principles and applications. *The Journal of Chemical Physics, 124*(24), 244704. DOI: 10.1063/1.2210932 PMID: 16821993

Okunev, A. G., Parkhomchuk, E. V. E., Lysikov, A. I., Parunin, P. D., Semeykina, V. S., & Parmon, V. N. (2015). Catalytic hydroprocessing of heavy oil feedstocks. *Russian Chemical Reviews, 84*(9), 981–999. DOI: 10.1070/RCR4486

Oviedo, F., Lavista Ferres, J., Buonassisi, T., & Butler, K. (2022). Interpretable and explainable machine learning for materials science and chemistry. *Accounts of Materials Research*, *3*(6), 597–607. Advance online publication. DOI: 10.1021/accountsmr.1c00244

Özkan, V., & Özkan, A. (2022). Adsorptive desulfurization of crude oil with clinoptilolite zeolite. *Natural and Engineering Sciences*, *7*(3), 284–293. DOI: 10.28978/nesciences.1222495

Pietsch, W. B. (2004). *Agglomeration in industry: Occurrence and applications* (Vol. 1). John Wiley & Sons. DOI: 10.1002/9783527619795

Potyrailo, R. A., Rajan, K., Stoewe, K., Takeuchi, I., Chisholm, B., & Lam, H. (2011). Combinatorial and high-throughput screening of materials libraries: A tutorial. *ACS Combinatorial Science*, *13*(6), 579–633. DOI: 10.1021/co200007w PMID: 21644562

Pouladi Borj, B., Fanaei, M. A., Esfandyari, M., Naddaf, A., Jafari, D., & Baghmisheh, G. (2024). Machine learning-assisted methods for prediction and optimisation of oxidative desulfurization of gas condensate via a novel oxidation system. *Journal of Sulfur Chemistry*, *45*(1), 84–100. DOI: 10.1080/17415993.2023.2257827

Qu, J., Shi, L., He, C., Gao, F., Li, B., Zhou, Q., Hu, H., Shao, G., Wang, X., & Qiu, J. (2014). Highly efficient synthesis of graphene/MnO_2 hybrids and their application for ultrafast oxidative decomposition of methylene blue. *Carbon*, *66*, 485–492. DOI: 10.1016/j.carbon.2013.09.025

Raccuglia, P., Elbert, K. C., Adler, P. D. F., Falk, C., Wenny, M. B., Mollo, A., & Norquist, A. J. (2016). Machine-learning-assisted materials discovery at record speed. *Nature*, *533*(7601), 73–76. DOI: 10.1038/nature17439 PMID: 27147027

Ramprasad, R., Batra, R., Pilania, G., Mannodi-Kanakkithodi, A., & Kim, C. (2017). Machine learning in materials informatics: Recent applications and prospects. *npj Computational Materials*, *3*, 54. DOI: 10.1038/s41524-017-0056-5

Rana, M. S., Sámano, V., Ancheyta, J., & Díaz, J. A. I. (2007). A review of recent advances on process technologies for upgrading of heavy oils and residue. *Fuel*, *86*(9), 1216–1231. DOI: 10.1016/j.fuel.2006.08.004

Roman, F. F., Diaz de Tuesta, J. L., Silva, A. M. T., Faria, J. L., & Gomes, H. T. (2021). Carbon-based materials for oxidative desulfurization and denitrogenation of fuels: A review. *Catalysts*, *11*(10), 1239. DOI: 10.3390/catal11101239

Sabbe, M. K., Reyniers, M. F., & Reuter, K. (2012). First-principles kinetic modeling in heterogeneous catalysis: An industrial perspective on best practice, gaps and needs. *Catalysis Science & Technology*, *2*(10), 2010–2024. DOI: 10.1039/c2cy20261a

Saha, B., Vedachalam, S., & Dalai, A. (2020). Review on recent advances in adsorptive desulfurization. *Fuel Processing Technology*, *214*, 106685. DOI: 10.1016/j.fuproc.2020.106685

Saleh, T. A. (2020). Characterisation, determination and elimination technologies for sulfur from petroleum: Toward cleaner fuel and a safe environment. *Trends in Environmental Analytical Chemistry*, *25*, e00080. DOI: 10.1016/j.teac.2020.e00080

Sanchez-Lengeling, B., & Aspuru-Guzik, A. (2018). Inverse molecular design using machine learning: Generative models for matter engineering. *Science*, *361*(6400), 360–365. DOI: 10.1126/science.aat2663 PMID: 30049875

Schleder, G. R., Padilha, A. C. M., Acosta, C. M., Costa, M., & Fazzio, A. (2019). From DFT to machine learning: Recent approaches to materials science—A review. *JPhys Materials*, *2*(3), 032001. DOI: 10.1088/2515-7639/ab084b

Schneider, G. (2018). Automating drug discovery. *Nature Reviews. Drug Discovery*, *17*(2), 97–113. DOI: 10.1038/nrd.2017.232 PMID: 29242609

Seriket, H., Bougzime, O., Song, Y., Qi, H. J., & Demoly, F. (2025). Reinforcement learning-enabled design of topological interlocking materials for sustainable multi-material additive manufacturing. *Additive Manufacturing*, *104992*, 104992. Advance online publication. DOI: 10.1016/j.addma.2025.104992

Shi, B., Lookman, T., & Xue, D. (2023). Multi-objective optimization and its application in materials science. *Materials Genome Engineering Advances*, *1*(2), e14. Advance online publication. DOI: 10.1002/mgea.14

Sholl, D. S., & Lively, R. P. (2016). Seven chemical separations to change the world. *Nature*, *532*(7600), 435–437. DOI: 10.1038/532435a PMID: 27121824

Singh, A., Nanade, A., Bandiwadekar, P., Patil, P., Nadar, S., Nanade, S., & Tiwari, M. (2025). Machine learning models for the prediction of CO_2 adsorption capacity of metal–organic frameworks. *Journal of the Indian Chemical Society*, *102*(9), 101904. DOI: 10.1016/j.jics.2025.101904

Skinner, E. N., Mason, J. F., & Moran, J. J. (1960). High temperature corrosion in refinery and petrochemical service. *Corrosion*, *16*(12), 593t–600t. DOI: 10.5006/0010-9312-16.12.85

Sofiev, M., Winebrake, J. J., Johansson, L., Carr, E. W., Prank, M., Soares, J., Vira, J., Kouznetsov, R., Jalkanen, J.-P., & Corbett, J. J. (2018). Cleaner fuels for ships provide public health benefits with climate trade-offs. *Nature Communications*, *9*(1), 406. DOI: 10.1038/s41467-017-02774-9 PMID: 29410475

Song, C. (2003). An overview of new approaches to deep desulfurization for transportation fuels. *Catalysis Today, 86*(1–4), 211–263. DOI: 10.1016/S0920-5861(03)00412-7

Speight, J. G. (2014). *The chemistry and technology of petroleum.* CRC Press., DOI: 10.1201/b16559

Srivastava, V. (2012). An evaluation of desulfurization technologies for sulfur removal from liquid fuels. *RSC Advances, 2*(3), 759–783. DOI: 10.1039/C1RA00309G

Stach, E., DeCost, B., Kusne, A. G., Hattrick-Simpers, J., Brown, K. A., Reyes, K. G., Schrier, J., Billinge, S., Buonassisi, T., Foster, I., Gomes, C. P., Gregoire, J. M., Mehta, A., Montoya, J., Olivetti, E., Park, C., Rotenberg, E., Saikin, S. K., Smullin, S., & Maruyama, B. (2021). Autonomous experimentation systems for materials development: A community perspective. *Matter, 4*(9), 2702–2726. DOI: 10.1016/j.matt.2021.06.036

Stanev, V., Oses, C., Kusne, A. G., et al. (2018). Machine learning modeling of superconducting critical temperature. *npj Computational Materials, 4*, 29. DOI: 10.1038/s41524-018-0085-8

Stanislaus, A., Marafi, A., & Rana, M. S. (2010). Recent advances in the science and technology of ultra-low sulfur diesel (ULSD) production. *Catalysis Today, 153*(1–2), 1–68. DOI: 10.1016/j.cattod.2010.05.011

Teo, W. L., Zhou, W., Qian, C., & Zhao, Y. (2021). Industrializing metal–organic frameworks: Scalable synthetic means and their transformation into functional materials. *Materials Today, 47*, 170–186. DOI: 10.1016/j.mattod.2021.01.010

Topsøe, H., Clausen, B. S., & Massoth, F. E. (1996). Hydrotreating catalysis. In *Catalysis: Science and Technology* (pp. 1–269). Springer. DOI: 10.1007/978-3-642-61040-0_1

Tran, K., & Ulissi, Z. W. (2018). Active learning across intermetallics to guide discovery of electrocatalysts for CO_2 reduction and H_2 evolution. *Nature Catalysis, 1*(9), 696–703. DOI: 10.1038/s41929-018-0142-1

U.S. Environmental Protection Agency. (2014). Control of air pollution from motor vehicles: Tier 3 motor vehicle emission and fuel standards. *Federal Register, 79*(81), 23414–23795. https://www.govinfo.gov/content/pkg/FR-2014-04-28/pdf/2014-06954.pdf

Velu, S., Watanabe, S., Ma, X., & Song, C. (2003). Regenerable adsorbents for the adsorptive desulfurization of transportation fuels for fuel cell applications. *ACS Division of Fuel Chemistry, Preprints, 48.*

Vosooghzadeh, B. (2018). Bitumen and heavy crude upgrading (review article). DOI: 10.13140/RG.2.2.25169.58721

Westermayr, J., Gastegger, M., Schütt, K. T., & Maurer, R. J. (2021). Perspective on integrating machine learning into computational chemistry and materials science. *The Journal of Chemical Physics, 154*(23), 230903. DOI: 10.1063/5.0047760 PMID: 34241249

Whitesides, G. M. (2005). Nanoscience, nanotechnology, and chemistry. *Small, 1*(2), 172–179. DOI: 10.1002/smll.200400130 PMID: 17193427

World Health Organization. (2021). *WHO global air quality guidelines: Particulate matter (PM2.5 and PM10), ozone, nitrogen dioxide, sulfur dioxide and carbon monoxide*. https://www.who.int/publications/i/item/9789240034228

Wu, F., Yuan, Q., Wang, J., Wang, X., Luo, J., Guo, Y., Xu, H., & Wei, X. (2025). Oxidative desulfurization catalysed by magnetically recoverable $CoFe_2O_4$ nanoparticles. *Arabian Journal of Chemistry, 18*(1), 106076. DOI: 10.1016/j.arabjc.2024.106076

Wu, Q. (2023). Acidic and basic catalytic cracking technologies and its development prospects for crude oil to chemicals. *Fuel, 332*, 126132. DOI: 10.1016/j.fuel.2022.126132

Xie, R., Xing, Q., Gao, J., Wang, X., Xu, W., Lv, Z., Yi, W., Zhao, J., Luo, Z., Wu, X., & Liu, H. (2024). Investigation on the effect of ship emissions on the air quality: A case study in Hainan Island, China. *Journal of Environmental Sciences (China), 156*, 114–125. Advance online publication. DOI: 10.1016/j.jes.2024.08.010 PMID: 40412918

Xu, F., Zhao, S., Li, B., Li, H., Ling, Z., Zhang, G., & Liu, M. (2024). Current status of zero liquid discharge technology for desulfurization wastewater. *Water (Basel), 16*(6), 900. DOI: 10.3390/w16060900

Zuo, Y., Qin, M., Chen, C., Ye, W., Li, X., Luo, J., & Ong, S. P. (2021). Accelerating materials discovery with Bayesian optimization and graph deep learning. *Materials Today, 51*, 126–135. DOI: 10.1016/j.mattod.2021.08.012

Compilation of References

Abas, N., Kalair, A., & Khan, N. (2015). Review of fossil fuels and future energy technologies. *Futures*, *69*, 31–49. DOI: 10.1016/j.futures.2015.03.003

Abboud, M., Alnefaie, R., & Alhanash, A. (2022). Synthesis and application of unsupported and silica-supported Ni nanoparticles in catalysis. *Journal of Nanoparticle Research*, *24*(2), 21. DOI: 10.1007/s11051-021-05350-x

Abd Zaid, S. M., AbdulRazak, A., & Abid, M. (2022). A review of nano-catalyst applications in kerosene desulfurization techniques. *Journal of Applied Sciences and Nanotechnology*, *2*(2), 86–102. DOI: 10.53293/jasn.2022.4302.1094

Abdali, A., Mahmoudian, M., & Nozad, E. (2021). Desulfurization of a model fuel using pervaporation membranes containing Zn-MOFs. *Journal of Polymer Research*, *28*(7), 271. DOI: 10.1007/s10965-021-02472-7

Abdu, A., Eljugmani, A. S., Alghadamsi, A., Ejdedhom, H., & Alowayb, R. (2024). Artificial neural network modelling techniques employed to study desulfurization of Zawia refinery diesel. *Proceedings of the 10th Jordan International Chemical Engineering Conference (JIChEC10).*

Abdullah, N., Zumra, N., & Ahmed, S. (2025). Recent innovations in nanocomposite beads for the removal of pollutants from water: A critical review. *Environmental Science. Water Research & Technology*, *11*(6), 1428–1445. DOI: 10.1039/D5EW00215J

Abdullah, T. A., Rasheed, R. T., Adnan, M., Qasim, B., Sajet, F., Mansoor, H. S., Talib, H., Kadhim, M. S., Kułacz, K., Meharban, F., & Abdul-Zahra, Z. (2024). A comprehensive review of the application of modified carbon nanotubes (CNTs) for the removal of metals from wastewater. *Materials Research Express*, *10*(12), 122005. DOI: 10.1088/2053-1591/ad1770

Abolore, R. S., Jaiswal, S., & Jaiswal, A. K. (2024). Green and sustainable pretreatment methods for cellulose extraction from lignocellulosic biomass and its applications: A review. *Carbohydrate Polymer Technologies and Applications*, *7*, 100396. DOI: 10.1016/j.carpta.2023.100396

Abou Hammad, A. B., Mansour, A. M., Elhelali, T. M., & El Nahrawy, A. M. (2022). Sol-Gel/Gel Casting Nanoarchitectonics of Hybrid Fe2O3–ZnO/PS-PEG Nanocomposites and Their Optomagnetic Properties. *Journal of Inorganic and Organometallic Polymers and Materials 2022 33:2*, *33*(2), 544–554. DOI: 10.1007/s10904-022-02519-2

Aceto, M., Calà, E., Gulino, F., Gullo, F., Labate, M., Agostino, A., & Picollo, M. (2022). The Use of UV-Visible Diffuse Reflectance Spectrophotometry for a Fast, Preliminary Authentication of Gemstones. *Molecules (Basel, Switzerland)*, *27*(15), 4716. DOI: 10.3390/molecules27154716 PMID: 35897892

Adams, Z. K. (2024). *Recycling of Rare Earth Magnets With Sulfur Based Chemistries and High Temperature Processing*. Massachusetts Institute of Technology.

Adassooriya, N. M., Manipura, A., & Husen, A. (2025). Industrial Scale-Up Production of Nanoparticles—Challenges and Opportunities. In Adassooriya, N. M., Manipura, A., & Husen, A. (Eds.), *Industrial Scale Production of Nanoparticles. Smart Nanomaterials Technology* (pp. 1–7). Springer., DOI: 10.1007/978-981-96-9176-0_1

Advani, S. G., & Hsiao, K. T. (2012). Introduction to composites and manufacturing processes. *Manufacturing Techniques for Polymer Matrix Composites (PMCs): A Volume in Woodhead Publishing Series in Composites Science and Engineering*, 1–12. DOI: 10.1533/9780857096258.1.1

Afolabi, O. A., & Ndou, N. (2024). Synergy of hybrid fillers for emerging composite and nanocomposite materials—A review. *Polymers*, *16*(13), 1907. DOI: 10.3390/polym16131907 PMID: 39000762

Afzal, S., Singh, A., Nicholson, S. R., Uekert, T., DesVeaux, J. S., Tan, E. C., Dutta, A., Carpenter, A. C., Baldwin, R. M., & Beckham, G. T. (2023). Techno-economic analysis and life cycle assessment of mixed plastic waste gasification for production of methanol and hydrogen. *Green Chemistry*, *25*(13), 5068–5085. DOI: 10.1039/D3GC00679D

Agbadaola, M. T., Adeojo, O. D., Michael-Agbadaola, O., & Babalola, J. O. (2025). Facile fabrication of a CuO/NiO-biomass heterostructure micro-/nano-composite for the adsorptive removal of methylene blue from wastewater. *Separation Science and Technology*, *60*(13), 1659–1672. DOI: 10.1080/01496395.2025.2514954

Aggarwal, S. K. (1998). A review of spray ignition phenomena: Present status and future research. *Progress in Energy and Combustion Science*, *24*(6), 565–600. DOI: 10.1016/S0360-1285(98)00016-1

Agrawal, A., & Choudhary, A. (2016). Perspective: Materials informatics and big data: Realisation of the "fourth paradigm" of science in materials science. *APL Materials*, *4*(5), 053208. DOI: 10.1063/1.4946894

Ahmad, W., & Ahmad, I. (2017). Mechanisms of desulfurization by nanomaterials. In *Nanotechnology in Oil and Gas Industries: Principles and Applications* (pp. 211-243). Springer.

Ahmed, H. R., & Karim, A. R. (2025). Structural insights into metal oxide nanocomposite graphene frameworks as catalysts for oxidative desulfurization of light gasoil. *Journal of Molecular Structure*, *1325*, 141081. DOI: 10.1016/j.molstruc.2024.141081

Ajien, A., Idris, J., Md Sofwan, N., Husen, R., & Seli, H. (2023). Coconut shell and husk biochar: A review of production and activation technology, economic, financial aspect and application. *Waste Management & Research*, *41*(1), 37–51. DOI: 10.1177/0734242X221127167 PMID: 36346183

Akhtar, M. S., Naseem, M. T., Ali, S., & Zaman, W. (2025). Metal-Based Catalysts in Biomass Transformation: From Plant Feedstocks to Renewable Fuels and Chemicals. *Catalysts*, *15*(1), 40. DOI: 10.3390/catal15010040

Akopyan, A., Polikarpova, P., Anisimov, A., Lysenko, S., Maslova, O., Sen'ko, O., & Efremenko, E. (2021). Oxidation of dibenzothiophene with the subsequent bioconversion of sulfone. *Theoretical Foundations of Chemical Engineering*, *55*(4), 778–785. DOI: 10.1134/S0040579521040035

Akter, T., Islam, A., Al Miad, A., Alam, M. K., Ahmed, S., & Sahadat Hossain, M. (2025). Morphological Modification of Metal Oxide Nanomaterials Using Different Types of Organic Modifiers. *ChemistryOpen*, *14*(12), 202500342. DOI: 10.1002/open.202500342 PMID: 40944616

Al Janabi, M. A. Y., El Houda Tiri, R. N., Cherif, A., Altuner, E. E., Lee, C. J., Sen, F., Dragoi, E. N., Karimi, F., & Kalikeri, S. (2024). Hydrogen Generation by Methanolysis of NaBH4 via Efficient CuFe2O4 Nanoparticle Catalyst: A Kinetic Study and DNN Model. *Topics in Catalysis*, *67*(9–12), 843–852. DOI: 10.1007/s11244-024-01904-0

Alabdullah, M. A., Gomez, A. R., Vittenet, J., Bendjeriou-Sedjerari, A., Xu, W., Abba, I. A., & Gascon, J. (2020). A viewpoint on the refinery of the future: Catalyst and process challenges. *ACS Catalysis*, *10*(15), 8131–8140. DOI: 10.1021/acscatal.0c02209

Alardhi, S. M., Jabbar, N. M., Al-Jadir, T., Ibrahim, N. K., Dakhil, A. M., Al-Saedi, N. D., Al-Saedi, H. D., & Adnan, M. (2022). Artificial neural network model for predicting the desulfurization efficiency of AL-Ahdab crude oil. In. AIP Conference Proceedings: Vol. 2443. *Issue 1* (p. 030033). AIP Publishing., DOI: 10.1063/5.0091975

Al-Attas, T. A., Ali, S. A., Zahir, M. H., Xiong, Q., Al-Bogami, S. A., Malaibari, Z. O., Razzak, S. A., & Hossain, M. M. (2019). Recent advances in heavy oil upgrading using dispersed catalysts. *Energy & Fuels*, *33*(9), 7917–7949. DOI: 10.1021/acs.energyfuels.9b01532

Albdiry, M. (2024). Effect of melt blending processing on mechanical properties of polymer nanocomposites: A review. *Polymer Bulletin*, *81*(7), 5793–5821. DOI: 10.1007/s00289-023-05012-z

Ali, A., & Zhao, C. (2020). Direct liquefaction techniques on lignite coal: A review. *Chinese Journal of Catalysis*, *41*(3), 375–389. DOI: 10.1016/S1872-2067(19)63492-3

Ali, G., Mirza, M., & Mustafa, G. M. (2025). CNT–metal oxide composites as cathode materials for Li-ion batteries. In *Nanostructured Lithium-Ion Battery Materials* (pp. 319–339). Elsevier., DOI: 10.1016/B978-0-443-13338-1.00015-0

Alshammari, A., & Kalevaru, V. N. (2016). Supported gold nanoparticles as promising catalysts. Catalytic Application of Nano-Gold Catalysts, 57-81.

Altuner, E. E., Arıkan, K., Burhan, H., Ozdemir, S., & Şen, F. (2021). Commercial aspects of direct alcohol fuel cells. In Nanomaterials for Direct Alcohol Fuel Cells (pp. 511–524). https://doi.org/ DOI: 10.1016/B978-0-12-821713-9.00012-3

Altuner, E. E., Bekmezci, M., & Sen, F. (2022). Manufacturing Techniques of Magnetic Polymer Nanocomposites. *Handbook of Magnetic Hybrid Nanoalloys and Their Nanocomposites*, 1–16. DOI: 10.1007/978-3-030-34007-0_12-1

Altuner, E. E., Bekmezci, M., & Sen, F. (2022). Manufacturing techniques of magnetic polymer nanocomposites. In *Handbook of Magnetic Hybrid Nanoalloys and Their Nanocomposites* (pp. 303–318). DOI: 10.1007/978-3-030-90948-2_12

Altuner, E. E., Bekmezci, M., Bayat, R., Akin, M., Isik, I., & Şen, F. (2021). Dendrimer-based nanocomposites for alcohol fuel cells. *Nanomaterials for Direct Alcohol Fuel Cells*, 337–352. DOI: 10.1016/B978-0-12-821713-9.00007-X

Altuner, E. E., Erduran, V., & Sen, F. (2023). Green synthesized nanomaterials for bioimaging. In Synthesis of Bionanomaterials for Biomedical Applications: Micro and Nano Technologies (pp. 265–286). Elsevier. https://doi.org/ DOI: 10.1016/B978-0-323-91195-5.00023-4

Altuner, E. E., Gur, T., & Şen, F. (2021). Ternary/quaternary nanomaterials for direct alcohol fuel cells. *Nanomaterials for Direct Alcohol Fuel Cells*, 157–172. DOI: 10.1016/B978-0-12-821713-9.00001-9

Altuner, E. E., Sen, F., & Dar, U. A. (2025). Commercialization of quinone-based drugs. *Quinone-Based Compounds in Drug Discovery: Trends and Applications*, 293–300. DOI: 10.1016/B978-0-443-24126-0.00017-2

Altuner, E. E. (2023). Electrochemical sensitivity, selectivity, and sensory properties of a natural polymer-supported titanium nanocomposite electrode towards lysine amino acid. *Muş Alparslan Üniversitesi Fen Bilimleri Dergisi*, *11*(1), 28–32. DOI: 10.18586/msufbd.1268440

Altuner, E. E. (2025). Synthesis, characterisation and photocatalytic activation of zinc nanoparticles via. biogenic methods. *International Journal of Environmental Analytical Chemistry*, *105*(19), 1–11. DOI: 10.1080/03067319.2025.2451356

Altuner, E. E., Akin, M., Bayat, R., Bekmezci, M., Burhan, H., & Sen, F. (2022). Challenges in commercialization of carbon nanomaterial-based sensors. In *Carbon Nanomaterials-Based Sensors* (pp. 381–392). Emerging Research Trends in Devices and Applications., DOI: 10.1016/B978-0-323-91174-0.00020-2

Altuner, E. E., Gulbagca, F., Tiri, R. N. E., Aygun, A., & Sen, F. (2023). Highly efficient palladium–zinc oxide nanoparticles synthesized by biogenic methods: Characterization, hydrogen production and photocatalytic activities. *Chemical Engineering Journal Advances*, *14*, 100465. DOI: 10.1016/j.ceja.2023.100465

Altuner, E. E., Ozalp, V. C., Yilmaz, M. D., Bekmezci, M., & Sen, F. (2022). High-efficiency application of CTS-Co NPs mimicking peroxidase enzyme on TMB(ox). *Chemosphere*, *292*, 133429. DOI: 10.1016/j.chemosphere.2021.133429

Altuner, E. E., Ozalp, V. C., Yilmaz, M. D., Sudagidan, M., Aygun, A., Acar, E. E., Tasbasi, B. B., & Sen, F. (2022). Development of electrochemical aptasensors detecting phosphate ions on TMB substrate with epoxy-based mesoporous silica nanoparticles. *Chemosphere*, *134077*, 134077. Advance online publication. DOI: 10.1016/j.chemosphere.2022.134077

Altuner, E. E., Tatli, H. N., Aldoori, B., & Dar, U. A. (2025). Waste Dye Removal with Nanotechnological Methods: Photocatalytic Processes. *Nanotechnology in the Life Sciences*, (Part F525), 249–270. DOI: 10.1007/978-3-031-82794-5_11

Alvarez, M. (2023). Challenges in scaling up nanomaterial synthesis for industry: Technical, economic, and environmental perspectives. American Journal of Materials Engineering, 4(3), 12–18. E-ISSN: 2688-1780.

Álvarez-Chimal, R., & Arenas-Alatorre, J. A. (2023). Green synthesis and characterization of nanoparticles. In Kumar, P. (Ed.), *Green Nanotechnology—Overview and Further Prospects* (pp. 1–27). IntechOpen., DOI: 10.5772/intechopen.110751

Alviar, K. A. T., Arbilo, A. N. V., Correa, S. S., & Choi, A. E. S. (2024). Advanced desulfurization via ultrasonic-assisted oxidative methods in gasoline and crude oil towards sustainability. *Chemical Engineering Transactions*, *114*, 55–60. DOI: 10.3303/CET24114010

Ameen, F., Altuner, E. E., Tiri, R. N. E., Gulbagca, F., Aygun, A., Sen, F., Majrashi, N., Orfali, R., & Dragoi, E. N. (2023). Highly active iron(II) oxide–zinc oxide nanocomposite synthesized using *Thymus vulgaris* plant as bioreduction catalyst: Characterization, hydrogen evolution and photocatalytic degradation. *International Journal of Hydrogen Energy*, *48*(55), 21139–21151. DOI: 10.1016/j.ijhydene.2022.11.229

Ameta, S. C. (2022). Nanomaterials: An Introduction. In The Science of Nanomaterials (pp. 1-18): Apple Academic Press.

Amthiou, H., Arioua, M., & Benbarrad, T. (2023). Digital twins in Industry 4.0: A literature review. *ITM Web of Conferences, 52*, 01002. DOI: 10.1051/itmconf/20235201002

Amulya, K., Morris, S., & Lens, P. N. (2023). Aquatic biomass as sustainable feedstock for biorefineries. *Biofuels, Bioproducts & Biorefining*, *17*(4), 1012–1029. DOI: 10.1002/bbb.2471

Anastas, P. T., & Warner, J. C. (1998). Green chemistry: Theory and practice. *Green Chemistry*, *1*(1), 1–12. DOI: 10.1039/A902355P

Ancheyta, J., Sánchez, S., & Rodríguez, M. A. (2005). Kinetic modelling of hydrocracking of heavy oil fractions: A review. *Catalysis Today*, *109*(1–4), 76–92. DOI: 10.1016/j.cattod.2005.08.015

Andreani, T., Nogueira, V., Gavina, A., Fernandes, S., Rodrigues, J. L., Pinto, V. V., Ferreira, M. J., Silva, A. M., Pereira, C. M., & Pereira, R. (2020). Ecotoxicity to freshwater organisms and cytotoxicity of nanomaterials: Are we generating sufficient data for their risk assessment? *Nanomaterials (Basel, Switzerland)*, *11*(1), 66. DOI: 10.3390/nano11010066 PMID: 33396620

Anekwe, I. M. S., & Isa, Y. M. (2025). Unlocking catalytic longevity: A critical review of catalyst deactivation pathways and regeneration technologies. *Energy Advances*, *4*(9), 1075–1113. DOI: 10.1039/D5YA00015G

Anshassi, M., & Townsend, T. G. (2021). Reviewing the underlying assumptions in waste LCA models to identify impacts on waste management decision making. *Journal of Cleaner Production*, *313*, 127913. DOI: 10.1016/j.jclepro.2021.127913

Areti, H. A., Hamda, A. S., Abo, L. D., Jabesa, A., & Muleta, M. D. (2025). Biowastes as sustainable catalysts for water treatment: A comprehensive overview. *Environmental Advances*, *19*, 100610. DOI: 10.1016/j.envadv.2025.100610

Arevalo-Basañez, E., Jiménez-García, G., Villalón-López, U. A., & Maya-Yescas, R. (2025). The HDS Process: Origin, Process Evolution, Reaction Mechanisms, Process Units, Catalysts, and Health Risks. *Processes (Basel, Switzerland)*, *13*(9), 2817. DOI: 10.3390/pr13092817

Armaroli, N., & Balzani, V. (2011a). The legacy of fossil fuels. *Chemistry, an Asian Journal*, *6*(3), 768–784. DOI: 10.1002/asia.201000797

Arvidsson, R., Molander, S., & Sandén, B. A. (2014). Review of life cycle assessment of nanomaterials. *Journal of Cleaner Production*, *63*, 1–12. DOI: 10.1016/j.jclepro.2013.09.038

Aschberger, K., Johnston, H. J., Stone, V., Aitken, R. J., Hankin, S. M., Peters, S. A. K., & Tran, C. L. (2010). Review of carbon nanotubes' toxicity and exposure. *Toxicology and Applied Pharmacology*, *243*(3), 316–325. DOI: 10.1016/j.taap.2009.11.020 PMID: 20860524

Ashok, B., Denis Ashok, S., & Ramesh Kumar, C. (2015). LPG diesel dual fuel engine – A critical review. *Alexandria Engineering Journal*, *54*(2), 105–126. DOI: 10.1016/j.aej.2015.03.002

Asif, M., & Kumar, A. (2022). Resistive switching in emerging materials and their characteristics for neuromorphic computing. *Materials Today Electronics*, *1*, 100004. DOI: 10.1016/j.mtelec.2022.100004

Aspuru-Guzik, A., Lindh, R., & Reiher, M. (2018). The matter of quantum chemistry. *ACS Central Science*, *4*(2), 144–152. DOI: 10.1021/acscentsci.7b00550 PMID: 29532014

ASTM International. (2019). *ASTM D5453–19: Standard test method for determination of total sulfur in light hydrocarbons, motor fuels, and oils by ultraviolet fluorescence.*

Ates, B., Koytepe, S., Ulu, A., Gurses, C., & Thakur, V. K. (2020). Chemistry, structures, and advanced applications of nanocomposites from biorenewable resources. *Chemical Reviews*, *120*(17), 9304–9362. DOI: 10.1021/acs.chemrev.9b00553 PMID: 32786427

Attaran, M., & Celik, B. (2023). Digital twin: Benefits, use cases, challenges, and opportunities. *Decision Analytics Journal*, *6*, 100165. DOI: 10.1016/j.dajour.2023.100165

Ausmeel, S., Eriksson, A., Ahlberg, E., Sporre, M. K., Spanne, M., & Kristensson, A. (2020). Ship plumes in the Baltic Sea Sulfur Emission Control Area: Chemical characterisation and contribution to coastal aerosol concentrations. *Atmospheric Chemistry and Physics*, *20*(15), 9135–9151. DOI: 10.5194/acp-20-9135-2020

Avtar, R., Tripathi, S., Aggarwal, A. K., & Kumar, P. (2019). Population–Urbanization–Energy Nexus: A Review. *Resources 2019, Vol. 8, Page 136, 8*(3), 136. DOI: 10.3390/resources8030136

Azeez, M. O., & Ganiyu, S. A. (2023). Review of biomass derived-activated carbon for production of clean fuels by adsorptive desulfurization: Insights into processes, modifications, properties, and performances. *Arabian Journal of Chemistry*, *16*(10), 105182. DOI: 10.1016/j.arabjc.2023.105182

Aziz, A. A. A., Yong, K. S., Ibrahim, S., & Pichiah, S. (2012). Enhanced magnetic separation and photocatalytic activity of nitrogen doped titania photocatalyst supported on strontium ferrite. *Journal of Hazardous Materials*, *199–200*, 1–9. DOI: 10.1016/j.jhazmat.2011.10.069

Babich, I. V., & Moulijn, J. A. (2003). Science and technology of novel processes for deep desulfurization of oil refinery streams: A review. *Fuel*, *82*(6), 607–631. DOI: 10.1016/S0016-2361(02)00324-1

Bagchi, A., & Srivastava, P. (2024). Genetic and metabolic engineering approaches for enhanced biodesulfurization. *Frontiers in Bioengineering and Biotechnology*, *12*, 1482270. DOI: 10.3389/fbioe.2024.1482270 PMID: 39530058

Bahari, N., Hashim, N., Abdan, K., Md Akim, A., Maringgal, B., & Al-Shdifat, L. (2023). Honey as reducing/capping agent in green nanoparticle synthesis. *Nanomaterials (Basel, Switzerland)*, *13*, 1244. DOI: 10.3390/nano13071244 PMID: 37049336

Bajjou, O., Bakour, A., Khenfouch, M., Baitoul, M., Mothudi, B., Maaza, M., & Faulques, E. (2018). Charge Carrier Dynamics and pH Effect on Optical Properties of Anionic and Cationic Porphyrin–Graphene Oxide Composites. *Journal of Electronic Materials*, *47*(5), 2897–2904. DOI: 10.1007/s11664-018-6139-3

Balusamy, S. R., Joshi, A. S., Perumalsamy, H., Mijakovic, I., & Singh, P. (2023). Advancing sustainable agriculture: A critical review of smart and eco-friendly nanomaterial applications. *Journal of Nanobiotechnology*, *21*(1), 372. DOI: 10.1186/s12951-023-02135-3 PMID: 37821961

Baptista, A., Silva, F., Porteiro, J., Míguez, J., & Pinto, G. (2018). Sputtering Physical Vapour Deposition (PVD) Coatings: A Critical Review on Process Improvement and Market Trend Demands. *Coatings 2018, Vol. 8, Page 402*, *8*(11), 402. DOI: 10.3390/coatings8110402

Barbhuiya, S., Das, B. B., Kapoor, K., Das, A., & Katare, V. (2025). Mechanical performance of bio-based materials in structural applications: A comprehensive review. *Structures (Oxford)*, *75*, 108726. DOI: 10.1016/j.istruc.2025.108726

Bartholomew, C. H. (2001). Mechanisms of catalyst deactivation. *Applied Catalysis A, General*, *212*(1–2), 17–60. DOI: 10.1016/S0926-860X(00)00843-7

Bartoli, M., Giorcelli, M., & Tagliaferro, A. (2023). A comprehensive overview on biochar-based materials for catalytic applications. *Catalysts*, *13*(10), 1336. DOI: 10.3390/catal13101336

Batchu, S. P., Hernandez, B., Malhotra, A., Fang, H., Ierapetritou, M., & Vlachos, D. G. (2022). Accelerating manufacturing for biomass conversion via integrated process and bench digitalization: A perspective. *Reaction Chemistry & Engineering*, *7*(4), 813–832. DOI: 10.1039/D1RE00560J

Batool, Z., Rehman, A. U., Sindhu, K., Hassan, R. U., Almufarij, R. S., Arayro, J., Jumanazarov, D., Abduvokhidov, A., Ashraf, G. A., & Jabeen, R. (2025). Structure–performance correlation in NiS–CNT hybrid electrodes for advanced electrochemical energy storage. *Journal of Electroceramics*, •••, 1–17. DOI: 10.1007/s10832-025-00445-6

Bazyari, A., Khodadadi, A. A., Haghighat Mamaghani, A., Beheshtian, J., Thompson, L. T., & Mortazavi, Y. (2016). Microporous titania–silica nanocomposite catalyst-adsorbent for ultra-deep oxidative desulfurization. *Applied Catalysis B: Environmental*, *180*, 65–77. DOI: 10.1016/j.apcatb.2015.06.011

Behler, J. (2016). Perspective: Machine learning potentials for atomistic simulations. *The Journal of Chemical Physics*, *145*(17), 170901. DOI: 10.1063/1.4966192 PMID: 27825224

Behrooznia, Z., & Nourmohammadi, J. (2024). Polysaccharide-based materials as an eco-friendly alternative in biomedical, environmental, and food packaging. *Giant (Oxford, England)*, *19*, 100301. DOI: 10.1016/j.giant.2024.100301

Bellanthudawa, B., Nawalage, N., Handapangoda, H., Suvendran, S., Wijayasenarathne, K., Rathnasuriya, M., Wickramasinghe, P. G. M. U., Aberathna, A. A. A. U., Tennakoon, A., & Perera, I. (2023). A perspective on biodegradable and non-biodegradable nanoparticles in industrial sectors: Applications, challenges, and future prospects. *Nanotechnology for Environmental Engineering*, *8*(4), 975–1013. DOI: 10.1007/s41204-023-00344-7

Beshtar, M., Ahmadi, S. M. M., Asgharinezhad, A. A., & Larimi, A. (2025). Catalysis and process parameters in photocatalytic oxidative desulfurization of liquid transportation fuels: An in-depth review. *Journal of Cleaner Production, 145462*. Advance online publication. DOI: 10.1016/j.jclepro.2025.145462

Bhagyaraj, S., & Krupa, I. (2023). Biopolymer-assisted synthesis of hybrid nanoflowers. *Inorganic Chemistry Communications*, *158*, 111644. DOI: 10.1016/j.inoche.2023.111644

Bhatia, T., & Sindhu, S. S. (2024). Sustainable management of organic agricultural wastes: Contributions in nutrients availability, pollution mitigation and crop production. *Discover Agriculture*, *2*(1), 130. DOI: 10.1007/s44279-024-00147-7

Bhushan, D., Shoran, S., Kumar, R., & Gupta, R. (2024). Plant biomass-based nanoparticles for remediation of contaminants from water ecosystems: Recent trends, challenges, and future perspectives. *Chemosphere*, *365*, 143340. DOI: 10.1016/j.chemosphere.2024.143340 PMID: 39278321

Bianchini Silva, M., Costa, U. O., Mattoso, L. H. C., Monteiro, S. N., de Souza, M. L., & Vitorazi, L. (2024). Magnetic Nanoparticles in Biopolymer Fibers: Fabrication Techniques and Characterization Methods. *Polymers*, *16*(19), 2805. DOI: 10.3390/polym16192805 PMID: 39408514

Bibri, S. E. (2022). The social shaping of the metaverse as an alternative to the imaginaries of data-driven smart Cities: A study in science, technology, and society. *Smart Cities*, *5*(3), 832–874. DOI: 10.3390/smartcities5030043

Bihari, P., Gróf, G., & Gács, I. (2010). Efficiency and cost modelling of thermal power plants. *Thermal Science*, *14*(3), 821–834. DOI: 10.2298/TSCI1003821B

Blumer, M., & Sass, J. (1972). Oil pollution: Persistence and degradation of spilled fuel oil. *Science*, *176*(4039), 1120–1122. DOI: 10.1126/science.176.4039.1120

Bo, L., Hu, Y., Zhang, Z., & Tong, J. (2019). Efficient photocatalytic degradation of Rhodamine B catalyzed by $SrFe_2O_4$/g-C_3N_4 composite under visible light. *Polyhedron*, *168*, 94–100. DOI: 10.1016/j.poly.2019.04.036

Boostani, F., Sharififard, H., Lashanizadegan, A., & Darvishi, P. (2024). Desulphurisation of liquid fuel using adsorption onto bio-based activated carbon modified with chitosan and Fe ions. *International Journal of Environmental Analytical Chemistry*, *104*(13), 2956–2979. DOI: 10.1080/03067319.2022.2076222

Borah, P., Kumar, M., & Devi, P. (2020). Types of inorganic pollutants: metals/metalloids, acids, and organic forms. *Inorganic Pollutants in Water*, 17–31. DOI: 10.1016/B978-0-12-818965-8.00002-0

Borj, B. P., Saeidi, S., As'adi, H., Naddaf, A., Fekri, H., & Sarkhoshkalat, M. (2025). Environmental life cycle assessment (LCA) of oxidative desulfurization of gas condensate in novel oxidation system. *Environmental Pollution*, *371*, 125958. DOI: 10.1016/j.envpol.2025.125958 PMID: 40024511

Boshagh, F., Rahmani, M., Rostami, K., & Yousefifar, M. (2021). Key factors affecting the development of oxidative desulfurization of liquid fuels: A critical review. *Energy & Fuels*, *36*(1), 98–132. DOI: 10.1021/acs.energyfuels.1c03396

Boshagh, F., Rahmani, M., & Zhu, W. (2022). Recent Advances and Challenges in Developing Technological Methods Assisting Oxidative Desulfurization of Liquid Fuels: A Review. *Energy & Fuels*, *36*(21), 12961–12985. DOI: 10.1021/acs.energyfuels.2c03218

Botin, A. A., Boldushevskii, R. E., Mozhaev, A. V., Ghambarian, M., Balar, M., Ghashghaee, M., & Nikulshin, P. A. (2023). Reactive adsorption desulfurization of model FCC gasoline on Ni-based adsorbents: Effect of active phase dispersion on activity and HDS/HYD selectivity. *Applied Catalysis B: Environmental*, *337*, 122946. DOI: 10.1016/j.apcatb.2023.122946

Bo, W., Sheng, H., & Wang, J. (2023). Orbital Polarization-Dependent Fragment Twist-Induced Intramolecular Electric-Field-Driven Charge Transfer. *Molecules (Basel, Switzerland)*, *28*(4), 1801. DOI: 10.3390/molecules28041801 PMID: 36838789

Bre, F., Giménez, J., & Fachinotti, V. (2017). Prediction of wind pressure coefficients on building surfaces using artificial neural networks. *Energy and Building*, *158*, 1429–1441. Advance online publication. DOI: 10.1016/j.enbuild.2017.11.045

Brook, R. D., Rajagopalan, S., Pope, C. A.III, Brook, J. R., Bhatnagar, A., Diez-Roux, A. V., Holguin, F., Hong, Y., Luepker, R. V., Mittleman, M. A., Peters, A., Siscovick, D., Smith, S. C.Jr, Whitsel, L., & Kaufman, J. D. (2010). Particulate matter air pollution and cardiovascular disease: An update to the scientific statement from the American Heart Association. *Circulation*, *121*(21), 2331–2378. DOI: 10.1161/CIR.0b013e3181dbece1 PMID: 20458016

Bumbrah, G. S., & Sharma, R. M. (2016). Raman spectroscopy – Basic principle, instrumentation and selected applications for the characterization of drugs of abuse. *Egyptian Journal of Forensic Sciences*, *6*(3), 209–215. DOI: 10.1016/j.ejfs.2015.06.001

Bunker, B., Dvorak, B., & Aly Hassan, A. (2023). Thermal regeneration of activated carbon used as an adsorbent for hydrogen sulfide (H2S). *Sustainability (Basel)*, *15*(8), 6435. DOI: 10.3390/su15086435

Burger, B., Maffettone, P. M., Gusev, V. V., Aitchison, C. M., Bai, Y., Wang, X., Li, X., Alston, B. M., Li, B., Clowes, R., Rankin, N., Harris, B., Sprick, R. S., & Cooper, A. I. (2020). A mobile robotic chemist. *Nature*, *583*(7815), 237–241. DOI: 10.1038/s41586-020-2442-2 PMID: 32641813

Burnham, A. K. (2017). Global Chemical Kinetics of Fossil Fuels: How to Model Maturation and Pyrolysis. *Global Chemical Kinetics of Fossil Fuels: How to Model Maturation and Pyrolysis*, 1–315. https://doi.org/DOI: 10.1007/978-3-319-49634-4/COVER

Busca, G. (2021). Production of gasolines and monocyclic aromatic hydrocarbons: From fossil raw materials to green processes. *Energies*, *14*(13), 4061. DOI: 10.3390/en14134061

Butler, K. T., Davies, D. W., Cartwright, H., Isayev, O., & Walsh, A. (2018). Machine learning for molecular and materials science. *Nature*, *559*(7715), 547–555. DOI: 10.1038/s41586-018-0337-2 PMID: 30046072

Cai, G., Yan, P., Zhang, L., Zhou, H.-C., & Jiang, H.-L. (2021). Metal–organic framework-based hierarchically porous materials: Synthesis and applications. *Chemical Reviews, 121*(20), 12278–12326. DOI: 10.1021/acs.chemrev.1c00243 PMID: 34280313

Campos-Martin, J. M., Capel-Sanchez, M. C., Perez, A., & Fierro, J. L. G. (2010). Oxidative desulfurization of diesel fuels. *Catalysis Today, 158*(1–2), 62–68. DOI: 10.1016/j.cattod.2010.05.005

Campos-Martin, J. M., Capel-Sanchez, M. C., Perez-Presas, P., & Fierro, J. L. G. (2010). Oxidative processes of desulphurization of liquid fuels. *Journal of Chemical Technology and Biotechnology, 85*(7), 879–890. DOI: 10.1002/jctb.2394

Candra, A., Ahmed, Y. W., & Kitaw, S. L.. (2024). Gold-decorated bacterial cellulose nanocomposites. *Journal of Drug Delivery Science and Technology, 94*, 105477. DOI: 10.1016/j.jddst.2024.105477

Castiello, C., Junghanns, P., Mergel, A., Jacob, C., Ducho, C., Valente, S., Rotili, D., Fioravanti, R., Zwergel, C., & Mai, A. (2023). GreenMedChem: The challenge in the next decade toward eco-friendly compounds and processes in drug design. *Green Chemistry, 25*(6), 2109–2169. DOI: 10.1039/D2GC03772F

Chadha, M., & Berry, S. (2025). A comparative investigation of bio-waste derived renewable catalysts for sustainable bis-enol derivative synthesis: Transforming waste biomass into value-added products. *Research on Chemical Intermediates*, 1–21. DOI: 10.1007/s11164-025-05813-z

Challa, A. A., Saha, N., Zhivkova, T., Alexandrova, R., & Saha, P. (2025). Bacterial Cellulose/Graphene Oxide/Hydroxyapatite Biocomposite: A Scaffold from Sustainable Sources for Bone Tissue Engineering. *ACS Applied Materials & Interfaces, 17*(1), 572–582. DOI: 10.1021/acsami.4c17306 PMID: 39698804

Chang, A. F., Pashikanti, K., & Liu, Y. A. (2013). *Refinery engineering: Integrated process modelling and optimisation*. John Wiley & Sons., DOI: 10.1002/9783527666836

Chen, S., Liu, Z., Du, H., Tang, C., Ji, C. Y., Quan, B., Pan, R., Yang, L., Li, X., Gu, C., Zhang, X., Yao, Y., Li, J., Fang, N. X., & Li, J. (2021). Electromechanically reconfigurable optical nano-kirigami. *Nature Communications 2021 12:1, 12*(1), 1–8. DOI: 10.1038/s41467-021-21565-x

Chen, H., Cong, T. N., Yang, W., Tan, C., Li, Y., & Ding, Y. (2009). Progress in electrical energy storage system: A critical review. *Progress in Natural Science, 19*(3), 291–312. DOI: 10.1016/j.pnsc.2008.07.014

Chen, J., Ma, X., Liang, M., Guo, Z., Cai, Y., Zhu, C., Wang, Z., Wang, S., Xu, J., & Ying, H. (2024). *Physical–chemical–biological pretreatment for biomass degradation and industrial applications: A review*. Waste. DOI: 10.3390/waste2040024

Chen, L., Tian, Y., Hu, X., Yao, S., Lu, Z., Chen, S., Zhang, X., & Zhou, Z. (2022). A universal machine learning framework for electrocatalyst innovation: A case study of discovering alloys for hydrogen evolution reaction. *Advanced Functional Materials, 32*(47), 2208418. Advance online publication. DOI: 10.1002/adfm.202208418

Chen, N., Zeng, Y., Li, T., Cui, P., Dionysiou, D. D., Wang, X., Liu, C., Fang, G., Ding, C., Zhao, Y., Gao, J., Wang, Y., & Zhou, D. (2023). Phosphorus doping significantly enhanced the catalytic performance of cobalt-single-atom catalyst for peroxymonosulfate activation and contaminants degradation. *Journal of Hazardous Materials, 454*, 131480. DOI: 10.1016/j.jhazmat.2023.131480 PMID: 37146341

Chen, T. M., Gokhale, J., Shofer, S., & Kuschner, W. G. (2007). Outdoor Air Pollution: Nitrogen Dioxide, Sulfur Dioxide, and Carbon Monoxide Health Effects. *The American Journal of the Medical Sciences, 333*(4), 249–256. DOI: 10.1097/MAJ.0b013e31803b900f

Chen, Y., Wang, T., Tian, H., Su, D., Zhang, Q., & Wang, G. (2021). Advances in lithium–sulfur batteries: From academic research to commercial viability. *Advanced Materials, 33*(29), 2003666. DOI: 10.1002/adma.202003666 PMID: 34096100

Chen, Y., Wu, Q., Bu, N., Wang, J., & Song, Y. (2019). Magnetic recyclable lanthanum–nitrogen co-doped titania/strontium ferrite/diatomite heterojunction composite for enhanced visible-light-driven photocatalytic activity and recyclability. *Chemical Engineering Journal, 373*, 192–202. DOI: 10.1016/j.cej.2019.05.047

Cherubala, R. B. (2021). Desulphurization of diesel fuel using carbon-based metal oxide nanocomposites. Vaal University of Technology (South Africa), Chi, N. T. L., Anto, S., Ahamed, T. S., Kumar, S. S., Shanmugam, S., Samuel, M. S., . . . Pugazhendhi, A. (2021). A review on biochar production techniques and biochar based catalyst for biofuel production from algae. *Fuel, 287*, 119411.

Chew, K. J. (2014). The future of oil: Unconventional fossil fuels. *Philosophical Transactions of the Royal Society A: Mathematical, Physical and Engineering Sciences, 372*(2006). https://doi.org/DOI: 10.1098/RSTA.2012.0324/58963

Chomiak, J., Longwell, J. P., & Sarofim, A. F. (1989). Combustion of low calorific value gases; Problems and prospects. *Progress in Energy and Combustion Science, 15*(2), 109–129. DOI: 10.1016/0360-1285(89)90012-9

Chu, F., Wei, X., Lu, B., Zhao, G., Li, Y., Yang, K., & Lü, H. (2024). Flexible encapsulation of polyoxometalate into metal–organic frameworks for ultra-deep aerobic oxidation desulfurization of diesel via a biomimetic approach in deep eutectic solvents. *Chemical Engineering Journal, 481*, 148549. DOI: 10.1016/j.cej.2024.148549

Coley, C. W., Eyke, N. S., & Jensen, K. F. (2019). Autonomous discovery in the chemical sciences, Part I: Progress. *Angewandte Chemie International Edition, 59*(51), 22858–22893. DOI: 10.1002/anie.201909987 PMID: 31553511

Correa-Baena, J.-P., Hippalgaonkar, K., Duren, J., Jaffer, S., Chandrasekhar, V., Stevanovic, V., Wadia, C., Guha, S., & Buonassisi, T. (2018). Accelerating materials development via automation, machine learning, and high-performance computing. *Joule, 2*(8), 1410–1420. Advance online publication. DOI: 10.1016/j.joule.2018.05.009

Cui, Y., & Lieber, C. M. (2001). Functional Nanoscale Electronic Devices Assembled Using Silicon Nanowire Building Blocks. *Science, 291*(5505), 851–853. DOI: 10.1126/science.291.5505.851

Curtarolo, S., Hart, G. L., Nardelli, M. B., Mingo, N., Sanvito, S., & Levy, O. (2013). The high-throughput highway to computational materials design. *Nature Materials, 12*(3), 191–201. DOI: 10.1038/nmat3568 PMID: 23422720

D'Adamo, I., Ippolito, N. M., Shalchian, H., Innocenzi, V., Ferella, F., & Veglio, F. (2023). A feasibility study for a circular approach in oil refining: Metals recovery from hydrodesulphurization catalysts. *Sustainable Materials and Technologies, 36*, e00634. DOI: 10.1016/j.susmat.2023.e00634

da Silva Júnior, A. H., de Oliveira, C. R. S., Pellenz, L., Moraes, P. A. D., Marques, W. P., Mazur, L. P., Costa, T. G., Horn, A.Jr, Guelli Ulson de Souza, S. M. A., de Souza, A. A. U., da Silva, L., & da Silva, A. (2024). Perovskite-type strontium ferrite-based catalyst: Characterization and antibiotic degradation approach. *Process Safety and Environmental Protection, 187*, 1403–1421. DOI: 10.1016/j.psep.2024.05.048

Dahlia, S., Singh, B., Singh, N., & Rai, A. K. (2021). Green nanotechnology: A sustainable approach in energy and environmental applications. *Environmental Nanotechnology, Monitoring & Management, 16*, 100504. DOI: 10.1016/j.enmm.2021.100504

Daligaux, V., Richard, R., & Manero, M. H. (2021). Deactivation and regeneration of zeolite catalysts used in pyrolysis of plastic wastes—A process and analytical review. *Catalysts, 11*(7), 770. DOI: 10.3390/catal11070770

Damian, C. S., Devarajan, Y., J, R., & T, R. (2025). Nanocatalysts in Biodiesel Production: Advancements, Challenges, and Sustainable Solutions. *ChemBioEng Reviews*, *12*(2), e202400055. DOI: 10.1002/cben.202400055

Danks, A. E., Hall, S. R., & Schnepp, Z. (2016). The evolution of 'sol–gel' chemistry as a technique for materials synthesis. *Materials Horizons*, *3*(2), 91–112. DOI: 10.1039/C5MH00260E

Darabi, R., Gu, Q., Mehrizi, A. A., Altuner, E. E., Alhrasishawi, W., Gulbagca, F., Tiri, R. N. E., Tekeli, Y., Seyrankaya, A., Kaynak, I., & Sen, F. (2023). Synthesis, characterization, and enhanced hydrogen generation from NaBH4 methanolysis of highly dispersed bimetallic PdNi nanoparticles supported on Vulcan carbon. *Molecular Catalysis*, *547*, 113245. DOI: 10.1016/j.mcat.2023.113245

Dave, V., Sharma, P., Kuila, A., & Sahay, S. S. (2024). Revolutionizing environmental sustainability: Exploring the transformative potential of nanocomposites/nanomaterials from recycled waste. *Environmental Science and Pollution Research International*, *31*(29), 41621–41623. DOI: 10.1007/s11356-024-33435-2 PMID: 38671270

De Clippel, F., Dusselier, M., Van De Vyver, S., Peng, L., Jacobs, P. A., & Sels, B. F. (2013). Tailoring nanohybrids and nanocomposites for catalytic applications. *Green Chemistry*, *15*(6), 1398–1430. DOI: 10.1039/c3gc37141g

de Conti, M. C. M. D., Dey, S., Pottker, W. E., & La Porta, F. A. (2023). An overview into advantages and applications of conventional and unconventional hydro(solvo) thermal approaches for novel advanced materials design. *Materials Today Sustainability*, *23*, 100458. DOI: 10.1016/j.mtsust.2023.100458

Deb, K., Pratap, A., Agarwal, S., & Meyarivan, T. (2002). A fast and elitist multi-objective genetic algorithm: NSGA-II. *IEEE Transactions on Evolutionary Computation*, *6*(2), 182–197. DOI: 10.1109/4235.996017

Dembaremba, T. O., Majodina, S., Walmsley, R. S., Ogunlaja, A. S., & Tshentu, Z. R. (2022). Perspectives on strategies for improving ultra-deep desulfurization of liquid fuels through hydrotreatment: Catalyst improvement and feedstock pre-treatment. *Frontiers in Chemistry*, *10*, 807225. DOI: 10.3389/fchem.2022.807225 PMID: 35936099

Demirbas, A. (2004). Combustion characteristics of different biomass fuels. *Progress in Energy and Combustion Science*, *30*(2), 219–230. DOI: 10.1016/j.pecs.2003.10.004

Demirbaş, A., & Demirbaş, A. H. (2004). Estimating the calorific values of lignocellulosic fuels. *Energy Exploration & Exploitation, 22*(2), 135–143. DOI: 10.1260/0144598041475198

Deng, C., Xu, L., Hu, K., Chen, X., Gao, R., Zhang, L., Wang, L., & Zhang, C. (2023). Research Advances on Nitrogen-Doped Carbon Materials in COx Hydrogenation. *Atmosphere (Basel), 14*(10), 1510. DOI: 10.3390/atmos14101510

Deng, J., Xiong, T., Xu, F., Li, M., Han, C., Gong, Y., Wang, H., & Wang, Y. (2015). Inspired by bread leavening: One-step synthesis of hierarchically porous carbons. *Green Chemistry, 17*(7), 4053–4060. DOI: 10.1039/C5GC00523J

Deshmukh, D. S., Singh, S., Badgujar, K. C., Humne, V. T., Korpe, G. V., & Bhanage, B. M. (2025). Green innovations in C–H bond functionalisation: Exploring homogeneous recyclable catalytic systems. *Green Chemistry, 27*(20), 5667–5708. DOI: 10.1039/D5GC00278H

Dessie, Y., Tilahun, E., & Wondimu, T. H. (2024). Functionalized carbon electrocatalysts in energy conversion and storage applications: A review. *Heliyon, 10*(20), e39395. DOI: 10.1016/j.heliyon.2024.e39395 PMID: 39492918

Dey, N., Bhardwaj, S., & Maji, P. K. (2025). Recent breakthroughs in the valorization of lignocellulosic biomass for advancements in the construction industry: A Review. *RSC Sustainability, 3*(8), 3307–3357. DOI: 10.1039/D5SU00142K

Dhaka, A., Mali, S. C., Sharma, S., & Trivedi, R. (2023). A review on biological synthesis of silver nanoparticles and their potential applications. *Results in Chemistry, 6*, 101108. DOI: 10.1016/j.rechem.2023.101108

Dhanyashree, S., Jahnavi, H., Harini, H., Prasad, S. R., Nagaswarupa, H., & Naik, R. (2024). Design and development of novel nanoferrite and metal oxide composites for energy applications: a review. Multifunctional Inorganic Nanomaterials for Energy Applications, 96-119.

Di Carmine, G., Abbott, A. P., & D'agostino, C. (2021). Deep eutectic solvents: Alternative reaction media for organic oxidation reactions. *Reaction Chemistry & Engineering, 6*(4), 582–598. DOI: 10.1039/D0RE00458H

Dickson, U. J., & Huang, C. (2025). 2050 UK shipping emissions of NOx, SOx, PM10, and PM2.5: Is zero pollution by 2050 feasible? *Marine Pollution Bulletin, 220*, 118371. DOI: 10.1016/j.marpolbul.2025.118371 PMID: 40644899

Dieterle, F. J. (2003). Multianalyte quantifications by means of integration of artificial neural networks, genetic algorithms and chemometrics for time-resolved analytical data.

Dignon, J. (1992). NOx and SOx emissions from fossil fuels: A global distribution. *Atmospheric Environment. Part A, General Topics*, *26*(6), 1157–1163. DOI: 10.1016/0960-1686(92)90047-O

Dillon, A. C. (2010). Carbon Nanotubes for Photoconversion and Electrical Energy Storage. *Chemical Reviews*, *110*(11), 6856–6872. DOI: 10.1021/cr9003314

Ding, Y., Shang, H., Yang, C., Zhao, L., & Duan, A. (2024). Identifying efficient and inexpensive hydrodesulfurization catalysts through machine-learning-assisted analysis of metal–sulfur bonds in transition metal sulfides. *Chemical Engineering Science*, *298*, 120337. DOI: 10.1016/j.ces.2024.120337

Doner, N., Burhan, H., Bayat, R., Esra Altuner, E., & Sen, F. (2023). Energy Generation through Thermopower Waves using Multi-walled Carbon Nanotube Yarn. *Fuel*, *331*, 125906. DOI: 10.1016/j.fuel.2022.125906

Dong, W., Xing, J., Chen, Q., Huang, Y., Wu, M., Yi, P., Pan, B., & Xing, B. (2024). Hydrogen bonds between the oxygen-containing functional groups of biochar and organic contaminants significantly enhance sorption affinity. *Chemical Engineering Journal*, *499*, 156654. DOI: 10.1016/j.cej.2024.156654

Đorđević, A. M., Milikić, J., Milanković, V., Bogdanović, D. B., Radinović, K., Kaninski, M. M., Relić, D., Stanković, D., & Šljukić, B. (2025). Efficient Removal of Nitrobenzene and Its Compounds by Coconut Shell-Derived Activated Carbon. *Processes (Basel, Switzerland)*, *13*(7), 2072. DOI: 10.3390/pr13072072

Duan, Y., Mehariya, S., Kumar, A., Singh, E., Yang, J., Kumar, S., Li, H., & Kumar Awasthi, M. (2021). Apple orchard waste recycling and valorization of valuable product-A review. *Bioengineered*, *12*(1), 476–495. DOI: 10.1080/21655979.2021.1872905 PMID: 33472503

Dzenis, Y. (2008). Structural nanocomposites. *Science*, *319*(5862), 419–420. DOI: 10.1126/science.1151434 PMID: 18218884

Eatough, D. J., Caka, F. M., & Farber, R. J. (1994). The Conversion of SO2 to Sulfate in the Atmosphere. *Israel Journal of Chemistry*, *34*(3–4), 301–314. DOI: 10.1002/ijch.199400034

El Kfoury, M.-T. (2022). Novel Process for Organic Peroxides Synthesis. Polytechnique Montréal, Eskandarinezhad, S., Khosravi, R., Amarzadeh, M., Mondal, P., & Magalhães Filho, F. J. C. (2021). Journal of Composites and Compounds. *Journal of Composites and Compounds*, *3*, 43–56.

Elanthamilan, E., Elizabeth, I. B., Wang, S.-F., & Lydia, I. S. (2023). Strontium hexaferrite microspheres: Synthesis, characterization and visible-light-driven photocatalytic activity towards the degradation of methylene blue dye. *Optical Materials, 137*, 113565. DOI: 10.1016/j.optmat.2023.113565

El-Fawal, E. M., El Naggar, A. M., El-Zahhar, A. A., Alghandi, M. M., Morshedy, A. S., El Sayed, H. A., & Mohammed, A. M. E. (2025). Biofuel production from waste residuals: Comprehensive insights into biomass conversion technologies and engineered biochar applications. *RSC Advances, 15*(15), 11942–11974. DOI: 10.1039/D5RA00857C PMID: 40264955

Elio, J., & Milcarek, R. J. (2022). Techno-economic analysis and case study of combined heat and power systems in a wastewater treatment plant. *Energy, 260*, 125106. DOI: 10.1016/j.energy.2022.125106

El-Khayatt, A. M., El Desouky, F. G., El-Khawas, E., & Azab, A. (2025). Nanoferrite-modified CuO/SnO_2 nanocomposites with tunable magnetic and electrical properties for prospective applications. *Inorganic Chemistry Communications, 180*, 115019. DOI: 10.1016/j.inoche.2025.115019

Environmental Protection Agency. (2023). *Integrated science assessment for particulate matter*. U.S. Environmental Protection Agency. https://www.epa.gov/

EPA. (2023). *National ambient air quality standards for sulfur oxides*. U.S. Environmental Protection Agency. https://www.epa.gov/sox-pollution

Epling, W. S., Campbell, L. E., Yezerets, A., Currier, N. W., & Parks, J. E.II. (2004). Overview of the fundamental reactions and degradation mechanisms of NOx storage/reduction catalysts. *Catalysis Reviews. Science and Engineering, 46*(2), 163–245. DOI: 10.1081/CR-200031932

Eskandari, N., Nabiyouni, G., Masoumi, S., & Ghanbari, D. (2019). Preparation of a new magnetic and photocatalyst $CoFe_2O_4$–$SrTiO_3$ perovskite nanocomposite for photodegradation of toxic dyes under short-time visible irradiation. *Composites. Part B, Engineering, 176*, 107343. DOI: 10.1016/j.compositesb.2019.107343

Esra Altuner, E., Yousif, M., & Al Janabi, A. (2025). Synthesis and Characterization of Iron-Cobalt@Activated Carbon Nanoparticles for Electrochemical Sensing of Dopamine Neurotransmitters. *Chemistry Africa, 2025*(3), 1–9. DOI: 10.1007/s42250-025-01197-0

European Commission. (2009). *Directive 2009/30/EC amending Directive 98/70/EC relating to the quality of petrol and diesel fuels*. Official Journal of the European Union.

European Parliament and Council. (2009). Directive 2009/30/EC of the European Parliament and of the Council of 23 April 2009 amending Directive 98/70/EC as regards the specification of petrol, diesel and gas-oil and introducing a mechanism to monitor and reduce greenhouse gas emissions. *Official Journal of the European Union, L, 140*(88). https://data.europa.eu/eli/dir/2009/30/oj

Fadel, M., Salvador Fernández, M., Pilati, V., Martínez-García, J. C., Alvarez-Alonso, P., & Rivas, M. (2025). A holistic view of magnetic nanoparticles for advanced water treatment and cleanup. *Journal of Physics. D, Applied Physics, 58*(40), 403001. DOI: 10.1088/1361-6463/ae01b2

Fahim, Y. A., Hasani, I. W., & Mahmoud Ragab, W. (2025). Promising biomedical applications using superparamagnetic nanoparticles. *European Journal of Medical Research, 30*(1), 441. DOI: 10.1186/s40001-025-02696-z PMID: 40452035

Fang, J., Wang, J., Gan, L. M., Ng, S. C., Ding, J., & Liu, X. (2000). Fine strontium ferrite powders from an ethanol-based microemulsion. *Journal of the American Ceramic Society, 83*(5), 1049–1055. DOI: 10.1111/j.1151-2916.2000.tb01329.x

Fard, N. E., Raeisi, I., Yousefipour, Z., & Mosavin, R. (2024). Review of metal sulfide-based MXene nanocomposites for environmental applications, gas sensing, energy storage, and photothermal therapy. *ACS Applied Nano Materials, 7*(6), 5698–5728. DOI: 10.1021/acsanm.3c05455

Feng, Q., Lin, Q., Gong, F., Sugita, S., & Shoya, M. (2004). Adsorption of lead and mercury by rice husk ash. *Journal of Colloid and Interface Science, 278*(1), 1–8. DOI: 10.1016/j.jcis.2004.05.030

Fernandes, S. C., Leo, P., Santos-Vieira, I., de Castro, B., Cunha-Silva, L., & Balula, S. S. (2024). Improving oxidative catalytic efficiency for fuels desulfurization using hybrid materials based on MOF-808@SBA-15. *ChemCatChem, 16*(16), e202400355. DOI: 10.1002/cctc.202400355

Ferreira, P., Neves, R. P., Miranda, F. P., Cunha, A. V., Havenith, R. W., Ramos, M. J., & Fernandes, P. A. (2024). DszA catalyzes C–S bond cleavage through N5–hydroperoxyl formation. *Journal of Chemical Information and Modeling, 64*(10), 4218–4230. DOI: 10.1021/acs.jcim.4c00301 PMID: 38684937

Ferry, J. G. (1995). CO dehydrogenase. *Annual Review of Microbiology, 49*(1), 305–334. https://go.gale.com/ps/i.do?p=AONE&sw=w&issn=00664227&v=2.1&it=r&id=GALE%7CA17715866&sid=googleScholar&linkaccess=fulltext. DOI: 10.1146/annurev.mi.49.100195.001513

Filková, I., & Mujumdar, A. S. (2020). Industrial Spray Drying Systems. *Handbook of Industrial Drying*, 263–307. DOI: 10.1201/9780429289774-9

Finkbeiner, M., Inaba, A., Tan, R., Christiansen, K., & Klüppel, H. J. (2010). The new international standards for life cycle assessment. *The International Journal of Life Cycle Assessment, 11*(2), 80–85. DOI: 10.1065/lca2006.02.002

Flagiello, D., Parisi, A., Lancia, A., & Di Natale, F. (2021). A Review on Gas-Liquid Mass Transfer Coefficients in Packed-Bed Columns. *ChemEngineering 2021, Vol. 5, Page 43, 5*(3), 43. DOI: 10.3390/chemengineering5030043

Fotiadis, K., Kostoglou, M., Baltzopoulou, P., Zaspalis, V., & Karagiannakis, G. (2024). Activated carbon modification for real diesel adsorptive deep desulfurization: Experiments and modeling. *Chemical Engineering Communications, 211*(9), 1319–1335. DOI: 10.1080/00986445.2024.2341265

Fowler, J., Kottwitz, M., Trask, N., & Dingreville, R. (2024). Beyond combinatorial materials science: The 100 prisoners problem. *Integrating Materials and Manufacturing Innovation, 13*(1), 83–91. Advance online publication. DOI: 10.1007/s40192-023-00330-6

Friedemann, A. J. (2021). *Life after Fossil Fuels. 81*. DOI: 10.1007/978-3-030-70335-6

Fu, J., & Turn, S. Q. (2025). Oxidation mechanism of sulfur-containing compounds and antioxidant depletion dynamics: Insights into interactions. *Fuel, 381*, 133341. DOI: 10.1016/j.fuel.2024.133341

Fulkerson, W., Judkins, R. R., & Sanghvi, M. K. (1990). Energy from Fossil Fuels. *JSTOR, 263*(3), 128–135. https://www.jstor.org/stable/24996937?seq=1

Furukawa, H., Cordova, K. E., O'Keeffe, M., & Yaghi, O. M. (2013). The chemistry and applications of metal–organic frameworks. *Science, 341*(6149), 1230444. DOI: 10.1126/science.1230444 PMID: 23990564

Gacem, M. A., & Abd-Elsalam, K. A. (2022). Strategies for scaling up of green-synthesized nanomaterials: Challenges and future trends. In *Green Synthesis of Silver Nanomaterials* (pp. 669–698). Elsevier., DOI: 10.1016/B978-0-12-824508-8.00008-3

Gallego-Ramírez, C., Chica, E., & Rubio-Clemente, A. (2023, September 4). Life Cycle Assessment of Raw and Fe-Modified Biochars: Contributing to Circular Economy. *Materials (Basel), 16*(17), 6059. DOI: 10.3390/ma16176059 PMID: 37687752

Gangwar, N., Gangwar, C., & Sarkar, J. (2024). A review on template-assisted approaches & self assembly of nanomaterials at liquid/liquid interface. *Heliyon, 10*(17), e36810. DOI: 10.1016/j.heliyon.2024.e36810 PMID: 39263084

Gao, H., Guo, C., & Jiang, X. (2020). Nanocomposite catalysts for oxidative desulphurization: Recent advances and perspectives. *Journal of Energy Chemistry, 46*, 1–14. DOI: 10.1016/j.jechem.2019.11.019

Gao, M., Wang, L., Yang, Y., Sun, Y., Zhao, X., & Wan, Y. (2023). Metal and Metal Oxide Supported on Ordered Mesoporous Carbon as Heterogeneous Catalysts. *ACS Catalysis, 13*(7), 4060–4090. DOI: 10.1021/acscatal.2c05894

Gao, Y., Lia, X., Huang, S., Zhao, N., & Zhao, J. (2025). Dicationic ionic liquid modified peroxophosphotungstate encapsulated in ordered macro-microporous 3DOM-ZIF-8 for ultra-deep desulfurization of simulated fuel. *Small, 2501667*(34), 2501667. Advance online publication. DOI: 10.1002/smll.202501667 PMID: 40590137

Gary, J. H., & Handwerk, G. E. (2019). *Petroleum refining: Technology and economics* (5th ed.). CRC Press.

George, D., Palanisamy, U. M., & Begum, K. M. S. (2023). Chitosan nanocomposite hydrogel carriers. *Canadian Journal of Chemical Engineering, 101*(7), 3937–3947. DOI: 10.1002/cjce.24782

Georgiou, O., Maksymenko, M., & Russo, S. (2025). Deep Insights and Advances in R&D Management. In *R&D Management and Technology Commercialization: Practical Concepts, Innovation, and Case Studies* (pp. 39-66). Springer. DOI: 10.1007/978-3-031-86789-7_5

Ghaffar, U., Habiba, U., Majeed, M. I., Nawaz, H., Rashid, N., Akhtar, N., & Hajab, H. (2024). Surface-enhanced Raman spectroscopy for real-time analysis of biodesulfurization of 4,6-DMDBT. *Analytical Letters, 57*(14), 2182–2199. DOI: 10.1080/00032719.2023.2288695

Ghamarpoor, R., Fallah, A., & Eghbali, T. (2024). Design of bifunctional sandwich-like Co@Si/Ox-MXene nanocomposite to increase the supercapacitor properties and removal of pollutants from wastewater. *Journal of Alloys and Compounds, 983*, 173920. DOI: 10.1016/j.jallcom.2024.173920

Ghanbari Ghozikali, M., Heibati, B., Naddafi, K., Kloog, I., Oliveri Conti, G., Polosa, R., & Ferrante, M. (2016). Evaluation of Chronic Obstructive Pulmonary Disease (COPD) attributed to atmospheric O3, NO2, and SO2 using Air Q Model (2011–2012 year). *Environmental Research, 144*, 99–105. DOI: 10.1016/j.envres.2015.10.030

Ghanbari Kermanshahi, M., & Bahrami, K. (2019). Fe 3 O 4 @BNPs@SiO 2 –SO 3 H as a highly chemoselective heterogeneous magnetic nanocatalyst for the oxidation of sulfides to sulfoxides or sulfones. *RSC Advances*, *9*(62), 36103–36112. DOI: 10.1039/C9RA06221A

Gheisari, M. (2022). Modelling of oxidative desulfurization process by artificial neural network. *EasyChair Preprint* No. 7943.

Ghotekar, S., Pagar, T., Pansambal, S., & Oza, R. (2020). A review on green synthesis of sulfur nanoparticles via plant extract, characterization and its applications. *Advanced Journal of Chemistry-Section B*, *2*(3), 128–143.

Gill, I., & Ballesteros, A. (2000). Bioencapsulation within synthetic polymers (Part 1): Sol-gel encapsulated biologicals. *Trends in Biotechnology*, *18*(7), 282–296. DOI: 10.1016/S0167-7799(00)01457-8

Glass, N. R. (1979). Environmental effects of increased coal utilization: Ecological effects of gaseous emissions from coal combustion. *Environmental Health Perspectives*, *33*, 249–272. DOI: 10.1289/ehp.7933249

Golab, E. G., Ghamarpoor, R., Kondori, F. J., Hosseini, S., & Al-Saedi, H. N. (2024). Synthesis of hydrophobic polymeric surfactant (polyacrylamide/zwitterionic) and its effect on enhanced oil recovery (EOR). *Chemical Physics Impact*, *9*, 100756. DOI: 10.1016/j.chphi.2024.100756

Gómez-Bombarelli, R., Wei, J. N., Duvenaud, D., Hernández-Lobato, J. M., Sánchez-Lengeling, B., Sheberla, D., Aguilera-Iparraguirre, J., Hirzel, T. D., Adams, R. P., & Aspuru-Guzik, A. (2018). Automatic chemical design using a data-driven continuous representation of molecules. *ACS Central Science*, *4*(2), 268–276. DOI: 10.1021/acscentsci.7b00572 PMID: 29532027

Gomez-Villalba, L. S., Salcines, C., & Fort, R.GOmez-Villalba. (2023). Application of inorganic nanomaterials in cultural heritage conservation, risk of toxicity, and preventive measures. *Nanomaterials (Basel, Switzerland)*, *13*(9), 1454. DOI: 10.3390/nano13091454 PMID: 37176999

Gonçalves, M. A., Mares, E. K. L., Zamian, J. R., da Rocha Filho, G. N., & da Conceição, L. R. V. (2021). Statistical optimization of biodiesel production from waste cooking oil using magnetic acid heterogeneous catalyst $MoO_3/SrFe_2O_4$. *Fuel*, *304*, 121417.

Gonçalves, M., Molina-Sabio, M., & Rodriguez-Reinoso, F. (2010). Modification of activated carbon hydrophobicity by pyrolysis of propene. *Journal of Analytical and Applied Pyrolysis*, *89*(1), 17–21. DOI: 10.1016/j.jaap.2010.04.009

Government of Canada. (2015, July 29). Regulations amending the sulphur in gasoline regulations. *Canada Gazette, 149*(15). https://gazette.gc.ca/rp-pr/p2/2015/2015-07-29/html/sor-dors187-eng.html

Government of India. (2016). *Bharat Stage (BS) VI Emission Standards—Automotive Industry Standards (AIS)*. Ministry of Road Transport & Highways.

Goyal, A., Meena, P. K., Shelare, S., & Kumar, S.GOyal. (2025). Advancing biofuel production through nanomaterials: A review of efficiency improvements, challenges, and future directions. *Nanocomposites, 11*(1), 151–172. DOI: 10.1080/20550324.2025.2518755

Grande, C., Araújo, J. D. P., Cavenati, S., Firpo, N., Basaldella, E., & Rodrigues, A. (2004). New π-complexation adsorbents for propane–propylene separation. *Langmuir, 20*(13), 5291–5297. DOI: 10.1021/la036400s PMID: 15986665

Greenberg, N., Carel, R. S., Derazne, E., Bibi, H., Shpriz, M., Tzur, D., & Portnov, B. A. (2016). Different effects of long-term exposures to SO2 and NO2 air pollutants on asthma severity in young adults. *Journal of Toxicology and Environmental Health. Part A., 79*(8), 342–351. DOI: 10.1080/15287394.2016.1153548

Grennfelt, P., Engleryd, A., Forsius, M., Hov, Ø., Rodhe, H., & Cowling, E. (2020). Acid rain and air pollution: 50 years of progress in environmental science and policy. *Ambio, 49*(4), 849–864. DOI: 10.1007/s13280-019-01244-4 PMID: 31542884

Grewal, J. K., Kaur, M., Sharma, R. K., Oliviera, A. C., Garg, V. K., & Sharma, V. K. (2022). Structural and photocatalytic studies on oxygen hyperstoichiometric titanium-substituted strontium ferrite nanoparticles. *Magnetochemistry (Basel, Switzerland), 8*(12), 120. DOI: 10.3390/magnetochemistry8100120

Guan, B., Jiang, H., Wei, Y., Liu, Z., Wu, X., Lin, H., & Huang, Z. (2021). Density functional theory research for atomic structure, properties prediction, and rational design of selective catalytic reduction catalysts: Current progresses and future perspectives. *Molecular Catalysis, 510*, 111704. DOI: 10.1016/j.mcat.2021.111704

Guelpa, E., & Verda, V. (2019). Thermal energy storage in district heating and cooling systems: A review. *Applied Energy, 252*, 113474. DOI: 10.1016/j.apenergy.2019.113474

Gu, H., Kong, Y., Huang, D., Wang, Y., Raghavan, V., & Wang, J. (2025). Scaling cultured meat: Challenges and solutions for affordable mass production. *Comprehensive Reviews in Food Science and Food Safety, 24*(4), e70221. DOI: 10.1111/1541-4337.70221 PMID: 40635127

Gundogdu, A., Bozbeyoglu, P., Imamoglu, M., Baltaci, C., Duran, C., & Bulut, V. N. (2022). Characterization of the adsorption mechanism of cadmium(II) and methylene blue upon corncobs activated carbon. *Analytical Letters*, *56*(3), 433–448. DOI: 10.1080/00032719.2022.2098310

Guo, X., Zhang, Q., Ding, X., Shen, Q., Wu, C., Zhang, L., & Yang, H. (2016). Synthesis and application of several sol–gel-derived materials via sol–gel process combining with other technologies: a review. *Journal of Sol-Gel Science and Technology 2015 79:2, 79*(2), 328–358. DOI: 10.1007/s10971-015-3935-6

Guo, H. Q., Xie, L. L., Wang, X. L., Liu, F. R., Wang, M. J., & Hu, R. S. (2014). Sulfur removal and release behaviors of sulfur-containing model compounds during pyrolysis under inert atmosphere by TG-MS connected with Py-GC. *Journal of Fuel Chemistry & Technology*, *42*(10), 1160–1166. DOI: 10.1016/S1872-5813(14)60047-5

Guosheng, L., Kunhong, L., Xiaohan, L., Xinrui, Y., Jiaxin, S., Rong, G., & Baojian, S. (2024). Acidity regulation of Fe-based catalysts and its effect on the selectivity of HDS reaction pathways. *Journal of Fuel Chemistry & Technology*, *52*(2), 234–247. DOI: 10.1016/S1872-5813(23)60389-5

Guo, Y., Xie, W., Li, H., Li, J., Hu, J., & Liu, H. (2022). Construction of hydrophobic channels on Cu (I)-MOF surface to improve selective adsorption desulfurization performance in presence of water. *Separation and Purification Technology*, *285*, 120287. DOI: 10.1016/j.seppur.2021.120287

Gupta, P. K. (2020). Solvents, Gasses, and Vapors. *Problem Solving Questions in Toxicology:*, 181–193. DOI: 10.1007/978-3-030-50409-0_13

Gupta, N., Mahur, B. K., Izrayeel, A. M. D., Ahuja, A., & Rastogi, V. K. (2022). Biomass conversion of agricultural waste residues for different applications: A comprehensive review. *Environmental Science and Pollution Research International*, *29*(49), 73622–73647. DOI: 10.1007/s11356-022-22802-6 PMID: 36071366

Habib, U., Ahmad, F., Awais, M., Naz, N., Aslam, M., Urooj, M., & Sajid, M. (2023). Sustainable catalysis: Navigating challenges and embracing opportunities for a greener future. *J. Chem. Environ*, *2*(2), 14–53. DOI: 10.56946/jce.v2i2.205

Hachhach, M., Bayou, S., El Kasmi, A., Saidi, M. Z., Akram, H., Hanafi, M., Achak, O., El Moujahid, C., & Chafik, T. (2025). Towards Sustainable Scaling-Up of Nanomaterials Fabrication: Current Situation, Challenges, and Future Perspectives. *Eng*, *6*(7), 149. DOI: 10.3390/eng6070149

Halvaeikhanekahdani, P., Zandi, S., Ahmad, Q., & Payravand, H. (2025). Smart magnetic nanoparticles under electromagnetic stimulation: A synergistic approach to targeted chemotherapy and hyperthermia. *Nanomedicine (London)*, *20*(16), 2143–2166. DOI: 10.1080/17435889.2025.2534324 PMID: 40685946

Hammer, B., & Nørskov, J. K. (1995). Electronic factors determining the reactivity of metal surfaces. *Surface Science*, *343*(3), 211–220. DOI: 10.1016/0039-6028(96)80007-0

Hamza, Z. A., Dawood, J. J., & Jabbar, M. A. (2024). Review of TiO_2 as desulfurization catalyst for petroleum. *Catalysts*, *14*(6), 381. DOI: 10.3390/catal14060381

Hanif, M. A., Ibrahim, N., & Abdul Jalil, A. (2020). Sulfur dioxide removal: An overview of regenerative flue gas desulfurization and factors affecting desulfurization capacity and sorbent regeneration. *Environmental Science and Pollution Research 2020 27:22*, *27*(22), 27515–27540. DOI: 10.1007/s11356-020-09191-4

Hanif, M. A., Ibrahim, N., Isa, K. M., Abdullah, T. A. T., & Jalil, A. A. (2021). Sulfur dioxide removal by mesoporous silica KCC-1 modified with low-coverage metal nitrates. *Materials Today: Proceedings*, *47*, 1323–1328. DOI: 10.1016/j.matpr.2021.02.807

Hannun, R. M., & Abdul Razzaq, A. H. (2022). Air Pollution Resulted from Coal, Oil and Gas Firing in Thermal Power Plants and Treatment: A Review. *IOP Conference Series. Earth and Environmental Science*, *1002*(1), 012008. DOI: 10.1088/1755-1315/1002/1/012008

Hao, L., Hurlock, M., Ding, G., & Zhang, Q. (2020). Metal–organic frameworks towards desulfurization of fuels. *Topics in Current Chemistry*, *378*(1), 17. DOI: 10.1007/s41061-020-0280-1 PMID: 31993845

Hedayati, K., Behesht-Ara, Z., & Ghanbari, D. (2017). Preparation and characterization of various morphologies of $SrFe_{12}O_{19}$ nanostructures: Investigation of magnetization and coercivity. *Journal of Materials Science Materials in Electronics*, *28*(1), 1–9. DOI: 10.1007/s10854-016-5486-1

He, G., He, H., He, H., & He, H. (2020). Water Promotes the Oxidation of SO2 by O2 over Carbonaceous Aerosols. *Environmental Science & Technology*, *54*(12), 7070–7077. DOI: 10.1021/acs.est.0c00021

He, H., Zhang, R., Zhang, P., Wang, P., Chen, N., Qian, B., Zhang, L., Yu, J., & Dai, B. (2023). Functional carbon from nature: Biomass-derived carbon materials and the recent progress of their applications. *Advancement of Science*, *10*(16), 2205557. DOI: 10.1002/advs.202205557 PMID: 36988448

Helgeson, N., Trueblood, D., Holden, B., Jack, J., & Griffin, G. (2010). *Diesel engine air emissions reduction technologies* (No. NFESCTR2328ENV).

Herman, R. A., Zhu, X., Ayepa, E., You, S., & Wang, J. (2023). Advances in the One-Step Approach of Polymeric Materials Using Enzymatic Techniques. *Polymers, 15*(3), 703. DOI: 10.3390/polym15030703 PMID: 36772002

Hernández-Maldonado, A., & Yang, R. (2007). Desulfurization of transportation fuels by adsorption. *Catalysis Reviews. Science and Engineering, 46*(2), 111–150. DOI: 10.1081/CR-200032697

Herold, F., Prosch, S., Oefner, N., Brunnengräber, K., Leubner, O., Hermans, Y., Hofmann, K., Drochner, A., Hofmann, J. P., Qi, W., & Etzold, B. J. M. (2021). Nanoscale hybrid amorphous/graphitic carbon as key towards next-generation carbon-based oxidative dehydrogenation catalysts. *Angewandte Chemie International Edition, 60*(11), 5898–5906. DOI: 10.1002/anie.202014862 PMID: 33497000

Hessel, V., Tran, N. N., Asrami, M. R., Tran, Q. D., Long, N. V. D., Escribà-Gelonch, M., & Sundmacher, K. (2022). Sustainability of green solvents–review and perspective. *Green Chemistry, 24*(2), 410–437. DOI: 10.1039/D1GC03662A

Hill, J., Mulholland, G., Persson, K., Seshadri, R., Wolverton, C., & Meredig, B. (2016). Materials science with large-scale data and informatics: Unlocking new opportunities. *MRS Bulletin, 41*(5), 399–409. DOI: 10.1557/mrs.2016.93

Himanen, L., Geurts, A., Foster, A. S., & Rinke, P. (2019). Data-driven materials science: Status, challenges, and perspectives. *Advancement of Science, 6*(21), 1900808. DOI: 10.1002/advs.201900808 PMID: 31728276

Hossain, M. A., & Jhung, S. H. (2024). Supported chromium nitride: A new catalyst for oxidative desulfurization of liquid fuel. *Fuel, 372*, 132178. DOI: 10.1016/j.fuel.2024.132178

Hosseini, A., Khoshsima, A., Sabzi, M., & Rostam, A. (2022). Toward application of ionic liquids to desulfurization of fuels: A review. *Energy & Fuels, 36*(8), 4119–4152. DOI: 10.1021/acs.energyfuels.1c03974

Hosseinzadeh, E., Foroumadi, A., & Firoozpour, L. (2023). DFT insights into phytochemical capping agents. *Inorganic Chemistry Communications, 147*, 110243. DOI: 10.1016/j.inoche.2022.110243

Houda, S., Lancelot, C., Blanchard, P., Poinel, L., & Lamonier, C. (2018). Oxidative desulfurization of heavy oils with high sulfur content: A review. *Catalysts, 8*(9), 344. DOI: 10.3390/catal8090344

Hou, J., Deng, H. K., Liu, Z. X., Xu, P., & Wang, L. J. (2023). Sulfur metabolism in Rhodococcus species and applications in fuel desulfurization. *Journal of Applied Microbiology, 134*(3), lxad048. Advance online publication. DOI: 10.1093/jambio/lxad048 PMID: 36893799

Hou, J., Yu, J., Li, W., He, X., & Li, X. (2022). The effects of chemical oxidation and high-temperature reduction on surface functional groups and the adsorption performance of biochar for sulfamethoxazole adsorption. *Agronomy (Basel), 12*(2), 510. DOI: 10.3390/agronomy12020510

Howarth, R. W. (2014). A bridge to nowhere: Methane emissions and the greenhouse gas footprint of natural gas. *Energy Science & Engineering*, *2*(2), 47–60. DOI: 10.1002/ese3.35

Hsu, C. S., & Robinson, P. (2019). Characterization techniques for sulfur compounds in petroleum fuels. *Energy & Fuels*, *33*(4), 3030–3045. DOI: 10.1021/acs.energyfuels.9b00345

Hsu, C. S., & Robinson, P. R. (2006). *Practical advances in petroleum processing*. Springer.

Hsu, C. S., & Robinson, P. R. (Eds.). (2017). *Petroleum science and technology* (2nd ed.). Springer.

Huang, L., He, G., Yuan, Y., Zhang, T. C., Wang, Y., & Yuan, S. (2024). Trivalent metal ion (Al, Ga, In)-doped TiO_2 for enhanced photocatalytic desulfurization of H_2S: Band structure regulation and mechanism. *Industrial & Engineering Chemistry Research*, *63*(16), 7154–7165. DOI: 10.1021/acs.iecr.4c00074

Huang, Y., Shen, H., Chen, Y., Zhong, Q., Chen, H., Wang, R., Shen, G., Liu, J., Li, B., & Tao, S. (2015). Global organic carbon emissions from primary sources from 1960 to 2009. *Atmospheric Environment*, *122*, 505–512. DOI: 10.1016/j.atmosenv.2015.10.017

Humadi, J. I., Hamad, K. I., Abdulkareem, H. A., Ismael, M. N., Jarullah, A. T., Ahmed, M. A., & Mujtaba, I. M. (2025). Eco-friendly oxidative–adsorptive desulfurization for real diesel fuel using green MnO_2 biowaste-extracted calcite in a digital basket reactor. *Processes (Basel, Switzerland), 13*(10), 3084. DOI: 10.3390/pr13103084

Hung, Y.-T., Wang, L. K., & Shammas, N. K. (2020). *Handbook of Environment and Waste Management. 03*. DOI: 10.1142/11470

Hu, Q., Lu, J., Hui, J., Rao, Z., Ren, Y., & Wang, H. (2025). Artificial intelligence-driven development in rechargeable battery materials: Progress, challenges, and future perspectives. *Advanced Functional Materials*, *08438*(52), e08438. Advance online publication. DOI: 10.1002/adfm.202508438

Husain, I. A. F., Alkhatib, M. F., Jami, M. S., Mirghani, M. E. S., Zainudin, Z. Bin, & Hoda, A. 2014. Problems, Control, and Treatment of Fat, Oil, and Grease FOGReview, A. (•••).. . *Journal of Oleo Science*, *63*(8), 747–752. DOI: 10.5650/jos.ess13182

Hu, Y., Li, N., Wang, M., Qiao, Z., Gu, D., Zhu, L., Yuan, D., & Wang, B. (2025). Overview of research status and development trends in diesel desulfurization technology. *Catalysts*, *15*(3), 251. DOI: 10.3390/catal15030251

Ighalo, J. O., Akaeme, F. C., Georgin, J., de Oliveira, J. S., & Franco, D. S. (2025). Biomass hydrochar: A critical review of process chemistry, synthesis methodology, and applications. *Sustainability (Basel)*, *17*(4), 1660. DOI: 10.3390/su17041660

Ikram, R., Mohamed Jan, B., Nagy, P. B., & Szabo, T. (2023). Recycling waste sources into nanocomposites of graphene materials: Overview from an energy-focused perspective. *Nanotechnology Reviews*, *12*(1), 20220512. DOI: 10.1515/ntrev-2022-0512

IMO. (2020). *IMO 2020 global sulphur cap*. International Maritime Organization. https://www.imo.org/en/MediaCentre/HotTopics/Pages/Sulphur-2020.aspx

Inman, M. (2013). The True Cost pf Fossil Fuels. *JSTOR, 308*(4), 58–61. https://www.jstor.org/stable/26018067

Inspektor, A., & Salvador, P. A. (2014). Architecture of PVD coatings for metal-cutting applications: A review. *Surface and Coatings Technology*, *257*, 138–153. DOI: 10.1016/j.surfcoat.2014.08.068

International Maritime Organisation. (2020). IMO 2020 – Cutting sulphur oxide emissions. IMO Media Centre. https://www.imo.org/en/MediaCentre/HotTopics/Pages/Sulphur-2020.aspx

International Maritime Organization. (2020). *IMO 2020 – Regulation 14: Sulphur oxides (SOx) and particulate matter*. International Maritime Organization. https://www.imo.org/

Ioannidou, O., & Zabaniotou, A. (2007). Agricultural residues as precursors for activated carbon production. *Renewable & Sustainable Energy Reviews*, *11*(9), 1966–2005. DOI: 10.1016/j.rser.2006.03.013

Io a, M.-A., Cursaru, L.-M., chiopu, A.-G., Tudor, I. A., Motoc, A.-M., & Piticescu, R. M. (2023). Fe_3O_4 Core–shell nanostructures with anticancer and antibacterial properties: A Mini-review. *Processes (Basel, Switzerland), 11*(7), 1882. DOI: 10.3390/pr11071882

Ippolito, N. M., Romano, P., Innocenzi, V., Ferella, F., & Giansante, G. (2023). The Circular Economy Approach in Oil Refining: Recycling of Spent FCC and HDS Catalysts. Paper presented at the Offshore Mediterranean Conference and Exhibition.

Iram, S., Qaisar, I., Shabbir, R., Pomee, M. S., Schmidt, M., & Hertig, E. (2025). Impact of air pollution and smog on human health in Pakistan: A systematic review. *Environments (Basel, Switzerland), 12*(2), 46. DOI: 10.3390/environments12020046

Iravani, S. (2011). Green synthesis of metal nanoparticles using plants. *Green Chemistry, 13*(10), 2638–2650. DOI: 10.1039/c1gc15386b

Iroegbu, A. O. C., & Ray, S. S. (2022). Recent developments and future perspectives of biorenewable nanocomposites for advanced applications. *Nanotechnology Reviews, 11*(1), 1696–1721. DOI: 10.1515/ntrev-2022-0105

Irshad, T., Saeed, M., Munir, M., Kousar, R., Intisar, A., Din, M. I., Ahsan, H., Abid, M. A., Irfan, M., & Osman, S. M. (2024). Excellent catalytic performance over reduced graphene-boosted novel nanoparticles for oxidative desulfurization of fuel oil. *Nanotechnology Reviews, 13*(1), 20240116. Advance online publication. DOI: 10.1515/ntrev-2024-0116

Ishihara, A., Wang, D., Dumeignil, F., Amano, H., Qian, E. W., & Kabe, T. (2005). Oxidative desulfurization and denitrogenation of a light gas oil using an oxidation/adsorption continuous flow process. *Applied Catalysis A, General, 279*(1–2), 279–287. DOI: 10.1016/j.apcata.2004.10.037

Ishizuki, N., Shimizu, R., & Hitosugi, T. (2023). Autonomous experimental systems in materials science. *Science and Technology of Advanced Materials: Methods, 3*(1), 2197519. DOI: 10.1080/27660400.2023.2197519

Jacob, A. G., Alisi, I. O., & Surajo, J. M. (2025). Kaolinite clay as green and sustainable raw material for zeolites production: A review. *FUDMA Journal of Sciences, 9*, 18–35. DOI: 10.33003/fjs-2025-09(AHBSI)-3496

Jadhav, R. M., & Sangwai, J. S. (2020). Interaction of heavy crude oil and nanoparticles for heavy oil upgrading. In *Nanotechnology for Energy and Environmental Engineering* (pp. 231–255). Springer. DOI: 10.1007/978-3-030-33774-2_10

Jafarinejad, S. (2016). Control and treatment of sulfur compounds especially sulfur oxides (SOx) emissions from the petroleum industry: A review. *Chemistry International, 2*(4), 242–253.

Jagadeesan, D., Vernekar, D., Gupta, S., & Jaiswal, G. (2018). New opportunities in heterogeneous catalysis. *Proceedings of the Indian National Science Academy*, 99. DOI: 10.16943/ptinsa/2018/49504

Jain, A., Balasubramanian, R., & Srinivasan, M. P. (2016). Hydrothermal conversion of biomass waste to activated carbon. *Chemical Engineering Journal, 283*, 789–805. DOI: 10.1016/j.cej.2015.08.014

Jain, A., Ong, S. P., Hautier, G., Chen, W., Richards, W. D., Dacek, S., Cholia, S., Gunter, D., Skinner, D., Ceder, G., & Persson, K. A. (2013). Commentary: The Materials Project: A materials genome approach to accelerating materials innovation. *APL Materials, 1*(1), 011002. DOI: 10.1063/1.4812323

Jamar, A., Majid, Z. A. A., Azmi, W. H., Norhafana, M., & Razak, A. A. (2016). A review of water heating system for solar energy applications. *International Communications in Heat and Mass Transfer, 76*, 178–187. DOI: 10.1016/j.icheatmasstransfer.2016.05.028

Jarullah, A. T., Hussein, A. K., Al-Tabbakh, B. A., Hameed, S. A., Mujtaba, I. M., Saeed, L. I., & Humadi, J. I. (2024). Production of green fuel using a synthetic magnetite mesoporous nano-silica composite catalyst for oxidative desulfurization: Experiments and process modeling. *Catalysts, 14*(8), 529. DOI: 10.3390/catal14080529

Jasni, M. Z., Abdullah, N. H., Abdullah, S., Razab, M. A., Noor, A. M., Mohamed, M., Yusuf, N. N., Amin, M. M., Rasat, M. M., & Amin, M. M. (2018). Preparation and characterization of activated carbon from cocos nucifera l.(coconut) shell and sugarcane bagasse. *Int. J. Cur. Res. Eng. Sci. Tech, 1*(Spl-1), 416–421. DOI: 10.30967/ijcrset.1.S1.2018.416-421

Jha, D., Gomey, A. K., Kumari, G., Maheswari, P., Haider, M. B., Kumar, R., Balathanigaimani, M., & Sahu, J. (2025). A review on the role of nanocomposites for desulfurization of liquid transportation fuels. *Journal of Environmental Management, 375*, 124255. DOI: 10.1016/j.jenvman.2025.124255 PMID: 39879928

Jha, M. K., Joshi, S., Sharma, R. K., Kim, A. A., Pant, B., Park, M., & Pant, H. R. (2021). Surface modified activated carbons: Sustainable bio-based materials for environmental remediation. *Nanomaterials (Basel, Switzerland), 11*(11), 3140. DOI: 10.3390/nano11113140 PMID: 34835907

Ji, D., Lin, Y., Guo, X., Ramasubramanian, B., Wang, R., Radacsi, N., Jose, R., Qin, X., & Ramakrishna, S. (2024). Electrospinning of nanofibres. *Nature Reviews Methods Primers 2024 4:1, 4*(1), 1-. DOI: 10.1038/s43586-023-00278-z

Jia, Y., Yang, C., Chen, X., Xue, W., Hutchins-Crawford, H. J., Yu, Q., Topham, P. D., & Wang, L. (2021). A review on electrospun magnetic nanomaterials: Methods, properties and applications. *Journal of Materials Chemistry. C, Materials for Optical and Electronic Devices*, *9*(29), 9042–9082. DOI: 10.1039/D1TC01477C

Ji, F. (2021). *Collective Behaviors of Small-Scale Magnetic Building Blocks in Fluid*. The Chinese University of Hong Kong.

Jiménez, M., Romero, L., Domínguez, I. A., Espinosa, M. D. M., & Domínguez, M. (2019). Additive Manufacturing Technologies: An Overview about 3D Printing Methods and Future Prospects. *Complexity*, *2019*(1), 9656938. DOI: 10.1155/2019/9656938

Jones, R. O. (2015). Density functional theory: Its origins, rise to prominence, and future. *Reviews of Modern Physics*, *87*(3), 897–923. DOI: 10.1103/RevModPhys.87.897

Jorjani, E., Chehreh Chelgani, S., & Mesroghli, S. (2007). Prediction of microbial desulfurization of coal using artificial neural networks. *Minerals Engineering*, *20*(14), 1285–1292. DOI: 10.1016/j.mineng.2007.07.003

Ju, H., Zhang, Y., Zhang, J., Yu, Z., Zhang, Y., Zhang, X., Guo, X., Liu, J., Mao, Q., Liu, Q., Zhao, Y., Qi, T., Jiang, X., Guo, Z., & Chen, S. (2025). Development of a novel Cu(I) π-complexation adsorbent for ultra-deep desulfurization from a carbon dioxide stream. *Smart Molecules*, *3*(1), e20240027. DOI: 10.1002/smo.20240027 PMID: 40625567

Junaid, M., Khir, M. H. M., Witjaksono, G., Tansu, N., Saheed, M. S. M., Kumar, P., Ullah, Z., Yar, A., & Usman, F. (2020). Boron-Doped Reduced Graphene Oxide with Tunable Bandgap and Enhanced Surface Plasmon Resonance. *Molecules (Basel, Switzerland)*, *25*(16), 3646. DOI: 10.3390/molecules25163646 PMID: 32796504

Kadhum, A. T., & Albayati, T. M. (2022). Desulfurization techniques process and future challenges for commercial of crude oil products. *AIP Conference Proceedings*, 030039. DOI: 10.1063/5.0092049

Kaeed, Z., Shahhosseini, S., & Rezaei, M. (2024). Preparation of the MoO_3–Fe_3O_4 nanocomposites for deep oxidative desulfurization of model fuel: Optimization and experimental design. *Journal of Industrial and Engineering Chemistry*, *139*, 267–280. DOI: 10.1016/j.jiec.2024.05.003

Kampouraki, Z.-C., Giannakoudakis, D. A., Nair, V., Hosseini-Bandegharaei, A., Colmenares, J. C., & Deliyanni, E. A. (2019). Metal–organic frameworks as desulfurization adsorbents of DBT and 4,6-DMDBT from fuels. *Molecules (Basel, Switzerland)*, *24*(24), 4525. DOI: 10.3390/molecules24244525 PMID: 31835616

Kanagaraj, K., Nangan, S., & Thirumalaivasan, N. (2025). Applications of carbon-based nanomaterials derived from plant extracts in composite materials. In *Nanostructured Carbon Materials from Plant Extracts* (pp. 351–380). Elsevier. DOI: 10.1016/B978-0-323-95126-5.00007-X

Kandav, G., & Sharma, T. (2024). Green synthesis overview. *Particulate Science and Technology*, *42*(5), 874–894. DOI: 10.1080/02726351.2023.2281452

Kappe, C. O. (2004). Controlled microwave heating in modern organic synthesis. *Angewandte Chemie International Edition*, *43*(46), 6250–6284. DOI: 10.1002/anie.200400655 PMID: 15558676

Kapusta, K., & Stańczyk, K. (2011). Pollution of water during underground coal gasification of hard coal and lignite. *Fuel*, *90*(5), 1927–1934. DOI: 10.1016/j.fuel.2010.11.025

Kargar, H., Ghahramaninezhad, M., Shahrak, M. N., & Balula, S. S. (2021). An effective magnetic catalyst for oxidative desulfurization of model and real fuels: Fe_3O_4/ZIF-8/TiO_2. *Microporous and Mesoporous Materials*, *317*, 110992. DOI: 10.1016/j.micromeso.2021.110992

Karimi, R., & Golmohammadi, B. (2024). Simulation and optimization of heavy fuel oil (HFO) refining platform to low sulfur marine fuel (LS-FO) by oxidative desulfurization. *Scientific Reports*, *14*(1), 29830. DOI: 10.1038/s41598-024-81242-z PMID: 39616214

Karpuz, Ö., Baltacı, C., & Gül, A. (2025). *ve diğerleri.* Karabuğday kabuğu atığının sulu özütünden yeşil demir ve gümüş nanopartiküllerinin sentezi: Antibakteriyel, sitotoksik ve boya giderici özellikleri. *Biomass Conversion and Biorefinery*, *15*, 27679–27711. DOI: 10.1007/s13399-024-06287-6

Kar, S., Dey, S., & Chowdhury, K. B.. (2023). CuO/biochar composites via phyto-assisted synthesis. *Environmental Research*, *236*, 116824. DOI: 10.1016/j.envres.2023.116824 PMID: 37549783

Karthik, A., Bhuvaneshwaran, M., Senthil Kumar, M., Palanisamy, S., Palaniappan, M., & Ayrilmis, N. (2024). A review on surface modification of plant fibers for enhancing properties of biocomposites. *ChemistrySelect*, *9*(21), e202400650. DOI: 10.1002/slct.202400650

Kawatra, S. K., & Eisele, T. C. (2001). *Coal desulfurization: High-efficiency preparation methods*. Taylor & Francis.

Kaya, D., Çanka Kılıç, F., & Öztürk, H. H. (2021). *Fuels and Combustion*. 227–263. DOI: 10.1007/978-3-030-25995-2_8

Keerthana, S. P., Yuvakkumar, R., Kumar, P. S., Ravi, G., & Velauthapillai, D. (2021). Rare earth metal (Sm) doped zinc ferrite ($ZnFe_2O_4$) for improved photocatalytic elimination of toxic dye from aquatic system. *Environmental Research, 197*, 111047. DOI: 10.1016/j.envres.2021.111047

Khaleghian, M., Sheikhi, M., Shahab, S., Yahyaei, H., & Ahmadianarog, M. (2024). Theoretical investigation of the adsorption behavior of dibenzothiophene molecule on pristine boron nitride nanosheet. *Main Group Chemistry*, *23*(4), 425–438. DOI: 10.3233/MGC-230002

Khamkeaw, A., Asavamongkolkul, T., Perngyai, T., Jongsomjit, B., & Phisalaphong, M. (2020). Interconnected micro, meso, and macro porous activated carbon from bacterial nanocellulose for superior adsorption properties and effective catalytic performance. *Molecules (Basel, Switzerland)*, *25*(18), 4063. DOI: 10.3390/molecules25184063 PMID: 32899569

Khan, M. K., Faheem, K., Khan, Z., & Qayyum, N. (2025). Green synthesis of polymeric materials and nanocomposites for effluent treatment. In Sustainable Effluent Treatment and Resource Recovery (Vol. 1, pp. 153–180). https://doi.org/ DOI: 10.1021/bk-2025-1517.ch005

Khikani, M., Isopencu, G. O., & Deleanu, I. M.. (2024). BC-based antibacterial nanocomposite membranes. *Journal of Composites Science*, *8*(11), 475. DOI: 10.3390/jcs8110475

Khvorostyanov, D. V., Krinner, G., Ciais, P., Heimann, M., & Zimov, S. A. (2008). Vulnerability of permafrost carbon to global warming. Part I: Model description and role of heat generated by organic matter decomposition. *Tellus, Series B: Chemical and Physical Meteorology, 60 B*(2), 250–264. https://doi.org/DOI: 10.1111/J.1600-0889.2007.00333.X;WGROUP:STRING:PUBLICATION

Kim, E., Tomczak, K., Chandrasiri, H., Pálmai, M., Ghaznavi, A., Gritsenko, D., Xu, J., & Snee, P. (2023). Nucleation control of quantum dot synthesis in a microfluidic continuous flow reactor. *Frontiers in Nanotechnology*, *4*, 1096267. DOI: 10.3389/fnano.2022.1096267

Kim, J. R., Hu, Y., Zavala, V. M., & Karthikeyan, K. G. (2022). Techno-economic analysis of pretreatments to dairy manure biomass for enhanced biogas production. *Bioresource Technology Reports, 20,* 101275. DOI: 10.1016/j.biteb.2022.101275

Kim, K., Wie, J., & Kim, J. (2020). Synergistic interaction of P and N co-doping EDTA with controllable active EDTA-cobalt sites as efficient electrocatalyst for oxygen reduction reaction. *Journal of Industrial and Engineering Chemistry, 83,* 252–259. DOI: 10.1016/j.jiec.2019.11.035

Kim, S. Y., Kumachang, C. C. F., & Doumon, N. Y. (2023). Characterization Tools to Probe Degradation Mechanisms in Organic and Perovskite Solar Cells. *Solar RRL, 7*(13), 2300155. Advance online publication. DOI: 10.1002/solr.202300155

Kirubakaran, D., Wahid, J. B. A., Karmegam, N., Jeevika, R., Sellapillai, L., Rajkumar, M., & SenthilKumar, K. J. (2026). A comprehensive review on the green synthesis of nanoparticles: Advancements in biomedical and environmental applications. *Biomedical Materials & Devices, 4*(1), 388–413. DOI: 10.1007/s44174-025-00295-4

Kitano, H. (2021). Nobel Turing Challenge: Creating the engine for scientific discovery. *npj Systems Biology and Applications, 7,* 29. DOI: 10.1038/s41540-021-00189-3

Klaine, S. J., Alvarez, P. J. J., Batley, G. E., Fernandes, T. F., Handy, R. D., Lyon, D. Y., Mahendra, S., McLaughlin, M. J., & Lead, J. R. (2012). Nanomaterials in the environment. *Environmental Toxicology and Chemistry, 27*(9), 1825–1851. DOI: 10.1897/08-090.1 PMID: 19086204

Klebe, S., Leigh, J., Henderson, D. W., & Nurminen, M. (2019). Asbestos, Smoking and Lung Cancer: An Update. *International Journal of Environmental Research and Public Health 2020, Vol. 17, Page 258, 17*(1), 258. DOI: 10.3390/ijerph17010258

Kobayashi, S. (2020). *Nanoparticles in catalysis.* DOI: 10.1007/978-3-030-56630-2

Kober, T., Schiffer, H.-W., Densing, M., & Panos, E. (2020). Global energy perspectives to 2060 – WEC's World Energy Scenarios 2019. *Energy Strategy Reviews, 31,* 100523. DOI: 10.1016/j.esr.2020.100523

Kocak, Y., Aygun, A., Altuner, E. E., Gonca, S., Berikten, D., Tiri, R. N. E., & Sen, F. (2024). *Eco-friendly production of platinum nanoparticles: physicochemical properties, evaluation of biological and catalytic activities. 21,* 51–62. https://doi .org/DOI: 10.1007/S13762-023-05232-W/FIGURES/7

Kojima, Y., Suzuki, K. I., Fukumoto, K., Sasaki, M., Yamamoto, T., Kawai, Y., & Hayashi, H. (2002). Hydrogen generation using sodium borohydride solution and metal catalyst coated on metal oxide. *International Journal of Hydrogen Energy, 27*(10), 1029–1034. DOI: 10.1016/S0360-3199(02)00014-9

Kozhevnikov, I. V. (1998). Catalysis by heteropoly acids and multicomponent polyoxometalates in liquid-phase reactions. *Chemical Reviews*, *98*(1), 171–198. DOI: 10.1021/cr960400y PMID: 11851502

Krishani, M., Sambudi, N. S., & Suhaimi, H. (2025). Green Catalysts in the Synthesis of Biomaterials for Biomedical Applications. In *Sustainable Green Biomaterials As Drug Delivery Systems* (pp. 217–252). Springer. DOI: 10.1007/978-3-031-79062-1_9

Kulak, M., & Gulmez Samsa, C. (2023). Flavonoid-mediated nanomaterial synthesis. In *Secondary metabolites based green synthesis* (pp. 49–65). Springer., DOI: 10.1007/978-981-99-0927-8_3

Kumari, P., Seth, R., Meena, A., & Sharma, D. (2023). Enzymatic synthesis of cellulose nanocrystals from lemongrass and its application in improving anti-cancer drug release, uptake and efficacy. *Industrial Crops and Products*, *192*, 115933. DOI: 10.1016/j.indcrop.2022.115933

Kumar, S., Singh, H., Singh, H., Soni, H., Bechelany, M., & Singh, J. (2025). Extraction of Metallic Nanoparticles from Electronic Waste for Biomedical Applications: Pioneering Sustainable Technological Progress. *Sustainability (Basel)*, *17*(5), 2100. DOI: 10.3390/su17052100

Kumar, V., Goyal, A., & Gupta, R. (2020). Sustainable approaches in carbon nanomaterial synthesis. *Materials Science for Energy Technologies*, *3*, 293–312. DOI: 10.1016/j.mset.2019.11.005

Kundalwal, S. I., & Rathi, A. (2020). Improved mechanical and viscoelastic properties of CNT-composites fabricated using an innovative ultrasonic dual mixing technique. *Journal of the Mechanical Behavior of Materials*, *29*(1), 77–85. DOI: 10.1515/jmbm-2020-0008

Kwao-Boateng, E., Ankudey, E. G., Darkwah, L., & Danquah, K. O. (2024). Assessment of diesel fuel quality. *Heliyon*, *10*(2), e24733. Advance online publication. DOI: 10.1016/j.heliyon.2024.e24733 PMID: 38312546

Lahiri, D., Nag, M., Ghosh, S., & Ray, R. R. (2021). Green synthesis of nanoparticles and their applications in the area of bioenergy and biofuel production. In *Nanomaterials* (pp. 195–219). Elsevier. DOI: 10.1016/B978-0-12-822401-4.00017-9

Lakhmir, M. A., Akhter, F., Ahmed, J., Mushtaq, S., Arain, H. J., & Ahsan, M. J. (2025). Impact of sulfur-ash reduction treatment on combustion performance, SOx emissions, and flue gas temperatures of lignite coals. *Environmental Science and Pollution Research International*, *32*(9), 5574–5588. DOI: 10.1007/s11356-025-36064-5 PMID: 39934577

Lam, N. L., Smith, K. R., Gauthier, A., & Bates, M. N. (2012). Kerosene: A review of household uses and their hazards in low-and middle-income countries. *Journal of Toxicology and Environmental Health. Part B, Critical Reviews*, *15*(6), 396–432. DOI: 10.1080/10937404.2012.710134

Lazcano-Ramírez, H. G., Garza-García, J. J., & Hernández-Díaz, J. A.. (2023). Plant-mediated selenium nanoparticles. *Antibiotics (Basel, Switzerland)*, *12*(1), 115. DOI: 10.3390/antibiotics12010115 PMID: 36671316

Leal, E., Torres-Mancera, P., Alonso, F., Luna, R., Muñoz, J. A., & Ancheyta, J. (2024). Optimization methodology of dual-bed catalyst stacking systems to produce ultralow-sulfur diesel. *Industrial & Engineering Chemistry Research*, *63*(42), 17857–17867. DOI: 10.1021/acs.iecr.4c02846

Le, D. L., Salomone, R., & Nguyen, Q. T. (2024). Sustainability assessment methods for circular bio-based building materials: A literature review. *Journal of Environmental Management*, *352*, 120137. DOI: 10.1016/j.jenvman.2024.120137 PMID: 38266527

Lee, W. S., Choi, J. H., & Kim, D. H. (2009). Analytical approaches for characterization of sulfur compounds in fuels. *Journal of Analytical Chemistry*, *64*(5), 423–435. DOI: 10.1021/ac900987t

Lehmann, J., Barrios, E., Devault, M., Li, L., Nelson, R., Six, J., & Trimmer, J. (2025). Biochar in the circular bionutrient economy. *Proceedings of the National Academy of Sciences of the United States of America*, *122*(33), e2503668122. DOI: 10.1073/pnas.2503668122 PMID: 40789023

Lellala, K., Behera, S. K., Srivastava, P., Saeed, W. S., Haidyrah, A. S., & Burile, A. N. (2024). Fe_3O_4 nanoparticles decorated on N-doped graphene oxide for heavy metal removal and desulfurization. *Diamond and Related Materials*, *150*, 111746. DOI: 10.1016/j.diamond.2024.111746

Lerotholi, L., Everson, R. C., Koech, L., Neomagus, H. W. J. P., Rutto, H. L., Branken, D., Hattingh, B. B., & Sukdeo, P. (2022). Semi-dry flue gas desulphurization in spray towers: a critical review of applicable models for computational fluid dynamics analysis. *Clean Technologies and Environmental Policy 2022 24:7*, *24*(7), 2011–2060. DOI: 10.1007/s10098-022-02308-y

Lewandowski, M., Janus, R., Wadrzyk, M., Jaroszewska, K., & Zaborowska, K. (2025). Activity of amorphous NiB/SiO_2 in hydrotreating model reactions: Hydrodesulfurization of 4,6-DMDBT. *ACS Omega*, *10*(34), 38968–38984. DOI: 10.1021/acsomega.5c04966 PMID: 40918321

Lewin, A. (2011). *Hospitalisation pour problèmes respiratoires et exposition aux émissions d'une aluminerie chez l'enfant* [Universite de Montreal]. DOI: 10.71781/725

Liang, Y., Zhang, W., Wu, D., Ni, Q. Q., & Zhang, M. Q. (2018). Interface Engineering of Carbon-Based Nanocomposites for Advanced Electrochemical Energy Storage. *Advanced Materials Interfaces*, *5*(14), 1800430. DOI: 10.1002/admi.201800430

Liao, M., Xiao, Y., Pi, X., Zhang, L., Zheng, M., Liu, J., & Guo, J. (2024). Zr-based UiO-66 doped with rare-earth elements supported polyoxometalate liquid for oxidative desulfurization of dibenzothiophene. *Energy & Fuels*, *38*(14), 12555–12568. DOI: 10.1021/acs.energyfuels.4c01125

Li, C., Zhao, D., & Shi, W. (2018). Metal–organic frameworks for the adsorptive desulphurization of fuels: A review. *Energy & Fuels*, *32*(12), 12439–12460. DOI: 10.1021/acs.energyfuels.8b03092

Li, H., Xiang, X., Zhang, Y., Cheng, H., Chen, Q., Li, X., Wu, F., & Liu, X. (2025). The green preparation of ZrO_2-modified WO_3–SiO_2 composite from rice husk and its excellent oxidative desulfurization performance. *Catalysts*, *15*(10), 996. DOI: 10.3390/catal15100996

Lim, X. B., & Ong, W.-J. (2021). A current overview of the oxidative desulfurization of fuels utilizing heat and solar light: From materials design to catalysis for clean energy. *Nanoscale Horizons*, *6*(8), 588–633. DOI: 10.1039/D1NH00127B PMID: 34018529

Linares, C. F., & Brunet, S. (2023). Improving the hydrogenation pathway from the dibenzothiophene (DBT) hydrodesulfurization (HDS) reaction using alloys of bulk NiXCoyborides. *Reaction Kinetics, Mechanisms and Catalysis*, *136*(2), 901–917. DOI: 10.1007/s11144-023-02393-0

Lin, X., Li, W., Nguyen, V., Wang, S., Yang, S., Ma, L., & Fan, Z. (2023). Fe single-atom catalyst nanocages supported by bacterial cellulose-derived carbon aerogel. *Chemical Engineering Journal*, *477*, 146977. DOI: 10.1016/j.cej.2023.146977

Li, P., Jiang, H., Barr, A., Ren, Z., Gao, R., Wang, H., Fan, W., Zhu, M., Xu, G., & Li, J. (2021). Reusable polyacrylonitrile-sulfur extractor of heavy metal ions from wastewater. *Advanced Functional Materials*, *31*(51), 2105845. DOI: 10.1002/adfm.202105845

Li, P., Lu, J., Cui, H., Ruan, S., & Zeng, Y.-J. (2021). The development, application, and performance of black phosphorus in energy storage and conversion. *Materials Advances*, *2*(8), 2483–2509. DOI: 10.1039/D0MA01016B

Liu, D., Zhang, G., Kang, X., Wu, W., Zhao, L., Gao, J., & Xu, C. (2025). Hydrodesulfurization and isomerization performance of model FCC naphtha over sulfided Co(Ni)–Mo/Y catalysts. *Chinese Journal of Chemical Engineering*, *77*, 185–194. DOI: 10.1016/j.cjche.2024.10.001

Liu, D., Zhang, Q., Chen, X., Zhu, P., Yan, F., Wang, X., Dai, H., Chen, J., Gong, G., Shang, C., Xie, L., & Zhai, X. (2022). The Lattice Distortion, Defect Evolution and Electrochemical Performance Improvement in Zn-VO2(B) Nanorods. *Nanomaterials (Basel, Switzerland)*, *12*(18), 3196. DOI: 10.3390/nano12183196 PMID: 36144984

Liu, J., Bunes, B. R., Zang, L., & Wang, C. (2018). Supported single-atom catalysts: Synthesis, characterization, properties, and applications. *Environmental Chemistry Letters*, *16*(2), 477–505. DOI: 10.1007/s10311-017-0679-2

Liu, J., Burciaga, R., Tang, S., Ding, S., Ran, H., Zhao, W., Wang, G., Zhuang, Z., Xie, L., Lyu, Z., Lin, Y., Du, A., Yuan, A., Fu, J., Song, B., Zhu, J., Sun, Z., Jin, X., Huo, Z. Y., & Zhu, W. (2024). Heterogeneous catalysis for the environment. *The Innovation Materials*, *2*(3), 100090. DOI: 10.59717/j.xinn-mater.2024.100090

Liu, J., Jiang, Y., Yao, W., Jiang, X., & Jiang, X. (2019). Molecular Characterization of Henan Anthracite Coal. *Energy & Fuels*, *33*(7), 6215–6225. DOI: 10.1021/acs.energyfuels.9b01061

Liu, M., Ye, Y., Ye, J., Gao, T., Wang, D., Chen, G., & Song, Z. (2023). Recent advances of Fe_3O_4-based magnetic materials in catalysis. *Magnetochemistry (Basel, Switzerland)*, *9*(4), 110. DOI: 10.3390/magnetochemistry9040110

Liu, P., Yang, S., & Wang, H. (2025). Mass transfer behavior and desulfurization characteristics in the three-dimensional wet flue gas desulfurization tower. *Energy*, *319*, 134840. DOI: 10.1016/j.energy.2025.134840

Liu, Q., Chang, Q., Liao, X., Jiang, Y., Lyu, X., Zhao, Q., Diao, J., Wang, X., Huang, X., & Lyu, S. (2024). Adsorption desulphurization of biochar derived from eucalyptus waste. *Powder Technology*, *448*, 120322. DOI: 10.1016/j.powtec.2024.120322

Liu, Q., Wu, X., Wang, X., Jin, Z., Zhu, D., Meng, Q., Huang, S., Liu, J., & Fu, Q. (2019). Carbon and hydrogen isotopes of methane, ethane, and propane: A review of genetic identification of natural gas. *Earth-Science Reviews*, *190*, 247–272. DOI: 10.1016/j.earscirev.2018.11.017

Liu, W., Gao, W., Zhang, H., Wei, B., Jiang, R., Liang, L., Zhang, L., & Xie, Y. (2025). A New Clean Completion Technology for Soluble Screen Tubes in Ultra-Deep Well with High Temperature and High Pressure. *International Petroleum Technology Conference, IPTC 2025*. DOI: 10.2523/IPTC-25094-MS

Liu, W., Xu, H., Liao, Y., Quan, Z., Li, S., Zhao, S., Qu, Z., & Yan, N. (2019). Recyclable CuS sorbent with large mercury adsorption capacity in the presence of SO2 from non-ferrous metal smelting flue gas. *Fuel*, *235*, 847–854. DOI: 10.1016/j.fuel.2018.08.062

Liu, Y., Biswas, B., Hassan, M., & Naidu, R. (2024). Green adsorbents for environmental remediation: Synthesis methods, ecotoxicity, and reusability prospects. *Processes (Basel, Switzerland)*, *12*(6), 1195. DOI: 10.3390/pr12061195

Livage, J. (1997). Sol-gel processes. *Current Opinion in Solid State and Materials Science*, *2*(2), 132–138. DOI: 10.1016/S1359-0286(97)80057-5

Li, W., Guo, Z., Yang, J., Li, Y., Sun, X., He, H., Li, S., & Zhang, J. (2022). Advanced strategies for stabilizing single-atom catalysts for energy storage and conversion. *Electrochemical Energy Reviews*, *5*(3), 9. DOI: 10.1007/s41918-022-00169-z

Li, W., Zhu, X., Chao, Y., & Wang, X. (2016). Adsorptive desulfurization of fuels using novel nanomaterials. *Energy & Fuels*, *30*(8), 6604–6621. DOI: 10.1021/acs.energyfuels.6b00874

Li, X., Yu, J., Wageh, S., Al-Ghamdi, A. A., & Xie, J. (2016). Graphene in Photocatalysis: A Review. *Small*, *12*(48), 6640–6696. DOI: 10.1002/smll.201600382

Li, X., Zhu, W., Lu, X., Zuo, S., Yao, C., & Ni, C. (2017). Integrated nanostructures of CeO2/attapulgite/g-C3N4 as efficient catalyst for photocatalytic desulfurization: Mechanism, kinetics and influencing factors. *Chemical Engineering Journal*, *326*, 87–98. DOI: 10.1016/j.cej.2017.05.131

Li, Z., Ma, X., Liao, Y., & Zhao, N. (2022). Characteristics analysis and parameters optimization of desulfurization wastewater evaporation in a rotary spray drying tower. *Powder Technology*, *399*, 117211. DOI: 10.1016/j.powtec.2022.117211

Lockwood, T. (2017). A Compararitive Review of Next-generation Carbon Capture Technologies for Coal-fired Power Plant. *Energy Procedia*, *114*, 2658–2670. DOI: 10.1016/j.egypro.2017.03.1850

Long, B., Zou, Y., Li, Z., Ma, Z., Jiang, W., Zou, H., & Chen, H. (2020). Effect of Phosphorus Doping on Conductivity, Diffusion, and High Rate Capability in Silicon Anode for Lithium-Ion Batteries. *ACS Applied Energy Materials*, *3*(6), 5572–5580. DOI: 10.1021/acsaem.0c00534

Longwell, J. P., Rubin, E. S., & Wilson, J. (1995). Coal: Energy for the future. *Progress in Energy and Combustion Science*, *21*(4), 269–360. DOI: 10.1016/0360-1285(95)00007-0

Lookman, T., Balachandran, P. V., Xue, D., & Yuan, R. (2019). Active learning in materials science with emphasis on adaptive sampling using uncertainties for targeted design. *npj Computational Materials, 5*, 21. DOI: 10.1038/s41524-019-0153-8

Loy, A. C. M., Ng, W. L., & Bhattacharya, S. (2024). Advanced characterization techniques for the development of Subatomic scale catalysts: One step closer to industrial scale fabrication. *Materials Today Catalysis, 4*, 100033. DOI: 10.1016/j.mtcata.2023.100033

Lundberg, J. O. N., Weitzberg, E., Lundberg, J. M., Alving, K., Lundberg, E. J. O. N., Weitzberg, J. M., & Lundberg, K. A. (1996). Nitric oxide in exhaled air. *The European Respiratory Journal, 9*(12), 2671–2680. DOI: 10.1183/09031936.96.09122671

Luo, S., Liu, Y., Zhu, Y., Niu, Q., Cheng, M., Ye, S., Yi, H., Shao, B., Shen, M., Wen, X., Zeng, G., & Liu, Z. (2021). Perspectives on palladium-based nanomaterials: Green synthesis, ecotoxicity, and risk assessment. *Environmental Science. Nano, 8*(1), 20–36. DOI: 10.1039/D0EN01048K

Lu, Z., Zeng, L., Song, W., Qin, Z., Zeng, D., & Xie, C. (2017). In situ synthesis of C-TiO2/g-C3N4 heterojunction nanocomposite as highly visible light active photocatalyst originated from effective interfacial charge transfer. *Applied Catalysis B: Environmental, 202*, 489–499. DOI: 10.1016/j.apcatb.2016.09.052

Lyon, R. E. (2000). *Heat Release Kinetics. Fire and Materials An International Journal.* Wiley., DOI: 10.1002/1099-1018(200007/08)24:4

M, A., & G, P. K. (2023). Hydrogen towards sustainable transition: A review of production, economic, environmental impact and scaling factors. *Results in Engineering, 20*, 101456. DOI: 10.1016/j.rineng.2023.101456

Mabena, K. G., Nomngongo, P. N., & Mketo, N. (2024). Selective and precise solid phase extraction based on a magnetic cellulose gold nanocomposite followed by ICP-OES analysis for monitoring of total sulfur content in liquid fuels. *RSC Advances, 14*(38), 27990–27998. DOI: 10.1039/D4RA04427D PMID: 39224633

Madani, M., Hosny, S., Alshangiti, D. M., Nady, N., Alkhursani, S. A., Alkhaldi, H., Al-Gahtany, S. A., Ghobashy, M. M., & Gaber, G. A. (2022). Green synthesis of nanoparticles for varied applications: Green renewable resources and energy-efficient synthetic routes. *Nanotechnology Reviews, 11*(1), 731–759. DOI: 10.1515/ntrev-2022-0034

Maginn, E. J. (2010). What to Do with CO2. *The Journal of Physical Chemistry Letters, 1*(24), 3478–3479. DOI: 10.1021/jz101582c

Maheshwaran, L., Nadarajah, L., & Senadeera, S. P. N. N.. (2024). Phytochemical screening methods. *Asian Plant Research Journal, 12*(5), 11–38. DOI: 10.9734/aprj/2024/v12i5267

Majewski, W. A. (2001). *Diesel particulate filters*. DieselNet, Ecopoint Inc., https://courses.washington.edu/cive494/DPF.pdf

Ma, L., Zhong, Y., Qu, L., Feng, X., Zhang, Y., Hou, H., & Yang, G. (2025). Conformational flexibility of a novel Cu(I)-based adsorbent enhancing the selectivity of adsorptive desulfurization. *Separation and Purification Technology, 358*, 130324. DOI: 10.1016/j.seppur.2024.130324

Malini, K., Selvakumar, D., & Kumar, N. (2023). Activated carbon from biomass: Preparation, factors improving basicity and surface properties for enhanced CO2 capture capacity–A review. *Journal of CO2 Utilization, 67*, 102318.

Mane, M. B. (2021). Studies in newer methodologies and materials/material modifications for environmental, pollution control applications specifically for water treatment and desulfurization.

Mansouri, M., Shayanmehr, M., & Ghaemi, A. (2025). Exploring the adsorption desulfurization efficiency using RSM and ANN methodologies. *Scientific Reports, 15*(1), 20869. DOI: 10.1038/s41598-025-05688-5 PMID: 40595999

Mao, X., Wang, Y., Ma, Z., Hao, R., Fu, D., Yuan, B., & Zhao, Y. (2025). Deep oxidation removal of NO and SO_2 via singlet oxygen and hydroxyl radicals in $LaCoO_3$–H_2O_2 system. *ACS Omega, 10*(39), 45297–45308. DOI: 10.1021/acsomega.5c04716 PMID: 41078817

Marafi, M., & Stanislaus, A. (2008). Hydrodesulfurization of refractory sulfur compounds in diesel fuels. *Catalysis Reviews, 50*(1), 1–33. DOI: 10.1080/01614940701777791

Marschall, R. (2014). Semiconductor composites: Strategies for enhancing charge carrier separation to improve photocatalytic activity. *Advanced Functional Materials, 24*(17), 2421–2440. DOI: 10.1002/adfm.201303214

Martins, F., Felgueiras, C., & Smitková, M. (2018). Fossil fuel energy consumption in European countries. *Energy Procedia, 153*, 107–111. DOI: 10.1016/j.egypro.2018.10.050

Mashkina, A. V. (1990). Catalytic Synthesis of Sulfoxides and Sulfones Via Oxidation of Sulfides by Molecular Oxygen. *Catalysis Reviews. Science and Engineering, 32*(1–2), 105–161. DOI: 10.1080/01614949009349941

Matloob, A. M., Abd El-Hafiz, D. R., Saad, L., & Mikhail, S. (2023). Cr/Fe–MOF/graphene nanocomposite for sulfur removal from diesel. *Journal of Inorganic and Organometallic Polymers and Materials, 33*(1), 254–265. DOI: 10.1007/s10904-022-02472-0

Matloob, A. M., Al-Hagar, O. E., Abol-Fotouh, D., & Salem, H. M. (2025). Green synthesis of ZnO–Zn–MOF/bacterial nanocellulose for ultra-oxidative desulfurization of actual diesel fuel. *Scientific Reports, 15*(1), 29246. DOI: 10.1038/s41598-025-14377-2 PMID: 40784888

Mattox, D. M. (2002). Physical vapor deposition (PVD) processes. *Metal Finishing, 100*(1, SUPPL.), 394–408. DOI: 10.1016/S0026-0576(02)82043-8

Ma, X., Sakanishi, K., & Isoda, T. (2002). A review of selective oxidative desulphurization of liquid fuels with hydrogen peroxide catalyzed by polyoxometalates. *Applied Catalysis A, General, 235*(1–2), 79–92. DOI: 10.1016/S0926-860X(02)00240-0

Ma, X., Sakanishi, K., & Mochida, I. (1994). Hydrodesulfurization reactivities of various sulfur compounds in diesel fuel. *Industrial & Engineering Chemistry Research, 33*(2), 218–222. DOI: 10.1021/ie00026a007

Mc Carthy, K., Gullapalli, H., Ryan, K. M., & Kennedy, T. (2022). Electrochemical impedance correlation analysis for the estimation of Li-ion battery state of charge, state of health and internal temperature. *Journal of Energy Storage, 50*, 104608. DOI: 10.1016/j.est.2022.104608

Mehdi, R., Khoja, A. H., Naqvi, S. R., Gao, N., & Amin, N. A. S. (2022). A review on production and surface modifications of biochar materials via biomass pyrolysis process for supercapacitor applications. *Catalysts, 12*(7), 798. DOI: 10.3390/catal12070798

Mehmood, A., Khan, F. S. A., Mubarak, N. M., Tan, Y. H., Karri, R. R., Khalid, M., Walvekar, R., Abdullah, E. C., Nizamuddin, S., & Mazari, S. A. (2021). Magnetic nanocomposites for sustainable water purification—A comprehensive review. *Environmental Science and Pollution Research International, 28*(16), 19563–19588. DOI: 10.1007/s11356-021-12589-3 PMID: 33651297

Mehri, M., Khosravi-Nikou, M. R., Aghajani, M., & Shariati, A. (2021). Process intensification of liquid hydrocarbon fuel desulfurization using catalytic–oxidative–adsorptive method. *Chemical Papers, 75*(7), 3667–3678. DOI: 10.1007/s11696-021-01618-7

Melnyk, O. M., Onishchenko, O. A., Shyshkin, O. V., Volkov, O. M., Volyanskyy, S. M., Maulevych, V. O., & Kreitser, K. O. (2024). Enhancing shipboard technical facility performance through the utilization of low-sulfur marine fuel grades. *Journal of Chemistry and Technologies*, *32*(1), 233–245. DOI: 10.15421/jchemtech.v32i1.297916

Metawea, R., Farag, H. A., El-Ashtoukhy, E. S., El-Latif, M. M. A., & El-Sayed, E. M. (2023). Ultrasound-assisted one-pot oxidative desulfurization using green-synthesized MIL-53(Al). *Scientific Reports*, *13*(1), 13728. DOI: 10.1038/s41598-023-40955-3 PMID: 37608052

Mignogna, D., Szabó, M., Ceci, P., & Avino, P. (2024). Biomass energy and biofuels: Perspective, potentials, and challenges in the energy transition. *Sustainability (Basel)*, *16*(16), 7036. DOI: 10.3390/su16167036

Mihai, S., Bondarev, A., & Necula, M. (2025). The potential of biogenic materials as sustainable and environmentally benign alternatives to conventional adsorbents for dyes removal: A review. *Processes (Basel, Switzerland)*, *13*(2), 589. DOI: 10.3390/pr13020589

Mikołajewska, E., Mikołajewski, D., Mikołajczyk, T., & Paczkowski, T. (2025). Generative AI in AI-based digital twins for fault diagnosis for predictive maintenance in Industry 4.0/5.0. *Applied Sciences (Basel, Switzerland)*, *15*(6), 3166. DOI: 10.3390/app15063166

Miramontes, J. A. J., Arvizu, J. L. D., Gutierrez, S., Zaragoza, M. J. M., Ortiz, L., & Martinez, C. (2017). Synthesis, characterization and photocatalytic evaluation of strontium ferrites towards H_2 production by water splitting under visible light irradiation. *International Journal of Hydrogen Energy*, *42*(51), 30257–30266. DOI: 10.1016/j.ijhydene.2017.09.162

Mirfakhraei, S., Hekmati, M., Eshbala, F. H., & Veisi, H. (2018). Fe3O4/PEG-SO3H as a heterogeneous and magnetically-recyclable nanocatalyst for the oxidation of sulfides to sulfones or sulfoxides. *New Journal of Chemistry*, *42*(3), 1757–1761. DOI: 10.1039/C7NJ02513K

Mishra, P. M., & Devi, A. P. (2023). Biogenic synthesis of metal oxide nanocomposites using plant specimens and their application towards treatment of wastewater. *Environmental Science and Pollution Research International*, *30*(50), 108512–108524. DOI: 10.1007/s11356-023-29989-2 PMID: 37775638

Mishra, R. K., Singh, B., & Acharya, B. (2024). A comprehensive review on activated carbon from pyrolysis of lignocellulosic biomass: An application for energy and the environment. *Carbon Resources Conversion*, *7*(4), 100228. DOI: 10.1016/j.crcon.2024.100228

Mitra, S., Racha, S. M., Shown, B., Mandal, S., & Das, A. K. (2023). New insights on oxidative desulfurization for low sulfur residual oil production. *Sustainable Energy & Fuels*, *7*(1), 270–279. DOI: 10.1039/D2SE01363K

Miwa, K., Aoki, M., Noritake, T., Ohba, N., Nakamori, Y., Towata, S. I., Züttel, A., & Orimo, S. I. (2006). Thermodynamical stability of calcium borohydride. *Physical Review B: Condensed Matter and Materials Physics*, *74*(15), 155122. DOI: 10.1103/PhysRevB.74.155122

Moghaddam, M. R. H., & Shahrak, M. N. (2024). Coupled one-pot oxidative extractive desulfurization of a model fuel catalyzed by ZIF-67/biomass-derived activated carbon composite. *Inorganic Chemistry Communications*, *165*, 112489. DOI: 10.1016/j.inoche.2024.112489

Mohan, N. S., Pandian, M., Vijayalakshmi, V., & Arulraj, A. (2025a). Improving catalytic performance of $BaTiO_3$ perovskite nanoparticles through Sm and Sm@Ag deposition for environmental sustainability. *Materials Science and Engineering B*, *311*, 117790. DOI: 10.1016/j.mseb.2024.117790

Mohan, N. S., Pandian, M., Vijayalakshmi, V., & Arulraj, A. (2025b). Enhancing catalytic efficiency of Sm and Sm@Ag doped strontium titanate for environmental and renewable energy applications. *Ceramics International*, *51*(11), 13803–13814. DOI: 10.1016/j.ceramint.2025.01.215

Mohd Basri, M. S., Abdul Karim Shah, N. N., Sulaiman, A., Mohamed Amin Tawakkal, I. S., Mohd Nor, M. Z., Ariffin, S. H., Abdul Ghani, N. H., & Mohd Salleh, F. S. (2021). Progress in the valorization of fruit and vegetable wastes: Active packaging, biocomposites, by-products, and innovative technologies used for bioactive compound extraction. *Polymers*, *13*(20), 3503. DOI: 10.3390/polym13203503 PMID: 34685262

Mohnen, V. A. (n.d.). The Challenge of Acid Rain on JSTOR. *Scientific American, JSTOR, 1988*, 30–39. Retrieved December 12, 2025, from https://www.jstor.org/stable/24989192

Montoya, J., & Persson, K. (2017). A high-throughput framework for determining adsorption energies on solid surfaces. *npj Computational Materials, 3*. DOI: 10.1038/s41524-017-0017-z

Moosavi, E. S., Dastgheib, S. A., & Karimzadeh, R. (2012). Adsorption of thiophenic compounds from model diesel fuel using copper and nickel impregnated activated carbons. *Energies*, *5*(10), 4233–4250. DOI: 10.3390/en5104233

Moroń, W., & Rybak, W. (2015). NOx and SO2 emissions of coals, biomass and their blends under different oxy-fuel atmospheres. *Atmospheric Environment*, *116*, 65–71. DOI: 10.1016/j.atmosenv.2015.06.013

Morozzi, P., Ballarin, B., Arcozzi, S., Brattich, E., Lucarelli, F., Nava, S., Gómez-Cascales, P. J., Orza, J. A. G., & Tositti, L. (2021). Ultraviolet–Visible Diffuse Reflectance Spectroscopy (UV–Vis DRS), a rapid and non-destructive analytical tool for the identification of Saharan dust events in particulate matter filters. *Atmospheric Environment*, *252*, 118297. DOI: 10.1016/j.atmosenv.2021.118297

Morshedy, A. S., Ali, H. R., Nada, A. A., Rabie, A. M., & El-Maghrabi, H. H. (2021). Highly efficient Imprinted Polymer Nanocomposites for photocatalytic desulfurization of real diesel fuel. *Environmental Technology & Innovation*, *21*, 101206. DOI: 10.1016/j.eti.2020.101206

Mosca, L. P., Gapan, A., Angeles, R., & Lopez, E. C. (2023). Stability of metal–organic frameworks: Recent advances and future trends. *Engineering Proceedings*, *56*, 146. DOI: 10.3390/ASEC2023-16280

Motaleb, K. A., Pranta, A. D., Islam, M. R., Islam, S., & Janutėnienė, J. (2025). Bio-waste as a resource for sustainable nanocomposites: Strategies and multifunctional applications. *Functional Composites and Structures*, *7*(2), 022002. DOI: 10.1088/2631-6331/add9e0

Mourdikoudis, S., Dutta, S., Kamal, S., Gómez-Graña, S., Pastoriza-Santos, I., Wuttke, S., & Polavarapu, L. (2025). State-of-the-Art, Insights, and Perspectives for MOFs-Nanocomposites and MOF-Derived (Nano) Materials. *Advanced Materials*, *37*(52), 2415399. DOI: 10.1002/adma.202415399 PMID: 40255059

Mourdikoudis, S., Pallares, R. M., & Thanh, N. T. K. (2018). Characterization techniques for nanoparticles: Comparison and complementarity upon studying nanoparticle properties. *Nanoscale*, *10*(27), 12871–12934. DOI: 10.1039/C8NR02278J PMID: 29926865

Mousavi, A. (2020). An Investigation of How Nitrogen Monoxide (NO) and Nitrogen Dioxide (NO2) Are Produced in Oxidations by Nitric Acid (HNO3). *Comments on Modern Chemistry. Part A, Comments on Inorganic Chemistry : a Journal of Critical Discussion of the Current Literature*, *40*(6), 307–313. DOI: 10.1080/02603594.2020.1803296

Mo, Y., Zhang, Y., & Zhang, Q. (2024). The pulmonary effects of nickel-containing nanoparticles: Cytotoxicity, genotoxicity, carcinogenicity, and their underlying mechanisms. *Environmental Science. Nano*, *11*(5), 1817–1846. DOI: 10.1039/D3EN00929G PMID: 38984270

Mozetič, M. (2019). Surface modification to improve properties of materials. *Materials (Basel)*, *12*(3), 441. DOI: 10.3390/ma12030441 PMID: 30709009

Mubarik, S., Qureshi, N., Sattar, Z., Shaheen, A., Kalsoom, A., Imran, M., & Hanif, F. (2021). Synthetic approach to rice waste-derived carbon-based nanomaterials and their applications. *Nanomanufacturing*, *1*(3), 109–159. DOI: 10.3390/nanomanufacturing1030010

Mudakkar, S. R., Zaman, K., Khan, M. M., & Ahmad, M. (2013). Energy for economic growth, industrialization, environment and natural resources: Living with just enough. *Renewable & Sustainable Energy Reviews*, *25*, 580–595. DOI: 10.1016/j.rser.2013.05.024

Mudoi, R., & Saikia, L. (2025). Heteroatom Doping, Defect Engineering, and Stability of Transition Metal Diselenides for Electrocatalytic Water Splitting. *Chemistry, an Asian Journal*, *20*(19), e00755. Advance online publication. DOI: 10.1002/asia.202500755 PMID: 40783792

Mugisha, A. I. (2024). Synthesis and application of multifunctional iron oxide-graphene oxide nanocomposite materials for sulfur pollutant removal. Université d'Ottawa| University of Ottawa, Murrieta-Rico, F. N., Antúnez-García, J., Yocupicio-Gaxiola, R. I., Galván, D. H., GOnzález, J. C., & Petranovskii, V. (2022). Zeolites as initial structures for the preparation of functional materials. *Journal of Applied Research and Technology*, *20*(1), 92–116.

Mugisha, A. I., Fauteux-Lefebvre, C., Thangarasu, P., & Huerta-Aguilar, C. A. (2025). 2D graphene oxide–iron oxide catalysts for oxidation of sulfides. *Journal of Molecular Structure*, *1339*, 142382. DOI: 10.1016/j.molstruc.2025.142382

Muhammad, T., Ahmad, W., Ahmad, I., & Yaseen, M. (2024). Integrated oxidative-adsorptive desulfurization of model and real fuel oils over a Zn-impregnated hydroxyapatite-activated carbon (Zn/HA-AC) composite. *Journal of Sulfur Chemistry*, *45*(2), 217–238. DOI: 10.1080/17415993.2023.2264439

Muzio, L. J., & Quartucy, G. C. (1997). Implementing NOx control: Research to application. *Progress in Energy and Combustion Science*, *23*(3), 233–266. DOI: 10.1016/S0360-1285(97)00002-6

Naaz, A., Sheikh, A. R., & Arham, M. (2022). Oxidative and adsorptive desulfurization of fuel by using metal–organic frameworks: Recent progress. *IOSR Journal of Applied Chemistry*, *15*, 34–53. DOI: 10.9790/5736-1510013453

Nahian, M. S., Pritom, R., & Islam, M. M. (2024). Mechanistic Insights into the Evolution of Graphitic Carbon from Sulfur Containing Polar Aromatic Feedstock. *ACS Applied Materials & Interfaces*, *16*(41), 55478–55489. DOI: 10.1021/acsami.4c12324 PMID: 39356548

Nakano, K., Ali, S., Kim, H.-J., Kim, T., Alhooshani, K., Park, J.-I., & Mochida, I. (2013). Deep desulfurization of gas oil over NiMoS catalysts supported on alumina-coated USY-zeolite. *Fuel Processing Technology*, *116*, 44–51. DOI: 10.1016/j.fuproc.2013.04.012

Nakashima, K., Toshima, Y., Kobayashi, Y., Ishikawa, Y., & Kakihana, M. (2019). Solvothermal synthesis and morphology control of NaNbO 3 nanocubes using a reaction medium of water and/or methanol. *Journal of Asian Ceramic Societies*, *7*(4), 544–550. DOI: 10.1080/21870764.2019.1683955

Naseer, A. (2024). Role of nanocomposites and nano adsorbents for heavy metals removal and dyes. An overview. *Desalination and Water Treatment*, *320*, 100662. DOI: 10.1016/j.dwt.2024.100662

Nath, P. C., Ojha, A., Debnath, S., Sharma, M., Sridhar, K., Nayak, P. K., & Inbaraj, B. S. (2023). Biogeneration of valuable nanomaterials from agro-wastes: A comprehensive review. *Agronomy (Basel)*, *13*(2), 561. DOI: 10.3390/agronomy13020561

Nawaf, A., & Abdul Majeed, B. (2024). Kinetics study of oxidative desulfurization of real diesel fuel over uncoated and coated nano-catalysts in an oscillatory helical baffled reactor. *Journal of Chemical and Petroleum Engineering*, *58*(2), 359–374. DOI: 10.22059/jchpe.2024.377539.1522

Nayak, S., Das, M., Pradhan, B., & Kundu, S. N. (2019). Environmental Impacts of Fossil Fuels. *Sustainable Energy and Environment*, 245–270. DOI: 10.1201/9780429430107-13

Neira, M. P. (2019). Air pollution and human health: A comment from the World Health Organisation. *Annals of Global Health*, *85*(1), 141. DOI: 10.5334/aogh.2712 PMID: 31871904

Nel, A., Xia, T., Mädler, L., & Li, N. (2006). Toxic potential of materials at the nanolevel. *Science*, *311*(5761), 622–627. DOI: 10.1126/science.1114397 PMID: 16456071

Nezafat, Z., Dong, Y., Nasrollahzadeh, M., Shafiei, N., Gharoubi, H., & Javanshir, S. (2024). Recent progresses in energy conversion and storage of agricultural waste-derived (carbon/nano) materials: A review. *Green Chemistry*, *26*(21), 10687–10717. DOI: 10.1039/D3GC04332K

Ng, S. J. C., & Choi, A. E. S. (2025). Advancements in New Desulfurization Technology for Industry Scale Applications. *Results in Engineering*, *27*, 106577. DOI: 10.1016/j.rineng.2025.106577

Nguyen, M. D., Tran, H.-V., Xu, S., & Lee, T. R. (2021). Fe_3O_4 nanoparticles: Structures, synthesis, magnetic properties, surface functionalization, and emerging applications. *Applied Sciences (Basel, Switzerland)*, *11*(23), 11301. DOI: 10.3390/app112311301 PMID: 35844268

Ni, K., Wu, Y., Karimi, F., Gulbagca, F., Seyrankaya, A., Esra Altuner, E., Kocak, Y., & Sen, F. (2023). Palladium based bimetallic nanocatalysts: Synthesis, characterization and hydrogen fuel production. *Fuel*, *341*, 127577. DOI: 10.1016/j.fuel.2023.127577

Nikolaidis, P., & Poullikkas, A. (2017). A comparative overview of hydrogen production processes. *Renewable & Sustainable Energy Reviews*, *67*, 597–611. DOI: 10.1016/j.rser.2016.09.044

Nimbagal, V., Banapurmath, N., Sajjan, A. M., Patil, A. Y., & Ganachari, S. V. (2021). Studies on hybrid bio-nanocomposites for structural applications. *Journal of Materials Engineering and Performance*, *30*(9), 6461–6480. DOI: 10.1007/s11665-021-05843-9

Norizan, M. N., Shazleen, S. S., Alias, A. H., Sabaruddin, F. A., Asyraf, M. R. M., Zainudin, E. S., Abdullah, N., Samsudin, M. S., Kamarudin, S. H., & Norrahim, M. N. F. (2022). Nanocellulose-based nanocomposites for sustainable applications: A review. *Nanomaterials (Basel, Switzerland)*, *12*(19), 3483. DOI: 10.3390/nano12193483 PMID: 36234612

Nørskov, J. K., Bligaard, T., Rossmeisl, J., & Christensen, C. H. (2009). Towards the computational design of solid catalysts. *Nature Chemistry*, *1*(1), 37–46. DOI: 10.1038/nchem.121 PMID: 21378799

Nour, M. M., Tony, M. A., Fouad, M. K., & Nabwey, H. A. (2025). Biowaste-to-catalyst: Magnetite-functionalized potato shell as green magnetic biochar catalyst (PtS200–Fe3O4). *Catalysts*, *15*(10), 997. DOI: 10.3390/catal15100997

Nur Tatli, H., Beniz Gunduz, S., Sahin, M., Esra Altuner, E., & Ali Dar, U. (2024). Fluorimetric methods for determination of aluminum in water resources utilizing newly synthesized N,N'-bis(2,5-dihydroxybenzylidene)-4,4′-diamino diphenyl ether. *Spectrochimica Acta. Part A: Molecular and Biomolecular Spectroscopy, 321*, 124726. DOI: 10.1016/j.saa.2024.124726

Nyambuu, U., & Semmler, W. (2014). Trends in the extraction of non-renewable resources: The case of fossil energy. *Economic Modelling, 37*, 271–279. DOI: 10.1016/j.econmod.2013.11.020

Oberdörster, G., Oberdörster, E., & Oberdörster, J. (2005). Nanotoxicology. *Environmental Health Perspectives, 113*(7), 823–839. DOI: 10.1289/ehp.7339 PMID: 16002369

Oganov, A. R., & Glass, C. W. (2006). Crystal structure prediction using *ab initio* evolutionary techniques: Principles and applications. *The Journal of Chemical Physics, 124*(24), 244704. DOI: 10.1063/1.2210932 PMID: 16821993

Okpala, C. C. (2013). Nanocomposites–an overview. *Int. J. Eng. Res. Dev, 8*(11), 17.

Oksel Karakus, C., Bilgi, E., & Winkler, D. A. (2021). Biomedical nanomaterials: Applications, toxicological concerns, and regulatory needs. *Nanotoxicology, 15*(3), 331–351. DOI: 10.1080/17435390.2020.1860265 PMID: 33337941

Okunev, A. G., Parkhomchuk, E. V. E., Lysikov, A. I., Parunin, P. D., Semeykina, V. S., & Parmon, V. N. (2015). Catalytic hydroprocessing of heavy oil feedstocks. *Russian Chemical Reviews, 84*(9), 981–999. DOI: 10.1070/RCR4486

Olaremu, A. G., Adedoyin, W. R., Ore, O. T., & Adeola, A. O. (2021). Sustainable development and enhancement of cracking processes using metallic composites. *Applied Petrochemical Research, 11*(1), 1–18. DOI: 10.1007/s13203-021-00263-1

Oleiwi, A. H., Jabur, A. R., Alsalhy, Q. F., & Zeinedini, A. (2025). Crude oil desulfurization by polystyrene/graphene nanocomposite membranes. *Periodica Polytechnica. Chemical Engineering, 69*(1), 1–10. DOI: 10.3311/PPch.38227

Oloyede, C. T., Itabiyi, O. E., Popoola, O. A., Jekayinfa, S. O., Olaniyan, M. A., Adebayo, A. O., Ogunkunle, O., Zamri, M. F. M. A., & Fattah, I. M. R. (2024). Navigating prospects and challenges for green fuels for achieving economical, environmental and ecological resilience: A scientific review. *Biofuels, 15*(7), 929–941. DOI: 10.1080/17597269.2023.2299090

Oparin, V. N., Kiryaeva, T. A., Gavrilov, V. Y., Tanashev, Y. Y., & Bolotov, V. A. (2017). Initiation of underground fire sources. *Journal of Mining Science 2016 52:3, 52*(3), 576–592. DOI: 10.1134/S1062739116030850

Orlova, E. V. (2022). Design technology and AI-based decision-making model for digital twin engineering. *Future Internet, 14*(9), 248. DOI: 10.3390/fi14090248

Otsuki, J., Ishihara, A., & Nakayama, T. (2000). Oxidative desulfurization of diesel fuel using hydrogen peroxide. *Energy & Fuels, 14*(2), 279–284. DOI: 10.1021/ef990125p

Otsuki, S., Nonaka, T., Takashima, N., Qian, W., Ishihara, A., Imai, T., & Kabe, T. (2000). Oxidative desulphurization of light gas oil and vacuum gas oil by oxidation and solvent extraction. *Energy & Fuels, 14*(6), 1232–1239. DOI: 10.1021/ef000096e

Oviedo, F., Lavista Ferres, J., Buonassisi, T., & Butler, K. (2022). Interpretable and explainable machine learning for materials science and chemistry. *Accounts of Materials Research, 3*(6), 597–607. Advance online publication. DOI: 10.1021/accountsmr.1c00244

Oyedeji, S., Patel, N., Krishnamurthy, R., & Fatoba, P. O. (2024). Agricultural wastes to value-added products: economic and environmental perspectives for waste conversion. *Biowaste to Value-added Products: Economics and Technologies*, 215-248.

Ozcan, D. O., Hendekci, M. C., & Ovez, B. (2024). Enhancing the adsorption capacity of organic and inorganic pollutants onto impregnated olive stone derived activated carbon. *Heliyon, 10*(12), e32792. DOI: 10.1016/j.heliyon.2024.e32792 PMID: 38975120

Özdemir, S., Turkan, Z., Kılınc, E., Altuner, E. E., & Sen, F. (2023). Bioaccumulation, resistance, and remediation of Mn(II) and Cu(II) and their impacts on antioxidant enzymes of Anoxybacillus flavithermus. *International Journal of Environmental Science and Technology 2023 20:10, 20*(10), 10823–10834. DOI: 10.1007/s13762-023-05133-y

Ozdemir, S., Turkan, Z., Kilinc, E., Altuner, E. E., & Sen, F. (2022). Anoxybacillus flavithermus loaded ɣ-Fe2O3 magnetic nanoparticles as an efficient magnetic sorbent for the preconcentrations of Cu(II) and Mn(II). *Food and Chemical Toxicology, 168*, 113334. DOI: 10.1016/j.fct.2022.113334

Özkan, V., & Bozbeyoğlu, P. (2025). Production of Pure PVC, Graphene and Carbon Nanotube-Doped PVC Electrospun Nanofiber Membranes: A Highly Effective Method for Removing Acid Blue 74 Dye from Wastewater. *Black Sea Journal of Engineering and Science, 8*(2), 517–523. DOI: 10.34248/bsengineering.1634974

Özkan, V., & Özkan, A. (2022). Adsorptive desulfurization of crude oil with clinoptilolite zeolite. *Natural and Engineering Sciences, 7*(3), 284–293. DOI: 10.28978/nesciences.1222495

Pandey, R. A., Biswas, R., Chakrabarti, T., & Devotta, S. (2005). Flue gas desulfurization: Physicochemical and biotechnological approaches. *Critical Reviews in Environmental Science and Technology, 35*(6), 571–622. DOI: 10.1080/10643380500326374

Panjan, P., Drnovšek, A., Gselman, P., Čekada, M., & Panjan, M. (2020). Review of Growth Defects in Thin Films Prepared by PVD Techniques. *Coatings 2020, Vol. 10, Page 447, 10*(5), 447. DOI: 10.3390/coatings10050447

Pantoja-Romero, W., Lavín-Flores, A., & Morell, G.. (2025). Solvent-free MOF–quantum dot composites. *Chemistry (Weinheim an der Bergstrasse, Germany), 31*(34), e202500655. DOI: 10.1002/chem.202500655 PMID: 40326208

Pan, Y., Huang, W., Liu, Z. H., Yang, Z. Z., Tan, J. J., & Xu, B. M. (2025). Recent progress of MOFs for liquid-fuel desulfurization. *Separation and Purification Reviews, 54*(2), 128–148. DOI: 10.1080/15422119.2024.2367646

Patil, A., Chaudhari, V., Patil, S. R., Borse, G. P., & Patil, V. (2023). A bio-waste derived sustainable heterogeneous catalyst for Biginelli reaction. *Journal of the Indian Chemical Society, 100*(9), 101080. DOI: 10.1016/j.jics.2023.101080

Pavithra, S., Sakunthala, A., Vinofia, S., Pramoda, K., Kathirvel, V., & Keri, R. S. (2024). Superior performance of hierarchical 3D microflower CuS grown on carbon cloth as binder free electrode for supercapacitor applications. *Materials Research Bulletin, 180*, 113010. DOI: 10.1016/j.materresbull.2024.113010

Pavlenko, V., Khosravi, H. S., Żółtowska, S., Haruna, A. B., Zahid, M., Mansurov, Z., Supiyeva, Z., Galal, A., Ozoemena, K. I., Abbas, Q., & Jesionowski, T. (2022). A comprehensive review of template-assisted porous carbons: Modern preparation methods and advanced applications. *Materials Science and Engineering R Reports, 149*, 100682. DOI: 10.1016/j.mser.2022.100682

Peixoto, R. S., Vermelho, A. B., & Rosado, A. S. (2011). Petroleum-Degrading Enzymes: Bioremediation and New Prospects. *Enzyme Research, 2011*(1), 475193. DOI: 10.4061/2011/475193

Peng, Y., Zhang, Y., Huang, H., & Zhong, C. (2018). Flexibility induced high-performance MOF-based adsorbent for nitroimidazole antibiotics capture. *Chemical Engineering Journal, 333*, 678–685. DOI: 10.1016/j.cej.2017.09.138

Perego, C., & Bosetti, A. (2011). Biomass to fuels: The role of zeolite and mesoporous materials. *Microporous and Mesoporous Materials, 144*(1-3), 28–39. DOI: 10.1016/j.micromeso.2010.11.034

Pérez Sánchez, A. J., Pérez Sánchez, E. J., & Segura Silva, R. M. (2016). Design of a packed-bed absorption column considering four packing types and applying MATLAB. *Nexo Revista Científica, ISSN-e 1995-9516, Vol. 29, No. 2, 2016 (Ejemplar Dedicado a: Nexo Revista Científica), Págs. 83-104, 29*(2), 83–104. DOI: 10.5377/nexo.v29i2.4577

Pérez, I. A., García, M. Á., Sánchez, M. L., Pardo, N., & Fernández-Duque, B. (2020). Key Points in Air Pollution Meteorology. *International Journal of Environmental Research and Public Health 2020, Vol. 17, Page 8349, 17*(22), 8349. DOI: 10.3390/ijerph17228349

Pérez-Botella, E., Valencia, S., & Rey, F. (2022). Zeolites in adsorption processes: State of the art and future prospects. *Chemical Reviews, 122*(24), 17647–17695. DOI: 10.1021/acs.chemrev.2c00140 PMID: 36260918

Pérez-Mayoral, E., Calvino-Casilda, V., & Soriano, E. (2016). Metal-supported carbon-based materials: Opportunities and challenges in the synthesis of valuable products. *Catalysis Science & Technology, 6*(5), 1265–1291. DOI: 10.1039/C5CY01437A

Periasamy, P., & Mohanta, Y. K. (2024). Global energy crisis: need for energy conversion and storage. In *Green Nanomaterials in Energy Conversion and Storage Applications* (pp. 45–73). Apple Academic Press.

Perot, G., & Guisnet, M. (1990). Advantages and disadvantages of zeolites as catalysts in organic chemistry. *Journal of Molecular Catalysis, 61*(2), 173–196. DOI: 10.1016/0304-5102(90)85154-A

Perraud, V., Horne, J. R., Martinez, A. S., Kalinowski, J., Meinardi, S., Dawson, M. L., Wingen, L. M., Dabdub, D., Blake, D. R., Gerber, R. B., & Finlayson-Pitts, B. J. (2015). The future of airborne sulfur-containing particles in the absence of fossil fuel sulfur dioxide emissions. *Proceedings of the National Academy of Sciences of the United States of America, 112*(44), 13514–13519. DOI: 10.1073/pnas.1510743112

Perreault, F., De Faria, A. F., & Elimelech, M. (2015). Environmental applications of graphene-based nanomaterials. *Chemical Society Reviews, 44*(16), 5861–5896. DOI: 10.1039/C5CS00021A PMID: 25812036

Pervaiz, S., Javed, M., Shah, A., Latif, A., Nasir, S., & Shah, I. (2025). Environmental applications of magnetic nanohybrid materials. *RSC Advances, 15*(25), 19899–19936. DOI: 10.1039/D5RA03470A PMID: 40503319

Piccinno, F., Hischier, R., Seeger, S., & Som, C. (2012). Scale-up framework for chemical processes. *Journal of Cleaner Production*, *28*, 108–116. DOI: 10.1016/j.jclepro.2011.12.035

Pietsch, W. B. (2004). *Agglomeration in industry: Occurrence and applications* (Vol. 1). John Wiley & Sons. DOI: 10.1002/9783527619795

Pirsaheb, M., Gholami, T., & Seifi, H.. (2024). Plant extracts in green nanomaterial synthesis. *Environmental Science and Pollution Research International*, *31*(17), 24768–24787. DOI: 10.1007/s11356-024-32983-x PMID: 38523214

Poopanya, P., Gadhavi, P. M., Talati, M., Sivalertporn, K., Som, N. N., & Mishra, A. K. (2025). Tailoring graphene for thermoelectric: Theoretical insights into position-dependent effects of BN Co-doping. *Computational Condensed Matter*, *44*, e01106. DOI: 10.1016/j.cocom.2025.e01106

Potyrailo, R. A., Rajan, K., Stoewe, K., Takeuchi, I., Chisholm, B., & Lam, H. (2011). Combinatorial and high-throughput screening of materials libraries: A tutorial. *ACS Combinatorial Science*, *13*(6), 579–633. DOI: 10.1021/co200007w PMID: 21644562

Pouladi Borj, B., Fanaei, M. A., Esfandyari, M., Naddaf, A., Jafari, D., & Baghmisheh, G. (2024). Machine learning-assisted methods for prediction and optimisation of oxidative desulfurization of gas condensate via a novel oxidation system. *Journal of Sulfur Chemistry*, *45*(1), 84–100. DOI: 10.1080/17415993.2023.2257827

Prasad, R. D., Prasad, R. S., Prasad, R. B., Prasad, S. R., Singha, S. B., Prasad, R. J., Sinha, P., Saxena, S., Vaidya, A. K. B., Teli, S. B., Saxena, U. R., Harale, A., Deshmukh, M. B., Padvi, M. N., & Navathe, G. J. (2024). *A Review on Modern Characterization Techniques for Analysis of Nanomaterials and Biomaterials*. ES Energy & Environment., DOI: 10.30919/esee1087

Prins, R., Bus, E., & van Bokhoven, J. A. (2006). Catalytic hydrodesulfurization. *Catalysis Today*, *117*(1–3), 177–189. DOI: 10.1016/j.cattod.2006.05.012

Priyadarshini, P., Sahoo, S. K., & Parida, K. (2025). Hierarchically structured MOF-on-MOF photocatalysts with engineered charge dynamics for sustainable green fuel generation. *RSC Sustainability*. Advance online publication. DOI: 10.1039/D5SU00595G

Priya, V. N., Rajkumar, M., Rajendran, V., Mobika, J., Sibi, S. P. L., Veena, B., Vijayalakshmi, V., & Ahila, P. (2025). Sustainable selenium ions adsorption of cyclodextrin- and cellulose-functionalized layered double hydroxide/reduced graphene oxide nanocomposites. *Journal of Water Process Engineering*, *69*, 106580. DOI: 10.1016/j.jwpe.2024.106580

Qasim, M., Clarkson, A. N., & Hinkley, S. F. (2023). Biomass-derived carbon nanoparticles. *International Journal of Molecular Sciences, 24*(2), 1023. DOI: 10.3390/ijms24021023 PMID: 36674532

Qiu, X., Wang, B., Wang, R., & Kozhevnikov, I. V. (2024). New adsorption materials for deep desulfurization of fuel oil. *Materials (Basel), 17*(8), 1803. DOI: 10.3390/ma17081803 PMID: 38673161

Qu, J., Shi, L., He, C., Gao, F., Li, B., Zhou, Q., Hu, H., Shao, G., Wang, X., & Qiu, J. (2014). Highly efficient synthesis of graphene/MnO_2 hybrids and their application for ultrafast oxidative decomposition of methylene blue. *Carbon, 66*, 485–492. DOI: 10.1016/j.carbon.2013.09.025

Rabbani, M. S., Wei, S., Anwar, M. I., Zafar, Z., Liu, Z.-Y., Wang, Y., & Yue, X.-Z. (2025). Al3+ Doping-Induced Electronic Structure Regulation in NiFeO for Accelerated Oxygen Evolution Kinetics and Superior Stability. *Journal of Alloys and Compounds, 182831*, 182831. Advance online publication. DOI: 10.1016/j.jallcom.2025.182831

Rabiee, N. (2025). Green Nanomaterials: Sustainable Innovations and Diverse Applications. ACS Symposium SeriesVol. 1520, Chapter 1pp 1-27, eISBN. DOI: 10.1021/bk-2025-1520

Raccuglia, P., Elbert, K. C., Adler, P. D. F., Falk, C., Wenny, M. B., Mollo, A., & Norquist, A. J. (2016). Machine-learning-assisted materials discovery at record speed. *Nature, 533*(7601), 73–76. DOI: 10.1038/nature17439 PMID: 27147027

Radišić, B. (2024).. . *Non-Renewable and Renewable Energy Sources., 2024*(1), 1–5. DOI: 10.3997/2214-4609.202449BGS49

Radulescu, D. M., Surdu, V. A., Ficai, A., Ficai, D., Grumezescu, A. M., & Andronescu, E. (2023). Green synthesis of metal and metal oxide nanoparticles: A review of the principles and biomedical applications. *International Journal of Molecular Sciences, 24*(20), 15397. DOI: 10.3390/ijms242015397 PMID: 37895077

Radwan, M. S., Aboutaleb, W. A., El Naggar, A. M., Medany, S. S., & Shehata, M. R. (2023). Synergetic effect of Fe_2O_3–CeO_2 nanocomposites prepared via different techniques on photocatalytic desulfurization of heavy gas oil. *Arabian Journal for Science and Engineering, 48*(12), 15837–15850. DOI: 10.1007/s13369-023-07633-y

Rafiee, E., Noori, E., & Eavani, S. (2018). Nanoadsorbents in fuel desulfurization. *Journal of Industrial and Engineering Chemistry, 61*, 1–16. DOI: 10.1016/j.jiec.2017.12.021

Ragazzi, M., Ionescu, G., & Cioranu, S. I. (2017). Assessment of environmental impact from renewable and non-renewable energy sources. *International Journal of Energy Production and Management*, *2*(1), 8–16. DOI: 10.2495/EQ-V2-N1-8-16

Rahman, K. M., Amiri, O., Ahmed, S. S., Ismael, S. J., Rasul, N. S., Babakrb, K. A., Dadkhah, M., & Jamal, M. A. (2024). Mechanism and kinetic of piezo-catalytic desulfurization of model and actual fuel samples over CexOy/SrO nanocomposite at room temperature. *Heliyon*, *10*(2), e24707. DOI: 10.1016/j.heliyon.2024.e24707 PMID: 38304796

Rajabi, F., Vessally, E., Luque, R., & Voskressensky, L. (2021). Highly efficient and selective aqueous aerobic oxidation of sulfides to sulfoxides or sulfones catalyzed by tungstate-functionalized nanomaterial. *Molecular Catalysis*, *515*, 111931. DOI: 10.1016/j.mcat.2021.111931

Rajendran, A., Cui, T.-y., Fan, H., Yang, Z., Feng, J., & Li, W.-y. (2020). A comprehensive review on oxidative desulfurization catalysts targeting clean energy and environment. *Journal of Materials Chemistry. A, Materials for Energy and Sustainability*, *8*(5), 2246–2285. DOI: 10.1039/C9TA12555H

Rajendran, S., Al-Samydai, A., Palani, G., Trilaksana, H., Sathish, T., Giri, J., Saravanan, R., Lalvani, J. I. J., & Nasri, F. (2025). Replacement of Petroleum Based Products With Plant-Based Materials, Green and Sustainable Energy—A Review. *Engineering Reports*, *7*(4), e70108. DOI: 10.1002/eng2.70108

Ramadan, A. E. K. (2004). *Acid Deposition Phenomena.*

Ramadan, B. S., Wibowo, Y. G., Anwar, D., & Maryani, A. T. (2024). A review of life cycle assessment of nanomaterials-based adsorbent for environmental remediation. *Global NEST Journal*, *26*(7), 06216. DOI: 10.30955/gnj.06216

Ramesh, M., Rajeshkumar, L., Balaji, D., & Bhuvaneswari, V. (2023). Sustainable and renewable nano-biocomposites for sensors and actuators: A review on preparation and performance. *Current Analytical Chemistry*, *19*(1), 38–69. DOI: 10.2174/1573411018666220421112916

Ramprasad, R., Batra, R., Pilania, G., Mannodi-Kanakkithodi, A., & Kim, C. (2017). Machine learning in materials informatics: Recent applications and prospects. *npj Computational Materials*, *3*, 54. DOI: 10.1038/s41524-017-0056-5

Rana, M. S., Sámano, V., Ancheyta, J., & Díaz, J. A. I. (2007). A review of recent advances on process technologies for upgrading of heavy oils and residue. *Fuel*, *86*(9), 1216–1231. DOI: 10.1016/j.fuel.2006.08.004

Rando, G., Sfameni, S., Galletta, M., Drommi, D., Cappello, S., & Plutino, M. R. (2022). Functional Nanohybrids and Nanocomposites development for the removal of environmental pollutants and bioremediation. *Molecules (Basel, Switzerland), 27*(15), 4856. DOI: 10.3390/molecules27154856 PMID: 35956804

Rani, N., Singh, P., Kumar, S., Kumar, P., Bhankar, V., & Kumar, K. (2023). Plant-mediated synthesis of nanoparticles and their applications: A review. *Materials Research Bulletin, 163*, 112233. DOI: 10.1016/j.materresbull.2023.112233

Rao, J. J., Leal, E., Trejo, F., & Ancheyta, J. (2023). Kinetic models of deep hydrotreating reactions to produce ultralow sulfur diesel. *Energy & Fuels, 37*(15), 11216–11247. DOI: 10.1021/acs.energyfuels.3c01466

Rashid, R., Shafiq, I., Akhter, P., Iqbal, M. J., & Hussain, M. (2021). A state-of-the-art review on wastewater treatment techniques: The effectiveness of adsorption method. *Environmental Science and Pollution Research International, 28*(8), 9050–9066. DOI: 10.1007/s11356-021-12395-x PMID: 33483933

Rashid, T., Raza, A., Saleh, H. A., Khan, S., Rahaman, S., Pandey, K., AlDamen, M. A., Sama, F., Ahmad, A., Shahid, M., & Ahmad, S. A. (2023). Green synthesis of Ni/Fe3O4/rGO nanocomposites for desulfurization of fuel. *ACS Applied Nano Materials, 6*(20), 18905–18917. DOI: 10.1021/acsanm.3c03270

Raza, Q., Bibi, I., Majid, F., Dahshan, A., Kamal, S., Jilani, K., Fatima, G., Taj, B., Alwadai, N., & Iqbal, M. (2024). Dielectric, ferroelectric, optical and photocatalytic properties of Al and Ag-doped strontium hexaferrite ($SrFe_{12}O_{19}$) synthesized via facile micro-emulsion approach. *Journal of Alloys and Compounds, 985*, 174098. DOI: 10.1016/j.jallcom.2024.174098

Reis Medeiros, K. A., da Costa, L. G., Bifano Manea, G. K., de Moraes Maciel, R., Caliman, E., da Silva, M. T., de Sena, R. C., & de Oliveira, E. C. (2024). Determination of total sulfur content in fuels: A comprehensive and metrological review. *Critical Reviews in Analytical Chemistry, 54*(8), 3398–3408. Advance online publication. DOI: 10.1080/10408347.2023.2249564 PMID: 37610291

Rene, E. R., Veiga, M. C., & Kennes, C. (2012). Combined biological and physicochemical waste-gas cleaning techniques. *Journal of Environmental Science and Health. Part A, Toxic/Hazardous Substances & Environmental Engineering, 47*(7), 920–939. DOI: 10.1080/10934529.2012.667289

Ren, J., Cao, J.-P., Yang, F.-L., Liu, Y.-L., Tang, W., & Zhao, X.-Y. (2021). Understandings of catalyst deactivation and regeneration during biomass tar reforming: A crucial review. *ACS Sustainable Chemistry & Engineering, 9*(51), 17186–17206. DOI: 10.1021/acssuschemeng.1c07483

Rezaei, B., Yari, P., Sanders, S. M., Wang, H., Chugh, V. K., Liang, S., & Gómez-Pastora, J. (2024). Magnetic nanoparticles: A review on synthesis, characterization, functionalization, and biomedical applications. *Small, 20*(5), 2304848. DOI: 10.1002/smll.202304848 PMID: 37732364

Riaz, F., Rasul, I., Azeem, F., ... (2024). Environmental impact and LCA of bionanocomposites. In Advances in Bionanocomposites (pp. 387–406). DOI: 10.1016/B978-0-323-91764-3.00009-7

Riseh, R. S., Vazvani, M. G., Hassanisaadi, M., & Thakur, V. K. (2024). Agricultural wastes: A practical and potential source for the isolation and preparation of cellulose and application in agriculture and different industries. *Industrial Crops and Products, 208*, 117904. DOI: 10.1016/j.indcrop.2023.117904

Ritchie, H., & Rosado, P. (2017). Fossil fuels. *Our World in Data*. https://ourworldindata.org/fossil-fuels

Rizwan, M., Ali, M. F., Nawaz, M. A., He, M., Song, Y., Jiang, P., Ullah, S., Hassan, M. M. A., & Zhou, X. (2025). Synergistic effect of NaOH/NaClO absorbent in a novel wet FGD scrubber to control SOx/NOx emissions. *Environmental Monitoring and Assessment, 197*(2), 170. DOI: 10.1007/s10661-024-13455-8 PMID: 39808324

Rocha, L. S., Sousa, É. M., Pereira, D., Gil, M. V., Otero-Irurueta, G., Gallo, M. J. H., & Calisto, V. (2021). Sustainable and recoverable waste-based magnetic nanocomposites used for the removal of pharmaceuticals from wastewater. *Chemical Engineering Journal, 426*, 129974. DOI: 10.1016/j.cej.2021.129974

Rogalinski, T., Ingram, T., & Brunner, G. (2008). Hydrolysis of lignocellulosic biomass in water under elevated temperatures and pressures. *The Journal of Supercritical Fluids, 47*(1), 54–63. DOI: 10.1016/j.supflu.2008.05.003

Rojas, L., Peña, Á., & Garcia, J. (2025). AI-driven predictive maintenance in mining: A systematic literature review on fault detection, digital twins, and intelligent asset management. *Applied Sciences (Basel, Switzerland), 15*(6), 3337. DOI: 10.3390/app15063337

Roman, F. F., Diaz de Tuesta, J. L., Silva, A. M. T., Faria, J. L., & Gomes, H. T. (2021). Carbon-based materials for oxidative desulfurization and denitrogenation of fuels: A review. *Catalysts, 11*(10), 1239. DOI: 10.3390/catal11101239

Rossoni, A. L., de Vasconcellos, E. P. G., & de Castilho Rossoni, R. L. (2024). Barriers and facilitators of university-industry collaboration for research, development and innovation: A systematic review. *Management Review Quarterly, 74*(3), 1841–1877. DOI: 10.1007/s11301-023-00349-1

Roy, A., Sharma, A., Yadav, S., Jule, L. T., & Krishnaraj, R. (2021). Nanomaterials for remediation of environmental pollutants. *Bioinorganic Chemistry and Applications, 2021*(1), 1764647. PMID: 34992641

Sabbe, M. K., Reyniers, M. F., & Reuter, K. (2012). First-principles kinetic modeling in heterogeneous catalysis: An industrial perspective on best practice, gaps and needs. *Catalysis Science & Technology, 2*(10), 2010–2024. DOI: 10.1039/c2cy20261a

Sabukuttan, D. (2025). Hydrothermal Synthesis and Physicochemical Characterization of Carbon Quantum Dots from Marine Cyanobacteria Leptolyngbya sp. nov. KMBMA-1. *BioNanoScience, 15*(3), 1–13.

Sadegh-Vaziri, R., & Babler, M. U. (2019). Removal of hydrogen sulfide with metal oxides in packed bed reactors—A review from a modeling perspective with practical implications. *Applied Sciences (Basel, Switzerland), 9*(24), 5316. DOI: 10.3390/app9245316

Sadrara, M., & Khorrami, M. K. (2023). Designing an efficient organic–inorganic hybrid nanocomposite for simultaneous oxidative/adsorptive desulfurization of model and real fuel oils. *Scientific Reports, 13*(1), 15134. DOI: 10.1038/s41598-023-42392-8 PMID: 37704702

Saeed, M., Firdous, A., Zaman, M. S., Izhar, F., Riaz, M., Haider, S., & Tariq, S. (2023). MOFs for desulfurization: Recent advances. *Journal of the Chinese Chemical Society (Taipei), 70*(4), 789–824. DOI: 10.1002/jccs.202200546

Saeed, M., Waseem, A., & Intisar, A. (2025). Catalytic activity of MnWO4@g-C3N4 and CuWO4@g-C3N4 nanocomposites for green fuel production. *Catalysis Surveys from Asia, 29*(1), 11–25. DOI: 10.1007/s10563-024-09437-y

Sagin, S., and Sagin, A. (2023). Managing risk factors for emergency situations when using low-sulfur content fuel in marine diesel engines. Technology Audit and Production Reserves, 5(1/73), 37–43. .DOI: 10.15587/2706-5448.2023.290198

Saha, B., Vedachalam, S., & Dalai, A. K. (2021). Review on recent advances in adsorptive desulfurization. *Fuel Processing Technology, 214*, 106685. DOI: 10.1016/j.fuproc.2020.106685

Sahraei, S. (2023). Assessment of reaction parameters in the oxidative desulfurization reaction. *Energy & Fuels, 37*(20), 15373–15393. DOI: 10.1021/acs.energyfuels.3c02169

Salazar Sandoval, S., Amenábar, A., Toledo, I., Silva, N., & Contreras, P. (2024). Advances in the sustainable development of biobased materials using plant and animal waste as raw materials: A review. *Sustainability (Basel)*, *16*(3), 1073. DOI: 10.3390/su16031073

Saleh, T. A. (2020). Characterisation, determination and elimination technologies for sulfur from petroleum: Toward cleaner fuel and a safe environment. *Trends in Environmental Analytical Chemistry*, *25*, e00080. DOI: 10.1016/j.teac.2020.e00080

Saleh, T. A. (2022). Global trends in technologies and nanomaterials for removal of sulfur organic compounds: Clean energy and green environment. *Journal of Molecular Liquids*, *359*, 119340. DOI: 10.1016/j.molliq.2022.119340

Saleh, T. A., Sulaiman, K. O., Al-Hammadi, S. A., Dafalla, H., & Danmaliki, G. I. (2017). Adsorptive desulfurization of thiophene, benzothiophene and dibenzothiophene over activated carbon manganese oxide nanocomposite: With column system evaluation. *Journal of Cleaner Production*, *154*, 401–412. DOI: 10.1016/j.jclepro.2017.03.169

Salih, Y. M. (2023). Adsorptive desulfurization of DBT using hybrid MOF–activated carbon. *Journal of Sulfur Chemistry*, *44*(6), 694–711. DOI: 10.1080/17415993.2023.2241594

Salman, M., Ur Rahman, A., Usman Farooq, M., Shakir, M., Subhan, F., Akram, F., Khan, K., Ullah, M., & Yaseen, M. (2024). Advances in the applications of mixed matrix membranes for desulfurization of transportation fuels. *Separation and Purification Reviews*, *53*(4), 448–474. DOI: 10.1080/15422119.2024.2322425

Sánchez, A. L., Urzay, J., & Liñán, A. (2015). The role of separation of scales in the description of spray combustion. *Proceedings of the Combustion Institute*, *35*(2), 1549–1577. DOI: 10.1016/j.proci.2014.08.018

Sanchez-Lengeling, B., & Aspuru-Guzik, A. (2018). Inverse molecular design using machine learning: Generative models for matter engineering. *Science*, *361*(6400), 360–365. DOI: 10.1126/science.aat2663 PMID: 30049875

Sang, P., Chen, Q., Wang, D.-Y., Guo, W., & Fu, Y. (2023). Organosulfur materials for rechargeable batteries: Structure, mechanism, and application. *Chemical Reviews*, *123*(4), 1262–1326. DOI: 10.1021/acs.chemrev.2c00739 PMID: 36757873

Sankar, M. K., Parvez, M., Rinawa, M. L., Chaudhary, V., Gupta, S., & Gupta, P. (2022). The aspect of green nanocomposites in green technology and sustainable development: state of the art and new challenges. *Nanomaterials for Innovative Energy Systems and Devices*, 583-599.

Sarbassov, Y., Duan, L., Manovic, V., & Anthony, E. J. (2018). Sulfur trioxide formation/emissions in coal-fired air- and oxy-fuel combustion processes: A review. *Greenhouse Gases: Science and Technology*, *8*(3), 402–428. DOI: 10.1002/ghg.1767

Sarkar, M., Sarkar, C., & Saha, S. (2025). Catalytic activity of biodegradable polymeric colloidal particles in ground water remediation: recent progress and challenges. Journal of Macromolecular Science, Part A, 1-36.

Saxena, P., & Shukla, P. (2023). A review on recent developments and advances in environmental gas sensors to monitor toxic gas pollutants. *Environmental Progress & Sustainable Energy*, *42*(5), e14126. DOI: 10.1002/ep.14126

Scala, F., D'Ascenzo, M., & Lancia, A. (2004). Modeling flue gas desulfurization by spray-dry absorption. *Separation and Purification Technology*, *34*(1–3), 143–153. DOI: 10.1016/S1383-5866(03)00188-6

Schleder, G. R., Padilha, A. C. M., Acosta, C. M., Costa, M., & Fazzio, A. (2019). From DFT to machine learning: Recent approaches to materials science—A review. *JPhys Materials*, *2*(3), 032001. DOI: 10.1088/2515-7639/ab084b

Schneider, G. (2018). Automating drug discovery. *Nature Reviews. Drug Discovery*, *17*(2), 97–113. DOI: 10.1038/nrd.2017.232 PMID: 29242609

Schnelle, S., & Favaro, F. M. (2023). ADS standardization landscape: Making sense of its status and of the associated research questions. arXiv preprint arXiv:2306.17682.

Schobesberger, S., Franchin, A., Bianchi, F., Rondo, L., Duplissy, J., Kürten, A., Ortega, I. K., Metzger, A., Schnitzhofer, R., Almeida, J., Amorim, A., Dommen, J., Dunne, E. M., Ehn, M., Gagné, S., Ickes, L., Junninen, H., Hansel, A., Kerminen, V. M., & Worsnop, D. R. (2015). On the composition of ammonia-sulfuric-acid ion clusters during aerosol particle formation. *Atmospheric Chemistry and Physics*, *15*(1), 55–78. DOI: 10.5194/acp-15-55-2015

Schwietzke, S., Sherwood, O. A., Bruhwiler, L. M. P., Miller, J. B., Etiope, G., Dlugokencky, E. J., Michel, S. E., Arling, V. A., Vaughn, B. H., White, J. W. C., & Tans, P. P. (2016). Upward revision of global fossil fuel methane emissions based on isotope database. *Nature 2016 538:7623*, *538*(7623), 88–91. DOI: 10.1038/nature19797

Schwietzke, S., Griffin, W. M., Matthews, H. S., & Bruhwiler, L. M. P. (2014). Natural Gas Fugitive Emissions Rates Constrained by Global Atmospheric Methane and Ethane. *Environmental Science & Technology*, *48*(14), 7714–7722. DOI: 10.1021/es501204c

Sedyaaw, P., Pandey, A., Kawade, S. S., Ade, P. J., Nimbarte, P. V., Bhaladhare, D. R., & Kamble, P. A. (2024). A review on acid rain it's causes, effects and management measures. [IJCRT]. *International Journal of Creative Research Thoughts, 12*(4), q959–q968.

Seewald, J. S., Benitez-Nelson, B. C., & Whelan, J. K. (1998). Laboratory and theoretical constraints on the generation and composition of natural gas. *Geochimica et Cosmochimica Acta, 62*(9), 1599–1617. DOI: 10.1016/S0016-7037(98)00000-3

Segovia-Hernández, J. G., Hernández, S., Cossío-Vargas, E., Juarez-García, M., & Sánchez-Ramírez, E. (2025). Green hydrogen production for sustainable development: A critical examination of barriers and strategic opportunities. *RSC Sustainability, 3*(1), 134–157. DOI: 10.1039/D4SU00630E

Sehgal, B., & Kunde, G. (2022). Recent Advances in the Catalytic Applications of Magnetic Nanomaterials. Emerging Applications of Low Dimensional Magnets, 9-31.

Sehhatigdiri, A., & Pourhossein, M. (2024). Application of aminated chitosan/functionalized CNT nanocomposite in adsorption desulfurization of fuels containing dibenzothiophene. *Fullerenes, Nanotubes, and Carbon Nanostructures, 32*(5), 483–492. DOI: 10.1080/1536383X.2023.2296124

Selvaraj, V., Sagadevan, S., & Karuppasamy, G. (2024). Scale-Up Process of Green Nanomaterials. In Sustainable Green Synthesised Nano-Dimensional Materials for Energy and Environmental Applications (pp. 107–122). CRC Press. DOI: 10.1201/9781003362241-6

Senthil, K. P. (2024). Employing carbon nanocomposites for the removal of hazardous contaminants from aqueous solution. *Malaysian NANO-An International Journal, 4*(1), 24–44. DOI: 10.22452/mnij.vol4no1.2

Serafin, J., Dziejarski, B., & Sreńscek-Nazzal, J. (2023). An innovative and environmentally friendly bioorganic synthesis of activated carbon based on olive stones and its potential application for CO2 capture. *Sustainable Materials and Technologies, 38*, e00717. DOI: 10.1016/j.susmat.2023.e00717

Seriket, H., Bougzime, O., Song, Y., Qi, H. J., & Demoly, F. (2025). Reinforcement learning-enabled design of topological interlocking materials for sustainable multi-material additive manufacturing. *Additive Manufacturing, 104992*, 104992. Advance online publication. DOI: 10.1016/j.addma.2025.104992

Sevilla, M., Díez, N., & Fuertes, A. B. (2021). More sustainable chemical activation strategies for the production of porous carbons. *ChemSusChem, 14*(1), 94–117. DOI: 10.1002/cssc.202001838 PMID: 33047490

Sevilla, M., & Fuertes, A. B. (2011). Sustainable porous carbons. *Energy & Environmental Science*, *4*(5), 1765–1771. DOI: 10.1039/c0ee00784f

Sevilla, M., Mokaya, R., & Fuertes, A. B. (2017). Nitrogen-doped porous carbons from algae. *Energy & Environmental Science*, *4*(8), 2930–2936. DOI: 10.1039/c1ee01608c

Sfeir, A., Teles, C. A., Ciotonea, C., Reddy, G. N. M., Marinova, M., Dhainaut, J., Löfberg, A., Dacquin, J.-P., Royer, S., & Laassiri, S. (2023). Enhancing ammonia catalytic production over spatially confined cobalt molybdenum nitride nanoparticles in SBA-15. *Applied Catalysis B: Environmental*, *325*, 122319. DOI: 10.1016/j.apcatb.2022.122319

Shadmehri, M. A., Housaindokht, M. R., & Nakhaei Pour, A. (2023). Metal oxides supported on GO for oxidative desulfurization. *Applied Organometallic Chemistry*, *37*(11), e7272. DOI: 10.1002/aoc.7272

Shafiee, S., & Topal, E. (2009). When will fossil fuel reserves be diminished? *Energy Policy*, *37*(1), 181–189. DOI: 10.1016/j.enpol.2008.08.016

Shafiee, S., & Topal, E. (2010). A long-term view of worldwide fossil fuel prices. *Applied Energy*, *87*(3), 988–1000. DOI: 10.1016/j.apenergy.2009.09.012

Shafiq, I., Shafique, S., Akhter, P., Abbas, G., Qurashi, A., & Hussain, M. (2022). Efficient catalyst development for deep aerobic photocatalytic oxidative desulfurization: Recent advances, confines, and outlooks. *Catalysis Reviews. Science and Engineering*, *64*(4), 789–834. DOI: 10.1080/01614940.2020.1864859

Shah, M. A., Hayder, G., Kumar, R., Kumar, V., Ahamad, T., Kalam, M. A., Soudagar, M. E. M., Mohamed Shamshuddin, S. Z., & Mubarak, N. M. (2023). Development of sustainable biomass residues for biofuels applications. *Scientific Reports*, *13*(1), 14248. DOI: 10.1038/s41598-023-41446-1 PMID: 37648719

Shaik, S., & Franchetti, M. J. (2022). Recycling and sustainability of nanocomposites. In *Applications of Nanocomposites* (pp. 236–259). CRC Press. DOI: 10.1201/9781003247074-11

Shakirzyanov, R. I., Kozlovskiy, A., Zdorovets, M. V., Zheludkevich, A., Shlimas, D., Abmiotka, N., Kazantsev, P. A., Zubar, T. I., Trukhanov, S. V., & Trukhanov, A. (2023). Impact of thermobaric conditions on phase content, magnetic and electrical properties of the $CoFe_2O4$ ceramics. *Journal of Alloys and Compounds*, *954*, 170083. DOI: 10.1016/j.jallcom.2023.170083

Sharifi, S., Javadpour, J., Rezaie, H., Jamshidi, M., & Ghamarpoor, R. (2024). Developing a deposited calcium-phosphate layer on zirconia surface by chemical grafting of L-serine molecules. *Journal of Materials Research and Technology, 28*, 66–77. DOI: 10.1016/j.jmrt.2023.11.285

Sharma, S., Kumar, R., and Kumar, K. (2024). Synthesis of biowaste-derived carbon nanomaterials: Efficiency, quality, and environmental impact. Indian Chemical Engineer, 1–20. .DOI: 10.1080/00194506.2024.2442719

Shen, P. C., Lin, Y., Wang, H., Park, J. H., Leong, W. S., Lu, A. Y., Palacios, T., & Kong, J. (2018). CVD Technology for 2-D Materials. *IEEE Transactions on Electron Devices, 65*(10), 4040–4052. DOI: 10.1109/TED.2018.2866390

Shetti, R. S., Shetti, S. M., Sreenivasulu, M., Alibrahim, K. A., Alodhayb, A. N., & Shetti, N. P. (2025). Microwave-assisted synthesis of Ni-doped carbon from bio-waste as a cost-effective electrocatalyst for effective and sustainable water electrolysis. *Materials Today Sustainability, 101236*, 101236. Advance online publication. DOI: 10.1016/j.mtsust.2025.101236

Shi, B., Lookman, T., & Xue, D. (2023). Multi-objective optimization and its application in materials science. *Materials Genome Engineering Advances, 1*(2), e14. Advance online publication. DOI: 10.1002/mgea.14

Shi, J. (2012). On the Synergetic Catalytic Effect in Heterogeneous Nanocomposite Catalysts. *Chemical Reviews, 113*(3), 2139–2181. DOI: 10.1021/cr3002752

Shinde, P. A., Shrestha, L. K., & Ariga, K. (2025). Heteroatom-functionalized carbon nanoarchitectonics: Unlocking the doping effects for supercapacitor electrode design. *Green Energy & Environment, 10*(9), 1838–1862. Advance online publication. DOI: 10.1016/j.gee.2025.02.007

Shi, Q., & Wu, J. (2021). Review on Sulfur Compounds in Petroleum and Its Products: State-of-the-Art and Perspectives. *Energy & Fuels, 35*(18), 14445–14461. DOI: 10.1021/acs.energyfuels.1c02229

Sholl, D. S., & Lively, R. P. (2016). Seven chemical separations to change the world. *Nature, 532*(7600), 435–437. DOI: 10.1038/532435a PMID: 27121824

Shrivastav, G., Prava Jyoti, T., Chandel, S., & Singh, R. (2025). Eco-friendly extraction innovations. *Separation and Purification Reviews, 54*(3), 241–257. DOI: 10.1080/15422119.2024.2381605

Shul'pin, G. B. (2010). Selectivity enhancement in functionalization of C–H bonds: A review. *Organic & Biomolecular Chemistry, 8*(19), 4217–4228. DOI: 10.1039/c004223d PMID: 20593075

SIDDIQUI, M. T. H. (2024). Fabrication of magnetic carbon nanocomposites using rice husk via one pot solvothermal carbonization coprecipitation synthesis. RMIT University, Singh, R., & Bhateria, R. (2021). Core–shell nanostructures: A simplest two-component system with enhanced properties and multiple applications. *Environmental Geochemistry and Health*, *43*(7), 2459–2482.

Singh, P., & Singh, J. S. (2025). Environmental Consequences of Crop Residue Burning and Its Threats to Urbanization.

Singh, A., Nanade, A., Bandiwadekar, P., Patil, P., Nadar, S., Nanade, S., & Tiwari, M. (2025). Machine learning models for the prediction of CO_2 adsorption capacity of metal–organic frameworks. *Journal of the Indian Chemical Society*, *102*(9), 101904. DOI: 10.1016/j.jics.2025.101904

Singh, G., Kaur, M., Garg, V. K., & Oliviera, A. C. (2022). Oxygen hyper-stoichiometric trimetallic titanium-doped magnesium ferrite: Structural and photocatalytic studies. *Ceramics International*, *48*(17), 24476–24484. DOI: 10.1016/j.ceramint.2022.05.067

Singh, H., Desimone, M. F., Pandya, S., Jasani, S., George, N., Adnan, M., Aldarhami, A., Bazaid, A. S., & Alderhami, S. A. (2023). Revisiting the green synthesis of nanoparticles: Uncovering influences of plant extracts as reducing agents for enhanced synthesis efficiency and its biomedical applications. *International Journal of Nanomedicine*, *18*, 4727–4750. DOI: 10.2147/IJN.S419369 PMID: 37621852

Singh, R., & Kumar, S. (2023). Regulatory and safety concerns regarding the use of active nanomaterials in food industry. In *Nanotechnology advancement in agro-food industry* (pp. 269–306). Springer. DOI: 10.1007/978-981-99-5045-4_8

Singh, S., Kulshrestha, M. J., Rani, N., Kumar, K., Sharma, C., & Aswal, D. K. (2023). An overview of vehicular emission standards. *MPAN. Journal of Metrology Society of India*, *38*(1), 241–263. DOI: 10.1007/s12647-022-00555-4

Singh, Y., Sharma, S., Kumar, U., Sihag, P., Balyan, P., Singh, K. P., & Dhankher, O. P. (2024). Strategies for economic utilization of rice straw residues into value-added by-products and prevention of environmental pollution. *The Science of the Total Environment*, *906*, 167714. DOI: 10.1016/j.scitotenv.2023.167714 PMID: 37832665

Sithole, T. (2024). A review on regeneration of adsorbent and recovery of metals: Adsorbent disposal and regeneration mechanism. *South African Journal of Chemical Engineering*, *50*(1), 39–50.

Sivalingam, S., Vishal, G., & Anush, B. (2024). Environmental and health effects of acid rain. *Health and Environmental Effects of Ambient Air Pollution: Volume 1: Air Pollution, Human Health, and the Environment, 1*, 91–107. DOI: 10.1016/B978-0-443-16088-2.00007-7

Skinner, E. N., Mason, J. F., & Moran, J. J. (1960). High temperature corrosion in refinery and petrochemical service. *Corrosion, 16*(12), 593t–600t. DOI: 10.5006/0010-9312-16.12.85

Sloan, M. E., Staubitz, A., Clark, T. J., Russell, C. A., Lloyd-Jones, G. C., & Manners, I. (2010). Homogeneous Catalytic Dehydrocoupling/Dehydrogenation of Amine–Borane Adducts by Early Transition Metal, Group 4 Metallocene Complexes. *Journal of the American Chemical Society, 132*(11), 3831–3841. DOI: 10.1021/ja909535a

Soeder, D. J. (2021). Fossil Fuels and Climate Change. *Fracking and the Environment*, 155–185. DOI: 10.1007/978-3-030-59121-2_9

Sofiev, M., Winebrake, J. J., Johansson, L., Carr, E. W., Prank, M., Soares, J., Vira, J., Kouznetsov, R., Jalkanen, J.-P., & Corbett, J. J. (2018). Cleaner fuels for ships provide public health benefits with climate trade-offs. *Nature Communications, 9*(1), 406. DOI: 10.1038/s41467-017-02774-9 PMID: 29410475

Song, C. (2003). An overview of new approaches to deep desulphurization for ultra-clean gasoline, diesel fuel and jet fuel. *Catalysis Today, 86*(1–4), 211–263. DOI: 10.1016/S0920-5861(03)00412-7

Song, C., Zhang, B., Hao, L., Min, J., Liu, N., Niu, R., Gong, J., & Tang, T. (2022). Converting poly(ethylene terephthalate) waste into N-doped porous carbon as CO2 adsorbent and solar steam generator. *Green Energy & Environment, 7*(3), 411–422. DOI: 10.1016/j.gee.2020.10.002

Sousa, J. P., Neves, R. P., Sousa, S. F., Ramos, M. J., & Fernandes, P. A. (2020). Reaction mechanism and determinants for efficient catalysis by DszB. *ACS Catalysis, 10*(16), 9545–9554. DOI: 10.1021/acscatal.0c03122

Sovilj, M. N., Nikolovski, B. G., Spasojević, M., & Mauhar, S. M. (2019). Packed bed absorption column: Hydrodynamics and mass transfer. *Acta Periodica Technologica, 50*(50), 260–267. DOI: 10.2298/APT1950260S

Speight, J. G. (2014). *Fuel chemistry and technology* (3rd ed.). CRC Press.

Speight, J. G. (2020). *Fuel science and technology handbook* (2nd ed.). CRC Press.

Speight, J. G. (2020). *The chemistry and technology of petroleum* (6th ed.). CRC Press.

Srivastava, R. K., Jozewicz, W., & Singer, C. (2001). SO2 scrubbing technologies: A review. *Environmental Progress*, *20*(4), 219–228. DOI: 10.1002/ep.670200410

Srivastava, V. C. (2012). Evaluation of desulfurization technologies. *RSC Advances*, *2*(3), 759–783. DOI: 10.1039/C1RA00309G

Stach, E., DeCost, B., Kusne, A. G., Hattrick-Simpers, J., Brown, K. A., Reyes, K. G., Schrier, J., Billinge, S., Buonassisi, T., Foster, I., Gomes, C. P., Gregoire, J. M., Mehta, A., Montoya, J., Olivetti, E., Park, C., Rotenberg, E., Saikin, S. K., Smullin, S., & Maruyama, B. (2021). Autonomous experimentation systems for materials development: A community perspective. *Matter*, *4*(9), 2702–2726. DOI: 10.1016/j.matt.2021.06.036

Stanev, V., Oses, C., Kusne, A. G., et al. (2018). Machine learning modeling of superconducting critical temperature. *npj Computational Materials, 4*, 29. DOI: 10.1038/s41524-018-0085-8

Stanislaus, A., Marafi, A., & Rana, M. S. (2010). Recent advances in the science and technology of ultra-low sulfur diesel (ULSD) production. *Catalysis Today*, *153*(1–2), 1–68. DOI: 10.1016/j.cattod.2010.05.011

Steijns, M., Derks, F., Verloop, A., & Mars, P. (1976). The mechanism of the catalytic oxidation of hydrogen sulfide: II. Kinetics and mechanism of hydrogen sulfide oxidation catalyzed by sulfur. *Journal of Catalysis*, *42*(1), 87–95. DOI: 10.1016/0021-9517(76)90094-4

Stejskal, V., Vendl, T., Aulicky, R., & Athanassiou, C. (2021). Synthetic and Natural Insecticides: Gas, Liquid, Gel and Solid Formulations for Stored-Product and Food-Industry Pest Control. *Insects 2021, Vol. 12, Page 590*, *12*(7), 590. DOI: 10.3390/insects12070590

Strassburg, S., Mayer, K., & Scheibel, T. (2022). Functionalization of biopolymer fibers with magnetic nanoparticles. *Physical Sciences Reviews*, *7*(10), 1091–1117. DOI: 10.1515/psr-2019-0118

Strielkowski, W., Lutsenko, E., & Pavlov, D. (2021). Fossil fuel industry development in the 21st century: a case of coal. *Fossil Fuel Industry Development in the 21st Century: A Case of Coal, 128*, 02004. DOI: 10.1051/shsconf/202112802004

Studart, A. R., Erb, R. M., & Libanori, R. (2015). Bioinspired hierarchical composites. In *Hybrid and Hierarchical composite materials* (pp. 287–318). Springer. DOI: 10.1007/978-3-319-12868-9_8

Subarmanian, K., Yuvakkumar, R., Ravi, G., Sehemi, A. A., & Velanthapillai, D. (2021). Synthesis of pure and lanthanum-doped barium ferrite nanoparticles for efficient removal of toxic pollutants. *Journal of Hazardous Materials*, *424*, 127604.

Suib, S. L. (2017). A Review of Nanoceramic Materials for Use in Ceramic Matrix Composites. *Sol-Gel Based Nanoceramic Materials: Preparation, Properties and Applications*, 185–230. DOI: 10.1007/978-3-319-49512-5_7

Sulis, D. B., Lavoine, N., Sederoff, H., Jiang, X., Marques, B. M., Lan, K., Cofre-Vega, C., Barrangou, R., & Wang, J. P. (2025). Advances in lignocellulosic feedstocks for bioenergy and bioproducts. *Nature Communications*, *16*(1), 1244. DOI: 10.1038/s41467-025-56472-y PMID: 39893176

Suman, S., Pravalika, J., Manjula, P., & Farooq, U. (2023). Gender and CVD- Does It Really Matters? *Current Problems in Cardiology*, *48*(5), 101604. DOI: 10.1016/j.cpcardiol.2023.101604

Süme, V., Daneshfaraz, R., Kerim, A., Abbaszadeh, H., & Abraham, J. (2024). Investigation of Clean Energy Production in Drinking Water Networks. *Water Resources Management 2024 38:6, 38*(6), 2189–2208. DOI: 10.1007/s11269-024-03752-9

Sun, M., Abazari, R., Chen, J., Hussain, C. M., Zhou, Y., & Kirillov, A. M. (2023). Encapsulation of H4SiW12O40 into amide-functionalized MOF for oxidative desulfurization of diesel. *ACS Applied Materials & Interfaces*, *15*(45), 52581–52592. DOI: 10.1021/acsami.3c12374 PMID: 37931039

SURESH, C. A. (2022). Nanomaterials: An Introduction. The Science of Nanomaterials: Basics and Applications, 1.

Süto, L. (2010). Mining: Extraction of Fossil Fuels. *Anthropogenic Geomorphology: A Guide to Man-Made Landforms*, 131–154. DOI: 10.1007/978-90-481-3058-0_10

Taiye, M. A., Hafida, W., Kong, F., & Zhou, C. (2024). A review of the use of rice husk silica as a sustainable alternative to traditional silica sources in various applications. *Environmental Progress & Sustainable Energy*, *43*(6), e14451. DOI: 10.1002/ep.14451

Tajdari, A., Babaei, A., Goudarzi, A., Partovi, R., & Rostami, A. (2021). Hybridization as an efficient strategy for enhancing the performance of polymer nanocomposites. *Polymer Composites*, *42*(12), 6801–6815. DOI: 10.1002/pc.26341

Tallon, C., & Franks, G. V. (2011). Recent trends in shape forming from colloidal processing: A review. *Journal of the Ceramic Society of Japan*, *119*(1387), 147–160. DOI: 10.2109/jcersj2.119.147

Tang, S., Wang, Y., He, P., Wang, Y., & Wei, G. (2024). Recent Advances in Metal–Organic Framework (MOF)-Based Composites for Organic Effluent Remediation. *Materials (Basel), 17*(11), 2660. DOI: 10.3390/ma17112660 PMID: 38893925

Tanimu, A., Tanimu, G., Ganiyu, S. A., Gambo, Y., Alasiri, H., & Alhooshani, K. (2022). Metal-free catalytic oxidative desulfurization of fuels— A review. *Energy & Fuels, 36*(7), 3394–3419. DOI: 10.1021/acs.energyfuels.1c04411

Temelso, B., Phan, T. N., & Shields, G. C. (2012). Computational Study of the Hydration of Sulfuric Acid Dimers: Implications for Acid Dissociation and Aerosol Formation. *The Journal of Physical Chemistry A, 116*(39), 9745–9758. DOI: 10.1021/jp3054394

Teo, W. L., Zhou, W., Qian, C., & Zhao, Y. (2021). Industrializing metal–organic frameworks: Scalable synthetic means and their transformation into functional materials. *Materials Today, 47*, 170–186. DOI: 10.1016/j.mattod.2021.01.010

Thakur, A., Kumar, A., & Somya, A. (2024). Forestry and Agricultural Residues-Based Wastes. *Biomass Wastes for Sustainable Industrial Applications: A Waste-to-Wealth Approach*, 95.

Thakur, A. K., Kurtyka, K., Majumder, M., Yang, X., Ta, H. Q., Bachmatiuk, A., Liu, L., Trzebicka, B., & Rummeli, M. H. (2022). Recent Advances in Boron- and Nitrogen-Doped Carbon-Based Materials and Their Various Applications. *Advanced Materials Interfaces, 9*(11), 2101964. Advance online publication. DOI: 10.1002/admi.202101964

Thakur, A., & Kumar, A. (2023). Ecotoxicity analysis and risk assessment of nanomaterials for environmental remediation. *Macromolecular Symposia, 410*(1), 2100438. DOI: 10.1002/masy.202100438

Thomas, S., & Nochehdehi, A. R. (2022). Handbook of magnetic hybrid nanoalloys and their nanocomposites: Springer Nature.

Titirici, M. M., & Antonietti, M. (2010). Sustainable carbon materials by hydrothermal carbonization. *Chemical Society Reviews, 39*(1), 103–116. DOI: 10.1039/B819318P PMID: 20023841

TONGDANG. I., & Kiatkittipong, W. (2023). *THE CATALYTIC PYROLYSIS OF COCONUT SHELL OVER Ni-DOPED ACTIVATED CARBON CATALYST FOR AROMATIC-RICH BIO-OIL PRODUCTION* Silpakorn University].

Topsøe, H., Clausen, B. S., & Massoth, F. E. (1996). Hydrotreating catalysis. In *Catalysis: Science and Technology* (pp. 1–269). Springer. DOI: 10.1007/978-3-642-61040-0_1

Torlak, Y., Altuner, E. E., Tekeli, Y., & Tekeli, T. (2025). Advanced Nanocomposite-based Electrochemical Biosensors for Sensitive Detection of Pharmaceutical Drugs. *Journal of Pharmaceutical Research International, 37*(11), 34–58. DOI: 10.9734/jpri/2025/v37i117765

Tovar-Lopez, F. J. (2023). Recent progress in micro-and nanotechnology-enabled sensors for biomedical and environmental challenges. *Sensors (Basel), 23*(12), 5406. DOI: 10.3390/s23125406 PMID: 37420577

Tran, K., & Ulissi, Z. W. (2018). Active learning across intermetallics to guide discovery of electrocatalysts for CO_2 reduction and H_2 evolution. *Nature Catalysis, 1*(9), 696–703. DOI: 10.1038/s41929-018-0142-1

Trimm, D. L. (2001). The regeneration or disposal of deactivated heterogeneous catalysts. *Applied Catalysis A, General, 212*(1-2), 153–160. DOI: 10.1016/S0926-860X(00)00852-8

Tripathi, K., Singh, R., & Pant, K. K. (2021). Tailoring the physicochemical properties of Mg promoted catalysts via one pot non-ionic surfactant assisted co-precipitation route for CO_2 Co-feeding syngas to methanol. *Topics in Catalysis, 64*(5), 395–413. DOI: 10.1007/s11244-020-01410-z

Trivedi, Y., Sharma, M., Mishra, R. K., Sharma, A., Joshi, J., Gupta, A. B., Achintya, B., Shah, K., & Vuppaladadiyamd, A. K. (2025). Biochar potential for pollutant removal during wastewater treatment: A comprehensive review of separation mechanisms, technological integration, and process analysis. *Desalination, 600*, 118509. DOI: 10.1016/j.desal.2024.118509

Tsai, M. L., Lo, A. Y., Liu, J. H., and Dai, Y. M. (2025). Adsorptive desulfurization using waste-tire-derived carbon. C, 11(3), 47. .DOI: 10.3390/c11030047

Tsoukleris, D. S., Gatou, M.-A., Lagopati, N., Sygellou, L., Christodouleas, D. C., Falaras, P., & Pavlatou, E. A. (2023). Chemically modified TiO2 photocatalysts as an alternative disinfection approach for municipal wastewater treatment plant effluents. *Water (Basel), 15*(11), 2052. DOI: 10.3390/w15112052

Turkay, M. (2018). *KARAYOLU ULAŞIMINDAN KAYNAKLANAN SERA GAZI EMİSYONUNUN (KARBON AYAK İZİNİN) HESAPLANMASI: ESKİŞEHİR İLİ ÖRNEĞİ* [Fen Bilimleri Enstitüsü]. https://tez.yok.gov.tr/UlusalTezMerkezi/tezDetay.jsp?id=h_oND-zFpEPniNFRp8KCtw&no=z7z8rI4ek1lfdPBfXH2JGQ

U.S. Environmental Protection Agency. (2014). Control of air pollution from motor vehicles: Tier 3 motor vehicle emission and fuel standards. *Federal Register, 79*(81), 23414–23795. https://www.govinfo.gov/content/pkg/FR-2014-04-28/pdf/2014-06954.pdf

United States Environmental Protection Agency (EPA). (2020). *Regulations for Fuels and Fuel Additives: Gasoline and Diesel Fuel Standards*. Retrieved from https://www.epa.gov

Upadhyayula, V. K. K., Meyer, D. E., Curran, M. A., & Gonzalez, M. A. (2012). LCA of carbon nanotube products. *Journal of Cleaner Production, 26*, 37–47. DOI: 10.1016/j.jclepro.2011.12.018

Ürge-Vorsatz, D., Cabeza, L. F., Serrano, S., Barreneche, C., & Petrichenko, K. (2015). Heating and cooling energy trends and drivers in buildings. *Renewable & Sustainable Energy Reviews, 41*, 85–98. DOI: 10.1016/j.rser.2014.08.039

Vahlas, C., Caussat, B., Serp, P., & Angelopoulos, G. N. (2006). Principles and applications of CVD powder technology. *Materials Science and Engineering R Reports, 53*(1–2), 1–72. DOI: 10.1016/j.mser.2006.05.001

van Droffelaar, B., & Jacobs, M. (2018). Nature-Based Training Program Fosters Authentic Leadership. *Journal of Leadership Studies, 12*(3), 7–18. DOI: 10.1002/jls.21569

van Thriel, C., Schäper, M., Kleinbeck, S., Kiesswetter, E., Blaszkewicz, M., Golka, K., Nies, E., Raulf-Heimsoth, M., & Brüning, T. (2010). Sensory and pulmonary effects of acute exposure to sulfur dioxide (SO2). *Toxicology Letters, 196*(1), 42–50. DOI: 10.1016/j.toxlet.2010.03.013

Velu, S., Watanabe, S., Ma, X., & Song, C. (2003). Regenerable adsorbents for the adsorptive desulfurization of transportation fuels for fuel cell applications. *ACS Division of Fuel Chemistry, Preprints, 48*.

Vermeersch, L., De Proft, F., Faulkner, V., & De Vleeschouwer, F. (2023). Unravelling the mechanism and governing factors in Lewis acid and noncovalent Diels–Alder catalysis. *International Journal of Molecular Sciences, 24*(5), 4938. DOI: 10.3390/ijms24054938 PMID: 36902369

Viell, J., Theis, S., Götz, R., Weber-Bernard, A., Echtermeyer, A. W. W., & Lindmeyer, M. (2025). Benefits and Challenges for Decentralized Use of Biomass as Feedstock for Chemicals. *Chemieingenieurtechnik (Weinheim), 97*(6), 00043. DOI: 10.1002/cite.202500043

Vijayalakshmi, K., Florence, J. A. K., & Arunachalam, P. (2023). Industrial wastewater sources and treatment strategies by using bionanomaterials. In *Nanobiomaterials* (pp. 255–266). CRC Press. DOI: 10.1201/9780429057038-21

von Glasow, R., Bobrowski, N., & Kern, C. (2009). The effects of volcanic eruptions on atmospheric chemistry. *Chemical Geology*, *263*(1–4), 131–142. DOI: 10.1016/j.chemgeo.2008.08.020

Vosooghzadeh, B. (2018). Bitumen and heavy crude upgrading (review article). DOI: 10.13140/RG.2.2.25169.58721

Wadi, V. S., Varghese, A. M., Kuppireddy, S., Shimoga, G., Karanikolos, G. N., & Alhassan, S. M. (2024). Three-dimensional graphene with the decoration of elemental sulfur-based sulfur-nitrogen dots for CO2 capture. *Journal of Environmental Chemical Engineering*, *12*(4), 113130. DOI: 10.1016/j.jece.2024.113130

Wadsley, J. W., Stadel, J., & Quinn, T. (2004). Gasoline: A flexible, parallel implementation of TreeSPH. *New Astronomy*, *9*(2), 137–158. DOI: 10.1016/j.newast.2003.08.004

Wang, C., Chen, S., Chen, Y., Zi, F., Hu, X., Qin, X., Zhang, Y., Yang, P., He, Y., He, P., Lin, Y., & Zhang, G. (2020). Modification of activated carbon by chemical vapour deposition through thermal decomposition of thiourea for enhanced adsorption of gold thiosulfate complex. *Separation and Purification Technology*, *241*, 116632. DOI: 10.1016/j.seppur.2020.116632

Wang, C., Lu, W., Wu, W., Zhang, L., Guo, W., Huang, G., & Xu, S. (2023). Mechanical grinding of FeNC nanomaterial with Fe_3O_4 to construct magnetic adsorbents for desulfurization. *Separation and Purification Technology*, *306*, 122574. DOI: 10.1016/j.seppur.2022.122574

Wang, J., Lou, F., Zhang, M., & Yuan, J. (2025). Recent progress in nitrogen-doped activated carbon: Microstructural regulation and enhanced gas adsorption. *Journal of Environmental Chemical Engineering*, *13*(5), 118107. DOI: 10.1016/j.jece.2025.118107

Wang, J., Tchapmi, L. P., Ravikumar, A. P., McGuire, M., Bell, C. S., Zimmerle, D., Savarese, S., & Brandt, A. R. (2020). Machine vision for natural gas methane emissions detection using an infrared camera. *Applied Energy*, *257*, 113998. DOI: 10.1016/j.apenergy.2019.113998

Wang, L. K., Williford, C., & Chen, W. Y. (2005). Desulfurization and emissions control. In *Advanced Air and Noise Pollution Control* (pp. 35–95). Humana Press. DOI: 10.1007/978-1-59259-779-6_2

Wang, M., Wang, X., Moni, P., Liu, A., Kim, D. H., Jo, W. J., Sojoudi, H., & Gleason, K. K. (2017). CVD Polymers for Devices and Device Fabrication. *Advanced Materials, 29*(11), 1604606. DOI: 10.1002/adma.201604606

Wang, P., Xi, B., & Xiong, S. (2024). Insights into the Optimization of Catalytic Active Sites in Lithium–Sulfur Batteries. *Accounts of Chemical Research, 57*(15), 2093–2104. DOI: 10.1021/acs.accounts.4c00244

Wang, Q., & Frenkel, A. I. (2017). *Characterization of Model Nanocatalysts by X-ray Absorption Spectroscopy.*, DOI: 10.1016/B978-0-12-805090-3.00005-X

Wang, Q., Huang, W., Li, X., Lin, S., Li, Z., & Ma, X. (2024). Electronic regulation of MoS_2 edge sites by d electron transfer of Ni or Co to improve the activity of CO sulfur-resistant methanation. *Chemical Engineering Journal, 480*, 148242. DOI: 10.1016/j.cej.2023.148242

Wang, Q., Wu, Y., Shan, Y., & Dong, W. (2025). Synergizing surface-bound radicals with singlet oxygen for degradation of emerging contaminants. *Environmental Science & Technology, 59*(42), 22901–22913. DOI: 10.1021/acs.est.5c04418 PMID: 41082611

Wang, X., Zhu, W., Wang, N., & Zuo, X. (2022). Nitrogen-doped oxidized activated carbon as efficient metal-free bifunctional material for bisphenol removal: A tuning of surface chemistry. *Journal of Environmental Chemical Engineering, 10*(6), 108509. DOI: 10.1016/j.jece.2022.108509

Wang, Y., Ding, L., Shi, Q., Liu, S., Qian, L., Yu, Z., Wang, H., Lei, J., Gao, Z., Long, H., & Charles Xu, C. (2022). Volatile organic compounds (VOC) emissions control in iron ore sintering process: Recent progress and future development. *Chemical Engineering Journal, 448*, 137601. DOI: 10.1016/j.cej.2022.137601

Wang, Y., Fens, M. H., Van Kronenburg, N. C., Shi, Y., Lammers, T., Heger, M., van Nostrum, C. F., & Hennink, W. E. (2022). Magnetic beads for the evaluation of drug release from biotinylated polymeric micelles in biological media. *Journal of Controlled Release, 349*, 954–962. DOI: 10.1016/j.jconrel.2022.07.044 PMID: 35931210

Wang, Y., Wang, Z., Lin, Y., Qin, Y., He, R., Wang, M., Sun, Q., & Peng, Y. (2024). Nanocellulose from agro-industrial wastes: A review on sources, production, applications, and current challenges. *Food Research International, 192*, 114741. DOI: 10.1016/j.foodres.2024.114741 PMID: 39147548

Wang, Y., Zhao, L., Otto, A., Robinius, M., & Stolten, D. (2017). A Review of Post-combustion CO2 Capture Technologies from Coal-fired Power Plants. *Energy Procedia, 114*, 650–665. DOI: 10.1016/j.egypro.2017.03.1209

Wei, W., Yu, D., Jiang, Z., & Zhou, J. (2020). Food waste-derived carbon materials. *Journal of Cleaner Production, 256*, 120678. DOI: 10.1016/j.jclepro.2020.120678

Westermayr, J., Gastegger, M., Schütt, K. T., & Maurer, R. J. (2021). Perspective on integrating machine learning into computational chemistry and materials science. *The Journal of Chemical Physics, 154*(23), 230903. DOI: 10.1063/5.0047760 PMID: 34241249

Whitesides, G. M. (2005). Nanoscience, nanotechnology, and chemistry. *Small, 1*(2), 172–179. DOI: 10.1002/smll.200400130 PMID: 17193427

WHO. (2021). *Air quality guidelines: Sulfur dioxide (SO_2)*. World Health Organization. https://www.who.int/publications/i/item/9789240034228

Wiatrowski, M., Klein, B., Do, T. N., Ou, L., Cai, H., Carlson, N., & Davis, R. (2025). Technology Case Study: Techno-Economic and Life Cycle Analysis for Microalgae Conversion Pathways to Fuels and Products (No. NREL/TP-5100-91854). National Renewable Energy Laboratory (NREL), Golden, CO (United States).

Winder, C. (2004). Toxicity of gases, vapours and particulates. *Occupational Toxicology*, 399–424. DOI: 10.1201/9781482289282-18

World Health Organization. (2021). *WHO global air quality guidelines: Particulate matter (PM2.5 and PM10), ozone, nitrogen dioxide, sulfur dioxide and carbon monoxide*. https://www.who.int/publications/i/item/9789240034228

World Health Organization. (2021). *WHO global air quality guidelines: Particulate matter, ozone, nitrogen dioxide, sulfur dioxide and carbon monoxide*. World Health Organization. https://www.who.int

Wuebbles, D. J., & Jain, A. K. (2001). Concerns about climate change and the role of fossil fuel use. *Fuel Processing Technology, 71*(1–3), 99–119. DOI: 10.1016/S0378-3820(01)00139-4

Wu, F., Yuan, Q., Wang, J., Wang, X., Luo, J., Guo, Y., Xu, H., & Wei, X. (2025). Oxidative desulfurization catalyzed by magnetically recoverable CoFe2O4 nanoparticles. *Arabian Journal of Chemistry, 18*(1), 106076. DOI: 10.1016/j.arabjc.2024.106076

Wu, P., Deng, C., Liu, F., Zhu, H., Chen, L., Liu, R., Zhu, W., & Xu, C. (2023). Magnetically recyclable high-entropy metal oxide catalyst for aerobic catalytic oxidative desulfurization. *Chinese Journal of Catalysis, 54*, 238–249. DOI: 10.1016/S1872-2067(23)64541-3

Wu, Q. (2019). The fabrication of magnetic recyclable nitrogen-modified titanium dioxide/strontium ferrite/diatomite heterojunction nanocomposite for enhanced visible-light-driven photodegradation of tetracycline. *International Journal of Hydrogen Energy, 44*(16), 8261–8272. DOI: 10.1016/j.ijhydene.2019.01.145

Wu, Q. (2023). Acidic and basic catalytic cracking technologies and its development prospects for crude oil to chemicals. *Fuel, 332*, 126132. DOI: 10.1016/j.fuel.2022.126132

Wu, S. R., Chen, C. H., Chung, I. L., & Lee, H. T. (2011). Combustion of low-calorific waste liquids in high temperature air. *Fuel, 90*(8), 2639–2644. DOI: 10.1016/j.fuel.2011.04.001

Wu, Z., Li, Y., Zhang, C., Huang, X., Peng, B., & Wang, G. (2022). Recent advances in metal-organic-framework-based catalysts for thermocatalytic selective oxidation of organic substances. *Chem Catalysis, 2*(5), 1009–1045. DOI: 10.1016/j.checat.2022.02.010

Xiao, C. K., Zou, Y. T., Li, D. Z., Wang, E. H., Wang, A. C., Gao, D. W., Duan, A.-J., Zheng, P., & Wang, X. L. (2023). Optimization of dendritic TS-1/silica micro–mesoporous composites for efficient hydrodesulfurization of dibenzothiophene and 4,6-dimethyldibenzothiophene. *Petroleum Science, 20*(4), 2521–2530. DOI: 10.1016/j.petsci.2023.07.004

Xiao, G., Amamoo-Otoo, C., Wang, T., Li, Q., & Biancardo, S. A. (2025). Evaluating the impact of ECA policy on sulfur emissions from the five busiest ports in America. *Frontiers in Marine Science, 12*, 1609261. DOI: 10.3389/fmars.2025.1609261

Xie, R., Xing, Q., Gao, J., Wang, X., Xu, W., Lv, Z., Yi, W., Zhao, J., Luo, Z., Wu, X., & Liu, H. (2024). Investigation on the effect of ship emissions on the air quality: A case study in Hainan Island, China. *Journal of Environmental Sciences (China), 156*, 114–125. Advance online publication. DOI: 10.1016/j.jes.2024.08.010 PMID: 40412918

Xi, Z., Xi, K., Lu, L., & Zhang, M. (2023). Study on oxidation characteristics and conversion of sulfur-containing model compounds in coal. *Fuel, 331*, 125756. DOI: 10.1016/j.fuel.2022.125756

Xue, L. L., He, Q., Li, Z. Z., Wu, J. Q., Li, S. W., & Zhao, J. S. (2025). Remarkable catalytic performance of supported P–Mo–W polyoxometalate on green-friendly fiber for ultra-deep oxidative desulfurization. *Applied Organometallic Chemistry*, *39*(11), e70410. DOI: 10.1002/aoc.70410

Xue, X.-X., Meng, H., Huang, Z., Feng, Y., & Qi, X. (2021). Black phosphorus-based materials for energy storage and electrocatalytic applications. *Journal of Physics: Energy*, *3*(4), 042002. DOI: 10.1088/2515-7655/abff6b

Xu, F., Zhao, S., Li, B., Li, H., Ling, Z., Zhang, G., & Liu, M. (2024). Current status of zero liquid discharge technology for desulfurization wastewater. *Water (Basel)*, *16*(6), 900. DOI: 10.3390/w16060900

Xu, W., & Zhang, R. (2012). Theoretical Investigation of Interaction of Dicarboxylic Acids with Common Aerosol Nucleation Precursors. *The Journal of Physical Chemistry A*, *116*(18), 4539–4550. DOI: 10.1021/jp301964u

Yaashikaa, P., Kumar, P. S., & Karishma, S. (2022). Review on biopolymers and composites–evolving material as adsorbents in removal of environmental pollutants. *Environmental Research*, *212*, 113114. DOI: 10.1016/j.envres.2022.113114 PMID: 35331699

Yadav, V. K., Mir, S. H., & Singh, J. K. (2019). A computational study of structural, electronic and carrier mobility of boron and phosphorus/nitrogen co-doped graphene. *Physica B, Condensed Matter*, *571*, 291–295. DOI: 10.1016/j.physb.2019.07.017

Yang, G., Liang, J., Hu, X., Liu, M., Zhang, X., & Wei, Y. (2021). Recent Advances on Fabrication of Polymeric Composites Based on Multicomponent Reactions for Bioimaging and Environmental Pollutant Removal. *Macromolecular Rapid Communications*, *42*(6), 2000563. DOI: 10.1002/marc.202000563 PMID: 33543565

Yang, J., Wei, Q., Tian, C., Li, D., Li, H., Qin, G., Hu, K., & Zhang, Q. (2023). Preparation of biomass carbon composites MgO@ ZnO@ BC and its adsorption and removal of Cu (II) and Pb (II) in wastewater. *Molecules (Basel, Switzerland)*, *28*(19), 6982. DOI: 10.3390/molecules28196982 PMID: 37836824

Yang, Z., Wu, Y., Zhang, Z., Li, H., Li, X., Egorov, R. I., Strizhak, P. A., & Gao, X. (2019). Recent advances in co-thermochemical conversions of biomass with fossil fuels focusing on the synergistic effects. *Renewable & Sustainable Energy Reviews*, *103*, 384–398. DOI: 10.1016/j.rser.2018.12.047

Yang, Z., Zhang, X., Zhang, J., Liu, H., & Yu, X. (2023). Application of Biomass-Based Nanomaterials in Energy. *Advanced Energy and Sustainability Research*, *4*(12), 2300141. DOI: 10.1002/aesr.202300141

Yao, J., Liu, D., Zhao, B., Zhou, Y., Li, R., Zhang, Y., & Peng, T. (2023). Insight into the effect of B-doping on electronic structure and photocatalytic H2 evolution performance of C3N5 nanosheets. *Journal of Photochemistry and Photobiology A Chemistry, 436*, 114359. DOI: 10.1016/j.jphotochem.2022.114359

Yeole, N. R., & Parthasarthy, V. (2022). Design of experiments (DOE) for adsorptive desulfurization (ADS) of liquid fuels–A review. *Materials Today: Proceedings, 57*, 1613–1618. DOI: 10.1016/j.matpr.2021.12.230

Yosefi, L., Khoshbin, R., & Karimzadeh, R. (2022). Beneficial incorporation of metal-sulfur interaction in adsorption capacity of boron nitride based adsorbents used in highly selective sulfur removal. *Fuel, 310*, 122277. DOI: 10.1016/j.fuel.2021.122277

Yousefpour, A., Hojjati, M., & Immarigeon, J. P. (2004). Fusion bonding/welding of thermoplastic composites. *Journal of Thermoplastic Composite Materials, 17*(4), 303–341. DOI: 10.1177/0892705704045187

You, X.-X., Gao, J., Duan, Y.-C., Geng, Y., Zhang, M., Zhao, L., & Su, Z.-M. (2022). A theoretical analysis on the electron and energy transfer between host and guest materials in phosphor–doped OLED. *Journal of Photochemistry and Photobiology A Chemistry, 432*, 114058. DOI: 10.1016/j.jphotochem.2022.114058

Yuan, Y., Huang, L., Yılmaz, M., Zhang, T. C., Wang, Y., & Yuan, S. (2023). $MgFe_2O_4$-loaded N-doped biochar for H_2S desulfurization. *Fuel, 339*, 127385. DOI: 10.1016/j.fuel.2022.127385

Yu, B., Mi, J., Cui, C., Xiang, S., & Qiu, K. (2025). Biochar-supported zinc ferrite for H2S removal at room temperature under anaerobic conditions. *Chemical Engineering Research & Design, 222*, 286–297. DOI: 10.1016/j.cherd.2025.09.015

Yue, X., Shan, Y., Jiao, W., Shen, K., & Zhang, Y. (2023). Direct current effects on peroxydisulfate activation: Singlet oxygen generation and accelerated degradation. *Journal of Hazardous Materials, 452*, 131187. DOI: 10.1016/j.jhazmat.2023.131187 PMID: 36996542

Yun, J., Zhu, C., Wang, Q., Hu, Q., & Yang, G. (2019). Catalytic conversions of atmospheric sulfur dioxide and formation of acid rain over mineral dusts: Molecular oxygen as the oxygen source. *Chemosphere, 217*, 18–25. DOI: 10.1016/j.chemosphere.2018.10.201

Yu, S., Wang, L., Li, Q., Zhang, Y., & Zhou, H. (2022). Sustainable carbon materials from the pyrolysis of lignocellulosic biomass. *Materials Today Sustainability, 19*, 100209. DOI: 10.1016/j.mtsust.2022.100209

Zakari, R. S. B., Guerrero Peña, G. D. J., Pasanaje, A. H., Elsayed, M., Kuppireddy, S., Hassan Ali, M. I., Raj, A., & Elkadi, M. (2024). Effect of phosphorus on the physicochemical and electrochemical properties of Ni/Carbon electrodes for glycerol electrooxidation. *Case Studies in Chemical and Environmental Engineering, 10*, 100768. DOI: 10.1016/j.cscee.2024.100768

Zaman, C. Z., Pal, K., Yehye, W. A., Sagadevan, S., Shah, S. T., Adebisi, G. A., Marliana, E., & Rafique, R. F. (2017). Pyrolysis: A sustainable way to generate energy. *Pyrolysis*, 1.

Zaman, S., Wang, M., Liu, H., Sun, F., Yu, Y., Shui, J., Chen, M., & Wang, H. (2022). Carbon-based catalyst supports for oxygen reduction in proton-exchange membrane fuel cells. *Trends in Chemistry*, *4*(10), 886–906. DOI: 10.1016/j.trechm.2022.07.007

Zanurin, A., Johari, N. A., Alias, J., Ayu, H. M., Redzuan, N., & Izman, S. (2022). Research progress of sol-gel ceramic coating: A review. *Materials Today: Proceedings*, *48*, 1849–1854. DOI: 10.1016/j.matpr.2021.09.203

Zare, N., Karimi, F., Altuner, E. E., Mashkoor, H., Tiri, R. N. E., Gulbagca, F., Bijad, M., Cherif, A., & Sen, F. (2024). Synthesis of Bimetallic Nickel-Cobalt Nanoparticles Supported by Functional Multi-Walled Carbon Nanotubes by Chemical Methods: Characterization and Hydrogen Energy Generation. *Topics in Catalysis*, *67*(9–12), 606–614. DOI: 10.1007/s11244-024-01907-x

Zbuzant, M. (2022). An overview of photocatalysts for desulfurization. Eurasian Journal of Chemical. *Medicinal and Petroleum Research*, *1*(1), 10–19. DOI: 10.5281/zenodo.7353454

Zbuzant, M. (2022). An overview of the use of photo catalysts for desulfurization. Eurasian Journal of Chemical. *Medicinal and Petroleum Research*, *1*(1), 10–19.

Zeng, X., Lin, C., Ren, Q., Shan, S., & Wang, H. (2024). CuO–MgO enhanced with $MgMoO_4$ for oxidative desulfurization. *Fuel*, *360*, 130630. DOI: 10.1016/j.fuel.2023.130630

Zhang, T., Tang, Q., Pu, C., & Yan, Y. (2022). Simulation on Evaporation and Motion of Atomized Droplets in Spray Dry Flue Gas Desulfurization Tower. *Journal of Thermal Science 2022 31:6*, *31*(6), 2252–2263. DOI: 10.1007/s11630-022-1612-y

Zhang, B., Jiang, Y., & Balasubramanian, R. (2021). Synthesis, formation mechanisms and applications of biomass-derived carbonaceous materials: A critical review. *Journal of Materials Chemistry. A, Materials for Energy and Sustainability*, *9*(44), 24759–24802. DOI: 10.1039/D1TA06874A

Zhang, E., Zhu, Q., Huang, J., Liu, J., Tan, G., Sun, C., Li, T., Liu, S., Li, Y., Wang, H., Wan, X., Wen, Z., Fan, F., Zhang, J., & Ariga, K. (2021). Visually resolving the direct Z-scheme heterojunction in CdS@ZnIn2S4 hollow cubes for photocatalytic evolution of H2 and H2O2 from pure water. *Applied Catalysis B: Environmental, 293*, 120213. DOI: 10.1016/j.apcatb.2021.120213

Zhang, H., Chen, L., Dong, F., Lu, Z., Lv, E., Dong, X., Li, H., Yuan, Z., Peng, X., Yang, S., Qiu, J., Guo, Z., & Wen, Z. (2024). Dynamic transformation of active sites in energy and environmental catalysis. *Energy & Environmental Science, 17*(18), 6435–6481. DOI: 10.1039/D4EE02365J

Zhang, H., Su, T., Yu, S., Liao, W., Ren, W., Zhu, Z., Yang, K., Len, C., Dong, G., Zhao, D., & Lü, H. (2023). Fabrication of Anderson-Polyoxometalates/TiO_2/g-C_3N_4 heterojunction composite for efficient visible-light-driven photooxidative desulfurization. *Molecular Catalysis, 536*, 112916. DOI: 10.1016/j.mcat.2023.112916

Zhang, H., Wang, Z., Luo, X., Lu, J., Peng, S., Wang, Y., & Han, L. (2020). Constructing hierarchical porous carbons with interconnected micro-mesopores for enhanced CO2 adsorption. *Frontiers in Chemistry, 7*, 919. DOI: 10.3389/fchem.2019.00919 PMID: 32010669

Zhang, Q., Uchaker, E., Candelaria, S. L., & Cao, G. (2018). Nanomaterials for energy conversion and storage. *Chemical Society Reviews, 42*(7), 3127–3171. DOI: 10.1039/c3cs00009e PMID: 23455759

Zhang, R., Du, X., Li, S., Guan, J., Fang, Y., Li, X., Dai, Y., & Zhang, M. (2022). Application of heteroatom doping strategy in electrolyzed water catalytic materials. *Journal of Electroanalytical Chemistry (Lausanne, Switzerland), 921*, 116679. DOI: 10.1016/j.jelechem.2022.116679

Zhang, T., Chao, Y., Liu, N., Thompson, J., Garcia, M., He, B. B., Gerpen, J. V., & Chen, S. (2011). Case study of biodiesel-diesel blends as a fuel in marine environment. *Advances in Chemical Engineering and Science, 1*(2), 65–71. DOI: 10.4236/aces.2011.12011

Zhang, T., Wang, Q., & Rui, Y. (2025). The impact of nanomaterials on plant health: A review of exposure, toxicity, and control. *Environmental Science. Nano, 12*(6), 2965–2982. DOI: 10.1039/D5EN00037H

Zhang, W., Lu, W.-C., Zhang, H.-X., Ho, K. M., & Wang, C. Z. (2016). Lattice distortion and electron charge redistribution induced by defects in graphene. *Carbon, 110*, 330–335. DOI: 10.1016/j.carbon.2016.09.031

Zhang, X., Cheng, X., Kang, J., Fan, F., Sun, H., Zhou, A., & Zhou, W. (2025). The influence of In-modified NiMoS active phase on the direct desulfurization reaction process of 4,6-dimethyldibenzothiophene. *Molecular Catalysis*, *581*, 115143. DOI: 10.1016/j.mcat.2025.115143

Zhang, X., & Wang, R. (2025). Phase-transfer catalysis for fuel desulfurization. *Catalysts*, *15*(8), 724. DOI: 10.3390/catal15080724

Zhang, Y., Lin, C., Li, J., & Wang, X. (2017). Low-temperature CVD of carbon nanomaterials. *Carbon*, *117*, 328–343. DOI: 10.1016/j.carbon.2017.02.063

Zhang, Y., Liu, F., Chen, W., Han, W., Zhai, W., Lu, Y., & Li, M. (2024). Effective reduction of hydrogen consumption in ultra-deep hydrodesulfurization of diesel: Deep insights into thermodynamic limitations during hydrotreating. *Fuel, 356*, 129640. DOI: 10.1016/j.fuel.2023.129640

Zhang, Y., Qu, M., Zhang, K., Liu, W., Wang, J., Ren, L., & Wang, F. (2025). Effect mechanisms of microwave on CO2 adsorption with cellulosic and non-cellulosic biochar. *Separation and Purification Technology*, *373*, 133610. DOI: 10.1016/j.seppur.2025.133610

Zhang, Z., Bai, L., Li, Z., Qu, Y., & Jing, L. (2019). Review of strategies for the fabrication of heterojunctional nanocomposites as efficient visible-light catalysts by modulating excited electrons with appropriate thermodynamic energy. *Journal of Materials Chemistry. A, Materials for Energy and Sustainability*, *7*(18), 10879–10897. DOI: 10.1039/C9TA02373A

Zhang, Z., Wu, G.-X., Tan, Z.-L., Chen, H.-Z., Yu, Z., Yu, Z.-R., Huang, X.-B., Yang, W.-J., & Huang, Y.-D. (2024). Lattice-Engineering modulated band structure facilitates enhanced stability of high-voltage LiCoO2 at 4.6 V. *Chemical Engineering Journal*, *502*, 158053. DOI: 10.1016/j.cej.2024.158053

Zhao, H., Guo, S., & Zhao, H. (2018). Impacts of GDP, Fossil Fuel Energy Consumption, Energy Consumption Intensity, and Economic Structure on SO2 Emissions: A Multi-Variate Panel Data Model Analysis on Selected Chinese Provinces. *Sustainability 2018, Vol. 10, Page 657, 10*(3), 657. DOI: 10.3390/su10030657

Zhao, C., Li, X., Cheng, J., & Yang, Y. (2018). Nitrogen-doped porous carbons for sulfur adsorption. *Fuel*, *225*, 446–455. DOI: 10.1016/j.fuel.2018.01.010

Zhao, M., Peng, H. J., Li, B. Q., & Huang, J. Q. (2024). Kinetic Promoters for Sulfur Cathodes in Lithium–Sulfur Batteries. *Accounts of Chemical Research*, acs.accounts.3c00698. Advance online publication. DOI: 10.1021/acs.accounts.3c00698

Zheng, H., Lian, J., Liu, Y., Tang, Z., & Ji, J. (2024). Molecular mechanism of competitive adsorption desulfurization involving sulfur-containing molecules. *Colloids and Surfaces. A, Physicochemical and Engineering Aspects, 690*, 133610. DOI: 10.1016/j.colsurfa.2024.133610

Zheng, S., Yang, S., Chen, J., Wu, D., Qi, B., Zhang, C., Deng, W., Li, J., Mei, T., Wang, S., & Wan, L. (2024). New progresses in efficient, selective, and environmentally friendly recovery of valuable metal from e-waste and industrial catalysts. *Advanced Sustainable Systems, 8*(6), 2300512. DOI: 10.1002/adsu.202300512

Zhou, W., Meng, X., Gao, J., Zhao, H., Zhao, G., & Ma, J. (2021). Electrochemical regeneration of carbon-based adsorbents: A review of regeneration mechanisms, reactors, and future prospects. *Chemical Engineering Journal Advances, 5*, 100083. DOI: 10.1016/j.ceja.2020.100083

Zhou, X., Jin, H., Xia, B. Y., Davey, K., Zheng, Y., & Qiao, S. Z. (2021). Molecular cleavage of metal-organic frameworks and application to energy storage and conversion. *Advanced Materials, 33*(51), 2104341. DOI: 10.1002/adma.202104341 PMID: 34605072

Zhou, X., Wang, T., Liu, H., Gao, X., Wang, C., & Wang, G. (2021a). Desulfurization through Photocatalytic Oxidation: A Critical Review. *ChemSusChem, 14*(2), 492–511. DOI: 10.1002/cssc.202002144

Zhou, X., Wang, T., Zhang, L., Che, S., Liu, H., Liu, S., Wang, C., Su, D., & Teng, Z. (2022). Highly efficient Ag_2O/Na-g-C_3N_4 heterojunction for photocatalytic desulfurization of thiophene in fuel under ambient air conditions. *Applied Catalysis B: Environmental, 316*, 121614. DOI: 10.1016/j.apcatb.2022.121614

Zhu, J., Guo, Y., Chen, N., & Chen, B. (2024). A Review of the Efficient and Thermal Utilization of Biomass Waste. *Sustainability (2071-1050), 16*(21).

Zou, H., Du, H., Wang, Y., Zhao, L., Mao, G., Zuo, J., Liu, Y., Liu, X., & Huisingh, D. (2017). A review of the first twenty-three years of articles published in the Journal of Cleaner Production: With a focus on trends, themes, collaboration networks, low/no-fossil carbon transformations and the future. *Journal of Cleaner Production, 163*, 1–14. DOI: 10.1016/j.jclepro.2017.04.157

Zou, Y., Sun, Z., Wang, Q., Ju, Y., Sun, N., Yue, Q., & Wang, Z. (2024). Core–Shell Magnetic Particles: Tailored Synthesis and Applications. *Chemical Reviews, 125*(2), 972–1048. DOI: 10.1021/acs.chemrev.4c00710 PMID: 39729245

Zuo, Y., Qin, M., Chen, C., Ye, W., Li, X., Luo, J., & Ong, S. P. (2021). Accelerating materials discovery with Bayesian optimization and graph deep learning. *Materials Today*, *51*, 126–135. DOI: 10.1016/j.mattod.2021.08.012

About the Contributors

Tawfik A. Saleh is currently a faculty member at King Fahd University of Petroleum and Minerals. He has published several papers as well as proceeding in well-known journals. He also published several books and book chapters and reviews. His research work focused largely in the area of materials technology; synthesis, characterization and applications of inorganic-organic nanocomposites, adsorption and separation. He is in the editorial board of several international journals.

***

Irshad Ali Khan is a Lecturer in the Department of Agriculture at the University of Swabi, Khyber Pakhtunkhwa, Pakistan. He holds a Ph.D. in Plant Pathology, which he earned in 2012 from Zhejiang University, China. His doctoral research focused on [insert your specific research topic, e.g. "the interaction between fungal pathogens and host resistance mechanisms in economically important crops"]. After completing his Ph.D., Dr. Khan returned to Pakistan to pursue academic and research work in the field of plant health, disease resistance, and crop protection. His research interests include plant pathology, disease diagnostics, biological control of plant diseases, molecular plant-pathogen interactions, and sustainable disease management strategies. At the University of Swabi, Dr. Khan is involved in both teaching and research. In his role as lecturer, he teaches courses related to plant pathology, plant disease management, and agricultural microbiology, mentoring undergraduate and postgraduate students. He also supervises student projects and contributes to curriculum development in the Department of Agriculture. Dr. Khan has published several research articles in peer-reviewed journals, presenting findings from his work on pathogen identification, host resistance, and disease control. He has also participated in national and international conferences, sharing his research with the scientific community. Committed to service and extension, Dr. Khan collaborates with local farmers, agricultural extension agencies, and governmental or non-governmental organizations to help implement disease mitigation strategies in crops of local importance. His aim is to contribute to food security and sustainable agriculture in the region. Beyond his academic and research work, Dr. Khan continues to stay updated in his field by attending workshops, trainings, and collaborations. Through his work, he strives to bridge the gap between scientific research and field application, helping farmers manage plant diseases and improve crop health in Khyber Pakhtunkhwa.

Mangaiyarkkarasi is currently working as an Associate Professor of Physics in Nadar Mahajana Sangam S. Vellaichamy Nadar College, Nagamalai, Madurai-625 019, Tamil Nadu, India. She has 25 years of teaching experience. Her major research interest includes synthesis of ceramic materials, powder X-ray analysis and electron density studies. She has published 11 research papers in the refereed International journals and 17 book chapters. She has prepared study materials (UG & PG Physics) for Madurai Kamaraj University Distance Education Course.

Abdullah Özkan is a researcher at the Faculty of Engineering and Natural Sciences, İskenderun Technical University. His interdisciplinary work centers on nanomaterials, green chemistry, spectroscopy, and nanocomposite technologies. He focuses on environmentally friendly synthesis methods, structure–property relationships, and advanced spectroscopic characterization techniques. His research contributes to the development of sustainable functional materials for applications in energy systems, catalysis, environmental remediation, and advanced material design. He actively collaborates in national and international research projects aiming to produce high-performance, eco-conscious materials for emerging technological needs.

Vildan Özkan completed her undergraduate studies in Chemical Engineering at Selçuk University, her master's degree in Chemistry at Mustafa Kemal University, and her doctorate in Mechanical Engineering at Iskenderun Technical University. She currently works as an Assistant Professor in the Biomedical Engineering Department at Iskenderun Technical University. Her research focuses primarily on polymer nanocomposites, nanofiber production via electrospinning, carbon nanotube and graphene-reinforced composites, electromagnetic shielding materials, adsorptive desulfurization, and environmental applications. Dr. Özkan has numerous scientific publications in internationally indexed journals and continues to conduct research in the fields of defense, energy, and environmental technologies.

Sokan-Adeaga Adewale Allen, is a Public/Environmental Health Consultant and Assistant Professor at Public Health Program, Bowen University, Iwo, Osun

State, Nigeria. He is formerly a University of Ibadan Postgraduate School Scholar from 2015 to 2018; Award Winner of the 2018 ISWA-SWIS Essay Competition, Honorary Fellow of Medical Research Council, USA (2019); & London Journal Press Quarterly Franklin Membership, UK (2017). His areas of interest include Environmental Sustainability, Biofuels, and Public Health Assessment.

Micheal Ayodeji Sokan-Adeaga, is a trained Microbiologist at the Obafemi Awolowo University (OAU) Ile-Ife, and also a trained and certified Medical Practitioner from University of Lagos, Lagos, Nigeria.He is an honorary Fellow of Medical Research Council (FMRC), USA. He has published widely and has been cited in both Local and International peer reviewed journals. His areas of interest are General Medicine and Surgery, General Microbiology and Public Health.

Index

F

G

H

L

N

O

P

R

S

www.ingramcontent.com/pod-product-compliance
Lightning Source LLC
LaVergne TN
LVHW062328300926
849161LV00007B/1610